MATHEMATICS ALIVE AND APPLIED

FOR BUSINESS, ECONOMICS, AND LIFE

MATHEMATICS ALIVE AND APPLIED

FOR BUSINESS, ECONOMICS, AND LIFE

Sherman Chottiner
Syracuse University

Author/publisher: Sherman Chottiner
                     104 Easterly Terrace
                     DeWitt, NY 13214
                     (315) 445-0654
                     (315) 383-3599
                     schot@som.syr.edu

Compositor/Printer: Sheridan Books, Inc.
                     613 E. Industrial Dr.
                     Chelsea, MI 48118

Copyright © 2004 by Sherman Chottiner
ISBN 0-9759145-0-2

All rights reserved. Printed in the United States of America.
No part of this book may be used or reproduced in any manner
whatsoever without written permission except in the case of
brief quotations embodied in critical articles and reviews. For
information, contact the author.

To my wife Carol

Contents

	Contents in Detail	ix
	Preface	xvii
MODULE 1	**The Preparation Module**	1
	1. Overview	3
	2. Algebra Review	23
MODULE 2	**Finite Mathematics**	55
	3. Progressions	57
	4. Mathematics of Finance	83
MODULE 3	**Modeling Variable Relationships**	123
	5. Functions	125
	6. Analytic Geometry	143
	7. Curve Fitting	207
MODULE 4	**Linear Systems**	253
	8. Linear Systems: Overview	255
	9. Linear System Solution Methods	275
	10. Matrix Algebra	307
	11. Linear Programming	367
MODULE 5	**Calculus**	423
	12. Introduction to Differential Calculus	425
	13. Derivative Rules	461
	14. Applications of the Derivative	495
	15. Optimization	535
	16. Introduction to Integral Calculus	581
	17. Integration Methods and their Application	603
	18. Advanced Integral Calculus	633
	Appendix	663
	Index	665

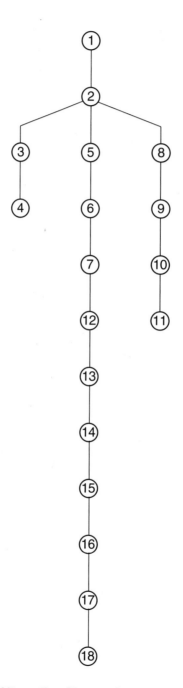

Alternative Chapter Sequences

Contents in Detail

Contents	vii
Preface	xvii

MODULE 1 The Preparation Module — 1

1. **Overview** — 3
   - 1.1 Introduction — 3
     Why Oh Why Do I Have to Learn Mathematics / Mathematics We Do and Math We Don't / Fermat's Last Theorem / Rule of 72
   - 1.2 Learning Plan — 6
     Exercises and Problems
   - 1.3 Mathematical Models — 7
   - 1.4 Applications Galore — 10
   - 1.5 The "Heart" of Mathematics — 11
     Sets / Numbers / Operations / Axioms / Logic, Proofs, and Theorems / Direct Proof / Indirect Proof / Expansion of the Mathematical World
   - 1.6 Summary — 19
     - Problems — 20

2. **Algebra Review** — 23
   - 2.1 Introduction — 23
     Algebra Quiz / Learning Plan
   - 2.2 Equations — 25
   - 2.3 Equality — 25
   - 2.4 Equation Manipulation to Isolate Unknowns — 26
   - 2.5 Exponents — 28
   - 2.6 Factoring — 30
   - 2.7 Fractions — 32
     Sign / Simplification / Common Denominator / Addition and Subtraction of Fractions / Multiplication and Division of Fractions / Compound Fractions
   - 2.8 Special Products — 37
     Binomial Expansion
   - 2.9 Simultaneous Equations — 40
   - 2.10 Logarithms — 42
     Logarithm Laws / Three Logarithm Laws / Richter Scales for Earthquakes

	2.11	Representation of Numbers	47
		Bases / Scientific Notation	
	2.12	The Magic of Algebra	51
		Answers to "Rust Detecting" Algebra Quiz	
		Problems	52

MODULE 2 Finite Mathematics 55

3. Progressions 57
   3.1 Introduction 57
       I.Q. Test for Aliens / Learning Plan
   3.2 Arithmetic Progressions 58
       Summing an Arithmetic Progression
   3.3 Some Applications of Arithmetic Progressions 62
       Theater-in-the-Round / Volume Discounts / Sky Diving / NBA Draft / Promotion Systems
   3.4 Geometric Progressions 67
       Summing a Geometric Progression
   3.5 Applications of Geometric Progressions 70
       Worker's Paradise / Yonkers NY Doubling Fines / Sure-Fire Gambling Scheme / Moore's Law of Computing Power / Sales Forecasting / Rule of 72 / Chain Letters / Population Explosion / March Madness
       Problems 78

4. Mathematics of Finance 83
   4.1 Introduction 83
       Learning Plan
   4.2 Compounding 84
       Compound Growth / Compound Interest / Basic Compound Interest Formula
   4.3 Compound Interest Applications 89
       Computing Bank Account Balances / The Indians Weren't So Dumb After All / Solving the World's Poverty Problem / Where to Stash Your Fortune / More Compounds Syndrome / Multiple Investments
   4.4 Present Values 93
       Savings Goals / Summing a Stream of Present Values / State Lottery
   4.5 Annuities 97
       Annuity for Junior's College Education / Sinking Fund for Fly Now Pay Later / More on Junior's College Fund / Ann Nudity Saves for Acting School / Retirement Annuities / Sports Annuities
   4.6 Capital Budgeting 104
   4.7 Depreciation Methods 107
       Sum-of-Years'-Digits Method / Double-Declining Balance Method
   4.8 Money Theory 111
   4.9 Oh Those Payments 112
   4.10 Cash Back Versus Low Financing 114
   4.11 Common Stock Valuation 115
       Problems 118

CONTENTS

MODULE 3 Modeling Variable Relationships — 123

5. Functions — 125
- 5.1 Introduction — 125
  Football Mathematics / Learning Plan
- 5.2 Numerical Functions — 129
  Equations of Functions / Univariate Functions / Multivariate Functions
- 5.3 Notation — 133
- 5.4 Graphical Representation — 135
- 5.5 Classification of Functions — 139
- Problems — 139

6. Analytic Geometry — 143
- 6.1 Introduction — 143
  Learning Plan
- 6.2 Linear Functions — 145
  i Is for Intercept / s Is for Slope / Point-Slope Formula / Graphing Linear Functions / Multivariate Linear Functions
- 6.3 Applications of Linear Functions — 153
  Demand Curves / Revenue Functions / Total Cost Function / Straight-Line Depreciation / Temperature Conversion / Cricket Chirps / Common Stock Betas / Postage Costs–Linear Step Function / Segmented Linear Functions / Income Tax Woes
- 6.4 Quadratic Equations — 162
  Quadratic Roots Formula / Identification
- 6.5 Applications of Quadratic Equations — 172
  Revenue Functions / Unit Cost Function / Sum of Arithmetic Progression / Bivariate Normal Distribution / Unitary Elasticity Demand Curve / Random Walk
- 6.6 Exponential Functions — 177
- 6.7 Applications of Exponential Functions — 180
  Worker's Paradise / Future and Present Values / Sales Forecasting / Exponential Probability Distribution / Normal Probability Distribution
- 6.8 Logarithmic Functions — 165
- 6.9 Applications of Logarithmic Functions — 188
  Learning Curves / Walking Pace
- 6.10 Relationship of Exponential and Logarithmic Functions — 190
- 6.11 Miscellaneous Functions — 191
  Power Functions / Polynomials / Mixed Functions
- 6.12 Applications of Miscellaneous Functions — 195
  Economic Order Quantity / Volume / Revenue with Parabolic Demand / Government Spending
- 6.13 Summary — 198
- Problems — 198

7. Curve Fitting — 207
- 7.1 Introduction — 207
  Basics of Slope / Learning Plan

	7.2	Curve-Fitting Models *Parameters*	210
	7.3	Constant Slope—Linear Model *Medical Bills / U.S. Crude Oil Production / Internal Revenue Service Blues*	219
	7.4	Constant Percentage Change—Exponential Model *Growth Companies / Population Growth / Car Depreciation*	222
	7.5	Increasing at Decreasing Rate—Logarithmic Model *Learning Curves / Yield Curves*	225
	7.6	Negative Slope with Lower Bound—Hyperbolic Model *Liquity Trap*	227
	7.7	Slope Reversal—Parabolic Model *Four-Minute Mile / Saving Gas*	
	7.8	The "Curvy" Road to Profits *Demand Curves / No-Extras / Warm Underwear / Kinked Television Demand / Revenue Functions / Constant Demand / Linear Demand / Hyperbolic Demand / Cost Curves*	
	7.9	Profit Curves—Breakeven Analysis *Linear Revenue and Cost Curves / Parabolic Revenue and Linear Cost*	242
		Problems	246

MODULE 4 Linear Systems 253

8. Linear Systems: Overview 255

	8.1	Introduction *Chapter and Module Learning Plan*	255
	8.2	Basic Concepts *Linear Family of Equations in Implicit Form / System Composition and Order / Solution / Variable Versus Unknown*	256
	8.3	Applications of Linear Systems *Investments / Gridiron Romance / Demand-and-Supply Equilibrium / Production / Least-Squares Regression / Input-Output Analysis*	260
		Problems	270

9. Linear System Solution Methods 275

	9.1	Introduction *Learning Plan*	275
	9.2	Solution Basics *Methods from the Past / Elimination Method / Constant Times an Equation / Addition and Subtraction of Equations / Solution Possibilities / Independent and Dependent Equations*	275
	9.3	General Solution of 2×2 Linear Systems	286
	9.4	General Solution of 3×3 Linear Systems	288
	9.5	Cramer's Rule *Cramer's Rule for 2×2 Systems / Cramer's Rule for 3×3 Systems / Generalization of Cramer's Rule*	293

	9.6	Cramer's Rule Solution of Applications	299
		Car Rental Problem / Gas Mileage Problem / Investments Problem / Gridiron Romance / Demand-and-Supply Equilibrium / Production Problem / Least-Squares Regression	
		Problems	305

10. Matrix Algebra 307

10.1	Introduction	307
	Learning Plan	
10.2	Matrix Basics	309
	What Is a Matrix? / Order / Transpose / Matrix Equality / Special Types of Matrices / Square / Vector / Identity / Elementary / Diagonal / Unity / Null	
10.3	Matrix Operations	315
	Multiplication of a Matrix by a Constant / Addition and Subtraction of Matrices / Combining Addition (Subtraction) and Constant Multiplication / Matrix Multiplication / Elements of Product Matrix / Matrix Multiplication is Not Commutative / Application of Matrix Multiplication / Multiplication by Identity Matrix / Multiplication of Matrix by Its Inverse	
10.4	$AX = B$ Really Is All Those Equations	330
10.5	Finding the Inverse of a Matrix	331
	Some Theory of Matrix Inversion / Elementary Row Operation Method / Application to Investments Problem	
10.6	Solving Linear Systems by Matrix Inversion	342
	Matrix Solution Flexibility	
10.7	Applications of Matrix Algebra	345
	Leontief Input-Output Model Revisited / Markov Chain Analysis of Market Share / Steady State / Student Load Model / Matrix Solution for Least Squares Regression	
10.8	Computer Matrix Operations	352
	Matrix Calculations with www.calc101.com / Matrix Calculations with Microsoft Excel / Multiplication of Matrices with Excel / Matrix Inverse on Excel / Excel with Large Matrices	
	Problems	359

11. Linear Programming 367

11.1	Introduction	367
	Learning Plan	
11.2	Inequalities	371
	Sense and Strength / Manipulation Rules / Graphical Perspectives / Formulating Real World Inequalities	
11.3	Applications of Linear Programming	378
	Vacation Planning / Production Scheduling / Diet Planning / Fuel Blending	
11.4	Basic Theory and Graphical Insights	383
	Feasible Solution Space / Corner Point Theorem	
11.5	Algebraic Solution Methods	388
	Slack Variables / Basic Solution / Moving on to Better Corners / Basis / Exchange Equations / Entering Variable / Limitations on Entering Variable / Departing Variable / Revision of Basis / Objective Function Value / Move on or Stop	

11.6	Simplex Method	399
	Step 1. Formulation of Initial Tableau Body / Step 2. Identify Variable to Enter Basis / Step 3. Identify Variable to Leave Basis / Step 4. Pivoting—Revise Tableau for New Basis / Step 5. Recycle Until Solution Found	
11.7	Simplex Solution of Applications	404
	Vacation Planning / Investments Problem / Production Scheduling / Diet Planning (A Minimization Case) / Fuel Blending	
11.8	Microsoft Excel Computer Methods	413
	Reference Problem / Fuel Blending Problem	
	Problems	417

MODULE 5 Calculus   423

12. Introduction to Differential Calculus   425

12.1	Introduction	425
	Rates Versus Totals / Rates of Change of Practical Significance / Learning Plan	
12.2	Rates of Change and Graphical Slope	428
	Average Rates of Change / Instantaneous Rates of Change	
12.3	Slope of Curved Functions	433
12.4	Limits	439
	Quick Fix: Set $\Delta x = 0$? / Limit Notation / Limit Laws / Limit Applications / The Important Constant called "e" / Banking (Continuous Compounding) Application	
12.5	Continuity	444
12.6	Programmed Learning of Calculus Basics	445
12.7	The Derivative	452
	Problems	454

13. Derivative Rules   461

13.1	Introduction	461
	Learning Plan	
13.2	Constant Rule	463
13.3	Power Rule	464
13.4	Logarithmic Rule	467
13.5	Exponential Rule	469
13.6	Constant Times a Function Rule	471
13.7	Sum of Functions Rule	473
13.8	Product Rule	475
13.9	Quotient Rule	476
13.10	Chain Rule	477
13.11	Summary of Derivative Rules	481
13.12	Partial Derivatives	483
13.13	Computer Differentiation	485
13.14	Whoa! Let's Stop and See What We Have Accomplished	489
	Problems	490

14. Applications of the Derivative   495

14.1	Introduction	495
	Learning Plan	

	14.2	Microeconomics	497
		Elasticity of Demand / Elasticity for Linear Demand / Relationship of Elasticity and Marginal Revenue / Unitary Elasticity / Constant Elasticity / Individual Consumption and Savings Functions	
	14.3	General Economic Models	508
		Simple Model of Economy / Multiplier / Harrod-Domar Growth Model	
	14.4	Sports	510
		Sports Statistics as Rates of Change / Batting Championship Dynamics	
	14.5	Public Administration	514
		State Budgets / Tax Equalization Rates	
	14.6	Velocity and Acceleration	518
		Falling Bodies / Propelled Bodies / Calculus I Rocket / Hot Rods	
	14.7	Probability	522
		Discrete Probability / Continuous Probability	
	14.8	Science and Engineering	525
		Chemical Reaction / Heat Transfer / Electric Current Flow	
	14.9	Graphing Aid	528
		Problems	530

15. Optimization — 535

	15.1	Introduction	535
		Learning Plan	
	15.2	Optimization Theory	537
		Second Derivative / Optimization Rule / Special Optimization Cases / Inflection Points / End-of-Domain Extremes / Optimization Rules for Functions of Two Variables	
	15.3	Overview of Optimization Applications	543
		Riches at Last for S. Lumlord / Applications of Optimization	
	15.4	Profit Maximization	546
		Maxi Taxi Company / Creaky Airlines No-Extras Fare / Profit Maximization–Breakeven Relationship	
	15.5	Net Return Maximization	550
		Student Recruiting / College and Big $	
	15.6	Cost Minimization	553
		Economic Order Quantity Model / Branch Office Model	
	15.7	Revenue Maximization	558
	15.8	Happiness Optimization	560
		Car Velocity Decision / Investment Portfolio Selection	
	15.9	Optimization in Statistics	565
		Sample Size Determination / Survey Efficiency	
	15.10	Student Study Optimization	568
	15.11	Physical Capacity Maximization	571
	15.12	Least-Squares Regression	573
		Problems	575

16. Introduction to Integral Calculus — 581

	16.1	Introduction	581
		Learning Plan	

	16.2	Area Under a Curve	582
		Midpoint Rectangle Method / "Worst" Methods—Smallest and Largest Rectangles / Intermediate Rectangle Method	
	16.3	Fundamental Theorem of Calculus	589
	16.4	Antidifferentiation	593
		Mystery Operation Revisited / Area Determination by Fundamental Theorem / The Constant of Integration	
		Problems	600

17. Integration Methods and Their Application — 603

	17.1	Introduction	603
		Learning Plan	
	17.2	Heuristic Development of Integration Rules	607
	17.3	Some Applications of Integration	612
		Sales Forecasting / Weather Forecasting / Market Research / Highway Design / Probability and Statistics / Averages—Central Tendency: Median and Mean / Variability: Variance and Standard Deviation / Economic Theory / Capital Budgeting	
	17.4	Integration Using www.calc101.com	624
		Problems	626

18. Advanced Integral Calculus — 633

	18.1	Introduction	633
		Learning Plan	
	18.2	Heuristic Development of Some More Integration Rules	633
		Change of Variable Integration / Integration by Parts	
	18.3	Integration Using www.calc101.com Software	641
	18.4	Applications of Integration	643
		Matt Tress's Capital Budgeting Problem Revisited / Learning Curve: Job Training / Physical Fitness / City Planning: Garbage Dump Forecasting / Ecology: Water Pollution / Ecology: Air Pollution / Probability—Exponential Distribution	
	18.5	Multiple Integration	653
		Single Independent Variable / Two Independent Variables / Probability—Exponential Distribution Application	
	18.6	The End!	658
		Problems	659

Appendix
	Table 1	Natural (base e) and Common (base 10) Logarithms	663
	Table 2	Powers of e and 10	664
Index			665

Preface

"Can we talk?", as comedian Joan Rivers would say. In other words, let's be brutally frank. Mathematics is not your favorite subject. Maybe you even hate it. You would prefer taking a course in your major. On the other hand, many mathematics professors would prefer teaching an advanced course to math majors.

So, how can one resolve the problem of neither student nor professor anxious to be here? Well there are three ingredients. First is your professor, who chose this rather unorthodox applied mathematics textbook. As such, your professor must be an innovative teacher who is student friendly and interested in motivating students. The second ingredient to solve this problem lies with you. You have to have an open mind toward the subject and its importance to your development. Think of it as medicine that doesn't taste good, but sure makes you healthy. The third necessary ingredient is this textbook. It is written in a student-friendly style with humor and a light conversational style to win you over. It covers a wide variety of interesting applications from business, economics, sports, and life in general. At the same time, the level of mathematics is high enough so that the instructor can feel that this is a "real" mathematics course. Overall, when this three-ingredient stew is cooked properly, you will get involved and gung-ho about applied mathematics. Your professor will sense this excitement and do an even better job of teaching. In turn, you will learn even more. And on and on this spiral of expanded learning will go on and on. What a synergism!

The ultimate purpose of this text is to enable the student to conveive of the real world in terms of mathematical concepts and models. This process of "thinking mathematically" involves taking an actual real world problem, fitting a mathematical model to it, solving that model, and then applying the results to better understand reality. The technical mathematical solution is an integral part of the process; yet it remains a means to an end and should not be overestimated. Since thinking mathematically can only come about within a total life experience, applications involving home, sports, probability and statistics, everyday life, as well as business and economics are incorporated into the text.

The presentation of this book is light and conversational, with some humor (OK, bad jokes) thrown in. It is hoped that the reader will even chuckle once in a while. However, there is no attempt to sacrifice mathematical rigor.

Mathematics is best learned by working increasingly more challenging problems. Reading the chapters—even several times—and nodding approvingly rarely helps. Accordingly, this text is written in an interactive mode, with new material, examples, and applications immediately followed by related exercises. The student should use the exercises to check if they have achieved the understanding of the previous work and to build problem-solving confidence. This ever-expanding problem-solving development continues at the end of the chapters. There, three categories of problems graded from easiest to most challenging are provided. Detailed answers to all the exercises and the odd-numbered end-of-chapter problems are provided in a separate

Solutions Manual. Be sure to give an all out effort to solve the problems before giving up and consulting the answers.

This book is organized into five modules: Preparation, Finite Mathematics, Modeling Variable Relationships, Linear Systems, and Calculus. The whole book can be covered in one semester at the master's level. Two semesters are recommended at the undergraduate level. The first semester could include the Preparation, Finite Mathematics, and Linear Systems modules, along with selective topics in the Modeling Variable Relationships module. The second semester could then include the Modeling Variable Relationships and Calculus modules.

Each chapter begins with an introductory section that gives the student an intuitive glimpse at the key concepts to be developed. That is followed by a Learning Plan that spells out the objectives of the chapter in terms of what the student is expected to learn.

Computers play an integral part of this textbook. Microsoft Excel and www.calc101.com software are employed. Excel is employed to perform matrix operations and solve linear programming problems. www.calc101.com is employed to perform matrix operations, solve linear systems, and carry out the calculus operations of finding derivatives and integrals.

I am indebted to many people who helped me write this book. Many of my students at Syracuse University, especially Tim Kallet, gave me valuable feedback and helped me make it a student-friendly book. Professor Erl Sorensen of Bentley College gave me many presentation tips and helped write some sections. As an often-winner of teaching awards at Syracuse, Northeastern, and Bentley, there is no better person to add relevance and student-friendliness to a book. Professor Fred Easton helped me develop the Microsoft Excel linear programming section. Matt Hiemstra, an all-around computer whiz, was instrumental in developing the various computer sections, as well as solving my own personal computer glitches. Mr. George Beck was kind enough to let me use the neat software, www.calc101.com, that he developed. My son, Marlund, helped me produce the *Solutions Manual.* The capable staff at Sheridan Books made the book production care free. I especially thank Mrs. Karen Wenk, who did an outstanding job as typesetter, project manager, and communicator. Finally, I thank my wife of 40 years, Carol, who helped me in the layout of the book and sacrificed much while I was "holed up" writing, rewriting, and rewriting some more.

I hope you will find this book worthwhile. I would appreciate suggestions for improvement, and will personally answer such letters addressed to me at: 104 Easterly Terrace, DeWitt, NY 13214.

Sherman Chottiner

MODULE ONE

The Preparation Module

The best advice you can get upon entering any mathematical study is to thoroughly learn the basics and that is the goal of this module. In Chapter 1, we see how and where mathematical models can be applied to solve real problems. Then we take a look at the basic elements of mathematics that underlie the entire textbook. Because the various mathematical topics and applications assume a good understanding of algebra, we present an algebra review in Chapter 2. This module is a foundation for the entire book; so don't be afraid to return to it whenever you feel that your mathematical building needs a little shoring up in the basement.

CHAPTER

Overview

1.1 INTRODUCTION
Why, Oh Why, Do I Have To Learn Mathematics?

Mathematics is a recent addition to the curriculum for students of management.[1] This contrasts with the physical sciences and engineering where extensive mathematical training has long been cherished. Since World War II, basic changes have led to the need for expanded mathematical training for management. One of these changes is the development of new and important mathematical techniques that have direct application to management. Another change is that economic theory, a fundamental tool of management, has been largely recast in mathematical language. To illustrate this point, several Nobel prizes in Economic Science have been awarded to economists for their work in econometrics, the mathematical expression of economic theory. A third factor is the development of the computer, which communicates best with those who speak its language—mathematics. Finally, the feats of the Space Age, which would not have been possible without mathematics, have provided a proper climate for and stimulated the use of mathematics in all disciplines.

Given the nature of management and mathematics, it is surprising that it took so long for this mathematical revolution to evolve. Although it is quite difficult to define mathematics,[2] suffice it to say it is a language for inquiring into and communicating about the interrelations of things that are measurable. Good management entails the understanding of complex interactions of variables that are both internal and external to the organization. Many of these decision-making variables can be measured: for example, profit, advertising expenditures, output, cash flow, investment, market share, and so forth. Inasmuch as mathematics is the only way such variables can be related in a systematic fashion, and so is an important means that management can profitably use to understand its environment, the mathematics revolution is long overdue.

At the same time that one realizes the importance of mathematics for management, one must also recognize its limitations. Some situations do not lend themselves to mathematical formulation. While some problems can be cast in mathematical terms, they can't be efficiently

---

[1] The terms *business* and *administration* can be substituted for *management* in this text.
[2] Kasner and Newman, in *Mathematics and the Imagination,* tried to define mathematics, but had to be content with the following: "Mathematics is the science which uses easy words for hard ideas."

solved mathematically. So in some cases the understanding of mathematics can either provide management with a powerful problem-solving tool or provide it with the basis for rejecting the use of mathematics if its isn't justified.

Having read the above, I know that many of you are still muttering, "I'll never have to use math; so, why must I learn it?" But wait, let me cite some specific situations where you will encounter mathematics in your studies and in your career.

Let's start with elasticity of demand. You'll study that concept in economics and maybe in several other courses. You would use that concept in marketing analysis in your career. For example, one always wants to know the effect on sales as a result of an increase in price. The answer to that question is whether the product has elastic or inelastic demand, as determined by elasticity of demand. Well, elasticity of demand is based on calculus!

How about compound interest, compound growth of IRA funds, etc.? Who wants to be ignorant of such everyday practical matters? Of course, without the mathematical understanding of the basic formulas behind compounding, you will be deficient in these important matters.

How about everyday terminology, such as "exponential growth" or "learning curve"? Again, mathematics gives you a depth of understanding that allows you to understand these concepts. And so, you will never be at a loss of words at your next cocktail party.

How about statistics, which you all must take? Did you realize that statistics is based completely on mathematics? Without a good understanding of the underlying mathematics, there goes your grade in the statistics course.

I could go on and on. Variance accounting in accounting, brand switch analysis in marketing, portfolio analysis in finance, etc. all require extensive mathematics for understanding.

OK, maybe you still are not convinced. How about this argument: "You will fail this course and ruin your whole career and life if you don't learn math!!!!!"

Mathematics We Do and Math We Don't

Mathematics books are written somewhere along the continuum of 100% applied to 100% theoretical. If you fear the theoretical end with abundant abstraction, humungous formulas, laced with strange symbols, devoid of friendly numbers, relax. We don't do that in this book. We leave that to the pure mathematicians and stay far away from them! At the same time, some of their work is very interesting. So, to see what you are not doing here in this book, consider the interesting history of Fermat's Last Theorem.

Fermat's Last Theorem

In 1637, the brilliant French mathematician, Pierre de Fermat, wrote the following formula in the margin of a book:

$x^n + y^n = z^n$

You probably recognize the $n = 2$ case from geometry as the Pythagorean Theorem:

$x^2 + y^2 = z^2$

For example, the right triangle with sides of $x = 3$ and $y = 4$ has a hypothesis of $z = 5$. So:

$3^2 + 4^2 = 5^2$

Or $9 + 16 = 25$

Fermat scribbled that he proved that one could not find whole numbers for x, y, and z to satisfy the formula when n was a whole number greater than 2; however, he didn't have enough space in the margin to show it. Fermat died without ever showing his proof, which since has been called Fermat's Last Theorem.

To illustrate this impossibility

$2^3 + 3^3 \neq 4^3$ or $8 + 27 \neq 64$

And no other sets of 3 whole numbers raised to the third power will ever yield an equality. Try a few to see!

Theoretical mathematicians have worked for over 350 years trying to prove Fermat's Last Theorem. That was one of the greatest challenges in pure mathematics. Many claims to a proof were offer, but they were all challenged and shown to be wrong. Finally, Dr. Andrew Wiles of Princeton University, after working many years in secret (so that those other mathematicians would not steal his prize) announced his success in 1993. Even then, other mathematicians discovered flaw in aspects of the proof. Finally in 1996, the world of pure mathematics was satisfied that Wiles's monumental 700 page proof, complete with seemingly unrelated topics such as elliptical curves was correct. Hallelujah!

Now back to this book and us. Breathe easy. 700 page proofs are banned from this book. Rather, we will focus on the applications of mathematics. Now mind you, we will sometimes have to do simple mathematical proofs to gain a solid understanding of mathematics. We may even have to use the results of a complicated proof (that thankfully we won't go through step by step) so that we can proceed to develop an application. But, our emphasis is always on the applications.

We illustrate our applications thrust and the use of selected mathematical theorems in the applications process with the so-called Rule of 72.

Rule of 72

Maybe as a result of some prior unpleasant mathematics course you may believe that mathematics makes things more complicated. But it can be just the opposite—mathematics can indeed make life easier and more successful. For example, consider the Rule of 72 (derived in Chapter 3).

$$n = \frac{72}{i}$$

This simple rule tells us how long (n) it would take for a quantity to double if it grows annually at a rate of i. For example, how long would it take your bank balance to double if the interest rate is 6%? The answer is $n = 72 \div 6 = 12$ years! Simple. This little rule alone can propel you to instant success.

Marketing V.P.	"Our new product should experience a doubling of sales in the next 4 years. On the other hand, profits from this product should grow smartly at an annual growth rate of 5%."
President of GM	"That sounds inconsistent. Can you explain further?"

Marketing V.P.	"Sir, that is all my staff person told me."
President	"You, new graduate of Managerial mathematics, what did you learn about all this?"
You (New intern)	"Sir, those projections would require an 18% annual growth rate for sales. That does seem high to me in consideration of the long 12 year time period for profits to double."
President	"I am amazed at you, son (daughter)*. Do you have a computer brain? I bet my daughter (son)* would like to meet such a talented person. She (He)* certainly would be interested in meeting our NEW V.P. of Marketing for Googles Manufacturing."

In Chapter 3, we derive the Rule of 72, using some mathematical proofs from the vast world of Fermat and Wiles. Much of this development is straightforward and easy to understand. One aspect is beyond the scope of this book. So in that case, we just use the theoretical result without any proof of our own. The end result, grounded in mathematical theory, led us to a wonderful application that we can profit from in the real world. The fact that we went through the steps of deriving the rule means that we won't misuse it in inappropriate situations. Also, we see that a complicated real world situation was reduced through mathematics to a simple usable formula. Yes, math can make complicated situations simple, as opposed to what some think—math makes things complicated.

1.2 LEARNING PLAN

Now that you know why you are studying mathematics, you should also know the nature and purpose of the various sections of this book in relation to your learning experience. This introductory learning plan will preface each chapter.

In this chapter, we first explore the nature of mathematical models using as an example the problem of how to choose the cheapest car rental. Once we understand the mathematics, these models enable us to investigate and solve specific problems. You will then see how mathematical models can be applied to help make more effective management decisions. We then list the many applications of mathematics in each chapter throughout the book. In the remainder of the chapter, we will take a brief look at the inner workings of mathematics by observing its heart—the system of sets, numbers, operations, axioms, logic, proof, and theorems. The goal of this latter section is not to make you a pure mathematician, but rather to give you an overview of mathematics so that you can appreciate its power and flexibility in investigating problems in many areas of life.

Exercises and Problems

"Plastics" was the tip for success for young Dustin Hoffman in the classic movie, *The Graduate*. Here, the tip is "exercises and problems". The exercises in this book are interactive and integrated into the chapters. You will find them immediately after a topic is explained or an application

---

*This book has been approved by the P.I.C. (Politically Incorrect Council) and E.O.D. (Equal Opportunity Discriminator)

is presented. These exercises are designed to augment your understanding of a recently discussed subject or application. The detailed answers to all the exercises can be found in a separate Solutions Manual.

At the end of each chapter, you will find problems organized into the following three categories:

1. Technical Triumphs
2. Confidence Builders
3. Mind Stretchers

The technical triumphs involve straight mathematical operations. They are designed to see if you can integrate the mathematical ideas of the chapter, without any potentially confusing real world complications. Mastering these mean you are at the C grade level. The latter two categories of problems are all word problems, and thus do incorporate real world complexity. Such "word" problems are the bugaboo of many students. So we approach them in a manner of increasing complexity. The Confidence Builders are designed to get your feet wet in fairly simple real situations. Mastering them mean that you are at about the B grade level. Then you graduate to the Mind Stretchers, which require you to think more and synthesize more than one idea from different sections of the chapter. Mastering these puts you at the A grade level. Getting 100% of them correct qualifies you to teach the course!

Detailed answers to all the odd-numbered problems are given in a separate Solutions Manual. It is your instructor's decision as to whether you can purchase the Solutions Manual or not.

We illustrate the nature of these different categories of exercises and problems, using the Rule of 72 below;

Rule of 72: $n = \dfrac{72}{i}$ where n = number of years for money to double

i = annual growth rate

Exercise: How long would it take for money in the bank to double if the bank paid a rate of 4%?

Technical Triumph: Solve for i in terms of n.

Confidence Builder: A saver is looking for a bank where their money will double in 10 years. What interest rate should the saver look for?

Mind Stretcher: Suppose one puts $1000 into the bank, if the bank pays an annual interest rate of 6%, how long will it take for the money to quadruple?

1.3 MATHEMATICAL MODELS

The power of mathematics is its ability to capture the essence of a real situation in terms of a set of mathematical statements called a *mathematical model*. This model can then be studied in such a way as to focus attention on the important aspects of the problem. Finally, with this better understanding, the original problem can be alleviated.

To illustrate this process, consider the plight of Matt Maddix, a man faced with a car-rental decision. He needs to rent a car for a one-day trip; only Gertz and Bavis are available to choose from. Considering cost as the only relevant factor, he wants to rent from the firm with better rates. In effect, Matt is saying that advertising claims as to which company "sells harder" are irrelevant.

For comparable models, Gertz charges $10 per day plus $.10 per mile driven, while Bavis charges $20 per day plus $.05 per mile driven. Matt could list all the possible mileage totals (1, 2, 3, 4, . . .) and calculate the costs for each company. For example, if he traveled 1 mile, Gertz would charge $10 + $.10 = $10.10, Bavis would charge $20 + $.05 = $20.05. If he traveled 100 miles, Gertz would charge $10 + 100($.10) = $20, Bavis would charge $20 + 100($.05) = $25, and so on. However, a mathematical equation for each company can serve to summarize all these possibilities. Thus,

$G = 10 + .1M$

$B = 20 + .05M$

where

G is the Gertz charge in dollars
B is the Bavis charge in dollars
M is the number of miles driven

This is a mathematical model. It relates each company's charge to the miles driven. The model makes no mention of such other factors as sales-person courtesy, cleanliness of the cars, mechanical condition of the cars, or convenience of the rental company's location. These variables are either assumed to be unimportant by Matt or they are the same for the two companies. If, for example, Gertz is 10 miles away and Bavis is next door, this would indirectly increase the cost of the Gertz car and should be included in the model.

As it stands, the model is sufficient for Matt's decision since it includes all the variables he considers relevant. Now the scene shifts to the mathematical world, where operations are performed and equations are manipulated. The original situation is not altered. Rather, all the mathematical operations merely serve to focus on, and thus provide better understanding of, the important aspects of the problem. In Matt's situation, the model must determine which rental company is cheaper.

The mathematical model is composed of two equations, each of a special type called a *linear equation*. Together, the two equations constitute a *linear system,* properties of which have been extensively studied. For such a linear system, mathematical theory states that the charges will be equal at one value of M, the indifference value; but when M is less than the indifference value, one charge will be lower and, conversely when M is greater than the indifference value, the other charge will be lower. Specifically, the indifference value, where $G = B,$ is $M = 200$ miles. For an M of less than 200, Gertz is cheaper; for an M greater than 200, Bavis is cheaper. This is shown graphically in Figure 1.1.

Fortunately, in Matt's case a mathematical model of the linear system type was able to duplicate the essence of the real situation (see Figure 1.2). The model has been solved on paper, which is more efficient than physically renting many cars and recording the charges. Once Matt estimates his total mileage, the more economical rental company is determined. Now we can turn our attention to other areas where mathematical models have been applied.

OVERVIEW

Figure 1.1

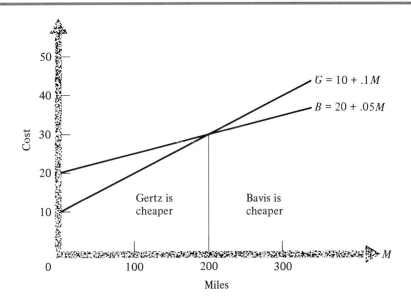

Figure 1.2

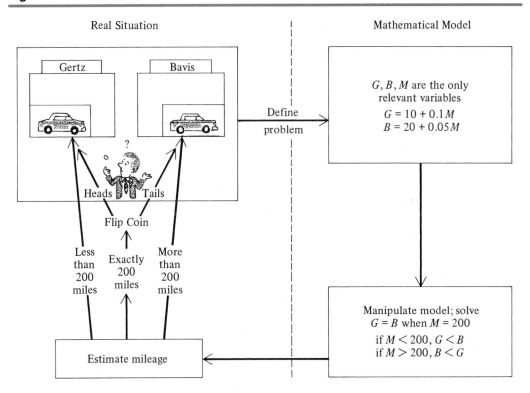

Figure 1.3

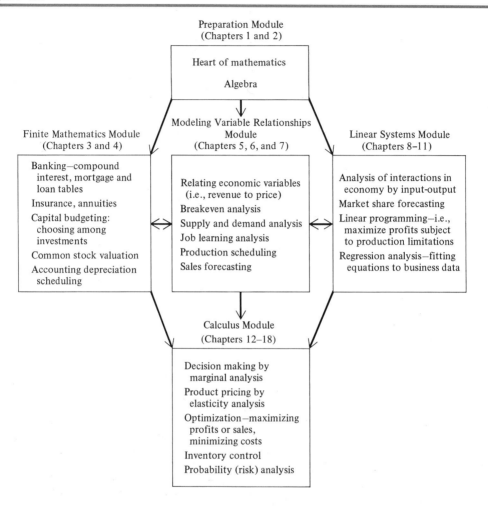

1.4 APPLICATIONS GALORE

Management students usually need to be convinced of the applicability of mathematics. Fortunately, mathematical models developed from many mathematical topics have been successfully applied in a wide variety of management areas. Figure 1.3 provides a partial list of the kinds of problems that have benefited from mathematical methods. The various applications are broken down into major mathematical areas of study, called modules, which are the basic building blocks of this text. However, since most applications require the integration of many topics in mathematics, one should be aware that the listing is not straightforward. For example, breakeven analysis requires mathematical tools from both the linear systems and the modeling variable relationships modules.

1.5 THE "HEART" OF MATHEMATICS

You may be wondering how mathematics is flexible enough to allow such widely varied applications. Any answer to this question can only be appreciated after the various mathematical topics and applications are understood. However, the flexibility and durability of the "heart" of mathematics has a lot to do with it.

The heart of mathematics contains the five basic components shown in the drawing.

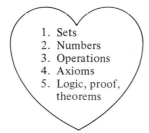

1. Sets
2. Numbers
3. Operations
4. Axioms
5. Logic, proof, theorems

Now let's carefully examine each of these components to see what makes the heart tick.

Sets

A set is a well-defined collection of objects, called elements. Normally we name sets with capital letters, reflecting their contents, if possible.

Examples[3]

$I = \{1, 2, 3, \ldots, 10\}$

$E = \{2, 4, 6, 8, \ldots, 100\}$

$V = \{a, e, i, o, u\}$

$F = \left\{\dfrac{1}{1}, \dfrac{1}{2}, \dfrac{1}{3}, \dfrac{1}{4}, \ldots\right\}$

$N = \{$All Nasdaq listed stocks as of Jan 1, 2003$\}$

$P = \{$All firms located on Lake Pollution$\}$

$B = \{$All dot com companies that went bankrupt in 2001$\}$

Collections of objects that are not precisely defined (jet set), or whose elements are a matter of opinion (set of good books), are not sets in the mathematical sense.

---

[3] ... means "and so on and so forth according to the established pattern." If the established pattern is terminated somewhere, the final element is inserted, as in sets I and E. If no terminal element is indicated, the pattern is assumed to continue for an infinite number of elements, as in set F.

A set can be defined by its individual elements, (sets *I, E, V, F*) or by a generic description of its elements (sets *N, P,* and *B*). A set can have a finite (all except set *F*) or an infinite (set *F*) number of elements.

Numbers

Numbers serve to measure. They answer "how many" and "how much" questions as expressed in some unit of measurement: 47° Fahrenheit, $82,000,000 sales, 150 hospital beds, 60 miles per hour, and so on. The set of numbers has expanded over time in response to both the theoretical advances of mathematics and the measurement requirements of practitioners. Here we trace the history of these developments.

Early man—the prehistoric hunter—and the agrarian man of earliest recorded history had need of only simple measurements. The former had to keep track of weapons, prey, and family members; the latter recorded sheep, crops, and wives. The *natural numbers* to which we now give the symbols

$\{1, 2, 3, 4, 5, \dots\}$

were sufficient for them.

This set of numbers is mathematically closed under the operation of addition. That is, one can take any two natural numbers, add them, and the result will be another natural number. So, if Tab the Hunter had two wives, and as a result of his fine hunting he was able to support three more, Tab knew that some natural number (in this case, five) would measure his new marital situation. Alan the Shepherd could even do all his subtraction problems with a set of natural numbers. But as man grew more commercial, a point was reached where the natural numbers were not sufficient for measurement purposes. This point occurred when a Babylonian trader wrote a check for more than he had in his account. At that point the banker said, "Ah, the natural numbers are not 'closed' under subtraction," meaning that the result of taking one natural number minus another need not be another natural number. At this time an expansion of the number system was in order and the set of *integers*

$\{\dots, -4, -3, -2, -1, 0, 1, 2, 3, 4, \dots\}$

resulted.

The integers include the natural numbers, zero, and negatives of natural numbers. With such numbers the trader could carry (for a while anyway) a negative bank balance, and the early "growth stocks" could show negative earnings.

Everything went nicely until a Greek manager of the Parthenon Construction Company tried to use cost accounting methods to determine the cost per man hour of labor. After finding a total cost of $87 for 1298 hours of labor (labor was cheap in those days), he summed up his plight by saying, "I can't get the answer because the integers are not closed under division." He meant that dividing one integer by another need not result in another integer. To correct this, the set of *rational* numbers was quickly invented. These numbers are of the form *p/q* where *p* and *q* are integers and *q* is not zero. For example,

$$\frac{3}{2}, \frac{7}{13}, \frac{-2}{181}, \frac{-11}{-3}, \frac{17}{-16}$$

are all rational numbers. The reader might be surprised to know that such numbers as 13.86 and even 17.1826403955 are rational numbers. This is because

$$13.86 = \frac{1386}{100} \quad \text{and} \quad 17.1826403955 = \frac{171{,}826{,}403{,}955}{10{,}000{,}000{,}000}$$

At this point it is difficult to think of a number that isn't a rational number. One could attach a trillion digits both before and after the decimal to the numbers above, and they still could be written as one integer divided by another. So, one can imagine the dismay and frustration of some Greek mathematicians when Pythagoras (fifth century B.C.) proved that $\sqrt{2}$, a number he encountered every day in his work with triangles, was not a rational number (see page 17 for this proof) since no two integers p and q could be found so that $p/q = \sqrt{2}$.

Defying rationality, $\sqrt{2}$ was called an *irrational* number. For about a century it was believed to be the only irrational number. But then several others ($\sqrt{3}$, $\sqrt{5}$, $\sqrt{6}$, $\sqrt{7}$) were shown not to be rational. As of a few centuries ago, it was known that such important numbers as π (of circle fame) and e (the base of the natural logarithmic system) were irrational; in fact, there are more irrational than rational numbers. The set of all rational and all irrational numbers is called the set of *real numbers*.

The distinction between rational and irrational is that rational numbers have a predictable decimal pattern, whereas irrational numbers have a non-predictable decimal pattern. For example,

$$\frac{4}{3} = 1.333\ldots$$

is, of course, rational because it is formed by dividing two integers, but note also that one can predict its decimal pattern as three's forever. The rational number

$$\frac{4}{11} = .363636\ldots$$

has a repeating 36 pattern forever; the rational number

$$\frac{1}{7} = .142857142857\ldots$$

has the 6-digit block, 142857, which repeats forever; and the rational number

$$\frac{60}{19} = 3.157894736842105263157894736842105263\ldots$$

has an 18-digit block which repeats forever. See if you can find the 16-digit repeating decimal block for

$$\frac{45}{17}$$

But

$$\pi = 3.14159\ldots$$

doesn't have a predictable decimal pattern. No one can forecast its one billionth decimal place. Even if its first one billion decimal places were calculated, this would not help in predicting the next billion places. It is as if the decimal is formed by random numbers, none of which nor any group of which will assist in predicting the digits yet to come.

It should be emphasized that the distinction between rational and irrational numbers is largely theoretical. For example, most practical problems involving the use of π can be satisfactorily worked by approximating π by the rational number 3.14159. However, since π has been calculated to a considerable number of decimal places, one can achieve any accuracy desired.[4]

Having reached the set of real numbers, one might believe that it contained all the numbers one would ever need. But, consider the equation

$$x^2 + 1 = 0.$$

The solution is

$$x^2 = -1 \text{ or } x = \pm\sqrt{-1}$$

the square root of -1. No such real number exists. Consequently, about three centuries ago, the so-called imaginary numbers, which allow for the above solution, came to be included in the number system. For many years these numbers were of theoretical interest only. Leibnitz, a famous mathematician, said of them in 1702, "They are a wonderful flight of God's spirit, they are almost an amphibian between being and not being." However, one should not let the term *imaginary number* and Leibnitz's statement mislead you. In the last century these numbers have proven to be essential for many applications in science, engineering (for example, alternating current network analysis), and even economics (multiplier-accelerator model of the economy). The real and imaginary numbers taken together form the set of *complex numbers*.

Today we have a number system expanded several times (as shown in Figure 1.4) to include new numbers essential to solve problems. Would anyone want to bet that some mathematician won't come along some day and uncover a whole new set of numbers essential to people in the future? In this book the real numbers will be sufficient to solve all our problems.

Operations

There are operations that can be performed on numbers, on matrices, on sets, on functions, and so on. For example, addition is an operation that can be performed on two numbers or two matrices; differentiation is an operation of calculus that can be performed on a function.

Operations involve doing something with mathematical entities. You are familiar with arithmetic operations of addition, subtraction, multiplication, and division that one can perform using two numbers. For example, $2 + 3 = 5$, $11 - 3 = 8$, $4 \cdot 5 = 20$, and $150 \div 15 = 10$. If only everything were that easy!!

We will study more advanced operations, such as taking the logarithm of a number, multiplying two matrices, taking the derivative of a function, etc. You may have never heard of some of these. But be assured that in time, they too will be easy!!

---

[4]The determination of the decimal places of π has received considerable attention over the ages. A landmark in this development was William Shanks' computation in 1873 of π to 707 decimal places, all of which, by his request, are inscribed on his tombstone. Today, computers have made possible the calculation of π out to over a trillion decimal places.

Figure 1.4

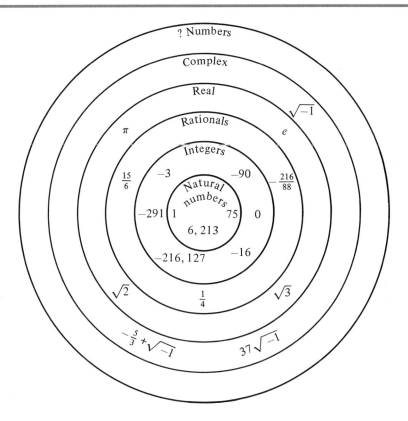

Axioms

With sets, numbers, and the arithmetic operations defined, the rules by which we operate on numbers (called *algebra*) can be considered. The rules of algebra, such as $5 \cdot 0 = 0$, $1/(1/10) = 10$, $xy + 2y = y(x + 2)$, and $\dfrac{1}{3} + \dfrac{3}{4} = \dfrac{1 \cdot 4 + 3 \cdot 3}{3 \cdot 4} = \dfrac{13}{12}$ have too often been memorized, but not really understood, by those who studied the "old math." However, each of these rules has a definite reason for existing; and they all follow from a few initial statements, call *axioms*. Since a mathematical system cannot evolve from nothing, the axioms provide necessary initial statements of truth that are accepted without proof. Then, using mathematical logic in direct, indirect, and inductive proofs, one can show that other mathematical statements called *theorems* follow from the axioms. These theorems are in turn used with the axioms to derive other theorems. This expansion of mathematics continues indefinitely. Through the use of the logic process, all the existing "truth" can be employed at any point in order to generate new statements (theorems), which are consistent with the previous body of knowledge. This process doesn't end with the development of algebra. The theorems of calculus and other advanced topics are built on top of those of algebra.

The choice of axiom set determines the nature of a mathematical system. Mathematicians have been able to determine the smallest set of axioms that will provide for a viable mathematical system. This set is listed below, where *a*, *b*, and *c* stand for any real numbers, and 0 and 1 are unique real numbers:

Type	Examples
Equality axioms	
1. $a = a$ (reflexivity)	$1 = 1$
2. If $a = b$ the $b = a$ (symmetry)	If $x = 2$, then $2 = x$
3. If $a = b$ and $b = c$, then $a = c$ (transitivity)	If $x = 3$ and $3 = y$, then $x = y$
Closure axioms	
4. (for addition) $a + b$ = unique real number	$4 + 5 = 9$
5. (for multiplication) $a \cdot b$ = unique real number	$4 \cdot 5 = 20$
Commutative axioms	
6. (for addition) $a + b = b + a$	$6 + 7 = 7 + 6$
7. (for multiplication) $a \cdot b = b \cdot a$	$6 \cdot 7 = 7 \cdot 6$
Associative axioms	
8. (for addition) $a + b + c = (a + b) + c = a + (b + c)$	$2 + 3 + 4 = (2 + 3) + 4 = 2 + (3 + 4)$
9. (for multiplication) $a \cdot b \cdot c = (a \cdot b) \cdot c = a \cdot (b \cdot c)$	$2 \cdot 3 \cdot 4 = (2 \cdot 3) \cdot 4 = 2 \cdot (3 \cdot 4)$
Distributive axiom	
10. $a \cdot (b + c) = a \cdot b + a \cdot c$	$2 \cdot (3 + 4) = 2 \cdot 3 + 2 \cdot 4$
Identity axioms	
11. (for addition) $a + 0 = a$	$5 + 0 = 5$
12. (for multiplication) $a \cdot 1 = a$	$5 \cdot 1 = 5$
Inverse axioms	
13. (for addition) $a + (-a) = 0$	$6 + (-6) = 0$
14. (for multiplication) $a \cdot \frac{1}{a} = 1$ ($a \neq 0$)	$6 \cdot \frac{1}{6} = 1$

While it is very possible that the reader is not too impressed by the above axiom set, you should consider that all the topics in this text follow from it, and, in fact, couldn't exist without it.

Logic, Proofs, and Theorems

Mathematical "truth" beyond the axioms is determined via certain logical steps, which, taken together, are called *proofs*. To illustrate this process, two proofs, one of the direct type and the other of the indirect type, are now presented.

Direct Proof

Show

$$a(b + c + d) = ab + ac + ad$$

where a, b, c, d stand for any real numbers. If, for example,

$$a = 2, b = 3, c = 4, d = 5$$

we can see that

$$2(3 + 4 + 5) = 2(12) = 24,$$

which is, indeed, equal to

$$2 \cdot 3 + 2 \cdot 4 + 2 \cdot 5 = 6 + 8 + 10 = 24.$$

But the mathematical proof must proceed on general lines; thus we must return to the letters.
First, notice that the distributive axiom shows how to expand a number multiplied by the sum of two numbers,

$$a(b + c) = ab + ac.$$

but no axiom describes the expansion of a number multiplied by the sum of three numbers:

$$a(b + c + d) = ?$$

But, with a series of logical steps, using the axioms, one can accomplish the proof as shown.

	Axiom or reason
$a(b + c + d) = a[(b + c) + d]$	Associative for addition
Since $(b + c)$ is equal to some unique real number, which we call β, or $(b + c) = \beta$	Additive closure
$\quad = a(\beta + d)$	
$\quad = a\beta + ad$	Distributive
$\quad = a(b + c) + ad$	Substituting back for β
$\quad = (ab + ac) + ad$	Distributive
Since $ab, ac,$ and ad are unique real numbers	Multiplicative closure
$\quad = ab + ac + ad$	Associative for addition

Indirect Proof

Show that $\sqrt{2}$ is not a rational number. This proof begins by supposing that $\sqrt{2}$ is a rational number. Then, after a series of *correct* mathematical steps, a point is reached where there is an obvious contradiction. Since the steps were correct, the only reason for a contradiction is that the first statement ($\sqrt{2}$ is rational) must be incorrect (this indirect way of proof is characteristic of the indirect proof method).

Suppose $\sqrt{2}$ is a rational number and, as such, can be expressed as the ratio of two integers.

$$\sqrt{2} = \frac{p}{q}$$

where p, q are positive integers and p/q is reduced to its lowest form, that is, no integer divides "evenly" into both p and q. Multiplying both sides of this equation by q gives

$$p = \sqrt{2}q.$$

then squaring both sides gives

$$p^2 = 2q^2$$

An odd integer squared is odd ($1^2 = 1$, $3^2 = 9$, $5^2 = 25$, $7^2 = 49$, etc.). An even integer squared is even ($2^2 = 4$, $4^2 = 16$, $6^2 = 36$, $8^2 = 64$). Two times any integer, whether even or odd, is always even ($2 \cdot 1 = 2$, $2 \cdot 2 = 4$, $2 \cdot 3 = 6$, $2 \cdot 4 = 8$, etc.). Since q is an integer, $2 \cdot q^2$ must be even, which of course, makes p^2 even. Being even, p^2 can be divided evenly by 2. It follows, therefore, that p must also be an even integer, since only an even integer could upon squaring give an even integer.

Inasmuch as any even integer can be written as 2 times another integer (i.e., $10 = 2 \cdot 5$, $12 = 2 \cdot 6$), we can let $p = 2r$, where r is some integer. Then

$$p^2 = 2q^2$$
$$(2r)^2 = 2q^2$$
$$4r^2 = 2q^2$$
$$q^2 = 2r^2$$

The same argument as above shows that $2r^2$ must be even, q^2 must be even, and q is even and can be divided evenly by 2.

So, we started with the statement that $\sqrt{2}$ is rational and, therefore, is equal to some integer p divided by some integer q, both of which are not divisible by the same integer. Lo and behold, we later find that both of them are divisible by 2. And, since no mistakes were made in the various steps (we hope not anyway), we can say that the initial supposition ($\sqrt{2}$ is a rational number) is wrong.

Expansion of the Mathematical World

In essence, our two proofs have shown that two mathematical statements (called *theorems*) are true, which means only that they are consistent with previously accepted mathematical truths. Each proof and resulting theorem thus expands the world of mathematics. Although you will not be expected to engage in extensive proving in this text, it is instructive to see the role of proofs in the development of mathematics.

The process, as illustrated in Figure 1.5, begins with the axioms, which are those mathematical statements accepted without proof. (We must start somewhere, and that set of state-

Figure 1.5

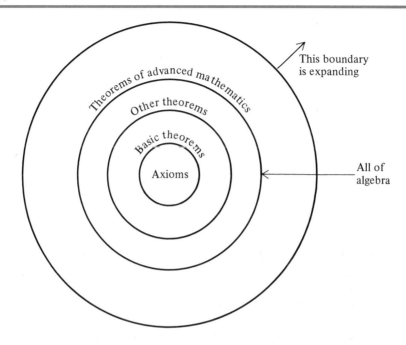

ments cannot be proven.) Then, using logic in the form of direct, indirect, or inductive proofs, other statements—basic theorems—can be shown to be consistent with the axioms. At this point in the development, our mathematical system would include the axioms and the basic theorems. From this base of expanded truths, other statements or theorems can be shown to be consistent, and at some point in this expansion process, all of algebra is developed. Then, additional proofs, aided by key definitions, open up whole new areas of mathematics. The process goes on and on because there is practically no limit to the expansion of mathematical truth. It is nice to know this, but don't let the thought of unlimited mathematical truth upset you. We will stay far away from the outer limits of this world.

1.6 SUMMARY

The usefulness of mathematical models in many diverse fields of inquiry is well documented. For many reasons, mathematics is just coming into its own as a means of solving management problems. With the growth of computers and mathematical models in managerial decision-making, today's student will have to understand more mathematics than did his or her predecessor. This understanding is helped by seeing the "big picture," including the basic components of the mathematical world and the process by which this world expands to include more truth. Providing this perspective was the purpose of Chapter 1. The next logical step is to review algebra, the foundation for all the advanced topics in this text.

PROBLEMS

TECHNICAL TRIUMPHS

1. How many elements are in the following sets?
   a. All presidents of the United States
   b. U.S. Senators at any given time.
   c. Planets in our solar system
   d. Courses you are taking this semester

2. What type of number is each of the following?
   a. 618
   b. 19.3
   c. .6666666 . . .
   d. 5.29185 . . . forever with no decimal pattern
   e. 0
   f. $\dfrac{1}{80,000,000}$ or Powerball chance of winning
   g. $\sqrt{-7}$

3. Find the set of repeating decimal places for the number $\dfrac{5}{13}$.

4. Which axiom is the bases for each of the following?
   a. $3x + 2 = y$ so $y = 3x + 2$
   b. $B(1 + i) = B \cdot 1 + Bi$
   c. $\dfrac{7x}{9x} = \dfrac{7}{9}$
   d. $x + (2y - 2y) = x$
   e. $x \cdot 5 = 5x$
   f. $4 + 200 = 204$
   g. If $x = y$ and $y = 10$, then $x = 10$
   h. $2x + (-2x) = 0$
   i. $2 + x + 3x = 2 + 4x$
   j. $9 \cdot x \cdot 7 = 9 \cdot 7 \cdot x$
   k. $40 = 40$

5. The Big Bang Theory that describes the beginning of our universe states that "the initial bang" took 10^{-39} seconds. This is a number that has 39 zeros after the decimal point followed by a 1. What type of number is this?

CONFIDENCE BUILDERS

1. The following operation

   $xy + xz = x(y + z)$

   is called "factoring", where the common term, x, is said to be "factored out". What axiom is factoring based on? Explain.

2. In baseball, batting average is defined as the number of base hits divided by the number of times at bat (excluding walks).
   a. What type of number is batting average?
   b. Is the set of possible batting averages finite or infinite? Why?

OVERVIEW

3. A baseball pitcher's earned run average, ERA, is calculated as the total number of earned runs allowed for the season divided by 9.
   a. What type of number is ERA?
   b. Is the set of possible ERA numbers finite or infinite? Why?

4. Consider the Rule of 72.
   a. Solve for the interest rate in terms of n, number of years to double.
   b. What interest rate would be necessary for your money to double in 8 years?

5. What axiom is the basis for cancellation? For example:

$$\frac{x(2 + y)}{x} = 2 + y$$

6. An ad for Bombardier offers a $1000 rebate or a 0% interest rate. What factors must be consider in order to make a decision?

7. Find another set of whole numbers (x, y, and z) that satisfy the equation

$$x^2 + y^2 = z^2$$

   (Note: find one other than $x = 3$, $y = 4$, and $z = 5$ given in the text)

MIND STRETCHERS

1. My wife and I both have medical insurance plans that include prescription drug benefits. Her plan has a $5 co-payment, which means that she has to pay $5 for each prescription. My plan pays 10% of the cost of the prescription. When filling a prescription, we have a choice as to which medical plan to use.
   a. Determine a model based on the prescription cost (C), which tells you when to use each one.
   b. Explain how this model is similar to the car-rental model. In what way is it different?
   c. Graph the model.

2. One internet service, LOA, offers unlimited access for $20 per month. Another, Marslink, offers 30 hours per month for $5, with an added charge of 50 cents for each hour over 30.
   a. Determine and graph the two cost equations.
   b. From the graph, what monthly usage would result in equal costs?
   c. Under what usage conditions would each service have a lower cost?

3. Two different stock brokerage firms have the following commissions formulas

   $y = 10 + .01n$ where $n =$ number of shares

   $y = 20$

   a. Graph both equations
   b. For how many shares would the two commission costs be equal?
   c. If you wanted the lowest commission, when would each firm be cheaper? (Note: when will be in terms of n, number of shares traded)

4. Use the axioms and the proof in the chapter

   $a(b + c + d) = ab + ac + ad$

   to prove

   $a(b + c + d + e) = ab + ac + ad + ae$

5. Consider Fermat's Last Theorem for $n = 3$. Also consider trying three consecutive numbers written in general as

$$x^3 + (x + 1)^3 \stackrel{?}{=} (x + 2)^3$$

A little bit of algebra (which you'll learn in the next chapter) requires the following equation

$$x^3 \stackrel{?}{=} 2x^2 + 9x + 7$$

to hold in order to have equality.
   a. Substitute in the values of $x = 2, 3, 4, 5$, and 6. In each case, determine the value of the left and right sides of the equation. You can see here that equality never occurred.
   b. Graph the difference (right side minus the left side).
   c. What value of x gave the closest to equality?
   d. From the trend in the difference, which side would you predict to be bigger when $x = 7$?

6. Consider the following Capital Budgeting model

$$P = \frac{20}{i}.$$

$$I = 2$$

where P is the present value of the future project returns, I is the initial investment, and i is the desired rate of return for the company.
   Whenever P is greater than I, one should invest in the project.
   a. Graph both equations, with i on the x-axis. (Note: $I = 2$ is a horizontal line)
   b. From the graph, what values of i result in P being greater than I?
   c. Solve algebraically for the interest rate when $P = I$.
   d. Construct a decision rule that tells when (what values of i) one should invest in the project.

7. 10-10-987, a long distance telephone company, charges 3 cents per minute, plus a 39 cent connect charge for calls in the USA. AT&T charges 7 cents per minute with no connect charge.
   a. Determine the telephone cost equations for each company.
   b. Determine the decision rule which gives the cheaper cost company for any telephone call time.

CHAPTER 2

Algebra Review

2.1 INTRODUCTION

Algebra is certainly not a glamorous subject. Nobody ever wrote home to his or her parents to brag that they are studying algebra. Can you just imagine the response if you did?

"Algebra! In college? The big-shot is studying algebra! Isn't that what our ninth-grader is studying? For that, I have to pay $15,000 tuition!"

For you MBA students, the humiliation would be even worse. Here is one come-on line that you would be well-advised not to use:

"Hi! I am in an MBA program where I am really excelling in algebra!" (Member of opposite sex exits immediately!)

Despite the lowliness of algebra's reputation, I guarantee you that there is no more important subject for insuring your mathematical success. For no matter how advanced the topic (especially calculus), algebra will be essential for your understanding. As verification of this statement, consider the problems at the end of this chapter. They were all gleaned from the various topics of this text. Without algebra, none of those advanced problems could be solved.

Everyone has taken algebra, most likely in high school years ago. Most likely, you haven't worn off the cover of your algebra book since then. So with lack of use, you have become a little rusty at it and need a review.

Before embarking on our review of algebra, we should assess the thickness of the rust you have undoubtedly accumulated about this subject. To accomplish this and to prescribe the necessary rust remover, take the following 10-minute algebra quiz (time yourself—no cheating)! Grade yourself with the answers provided at the end of the chapter and take the appropriate corrective action described in the Learning Plan.

Algebra Quiz

1. $\left(\dfrac{5}{4}\right)^2 =$
2. $3^{-2} =$
3. $10^0 =$
4. $\log_{10} 100 =$
5. Which is true? (circle one) $-\left(\dfrac{3}{5}\right) = \dfrac{-3}{+5}$ or $-\left(\dfrac{3}{5}\right) = \dfrac{-3}{-5}$
6. Simplify: $\dfrac{6x^2y + 9xy^2}{3xy} =$
7. Form a common denominator and simplify: $1 - \left(\dfrac{1+x}{1+y}\right) =$
8. Expand: $(2x + 3)^2 =$
9. Solve for x: $\quad ax - x = b$
10. Solve the system: $\quad \begin{array}{l} x + y = 2 \\ 2x - y = -5 \end{array}$

Learning Plan

How did you do? Were there questions that you almost got? Be tough! Give no partial credit! At this stage of the game it is better to err on the side of a stronger rust remover. With that philosophy in mind, here are the corrective actions you should take.

Number correct	Condition	Corrective action
10	Beautiful	Congratulations. You just finished this chapter.
7–9	Slightly tarnished	Proceed through this chapter.
4–6	Rusty	This chapter, plus Schaum's *Outline of College Algebra* (McGraw-Hill) (or any good algebra book) will help.
0–3	Corroded	Begin with Schaum's *Outline of Elementary Algebra,* and then proceed to take the corrective action for a rusty condition. "Sesame Street" wouldn't hurt either.

In the review process it is often wise to relate to the axioms. Ask yourself what axioms allow a certain algebraic step. When you do this, you are less likely to violate the law and end your days in the jail for mathematical sinners.

The algebra review in this chapter will proceed along the following topical order:

1. Equations
2. Equality
3. Equation manipulation to isolate unknowns
4. Exponents
5. Factoring
6. Fractions
7. Special products

ALGEBRA REVIEW

    8. Simultaneous equations
    9. Logarithms
  10. Representation of numbers

If your rust removal prescription includes doses of other texts, be sure to emphasize these topics there.

2.2 EQUATIONS

An equation has a left (L) and a right (R) side separated by an equal sign:

$$L = R$$

The equal sign means that the number L is the same as the number R. These letters, L and R, may stand for complicated expressions. For example, if

$$\frac{36 + 4}{10} + 6 = 1 + 2 + 3 + 4$$

then

$$L = \frac{36 + 4}{10} + 6 = 10$$

$$R = 1 + 2 + 3 + 4 = 10$$

L may be unknown, but determinable once the value of R is found. For example,

$$L = (2 \cdot 3) + 4 = 10$$

Part of L may be unknown. For example,

$$x + 3 = 8$$

Since $L = R$, $L = 8$ here. But more than likely, the value of x, not L, would have been sought.

Both L and R could be unknown. If x and y are variables that can assume many different values as defined by the equation

$$x = 2y$$

then L and R can have many possible values.

2.3 EQUALITY

An equation can be viewed as a two-pan scale with the same weight (number) on each pan. The same weight can be added or subtracted from both sides of a two-pan scale without affecting the balance. Even 6 times or $\frac{1}{3}$ of the weights can be substituted without affecting the scale balance. So, too, with equations. Adding, subtracting, multiplying, or dividing both sides of an equation by the same number doesn't affect the balance and thus the equality. In fact, any legitimate mathematical operation (such as taking logarithms or taking square roots performed on *both* sides of the equation) does not violate the equality.

News Bulletin

Mathachusetts has just passed the ERA (Equation Rights Amendment), making it illegal to treat one side of an equation differently from the other side. Senator Joe NoMath, who cast the only negative vote, proclaimed that giving the Left equal rights would destroy all our precious freedoms.

AP—"Algebra Press"

2.4 EQUATION MANIPULATION TO ISOLATE UNKNOWNS

Doing the same thing to both sides of an equation (equation manipulation) allows for isolation of the unknown and, consequently, solving for its numerical value. Let's illustrate with three examples. In these examples, we will employ the axioms.

Examples 1

1. $x + 3 = 8$

Note that the unknown, x, could be isolated on the left side if the 3 were somehow removed. This can be done by adding -3 to both sides.

$x + 3 + (-3) = 8 + (-3)$

$3 + (-3) = 0$ *Inverse axiom for addition*

$x + 0 = x$ *Identity axiom for addition*

So

$x = 8 - 3 = 5$

ALGEBRA REVIEW

2. $10 = \dfrac{x+4}{5}$

In order to solve for x, the equation must be manipulated to isolate x. This can be accomplished as follows. First both sides are multiplied by 5.

$5 \cdot 10 = \dfrac{5(x+4)}{5}$

Since dividing by 5 is the same as multiplying by $\dfrac{1}{5}$, we have

$50 = \dfrac{1}{5} \cdot 5(x+4)$

$ = 1(x+4)$ *Inverse axiom for multiplication*

$50 = (x+4)$ *Identity axiom for multiplication*

This puts us at the starting point of our first example. So by judicious use of the additive inverse and identity axioms, we get

$46 = x$ or $x = 46$

3. $\dfrac{3}{(x-2)} = 4$

First, we multiply both sides of the equation by $(x-2)$:

$\dfrac{3}{(x-2)} \cdot (x-2) = 4(x-2)$

$\qquad\qquad 3 \cdot 1 = 4(x-2)$ *Inverse axiom for multiplication*

$\qquad\qquad 3 = 4(x-2)$ *Identity axiom for multiplication*

$\qquad\qquad 3 = 4x - 8$ *Distributive axiom*

$\qquad 3 + 8 = 4x - 8 + 8$ *Add 8 to both sides*

$\qquad\qquad 11 = 4x$ *Inverse and identity axioms for addition*

$\qquad \dfrac{1}{4} \cdot 11 = \dfrac{1}{4} \cdot 4x$ *Multiply both sides by $\dfrac{1}{4}$*

$\qquad\qquad \dfrac{11}{4} = x$

You may be thinking that this process to solve for x is too detailed, and that you could combine several steps without once mentioning an axiom. This may be so, but you just may be the person who can quickly violate the axioms and get the wrong answer. So for a while, you should get into the habit of referring to the axioms and taking the steps one by one. Try it on these exercises.

Exercises 1

Manipulate the equations to solve for x.

1. $3 - x = 7$
2. $\dfrac{(x-1)}{4} = 2$
3. $2 = \dfrac{4}{(x+3)}$
4. $2x + 1 = x - 5$

2.5 EXPONENTS

In mathematical applications we often are called upon to multiply a number by itself several times. To simplify this process, the following definitions have been adopted.

$$\underbrace{a \cdot a \cdot a \cdot \ldots \cdot a}_{m\ a\text{'s}} \equiv a^m$$

$$\dfrac{1}{\underbrace{a \cdot a \cdot a \cdot \ldots \cdot a}_{m\ a\text{'s}}} \equiv a^{-m}$$

(Note that $\equiv$ means "is defined as.")

Here a (called the base) is any real number, and m (called the exponent) is any natural number.

Examples 2

1. $2 \cdot 2 \cdot 2 = 2^3$
2. $\left(\dfrac{4}{5}\right)\left(\dfrac{4}{5}\right) = \left(\dfrac{4}{5}\right)^2$
3. $\dfrac{1}{3 \cdot 3 \cdot 3 \cdot 3} = 3^{-4}$
4. $(-1)(-1)(-1)(-1) = (-1)^4$
5. $\dfrac{1}{\left(\dfrac{1}{5}\right)\left(\dfrac{1}{5}\right)} = \left(\dfrac{1}{5}\right)^{-2}$

Exercises 2

1. $10 \cdot 10 \cdot 10 \cdot 10 =$
2. $\dfrac{1}{6 \cdot 6 \cdot 6} =$
3. $(-4)(-4) =$
4. $\dfrac{1}{2 \cdot 2 \cdot 2 \cdot 2 \cdot 2 \cdot 2} =$
5. $(1.05)(1.05)(1.05)(1.05) =$

Additionally, we will need to use numbers with rational exponents, defined as $a^{m/n} \equiv n^{th}$ root of a^m, or the number which, if multiplied by itself n times, equals a^m. Alternatively this may be written as $\sqrt[n]{a^m}$. For example

ALGEBRA REVIEW

$$4^{3/2} = \sqrt[2]{4^3} = \sqrt[2]{4 \cdot 4 \cdot 4} = \sqrt[2]{64} = \pm 8$$

8 is the number, which, multiplied by itself two times (8 · 8) gives 4^3, or 64 (same as for -8)

Using this notation, the square root of a number can be symbolically represented in two ways.

$$\sqrt[2]{a} \quad \text{or} \quad a^{1/2}$$

In most cases the latter notation is preferred as it allows easier algebraic manipulation.

Examples 3

1. $25^{1/2} = \pm 5$
2. $8^{1/3} = \sqrt[3]{8} = 2$
3. $3^{5/2} = \sqrt[2]{3^5} = \sqrt[2]{243} \cong \pm 15.6$
4. $(-1)^{4/6} = \sqrt[6]{(-1)^4} = \sqrt[6]{1} = \pm 1$

Exercises 3

1. $36^{1/2} =$
2. $27^{1/3} =$
3. $25^{5/2} =$
4. $1^{9/7} =$

Finally, our last exponent definition is

$$a^0 \equiv 1$$

or any real number raised to the zero power equals 1.

Examples 4

1. $10^0 = 1$
2. $(-6)^0 = 1$
3. $\left(\dfrac{7}{5}\right)^0 = 1$

Exercises 4

1. $3^0 =$
2. $\left(\dfrac{1}{2}\right)^0 =$
3. $(-4)^0 =$

Wouldn't it be nice if all the exercises were this easy!

Simplification of numbers with exponents in a fraction often involves moving the number from numerator to denominator or vice versa. Let's see how that works.

We already know that

$$a^{-m} = \dfrac{1}{a^m} \quad \text{So,} \quad \dfrac{a^{-m}}{1} = \dfrac{1}{a^m}$$

so a negative exponent in the numerator becomes a positive exponent in the denominator, or vice versa.

Now consider a negative exponent in the denominator.

$$\frac{1}{a^{-m}} = \frac{1}{\frac{1}{a^m}} = \frac{a^m \cdot 1}{a^m \cdot \frac{1}{a^m}} = a^m$$

So a negative exponent in the denominator becomes a positive exponent in the numerator and vice versa.

In summary, the moral of this story is that you can move a number with an exponent from numerator to denominator or vice versa by merely changing the sign of the exponent. Easy!

Examples 5

1. $7^{-3} = \frac{1}{7^3}$
2. $\frac{1}{4^2} = 4^{-2}$
3. $\frac{x^5}{x^3} = x^5 \cdot x^{-3} = x^2$

Exercises 5

1. $10^{-4} =$
2. $\frac{1}{6^3} =$
3. $\frac{x^2}{x^5} =$

From these definitions, other rules or laws of exponents follow. We will develop each with examples and exercises in Table 1. Although we illustrated the basic definitions with natural or rational number exponents, the laws apply for all real number bases and exponents.

2.6 FACTORING

Factoring, a technique for manipulating algebraic expressions, is based on the distributive axiom $a(b + c) = ab + ac$. Or, in reverse, $ab + ac = a(b + c)$. Thus, if a sum of two terms, each being a product of two numbers, has a factor in common (a number common to both terms), this factor can be brought out or, as it is called, factored out.

Let's illustrate.

$3 \cdot 2 + 3 \cdot 5 = 3(2 + 5)$

Here, 3 corresponds to *a*, 2 corresponds to *b*, and 5 corresponds to *c* in the reverse distributive axiom.

$B + Bg = B \cdot 1 + Bg$
$ = B(1 + g)$

Here, *B* corresponds to *a*, 1 (which we included via the identity axiom for multiplication) corresponds to *b*, and *g* corresponds to *c* in the reverse distributive axiom.

Be especially cautious in cases such as the second illustration. Without explicitly incorporating the 1 as a multiple of *B*, students have often been known to do this:

$B + Bg = B(g)$ *Warning: This is a No-No!*

ALGEBRA REVIEW

Table 1. Exponent Laws

	Examples 6	Exercises 6
Law 1 $a^m a^n = \underbrace{(a \cdot a \cdot a \cdot \ldots \cdot a)}_{m \text{ a's}} \underbrace{(a \cdot a \cdot \ldots \cdot a)}_{n \text{ a's}}$ $= \text{product of } (m+n) \text{ a's}$ $= a^{m+n}$	$11^2 \cdot 11^3 = 11^{2+3} = 11^5$ $5^2 \cdot 5^{-1} = 5^{2-1} = 5^1$ $(1.8)^2 (1.8)^4 = (1.8)^{2+4}$ $= (1.8)^6$ $(-3)^3(-3)^{-4} = (-3)^{3-4}$ $= -3^{-1}$ $2^{1/2} \cdot 2^{1/2} = 2^{1/2+1/2} = 2^1$	$46^2 \cdot 46^1 =$ $\left(-\dfrac{1}{2}\right)^1 \left(-\dfrac{1}{2}\right)^3 =$ $(6.7)^3 (6.7)^4 =$ $\left(\dfrac{3}{4}\right)^{-2} \left(\dfrac{3}{4}\right)^2 =$ $3^{1/3} \cdot 3^{2/3} =$
Law 2 $(a^m)^n = \underbrace{a^m \cdot a^m \cdot a^m \cdot \ldots \cdot a^m}_{n \text{ of these } a^m}$ $= \text{product of } nm \text{ a's}$ $= a^{mn}$	$(2^2)^3 = 2^{2 \cdot 3} = 2^6$ $[(-3)^3]^2 = (-3)^6$ $((-4)^2)^{-4} = (-4)^{-8}$ $\left[\left(\dfrac{1}{5}\right)^2\right]^{-3} = \left(\dfrac{1}{5}\right)^{-6}$	$[(-7)^4]^{-2} =$ $\left[\left(\dfrac{1}{2}\right)^2\right]^2 =$ $[(403)^{-1}]^3 =$ $[(8.3)^3]^4 =$ $(2^4)^{1/2} =$
Law 3 $(ab)^m = (ab)(ab)(ab)\ldots(ab)$ $= \underbrace{(a \cdot a \cdot a \cdot \ldots \cdot a)}_{m \text{ a's}} \underbrace{(b \cdot b \cdot b \cdot \ldots \cdot b)}_{m \text{ b's}}$ $= a^m b^m$	$(3 \cdot 4)^2 = 3^2 \cdot 4^2$ $[2 \cdot (-5)]^3 = 2^3 \cdot (-5)^3$ $(3x)^4 = 3^4 x^4$ $(2y)^{-1} = 2^{-1} y^{-1}$ $\left(\dfrac{8}{x}\right)^2 = \left(8 \cdot \dfrac{1}{x}\right)^2 = 8^2 \cdot \left(\dfrac{1}{x}\right)^2$ $(xy)^{1/2} = x^{1/2} \cdot y^{1/2}$	$(2 \cdot 3)^4 =$ $[(-1)(-6)]^2 =$ $[(8.4)(9)]^3 =$ $(xy)^5 =$ $(8x)^{1/3} =$
Law 4 $\left(\dfrac{a}{b}\right)^m = \underbrace{\left(\dfrac{a}{b}\right)\left(\dfrac{a}{b}\right)\left(\dfrac{a}{b}\right)\ldots\left(\dfrac{a}{b}\right)}_{m \text{ of these}}$ $= \dfrac{a \cdot a \cdot a \cdot \ldots \cdot a}{b \cdot b \cdot b \cdot \ldots \cdot b}$ $= \dfrac{a^m}{b^m} = a^m b^{-m}$	$\left(\dfrac{1}{3}\right)^2 = \dfrac{1^2}{3^2} = \dfrac{1}{9}$ $\left(\dfrac{-2}{3}\right)^3 = \dfrac{-2^3}{3^3} = \dfrac{-8}{27}$ $\left(\dfrac{7}{x}\right)^4 = \dfrac{7^4}{x^4} = 7^4 x^{-4}$ $\left(\dfrac{y}{z}\right)^2 = \dfrac{y^2}{z^2} = y^2 z^{-2}$ $\left(\dfrac{1}{4}\right)^{1/2} = \dfrac{1^{1/2}}{4^{1/2}} = \dfrac{1}{2}$	$\left(\dfrac{1}{2}\right)^3 =$ $\left(\dfrac{4}{3}\right)^{-2} =$ $\left(\dfrac{x}{6}\right)^5 =$ $\left(\dfrac{9.2}{z}\right)^2 =$ $\left(\dfrac{1}{8}\right)^{1/3} =$

Note: a, b, m, n stand for any real numbers.

Of course, you won't do this!!

Now consider the following examples; once they are mastered, try the exercises.

Examples 7

1. $xy + xz = x(y + z)$
2. $3x + 6x^2 = 3x \cdot 1 + 3x \cdot 2x$
   $= 3x(1 + 2x)$
3. $\dfrac{x}{y} + \dfrac{z}{y} = \dfrac{1}{y}x + \dfrac{1}{y}z$
   $= \dfrac{1}{y}(x + z)$
4. $\dfrac{x}{2} + \dfrac{x}{4} = \dfrac{x}{2} \cdot 1 + \dfrac{x}{2} \cdot \dfrac{1}{2}$
   $= \dfrac{x}{2}\left(1 + \dfrac{1}{2}\right) = \dfrac{x}{2} \cdot \dfrac{3}{2} = \dfrac{3x}{4}$

Exercises 7

1. $x + xy =$
2. $4 \cdot 3 + 5 \cdot 4 =$
3. $2x^2y + 4xy^2 =$
4. $\dfrac{1}{x} + \dfrac{2}{x} =$

On page 17 we extended the distributive axiom by showing that $a(b + c + d) = ab + ac + ad$. If $b_1, b_2, b_3, \ldots, b_n$ stand for any n real numbers, this axiom can be further extended to read,

$$a(b_1 + b_2 + b_3 + b_4 + \cdots + b_n) = ab_1 + ab_2 + ab_3 + ab_4 + \cdots + ab_n.$$

So a sum of products, all having a common factor (as has the right side), can be "factored" to look like the left side.

Examples 8

1. $2x + 2y + 2z = 2(x + y + z)$
2. $x + x^2 + x^3 = x \cdot 1 + x \cdot x + x \cdot x^2$
   $= x(1 + x + x^2)$
3. $xy + xy^2 + xy^3 + xy^4 + xy^5$
   $= xy \cdot 1 + xy \cdot y + xy \cdot y^2 + xy \cdot y^3 + xy$
   $= xy(1 + y + y^2 + y^3 + y^4)$
4. $\dfrac{x}{y} + \dfrac{x}{z} + 2x = x \cdot \dfrac{1}{y} + x \cdot \dfrac{1}{z} + x \cdot 2$
   $= x\left(\dfrac{1}{y} + \dfrac{1}{z} + 2\right)$

Exercises 8

1. $2 \cdot 3 + 2 \cdot 4 + 2 \cdot 5 =$
2. $t^4y + t^3y^2 + t^2y^3 + ty^4 =$
3. $\dfrac{1}{5} + \dfrac{2}{5} + \dfrac{3}{5} =$
4. $\dfrac{2x}{y} + \dfrac{4x}{y} + \dfrac{6x}{y} =$

2.7 FRACTIONS

A fraction, such as $\dfrac{n_1}{n_2}$, has a number in the numerator (top) divided by a number in the denominator (bottom). The sign; simplification; finding a common denominator; addition and subtrac-

ALGEBRA REVIEW

tion of fractions; multiplication and division of fractions; and compound fractions will be reviewed in this section.

Sign

$$+\left(\frac{n_1}{n_2}\right) = \frac{+n_1}{+n_2} = \frac{-n_1}{-n_2}$$

and

$$-\left(\frac{n_1}{n_2}\right) = \frac{-n_1}{+n_2} = \frac{+n_1}{-n_2}$$

Examples 9

1. $+\left(\dfrac{3}{5}\right) = \dfrac{+3}{+5} = \dfrac{-3}{-5}$
2. $-\left(\dfrac{9}{7}\right) = \dfrac{-9}{+7} = \dfrac{+9}{-7}$
3. $\dfrac{-6}{-4} = +\left(\dfrac{6}{4}\right) = +\dfrac{3}{2}$
4. $\dfrac{-10}{+5} = -\left(\dfrac{10}{5}\right) = -2$

Exercises 9

1. $-\left(\dfrac{2}{7}\right) =$
2. $\dfrac{-6}{+3} =$
3. $+\left(\dfrac{8}{5}\right) =$
4. $\dfrac{-9}{-4} =$

Simplification

If a common factor exists in n_1 and n_2, it can be cancelled out from numerator and denominator by using the inverse axiom for multiplication.

Examples 10

1. $\dfrac{10}{15} = \dfrac{\cancel{5} \cdot 2}{\cancel{5} \cdot 3} = \dfrac{2}{3}$
2. $\dfrac{2x}{6x} = \dfrac{\cancel{2x} \cdot 1}{\cancel{2x} \cdot 3} = \dfrac{1}{3}$
3. $\dfrac{x^5}{x^2} = \dfrac{\cancel{x^2} \cdot x^3}{\cancel{x^2}} = x^3$
4. $\dfrac{x^2 y}{xy^2} = \dfrac{\cancel{x}\cancel{x}x}{\cancel{x}\cancel{y}y} = \dfrac{x}{y}$
5. $\dfrac{x + x^2}{x^2} = \dfrac{\cancel{x}(1 + x)}{\cancel{x} \cdot x} = \dfrac{1 + x}{x}$

Exercises 10

1. $\dfrac{3y}{9y} =$
2. $\dfrac{x^3}{x^5} =$
3. $\dfrac{xy^2}{x^2 y} =$
4. $\dfrac{2 + 2x^3}{2} =$
5. $\dfrac{x^2 y^3 + y^3 x^2}{x^2 y^2} =$

Common Denominator

Two fractions, a/b and c/d, can be made to have the same denominator, which is called a *common denominator*. This could be accomplished if the first fraction had the factor d in the denominator, and the second fraction had the factor b in its denominator. Then, bd is a common denominator.

$$\frac{a}{b} = \frac{a}{b} \cdot 1 = \frac{a}{b} \cdot \frac{d}{d} = \frac{ad}{bd}$$

$$\frac{c}{d} = \frac{c}{d} \cdot 1 = \frac{c}{d} \cdot \frac{b}{b} = \frac{cb}{db} = \frac{cb}{bd}$$

Examples 11

A common denominator for

1. $\frac{1}{2}$ and $\frac{2}{3}$ is $2 \cdot 3 = 6$
2. $\frac{4}{5}$ and $\frac{5}{3}$ is $5 \cdot 3 = 15$
3. $\frac{3}{6.5}$ and $\frac{4}{2.81}$ is $(6.5)(2.81)$

Exercises 11

Find a common denominator for the following.

1. $\frac{1}{2}$ and $\frac{2}{5}$
2. $\frac{2}{3}$ and $\frac{3}{7}$
3. $\frac{3}{5.5}$ and $\frac{5}{6.5}$

Addition and Subtraction of Fractions

Addition and subtraction of fractions involves finding a common denominator and then using the distributive axiom. Lets illustrate with addition of fractions.

$$\frac{a}{b} + \frac{c}{d} = \frac{a}{b} \cdot 1 + \frac{c}{d} \cdot 1$$

$$= \frac{a}{b} \cdot \frac{d}{d} + \frac{c}{d} \cdot \frac{b}{b}$$

$$= \frac{ad}{bd} + \frac{cb}{bd}$$

Since each term has a common factor $1/bd$

$$= \frac{1}{bd}(ad + cb)$$

$$= \frac{ad + cb}{bd}$$

The rule for subtracting fractions is

$$\frac{a}{b} - \frac{c}{d} = \frac{ad - cb}{bd}$$

ALGEBRA REVIEW

Examples 12

1. $\dfrac{1}{2} + \dfrac{2}{3} = \dfrac{1 \cdot 3 + 2 \cdot 2}{2 \cdot 3} = \dfrac{7}{6}$

2. $\dfrac{6}{5} + \dfrac{8}{11} = \dfrac{6 \cdot 11 + 8 \cdot 5}{5 \cdot 11} = \dfrac{106}{55}$

3. $\dfrac{x}{2} + \dfrac{y}{5} = \dfrac{x \cdot 5 + y \cdot 2}{2 \cdot 5} = \dfrac{5x + 2y}{10}$

4. $\dfrac{3}{2} - \dfrac{4}{5} = \dfrac{3 \cdot 5 - 4 \cdot 2}{2 \cdot 5} = \dfrac{7}{10}$

5. $\dfrac{x}{3} - \dfrac{1}{2} = \dfrac{x \cdot 2 - 1 \cdot 3}{3 \cdot 2} = \dfrac{2x - 3}{6}$

Exercises 12

1. $\dfrac{1}{3} + \dfrac{3}{4} =$

2. $\dfrac{x}{2} + \dfrac{2}{3} =$

3. $\dfrac{x}{y} + \dfrac{4}{5} =$

4. $\dfrac{5}{6} - \dfrac{1}{3} =$

5. $\dfrac{x}{y} - \dfrac{y}{x} =$

Multiplication and Division of Fractions

Multiplication

$$\left(\dfrac{a}{b}\right)\left(\dfrac{c}{d}\right) = \dfrac{ac}{bd}$$

Division

$$\dfrac{(a/b)}{(c/d)} = \dfrac{ad}{cb}$$

Proof:

$$\dfrac{\dfrac{a}{b}}{\dfrac{c}{d}}(1)(1) = \dfrac{\dfrac{a}{b}}{\dfrac{c}{d}} \cdot \dfrac{b}{b} \cdot \dfrac{d}{d} = \dfrac{\dfrac{a}{\cancel{b}} \cdot \cancel{b} \cdot d}{\dfrac{c}{\cancel{d}} \cdot b \cdot \cancel{d}} = \dfrac{ad}{cb}$$

Examples 13

1. $\left(\dfrac{1}{2}\right)\left(\dfrac{3}{5}\right) = \dfrac{1 \cdot 3}{2 \cdot 5} = \dfrac{3}{10}$

2. $\dfrac{(5/4)}{(3/7)} = \dfrac{5 \cdot 7}{3 \cdot 4} = \dfrac{35}{12}$

3. $\left(\dfrac{-9}{4}\right)\left(\dfrac{5}{7}\right) = \dfrac{-9 \cdot 5}{4 \cdot 7} = \dfrac{-45}{28}$

4. $\dfrac{(1/2)}{3} = \dfrac{(1/2)}{(3/1)} = \dfrac{1 \cdot 1}{2 \cdot 3} = \dfrac{1}{6}$

5. $\dfrac{1}{(1/a)} = \dfrac{(1/1)}{(1/a)} = \dfrac{1 \cdot a}{1 \cdot 1} = a$

Exercises 13

1. $\left(\dfrac{2}{3}\right)\left(\dfrac{4}{5}\right) =$

2. $\dfrac{(3/4)}{(5/6)} =$

3. $\left(\dfrac{7}{3}\right)\left(\dfrac{-2}{3}\right) =$

4. $\dfrac{4}{(3/5)} =$

5. $\dfrac{1}{(1/2)} =$

Compound Fractions

Compound fractions are characterized by numerators or denominators or both having more than one term. Some examples are shown as follows.

$$\frac{x+y}{2} \quad \frac{D}{i-g} \quad \frac{5+x}{10+x} \quad \frac{5+x}{10+x+y}$$

One must be careful to use the laws of algebra to manipulate or simplify compound fractions. For example, the distributive axiom allows

$$\frac{a+b}{c} = \frac{1}{c}(a+b) = \frac{1}{c}a + \frac{1}{c}b = \frac{a}{c} + \frac{b}{c}$$

For example,

$$\frac{8+x}{2} = \frac{8}{2} + \frac{x}{2} = 4 + \frac{x}{2}$$

But NO axiom allows

$$\frac{a}{b+c} = \frac{a}{b} + \frac{a}{c}$$

For example,

$$\frac{1}{2+3} = \frac{1}{5} \neq \frac{1}{2} + \frac{1}{3} = \frac{5}{6} \qquad \textit{Warning: another No-No!}$$

To simplify compound fractions, the distributive axiom along with the inverse and identity axioms for multiplication are useful. Let's illustrate.

$$\frac{2x}{2x+4} = \frac{2x}{2(x+2)} \qquad \textit{Distributive axiom}$$

$$= \frac{1}{2} \cdot 2 \cdot \frac{x}{x+2}$$

$$= 1 \cdot \frac{x}{x+2} \qquad \textit{Inverse axiom for multiplication}$$

$$= \frac{x}{x+2} \qquad \textit{Identify axiom for multiplication}$$

You may have been able to do the above "cancellation" in one step. But, without the guidance of the axioms, it is more likely that you would perpetrate the following no-no in one easy step.

$$\frac{2x}{2x+4} = \frac{x}{x+4} \qquad \textit{Warning: another No-No!}$$

ALGEBRA REVIEW

Consider the following examples of compound fraction simplification, and then try the accompanying exercises.

Examples 14

1. $\dfrac{2x}{4x-6} = \dfrac{2x}{2(2x-3)} = \dfrac{x}{2x-3}$
2. $\dfrac{x^2+x}{xy} = \dfrac{x(x+1)}{xy} = \dfrac{x+1}{y}$
3. $\dfrac{p(1+i)^3}{i+i^2} = \dfrac{p(1+i)^3}{i(1+i)} = \dfrac{p(1+i)^2}{i}$
4. $\dfrac{(x+1)^2}{2x+2} = \dfrac{(x+1)^2}{2(x+1)} = \dfrac{x+1}{2}$
5. $\dfrac{a^2x^2y - xy}{axy} = \dfrac{xy(a^2x-1)}{xy \cdot a} = \dfrac{a^2x-1}{a}$

Exercises 14

1. $\dfrac{x}{x+x^2} =$
2. $\dfrac{3+9y}{6xy} =$
3. $\dfrac{B+Bg}{1+g} =$
4. $\dfrac{4y+16y^2-8y^3}{2y-4y^2} =$
5. $\dfrac{a+a^{1/2}}{a^{1/2}} =$

2.8 SPECIAL PRODUCTS

Certain products appear frequently in mathematics, and thus deserve some attention. Especially important is the multiplication of two binomial (sum or difference of two terms) expressions. (*Note: a, b, c,* and *d* stand for any real numbers.)

$(a+b)^2 = (a+b)(a+b) = a^2 + 2ab + b^2$
$(a-b^2) = (a-b)(a-b) = a^2 - 2ab + b^2$
$(a+b)(a-b) = a^2 - b^2$
$(a+b)(c+d) = ac + ad + bc + bd$

As you might expect, these results have their basis in the axioms. Let's illustrate by proving the first result. (*Note:* The purpose here is to become more aware of the axioms, rather than to memorize a proof.)

	Axiom or reason
$(a+b)^2 = (a+b)(a+b)$	*Exponent definition*
Since $(a+b)$ is a unique real number, which we can call c	*Closure for addition*
$= c(a+b)$	
$= ca + cb$	*Distributive*
$= ac + bc$	*Commutative for multiplication*
$= a(a+b) + b(a+b)$	
$= a^2 + ab + ba + b^2$	*Distributive*
Since $ba = ab$ and $ab + ab = 2ab$	*Commutative for multiplication*
$= a^2 + 2ab + b^2$	

Now, using the four binomial multiplication rules, see if you can understand the following examples, then go on to the exercises.

Examples 15

1. $(3 + 4)^2 = 7^2 = 49 = 3^2 + 2 \cdot 3 \cdot 4 + 4^2 = 49$
2. $(x + 1)^2 = x^2 + 2 \cdot x \cdot 1 + 1^2 = x^2 + 2x + 1$
3. $(x + 2y)^2 = x^2 + 2 \cdot x \cdot 2y + (2y)^2 = x^2 + 4xy + 4y^2$
4. $\left(\dfrac{5}{3} + \dfrac{x}{2}\right)^2 = \left(\dfrac{5}{3}\right)^2 + 2 \cdot \dfrac{5}{3} \cdot \dfrac{x}{2} + \left(\dfrac{x}{2}\right)^2 = \dfrac{25}{9} + \dfrac{5}{3}x + \dfrac{x^2}{4}$
5. $(y - 6)^2 = y^2 - 2 \cdot y \cdot 6 + 6^2 = y^2 - 12y + 36$
6. $(w + 2u)(w - 2u) = w^2 - (2u)^2 = w^2 - 4u^2$
7. $\left(3u + \dfrac{v}{2}\right)\left(3u - \dfrac{v}{2}\right) = (3u)^2 - \left(\dfrac{v}{2}\right)^2 = 9u^2 - \dfrac{v^2}{4}$
8. $x^2 + 4x + 4) = (x + 2)^2$
9. $(w^4 - 9) = (w^2 + 3)(w^2 - 3)$
10. $(3 + x)(y + 4) = 3y + 12 + xy + 4x$

Exercises 15

1. $(x - 1)^2 =$
2. $(2a + 3b)^2 =$
3. $(4x + y)(4x - y) =$
4. $25 - x^2 =$
5. $(x + 2)(y + 4) =$

Binomial Expansion

Finding the sum of two numbers raised to a power, or

$(a + b)^n$

is referred to as a binomial expansion. Earlier, we found the simplest case ($n = 2$) of a binomial expansion

$(a + b)^2 = a^2 + 2ab + b^2$

The case of a binomial expansion for $n = 3$ can be found as

$(a + b)^3 = (a + b)(a + b)^2 = (a + b)(a^2 + 2ab + b^2)$
$ = (a + b)(a^2 + 2ab + b^2)$
$ = a(a^2 + 2ab + b^2) + b(a^2 + 2ab + b^2)$
$ = a^3 + 2a^2b + ab^2 - a^2b + 2ab^2 + b^3$
$ = a^3 + 3a^2b + 3ab^2 + b^3$

In a similar fashion, you show that

$(a + b)^4 = (a + b)(a + b)^3$
$ = a^4 + 4a^3b + 6a^2b^2 + 4ab^3 + b^4$

Reviewing the results of binomial expansions for $n = 2, 3$, and 4, can you detect the following patterns?

ALGEBRA REVIEW

1. The exponent for a starts at n and drops by one for each term.
2. The exponent for b starts at 0 and increases by one for each term.
3. There are $n + 1$ terms in the expansion.

The coefficients for the terms are mysterious at first. However, some of you may recognize them from Pascal's triangle; others may have learned them as binomial coefficients. But, regardless of your previous exposure to these coefficients, you can all learn them now. The coefficient for the tth term is:

$$\frac{n!}{k!(n-k)!} \quad \text{where} \quad k = t - 1$$

where ! is the factorial symbol. $x!$ simply means to multiply all the natural numbers from x down to 1. For example, $4! = 4 \cdot 3 \cdot 2 \cdot 1 = 24$.

Let's illustrate the use of this formula for the expansion of $(a + b)^4$. (Note: $n = 4$)

Term t	$k = t - 1$	Coefficient	
1	0	$\frac{4!}{0!4!} = 1$	(Note $0!$ is defined as 1)
2	1	$\frac{4!}{1!3!} = \frac{4 \cdot 3 \cdot 2 \cdot 1}{(1)(3 \cdot 2 \cdot 1)} = 4$	
3	2	$\frac{4!}{2!2!} = \frac{4 \cdot 3 \cdot 2 \cdot 1}{(2 \cdot 1)(2 \cdot 1)} = 6$	
4	3	$\frac{4!}{3!1!} = 4$	(Same as for $k = 1$)
5	4	$\frac{4!}{4!0!} = 1$	

So $(a + b)^4 = 1a^4 + 4a^3b + 6a^2b^2 + 4ab^3 + 1b^4$

Examples 16

1. $(a + b)^5 = \frac{5!}{0!5!}a^5 + \frac{5!}{1!4!}a^4b + \frac{5!}{2!3!}a^3b^2 + \frac{5!}{3!2!}a^2b^3 + \frac{5!}{4!1!}a^1b^4 + \frac{5!}{5!0!}b^5$

    $= 1a^5 + 5a^4b + 10a^3b^2 + 10a^2b^3 + 5ab^4 + 1b^5$

2. $(x + 2)^3 = \frac{3!}{0!3!}x^3 2^0 + \frac{3!}{1!2!}x^2 2^1 + \frac{3!}{2!1!}x^1 2^2 + \frac{3!}{3!0!}x^0 2^3$

    $= 1x^3 1 + 3x^2 2 + 3x^1 4 + 1 \cdot 2^3$
    $= x^3 + 6x^2 + 12x + 8$

3. $(y + 1)^4 = 1y^4 1^0 + 4y^3 1^1 + 6y^2 1^2 + 4y^1 1^3 + 1y^0 1^4$
    $= y^4 + 4y^3 + 6y^2 + 4y + 1$

Exercises 16

1. $(3 + x)^3 =$

2. $\left(y + \frac{1}{2}\right)^4 =$

3. $(a + b)^6 =$

2.9 SIMULTANEOUS EQUATIONS

Many mathematical applications require the simultaneous solution of equations—for example, Matt Maddix car rental problem in Chapter 1. Although we will develop this subject in depth in the linear systems module of the text, in the interim we will need to solve two equations simultaneously. A review of this process, which you undoubtedly learned somewhere in your mathematical past is in order.

Consider an equation with two unknowns—for example $y = 2x$. Many pairs of x and y values would satisfy (solve) such an equation. For example, $x = 0$ and $y = 0$, $x = 1$ and $y = 2$, $x = 2$ and $y = 4$ all satisfy the above equation. Many other pairs would work also. Now consider another equation relating the same two unknowns, say, $y = x + 1$. This equation also has many solutions: $x = 0$ and $y = 1$, $x = 1$ and $y = 2$, $x = 2$ and $y = 3$ are three of them.

Now consider the two equations taken together. A pair of values for x and y that satisfy both equations is said to be a solution to the system of two equations. From above you can see that $x = 1$ and $y = 2$ is such a point.

Listing the pairs of unknown values that will work in each equation and comparing them to find a pair that will work in both equations is a very inefficient process for solving simultaneous equations. More rewarding methods will be developed later, but until then, the substitution of variable method described below will do the job.

Consider the above system of equations:

$y = 2x$

$y = x + 1$

The goal of the method is to obtain one of the equations with only one variable present. Since $y = 2x$ from the first equation, this goal can be accomplished by substituting $2x$ for y in the second equation.

$2x = x + 1$

After some equation manipulations, we solve for x as

$2x - x = (x - x) + 1$

$1x = 0 + 1$

$x = 1$

Substituting this value of x into either equation gives the solution value of y:

$y = 2 \cdot 1 = 2 \quad or \quad y = 1 + 1 = 2$

Alternatively, we could have solved for x in the second equation,

$x = y - 1$

substituted this in the first equation,

$y = 2x = 2(y - 1) = 2y - 2$

ALGEBRA REVIEW

carried out some equation manipulations to find y,

$$y - 2y = (2y - 2y) - 2$$
$$-1y = -2$$
$$y = 2$$

and then solved for x in either equation.

There are two other ways to solve, namely, (1) Substitute $(x + 1)$ for y in the first equation, and (2) Substitute $\frac{1}{2}y$ for x in the second equation. Carry out these alternative substitution methods and show that you obtain the same solution.

There are always four ways to solve a two-equation simultaneous system by the substitution method. Since they all give the same solution, we pick the one that appears easiest. Let's illustrate with two examples.

Examples 17

1. $2x = y - 5$

   $x = y - 4$

   Since x is given directly in terms of y in the second equation, it seems best to substitute this for x in the first equation.

   $$2(y - 4) = y - 5$$
   $$2y - 8 = y - 5$$
   $$2y - y = 8 - 5$$
   $$y = 3$$

   It is then simplest to substitute this value of y in the second equation and solve for x:

   $$x = y - 4$$
   $$= 3 - 4 = -1$$

2. $2x - 3y = 3$

   $-x + y = -\frac{3}{2}$

   Here it appears easiest to solve for y in terms of x in the second equation and substitute this in the first equation.

   $$y = -\frac{3}{2} + x$$
   $$2x - 3\left(-\frac{3}{2} + x\right) = 3$$

$$2x + \frac{9}{2} - 3x = 3$$

$$-1x = -\frac{3}{2}$$

$$x = \frac{3}{2}$$

Substituting this value of x in the first equation quickly finds y:

$$2\left(\frac{3}{2}\right) - 3y = 3$$

$$3 - 3y = 3$$

$$y = 0$$

Exercises 17

1. $2x = 6 - y$
   $x = 10 - 3y$
2. $10x - y = -10$
   $2x + 2y = 20$
3. $x + 4y = 7$
   $5x - y = -7$
4. $-\left(\frac{2}{3}\right)x + \frac{1}{7}y = -5$
   $x + \frac{1}{2}y = \frac{5}{2}$

2.10 LOGARITHMS

For some reason,[1] the mention of logarithms sends chills through many students. But actually, the only difficulty anyone should have with them is remembering how to spell the word. We can solve that problem if we refer to them symbolically, as $\log_b N$.

In the good old days of my high school, we didn't have calculators or computers. Do you know what we used to multiply or divide massive multidigit monster numbers? Can you believe that it was logarithms? Today, we don't need logarithms for that purpose. But we do need them for modeling the real world and solving equations. To accomplish this goal, we need to find out what makes them tick. We do this as follows: first we define a logarithm, then we see how their values are determined, then we develop the three laws of logarithms, and some special relationships among logs, finally we breathe a sigh of relief.

Definition: The symbol, $\log_b N$, which in words means the logarithm of N for the base b, represents that exponent to which the base b (b is a positive number) must be raised in order to equal N. For example,

---

[1] It all began in Mr. Oldmath's class in high school, where endless log tables, mantissas, and characteristics coalesced in a big cloud heading toward oblivion.

ALGEBRA REVIEW

Examples 18

$\log_{10} 100 = 2$ since $10^2 = 100$

$\log_2 8 = 3$ since $2^3 = 8$

$\log_3 81 = 4$ since $3^4 = 81$

$\log_{100} 100 = 1$ since $100^1 = 100$

$\log_5 125 = 3$ since $5^3 = 125$

$\log_{1/2} \dfrac{1}{4} = 2$ since $\left(\dfrac{1}{2}\right)^2 = \dfrac{1}{4}$

Notice that two different numbers can have the same logarithm if the bases are different ($\log_2 8 = \log_5 125 = 3$). Also notice that $\log_b b = 1$. Why?

Don't be misled by all the positive logarithms in the examples. Logarithms can also be negative.

Examples 18 continued

$\log_{10} \dfrac{1}{10} = -1$ since $10^{-1} = \dfrac{1}{10}$

$\log_2 \dfrac{1}{32} = -5$ since $2^{-5} = \dfrac{1}{32}$

$\log_3 \dfrac{1}{27} = -3$ since $3^{-3} = \dfrac{1}{27}$

$\log_{1/3} 9 = -2$ since $\left(\dfrac{1}{3}\right)^{-2} = 9$

The value of $\log_b 0$ is $-\infty$ since only $L = -\infty$ satisfies the equation

$$b^L = \dfrac{1}{b^{-L}} = \dfrac{1}{b^{\infty}} = 0$$

Logarithms are not defined for negative values of N. For example, $\log_b -10$ is not defined since no real number (L) can be found so that $b^L = -10$. (Note that with positive b, b^L is always positive even if L is negative.)

We can find a unknown base too. For example:

$\log_b 25 = 2$ causes us to ask what value of b gives $b^2 = 25$

since $5^2 = 25$, then the base, b, must be 5

Exercises 18

1. Find
   (a) $\log_4 64$
   (b) $\log_{11} 121$
   (c) $\log_{1/3} 9$
   (d) $\log_2 16$
   (e) $\log_9 \dfrac{1}{81}$
   (f) $\log_{6.3} 6.3$
   (g) $\log_{1/7} 7$
   (h) $\log_{3/2} \dfrac{16}{81}$

2. Determine the missing base.
   (a) $\log_b 27 = 3$
   (b) $\log_b \dfrac{1}{16} = -4$

Logarithms Laws

With the log basics in hand, we are now ready to develop the three laws of logarithms. It all begins with

$$b^{\log_b N}$$

What do you suppose this equals? How about N? Why? Well, a careful reading of the definition shows that if $\log_b N$ is the number one raises the base b to equal N, then the mere act of doing that should give N.

This simple truth (it really is only a definition) allows us to prove the three laws of logarithms.

Three Logarithm Laws

1. $\log_b NM = \log_b N + \log_b M$ where b, N, and M are positive real numbers
2. $\log_b \dfrac{N}{M} = \log_b N - \log_b M$
3. $\log_b N^M = M \log_b N$

Let's see how easy it is to prove the first law. Since from the definition of a logarithm

$b^{\log_b N} = N$

$b^{\log_b M} = M$

$b^{\log_b NM} = NM$

And obviously

ALGEBRA REVIEW

$(N)(M) = (NM)$

Then

$(b^{\log_b N})(b^{\log_b M}) = b^{\log_b NM}$

Then the following step is brought to you by the first exponent law in Table 1 on page 31.

$b^{(\log_b N + \log_b M)} = b^{\log_b NM}$

In order for equality to hold, the exponents must be equal. Thus,

$\log_b NM = \log_b N + \log_b M$

This law is the gem that allowed me to multiply two monstrous numbers back in high school merely by adding their logarithms.

The second law of logarithms has a proof that follows the same pattern as the above proof. Why don't you see if you can follow through.

The third law of logarithms is merely an application of the first law. The first law generalizes to accommodate a product of more than two numbers, as

$\log_b(N \cdot N \cdots N) = \log_b N + \log_b N + \cdots + \log_b N$

so

$$\log_b N^M = \overbrace{\log_b N + \log_b N + \log_b N + \cdots + \log_b N}^{M \text{ of these}}$$

$$= M \cdot \log_b N$$

We will often need these log laws in later chapters to solve equations. To give you a hint of that process, consider the application of finding how long it would take for the Earth's population, currently 6 billion, to triple if population grows at an annual rate of 2%. We will see in Chapter 3 that the relevant starting equation is

$6(1.02)^x = 18$

Simplifying by dividing by 6 gives

$(1.02)^x = 3$

Notice how the unknown, x, is in the exponent part of the equation. There is only one way in the world (you guessed it, logarithms) to get that unknown isolated on the main line of the equation. So we take the logarithm of both sides

$\log_{10}(1.02)^x = \log_{10} 3$

Then the third law of logarithms allows us to bring the unknown down from the exponent

$x \log_{10}(1.02)^x = \log_{10} 3$

So

$$x = \frac{\log_{10} 3}{\log_{10}(1.02)}$$

The final numerical solution for x requires us to calculate 2 logarithms. These logs are not the easy "do in your head" ones that we previously encountered. So how do we do it?

There are two ways to find the numerical value of a logarithm. First, calculators, except for those real cheapo ones, have log keys for base 10 and/or base 2.718 . . . , so called natural logarithms symbolized as ln. Second, you can use Table 1 in the Appendix. The latter only lists log values for numbers under 10 with one decimal place.

Using a calculator, we find

$$x = \frac{\log_{10} 3}{\log_{10} 1.02} = \frac{.4771}{.0086} = 55.48$$

Sometimes, we will have to deal with logarithms with bases other than 10 or 2.718, say base β. We can find their values in relation to the "standard" two bases by the formulas

$$\log_\beta x = \frac{\log_{10} x}{\log_{10} \beta} = \frac{\ln x}{\ln \beta}$$

Examples 19   *Exercises 19*

Log Laws

1. $\log_{10}(9.5)(6.3) = \log_{10} 9.5 + \log_{10} 6.3$
$= .973 + .799$
$= 1.772$

2. $\log_{10} \frac{7.8}{9.8} = \log_{10} 7.8 - \log_{10} 9.8$
$= .892 - .991 = -.099$

3. $\log_{10} \frac{(5)(3.2)}{4.9} = \log_{10} 5 + \log_{10} 3.2 - \log_{10} 4.9$
$= .699 + .505 - .69$
$= .514$

4. $\ln 6^5 = 5 \ln 6 = 5(1.792)$
$= 8.66$

Logs of "Unusual" Bases

5. $\log_3 8 = \frac{\log_{10} 8}{\log_{10} 3} = \frac{.903}{.477} = 1.89$

6. $\log_3 8 = \frac{\ln 8}{\ln 3} = \frac{2.079}{1.099} = 1.89$

7. $\log_{1.5} 7.2 = \frac{\ln 7.2}{\ln 1.5} = \frac{1.974}{.405} = 4$

1. $\log_{10}(6.8)(4.5) =$

2. $\log_{10} \frac{8.8}{7.7} =$

3. $\log_{10} \frac{(1.9)(9.3)}{3.6} =$

4. $\ln 7^6 =$

5. $\log_5 7 =$

6. $\log_{.4} 3 =$

7. $\log_8 3 =$

Richter Scale for Earthquakes

Scientists have devised the Richter Scale for measuring the intensity of earthquakes. This scale value (R) is related to earthquake intensity (Q) by the base 10 logarithmic formula:

$$R = \log_{10} Q$$

We can solve for the intensity by raising both sides to the power of 10, or

$$10^R = 10^{\log_{10} Q} = Q$$

So, for each point increase in the Richter Scale, the intensity increases by a factor of 10, as seen by the following. Consider an earthquake with intensity Q_0 and Richter Scale reading of R. Suppose another earthquake had a Richter Scale reading that was one point greater. The ratio of the intensities is then:

$$\frac{Q}{Q_0} = \frac{10^{R+1}}{10^R} = 10$$

Example 20

As application of this, consider the 1999 earthquake in Turkey that had $R = 7.4$ and the 1906 San Francisco earthquake with $R = 8.3$. With the latter being nearly 1 point higher, we can say that the San Francisco earthquake was nearly 10 times as strong as the Turkish earthquake.

Exercises 20

1. Consider two earthquakes with Richter scale values that were 2 points apart. How many times more intense is the stronger earthquake?
2. In 1989, an earthquake with $R = 5.3$ hit Montana. How many times more intense was the
   a. San Francisco earthquake?
   b. Turkish earthquake?

2.11 REPRESENTATION OF NUMBERS

Bases

Have you ever thought of what a number representation, for example 27, really means? Perhaps you recall your teacher in grade school saying that the 2 is in the ten's column and the 7 is in the one's column. Your teacher meant that you have 2 tens and 7 ones, or

$$27 = 20 + 7$$
$$= 2 \cdot 10 + 7 \cdot 1$$

Now consider a bigger number, say 8573. This could be written as

$8573 = 8000 + 500 + 70 + 3$

or

$8573 = 8 \cdot 1000 + 5 \cdot 100 + 7 \cdot 10 + 3 \cdot 1$

Since $10^0 = 1$, $10^1 = 10$, $10^2 = 100$, *and* $10^3 = 1000$, $8573 = 8 \cdot 10^3 + 5 \cdot 10^2 + 7 \cdot 10^1 + 3 \cdot 10^0$.

The digits in the representation of a number (base 10) are the multiples of 10^k (where $k = 0$ for the right-most digit, $k = 1$ for the second from the right digit, $k = 2$ for the third from the right digit, and so on).

Examples 21

1. $908 = 900 + 00 + 8 = 9 \cdot 10^2 + 0 \cdot 10^1 + 8 \cdot 10^0$
2. $28,376 = 20,000 + 8,000 + 300 + 70 + 6$
   $= 2 \cdot 10^4 + 8 \cdot 10^3 + 3 \cdot 10^2 + 7 \cdot 10 + 6 \cdot 10^0$
3. $4,600,000 = 4,000,000 + 600,000$
   $= 4 \cdot 10^6 + 6 \cdot 10^5$

Exercises 21

Express the following numbers as powers of 10

1. $444 =$
2. $38,204 =$
3. $9,150,022 =$

Ten is not the only base that could be used in representing numbers. In fact, computers extensively use bases of 2, 8, and 16.

To familiarize yourself with another base, consider base 2. The only difference between this and the base 10 case is that the digits will represent the multiples of 2^0, 2^1, 2^2, and so on. For example, let's represent the number 5 (base 10) in base 2 notation.

$5 = 4 + 1$
$ = 2^2 + 2^0$
$ = 1 \cdot 2^2 + 0 \cdot 2^1 + 1 \cdot 2^0$

5 (*base* 10) = 101 (*base* 2)

The 101 here is not 101 (base 10) which we know as one more than one hundred. Rather, the three digits are merely the multiples of 2^2, 2^1, and 2^0, respectively. So, although 5 and 101 are different numerals, they represent the same number.

Examples 22

1. $7 = d_2 \cdot 2^2 + d_1 \cdot 2^1 + d_0 \cdot 2^0$

   where $d_2 d_1 d_0$ is the number in base 2 representation.

   $7 = 4 + 2 + 1$
   $ = 1 \cdot 2^2 + 1 \cdot 2^1 + 1 \cdot 2^0$

   Thus 7 (base 10) = 111 (base 2).

ALGEBRA REVIEW

2. $24 \ (base\ 10) = d_4 \cdot 2^4 + d_3 \cdot 2^3 + d_2 \cdot 2^2 + d_1 \cdot 2^1 + d_0 \cdot 2^0$

$\qquad = 16 + 8 + 0 + 0 + 0$

$\qquad = 1 \cdot 2^4 + 1 \cdot 2^3 + 0 \cdot 2^2 + 0 \cdot 2^1 + 0 \cdot 2^0$

$\qquad = 11{,}000 \ (base\ 2)$

Exercises 22

Write the following base 10 numerals in base 2 notation.
1. 3
2. 10
3. 20

Scientific Notation

Scientists, who need a convenient easy means of expressing the very large and very small numbers with which they work, have devised what is known as *scientific notation*. This is nothing more than expressing numbers as multiples of powers, just as we did in the previous section.

To continue with base 10, some examples of writing large numbers in scientific notation are as follows:

1. $4000 = 4 \cdot 1000 = 4 \cdot 10^3$
2. $80{,}000 = 8 \cdot 10{,}000 = 8 \cdot 10^4$
3. $17{,}500{,}000{,}000 = 1.75 \cdot 10{,}000{,}000{,}000 = 1.75 \cdot 10^{10}$

Notice that we could have written the last example as

$1 \cdot 10^{10} + 7 \cdot 10^9 + 5 \cdot 10^8$

But, this isn't as understandable as the former representation.

Note that

$\qquad 1 = 10^0$

$\qquad 10 = 10^1$

$\qquad 100 = 10^2$

$\qquad 1{,}000 = 10^3$

$\qquad 10{,}000 = 10^4$

$\qquad 100{,}000 = 10^5,$ and so on.

So, the number of zero digits equals the power to which 10 must be raised. For example, $1{,}000{,}000{,}000{,}000{,}000 = 10^{15}$, since there are 15 zero digits after the 1. This observation makes it easy to write

$$9{,}800{,}000{,}000 = 9.8 \cdot 1{,}000{,}000{,}000$$
$$= 9.8 \cdot 10^9$$

Very small numbers can also be written effectively in scientific notation. This will be apparent when one realizes that $.1 = 10^{-1}$, $.01 = 10^{-2}$, $.001 = 10^{-3}$, and so on. You can see that the number of digits to the right of the decimal is the exponent (there is a negative sign attached to it) of 10. For example, $.0000001 = 10^{-7}$. We can express multiples of 10^{-k} just as we did with multiples of 10^k. For example, $.0007 = 7(.0001) = 7 \cdot 10^{-4}$.

Examples 23

1. $.00000003 = 3(.00000001) = 3 \cdot 10^{-8}$
2. $.0094 = 94(.0001)$
   $= 94 \cdot 10^{-4}$
   $= (9.4 \cdot 10^1) \cdot 10^{-4}$
   $= 9.4 \cdot 10^{-3}$
3. $-.002 = -2(.001) = -2 \cdot 10^{-3}$

Exercises 23

Express the following numbers in scientific notation
1. .000007
2. .00000033
3. −.00000000246
4. Astronomers in 2003 estimated that the total number of stars, planets, moons, etc., in the universe is 70 followed by 21 zeros. Write this number in scientific notation.

Some well known big or small numbers in science are

1. Avogadro's number $6.025 \cdot 10^{23}$ — This is the number of molecules in a mole of any chemical.
2. Big Bang seconds: 10^{-35} — This is the amount of time that "Big Bang" theorists think the Universe was born in.
3. Planck's length 10^{-35} meters — The smallest width that any object can be
4. Planck's time 10^{-43} seconds — The time it takes for light to traverse Planck's length

James R. Newman has a very interesting discussion of large numbers in his fascinating book entitled, *Mathematics and the Imagination*. There he reports the following large number estimates of others:

Ice crystals needed to form ice age: 10^{30}
Grains of sand on Coney Island beach: 10^{20}
Number of words spoken since beginning of time: 10^{16}

Newman goes on to name the epitome of large numbers as the google (actually his 9-year old nephew picked that name). A google is 1 followed by 100 zeros, or

google = 10^{100} (I wonder if the search engine got their name from this.)

You might think that numbers that large would never ever have any application. But you are

ALGEBRA REVIEW

wrong. When we study sampling later, you will see that the number of different samples of size 36 from a population of 15,000 is, you guessed it, a google!

2.12 THE MAGIC OF ALGEBRA

We have all experienced "magical games" by which someone can predict how many times we eat out per week (or something even sexier) and your age. You might be surprised to know that algebra is behind all that "magic". Soon with our algebraic power, the "magic" will be gone!

Let's illustrate with one such "magical game" that went around the internet in 2002. We list the various steps of the procedure, along with the algebra that deglamorizes them.

Step	"Magical Instruction"	Algebra
1	Think about the number of times you eat out per week	Call this number x
2	Multiply that number by 2	$2x$
3	Add 5 to the result	$2x + 5$
4	Multiply this result by 50	$50(2x + 5) = 100x + 250$
5	Add 1752	$100x + \underbrace{250 + 1752}_{2002 \uparrow \text{Current year!}}$
6	Subtract your year of birth (Let's assume you are 22 and were born in 1980)	$100x + \underbrace{(2002 - 1980)}_{22 \uparrow \text{Note your age of 22}}$

Now, let's analyze the final result of $100x + 22$.

Assuming that you eat out less than 10 times per week, this is a 3-digit number, since x must be 9 or less. For example, if you eat out 4 times per week, the result will be $100(4) + 22 = 400 + 22 = 422$. Notice how the number of dinners out becomes the left-most digit in the 3-digit number and the other two digits become your age at the end of the calendar year.

If you eat out 10 or more times per week, the "magic" still works, except that the number will be 4 digits, with the left 2 digits being how often you eat out. For example, suppose you eat out 14 times per week, then

$100(x) + 22 = 100(14) + 22 = 1400 + 22 = 1422$

Exercises 24

1. Explain how you would change the "magical game" so that it would work in the year 2004.
2. Show that a new procedure of multiplying the original number by 4, then adding 5, then multiplying by 25 works also.

Answers to "Rust Detecting" Algebra Quiz

1. $\dfrac{25}{16}$
2. $\dfrac{1}{9}$
3. 1
4. 2
5. $\dfrac{-3}{+5}$
6. $2x + 3y$
7. $\dfrac{y - x}{1 + y}$
8. $4x^2 + 12x + 9$
9. $x = \dfrac{b}{a - 1}$
10. $x = -1, y = 3$

PROBLEMS

Algebra is essential for all the advanced topics in this textbook. To impress this point on you and to provide you with a chance to work with algebraic rules in an applied context, the following algebra problems encountered in the solution of applications of widely varied topics in this book are offered.

Normally in this book, we employ word problems in the Confidence Builder and Mind Stretcher sections. This chapter is the only exception as we completely focus on algebraic operations. However, we do use the three problem sections to indicate "degree of difficulty"

TECHNICAL TRIUMPHS

1. Show that $B + B\dfrac{i}{c} = B\left(1 + \dfrac{i}{c}\right)$.

2. Consider the 2 equations:

    $C = 6.40 + 2P$
    $C = 12.00 + 1.3P$

    Substitute the formula for C in the first equation in for C in the second equation. Then, simplify and show that $P = 8$

3. Show that $(a + 2d) + (a + (n - 3)d) = 2a + (n - 1)d$

4. Show that $(1.01)^4 \cdot (1.01^4)^{17} = 1.01^{72}$

5. Show that $\log_b kx^{-c} = \log_b k + c \log_b x$

6. Solve the following equation for Q

    $PQ - F - vQ = 0$

ALGEBRA REVIEW

7. Consider the following set of simultaneous equations

$$b - 500c = 50{,}000$$
$$b - 460c = 55{,}200$$

Subtract the second equation from the first equation. In doing so, the resulting equation will not have the b variable. Then solve for c.

8. Solve the simultaneous equations.

$$C = 6.40 + 2P$$
$$C = 12.00 + 1.3P$$

9. Isolate T if $T = .3M + .2T + .4F + .1C + 3$.
10. Find E in terms of λ: $\lambda - 2(E-5) = 0$
11. Simplify $k(\tfrac{1}{2}) - k(\tfrac{1}{2})^2$

CONFIDENCE BUILDERS

1. Consider the Matt Maddix car rental problem of Chapter 1. Solve the two cost equations simultaneously to show that they intersect at the point where cost is \$30 and mileage is 200 miles.

2. Show that $\dfrac{10^k(10^k + 1)}{2} = 5 \cdot 10^{2k-1} + 5 \cdot 10^{k-1}$.

3. Find k if $1023 = 2^k - 1$.

4. Show that $\dfrac{B\left(1 + \dfrac{i}{c}\right)^e - B}{B} = \left(1 + \dfrac{i}{c}\right)^e - 1$.

5. Show that

$$\dfrac{\dfrac{R}{1+i}\left[\left(\dfrac{1-g}{1+i}\right)^n - 1\right]}{\left(\dfrac{1-g}{1+i}\right) - 1} = \dfrac{R\left[1 - \left(\dfrac{1-g}{1+i}\right)^n\right]}{g + i}$$

6. Simplify

$$\dfrac{\dfrac{2}{n}\left[1 - \left(1 - \dfrac{2}{n}\right)^n\right]}{1 - \left(1 - \dfrac{2}{n}\right)}$$

7. Solve for t if $1.2^t = 5(1.1)^t$

8. Show that $B + B\dfrac{i}{c} + \left(B + B\dfrac{i}{c}\right)\dfrac{i}{c} = B\left(1 + \dfrac{i}{c}\right)^2$

MIND STRETCHERS

1. If $y_1 = sx_1 + i$
   $y_2 = sx_2 + i$ find $y_2 - y_1$ in simplest form

2. Simplify $\dfrac{e^x(1 + 6x) - (5 + x + 3x^2)e^x}{e^{2x}}$

3. Show that $\dfrac{-P}{kp^{-n}}[k(-n)p^{-n-1}] = n$

4. Solve for n if $\dfrac{-kS}{2n^2} + K = 0$

5. If $n = \dfrac{F}{v}$ and $C = F + vn$, show that (a) $Fn^{-1/2} - vn^{1/2} = 0$, and (b) $C = 2F$.

6. Solve for S if $12S^{-1/2} - \dfrac{3}{2}S^{1/2} = 0$.

7. Solve for x if $\dfrac{.3}{8}x^{-1/2} - \dfrac{.03}{4} = 0$.

8. If $C = kn^{-1/2} + Kn$
   where $n = \left(\dfrac{k}{2K}\right)^{2/3}$, find C in terms of k and K.

9. Solve for P if $1 = \dfrac{-sP}{sP + i}$.

10. Simplify the expression $\dfrac{-P}{kP^{-1}}\left(\dfrac{-k}{P^2}\right)$.

11. Solve for x if $70{,}000\left(\dfrac{1}{x+1}\right) - 10{,}000 = 0$.

12. Simplify the expression $\dfrac{(P^2 + 1)(50{,}000) - (50{,}000P)(2P)}{(P^2 + 1)^2}$.

13. Multiply out: $V = (10 - 2x)(8 - 2x)(x)$.

MODULE TWO

Finite Mathematics

When observations are recorded at discrete points in time, variables change by finite (thus the name of the module) amounts. Certain patterns of such observations, called progressions, have wide applicability. The theory of progressions and some of their applications in life are discussed in Chapter 3. Progressions have many important applications in finance, investments, banking, and accounting. The study of such applications is the topic of Chapter 4. Since this module requires extensive use of algebra, it should provide an early test of how well you have mastered the preparation module.

CHAPTER

Progressions

3.1 INTRODUCTION

Can you fill in the blanks?

1. 1 2 3 ___
2. 2 4 8 ___
3. 9 5 1 ___
4. 3 1 $\frac{1}{3}$ ___
5. 2 0 0 ___

Your answers were probably 4, 16, −3, and $\frac{1}{9}$ for the first four sets. Check the calendar if you missed the last one. Except for the tricky one, each set of numbers has a perceptible pattern, which allows for the determination of numbers "down the line." So for set 1, each number is one more than its predecessor; for set 2, each number is two times its predecessor; for set 3, each number is four less than its predecessor; for set 4, each number is one-third times its predecessor.

Now try your hand at these.

6. 5   −10   20   −40   ___
7. 0 $-\frac{1}{2}$ ___ $-\frac{3}{2}$ −2 $-\frac{5}{2}$
8. 18   8   10   9   −1   ___
9. 100   ___   25   12.5   6.25   ___
10. −4   −2   ___   2   ___   6

The above ten sets fall into three categories, depending on the pattern of their numbers. Two of these categories, which are progressions, are now delineated.

Type of pattern	Operation	Description	Reference Set	Specific pattern
Arithmetic progression	+ or −	Each number is a constant amount *more (or less)* than its predecessor	1 3 7 10	+1 −4 $-\frac{1}{2}$ +2

57

Geometric progression	× or ÷	Each number is a constant amount *times (or divided into)* its predecessor	2 4 6 9	×2 ×$\frac{1}{3}$ ×(−2) ×$\frac{1}{2}$
Irregular	?	Neither arithmetic nor geometric	5 8	year ?

The one you couldn't get (set 8) is reminiscent of the ones on standardized tests (I.Q. test, S.A.T.'s, etc.) that seem as though they have no solution.

I.Q. Test for Aliens

Our National Aeronautics and Space Administration (NASA) enclosed a sequential pattern quiz inside Pioneer 10, a spaceship which was sent in search of extraterrestrial life. Enclosed in the spaceship are many things describing our civilization. For instance, a depiction of Leonardo da Vinci's perfect man is enclosed, as well as indications of our alphabet and number systems. One item included is this sequence of numbers with some blank spaces at the end. It is hoped that if intelligent life forms see this, they will be able to communicate to us the missing numbers, thus indicating in a way we can comprehend that they have received our message. The sequence is 3 1 4 1 5 9 _ _ _.

Could you pass this I.Q. test designed for aliens? (Hint: This sequence could drive one in circles. Now it should be as easy as pie.)

Learning Plan

Arithmetic and geometric progressions lend themselves to systematic understanding; irregular patterns, although they occur frequently in the real world, do not. Thus, the best we can do here is to develop the theory of the first two types of progressions. Fortunately, arithmetic and geometric progressions have important applications in many fields of human endeavor.

In this chapter, we first discuss arithmetic progressions and then go on to geometric progressions. For each type you will be introduced first to its general expression and then to its parameters. Given such parameters we can derive formulas for finding the sum of the terms in the progression. (This is not just a mathematical exercise; most applications require knowing how to find these sums.) With the theory well in hand, you will then be in a good position to understand the various applications that follow.

A side benefit of this instruction module is that it requires a great deal of algebra. Thus, you will have the opportunity to exhibit your new or relearned skill. Be sure to refer back to the axioms, exponent laws, logarithm laws, and so on when you are unsure of a particular algebraic step. (Because there are complicated calculations in this chapter, you will need a good calculator or access to a computer.)

3.2 ARITHMETIC PROGRESSIONS

The general expression for the terms in an arithmetic progression is

Term Number:	1	2	3	⋯	x	⋯	n
Value of Term:	a	$a + d$	$a + 2d$	⋯	$a + (x − 1)d$	⋯	$a + (n − 1)d$

PROGRESSIONS

where

> a = numerical value of first term
> d = common difference
> x = term number
> n = number of terms

Unique to arithmetic progressions is that each term (other than the first) is d more than its predecessor. Thus, the third term is the second term plus d:

$$(a + d) + d = a + (d + d) = a + d(1 + 1) = a + 2d$$

The fourth term is the third term plus d:

$$(a + 2d) + d = a + (2d + d) = a + d(2 + 1) = a + 3d$$

Notice that the number of d's for a term is always one less than the term number. For example, the second term has $1d$; the third term has $2d$; the fourth term has $3d$; the last or nth term has $(n - 1)d$. The value of d can be negative so that successive terms are smaller and smaller.

Examples

	a	d	n
1. 2 3 4 5 6 7 8 9	2	1	8
2. −6 −8 −10 −12 −14	−6	−2	5
3. .2 .1 .0 −.1 −.2 $\cdots$ −1.1	.2	−.1	14
4. 0 100 200 300 $\cdots$	0	100	∞
5. 5 5 5 5 5 5	5	0	6

Exercise 1

Provide the missing information.

	a	d	n
a. 14 10 5 $\cdots$ −20			
b. −100 −113 $\cdots$ −139			
c.	62	−4	7
d.	0	.6	10
e. $\frac{9}{8}$ $\frac{8}{8}$ $\frac{7}{8}$ $\cdots$ $\frac{2}{8}$			

Summing An Arithmetic Progression

Finding the sum of the terms in an arithmetic progression is necessary for many applications.

Karl Gauss first discovered how to arrive at such a sum in the 18th century. That such an eminent mathematician (he was called the Prince of Mathematicians) was the discoverer isn't startling. We would expect such discoveries from him. However, that he did it at the age of 6 is indeed humbling!

Legend is that his first grade teacher in Germany, just wanting to survive a day in the classroom with a headache, asked the class to add all the natural numbers from 1 to 100. Think about it. Could you have done that in first grade? Expecting a few hours of peace and quite, can you imagine the look on her face when young Karl came to her desk after a short while with the correct answer and a mathematical derivation for the general case? Migraine!!

But, how did this six year old add those numbers? What he did was to notice a very important relationship in the sequence which greatly simplified the task. Look at the task as he saw it.

Karl's teacher saw the problem as

$1 + 2 + 3 + \cdots + 98 + 99 + 100$

Rather than adding the numbers in order $1 + 2 + 3 + \cdots$, Karl added the first to the last, and then the second to the second last, etc. to form subtotals. In doing so, each subtotal was the same.

$1 + 2 + 3 + \cdots + 98 + 99 + 100$

101

101

101

So he quickly saw that in pairing the numbers like this he only needed to sum all the subtotals of 101. If there were 100 numbers to start with, there would be $\frac{100}{2}$ or 50 pairs or subtotals and, therefore, the answer would be 50 times 101 which equals 5050.

The logic which young Gauss discovered that morning is exactly what we use today to sum the terms of any arithmetic progression. Let's take a look at a few examples.

Examples

1. $1 + 2 + 3 + 4 + \cdots + 9 + 10$ $= \frac{10}{2}(1 + 10)$ $= 55$
2. $1 + 2 + 3 + 4 + \cdots + 19 + 20$ $= \frac{20}{2}(1 + 20)$ $= 210$
3. $2 + 4 + 6 + \cdots + 18 + 20$ $= \frac{10}{2}(2 + 20)$ $= 110$
4. $5 + 10 + 15 + 20 + \cdots + 95 + 100$ $= \frac{20}{2}(5 + 100)$ $= 1{,}050$
5. $20 + 40 + 60 + 80 + 100$ $= \frac{5}{2}(20 + 100)$ $= 300$

Exercise 2

Find the sums of the following arithmetic progressions using this same logic.
a. $1 + 2 + 3 + \cdots + 49 + 50$ b. $3 + 6 + 9 + \cdots + 297 + 300$
c. $7 + 5 + 3 + 1 + (-1) + (-3)$ d. $5 + 10 + 15 + 20 + 25$
e. $8 + 16 + 24 + \cdots + 8{,}000$
f. $a = 4, d = 2, n = 7$ (Hint: Remember, the last term is always equal to the first term plus $(n - 1)$ common differences [i.e. $a + (n - 1)d$].)

Note that if there is an even number of numbers in the progression, this pairing system works perfectly. If, however, there is an odd numbers, there will be one number left over, the middle-most number in the progression, but his number will be exactly one-half of the subtotal

PROGRESSIONS

for each of the pairs. So, using $\frac{n}{2}$ to multiply the common subtotal by will work perfectly each time.

Let's see, by considering the general expression for an arithmetic progression, if this phenomenon always holds.

General arithmetic progression

$$a \quad a+d \quad a+2d \quad \cdots \quad a+(n-2)d \quad a+(n-1)d$$

The subtotals would be

First term + Last term

$$a \quad + a + (n-1)d = 2a + (n-1)d$$

Second term + Second to last term

$$a + d \quad + a + (n-2)d = 2a + (n-1)d$$

And so forth . . .

All the subtotals are the same, $2a + (n-1)d$.

We can get some good algebra practice by verifying a few of these sums. For example, adding the second term to the second to last term goes like this,

Second term + Second to last term

$$[a + d] \quad + \quad [a + (n-2)d]$$
$$= a + d + a + (n-2)d$$
$$= a + a + d + (n-2)d$$
$$= 2a + d[1 + (n-2)]$$
$$= 2a + d[n-1]$$
$$= 2a + (n-1)d$$

Exercises 3

a. Showing the algebraic steps, add the third term of the general expression to the third to last term.
b. Now try adding the fourth term to the fourth to last term.

There will be $n/2$ of these sums which means the overall sum will be $\frac{n}{2}[2a + (n-1)d]$. So, using S_A as our symbol for the sum of an arithmetic progression, we have

$$S_A = \frac{n}{2}[2a + (n-1)d].$$

Remember the logic which gave us this formula was adding the $n/2$ common subtotals found by adding the first term to the last term.

Alternatively, if we call the last term L, we could show the formula for the sum of an arithmetic progression as

$$S_A = \frac{n}{2}(a + L).$$

Now let's make use of these concepts and formulas in various daily situations.

3.3 SOME APPLICATIONS OF ARITHMETIC PROGRESSIONS

Theater-in-the-Round

In Hyannis, Cape Cod, at the Melody Tent theater, I had the pleasure of seeing "The Bugs Bunny Children's Show with Batman and Robin" in the company of what seemed like an infinite number of screaming kids.

When Robin was frozen by Dr. Danger, the announcer pleaded that only an all-out shout of "Wake up Robin!" would save our boy wonder. (What a great plot!) The first two ear-splitting screams did no good. Thank goodness, the third one worked. It was then that I began to wonder just how many kids were necessary to inflict that much eardrum damage.

Melody Tent has a circular seating arrangement around the stage as shown in Figure 3.1. There are eight sections: four with 12 seats in row 1 and four with 4 seats in row 1. In each sec-

Figure 3.1

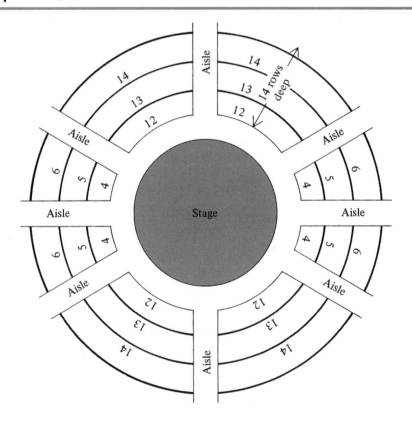

PROGRESSIONS

tion, each row has one more seat than the previous row. Thus, a section with 12 seats in the first row has 13 seats in the second row, 14 seats in the third row, and so on: a section with 4 seats in the first row has 5 seats in the second row, 6 seats in the third row, and so on. Each section has a total of 14 rows.

If Melody Tent was sold out for "Bugs and Co.," and if 75 percent of the seats were occupied by screaming kids, let's see how many of them were responsible for my ear troubles. First we must find the total number of seats in Melody Tent.

Notice that the entire first row has

$4 \cdot 12 + 4 \cdot 4 = 64$ seats.

The entire second row has

$4 \cdot 13 + 4 \cdot 5 = 72$ seats.

This is 8 more seats than the first row, which makes sense since each of the eight sections has 1 more seat in its second row than in its first row. Following this logic, the entire third row has 72 + 8 = 80 seats. the entire fourth row has 80 + 8 = 88 seats, and so on. So the total number of seats in the house is

$64 + 72 + 80 + 88 + \cdots$ and so on for 14 terms.

This is an arithmetic progression, with $a = 64$, $d = 8$, and $n = 14$, whose sum is

$$S_A = \tfrac{n}{2}[2a + (n-1)d] = \tfrac{14}{2}[2 \cdot 64 + 13 \cdot 8]$$
$$= 1624 \text{ seats}$$

So, 75 percent of 1624, or 1218, strong lungs equals one big earache.

Exercise 4

Given the same relationship among rows, how many seats would be in Melody Tent if the number of rows were doubled to 28?

The designers of Melody Tent probably started with a desired amount of seating and then worked backwards through the summation formula to determine the number of rows required. Let's illustrate this process by considering the design of an 8800 seat Melody Tent (Oh my ears!).

Here we know the sum of the terms in an arithmetic progression, but we don't know how many terms there are.

$$S_A = 8800 = 64 + 72 + 80 + 88 + \cdots \text{ and so on for } n \text{ terms}$$
$$8800 = \tfrac{n}{2}[2 \cdot 64 + (n-1)8]$$
$$= \tfrac{n}{2}[128 + 8n - 8]$$
$$8800 = 4n^2 + 60n$$

Exercise 5

Set up the equation (don't solve) to find the number of rows needed to have 5400 seats in Melody Tent.

Volume Discounts

The price of a product is often lower when the order is increased. You may have observed this phenomenon in your local A&P (Arithmetic & Progressions) supermarket. When Grandma's soup is priced at 2 for 99¢, you pay 50¢ for one can and get the second one for 49¢. Now just imagine a continuation of this pattern so that the third can costs 48¢, the fourth can costs 47¢, and so on. Then total cost, T, is the following arithmetic progression.

$$T = 50 + 49 + 48 + 47 + \cdots$$

If A&P limits its offer to a dozen cans, let's find the total and average cost.

$$T = \tfrac{n}{2}[2a + (n-1)d] = \tfrac{12}{2}[2 \cdot 50 + 11(-1)]$$
$$= 534¢, \quad \text{or} \quad \$5.34$$

Average cost is total cost (T) divided by the number of units purchased (12). For this calculation, let's use the alternative sum formula.

$$AC = \frac{T}{n} = \frac{\tfrac{n}{2}(a+L)}{n} = \frac{\tfrac{12}{2}(a+L)}{12} = \frac{a+L}{2}$$

$$= \frac{(50+39)}{2}$$

$$= 44.5¢$$

Note: $L = a + (n-1)d$
$$= 50 + 11(-1)$$
$$= 39$$

Notice how the average cost of all the units is merely the average of the first and last unit.

Exercise 6

Progression Car Rentals charge $1.00 for the first mile, $.98 for the second mile, $.96 for the third mile.... $.02 for the fiftieth mile. How much would they charge for a 50-mile trip? What would the average cost per mile be?

Sky Diving

The *Syracuse Herald Journal* of May 25, 1975, reported that Mr. George McCulloch of Seneca Falls, New York, celebrates his birthday each year by jumping out of an airplane. Each year he counts his age before pulling the cord. "It gets longer every year," he said. "I guess I'll have to jump from 14,000 feet." The count up to his age is so long that Mr. McCulloch, who is 70, will fall 12,000 feet before the parachute opens.

The "To Tell the Truth" show on CBS in the 1970's interviewed a sky diver who jumped from 3000 feet above Arizona. Despite the fact that his parachute didn't open, he lived (but lost $1\frac{1}{2}$ inches of height).

ABC's "Wide World of Sports" in the 1980's telecasted the diving competition at Acapulco, Mexico. There divers jump into the water (just missing the cliffs) from a height of 137 feet.

Mr. McCulloch, the Arizona diver, and the divers of Acapulco were all using earth's gravity to get their kicks. Scientists have determined that a body falling in earth's gravity will fall 16 feet during the first second, 48 feet during the second second, 80 feet during the third second, and so on.[1] So the drops per second follow the arithmetic progression

16 48 80 112 ···

where $a = 16$ and $d = 32$.

The total distance fallen, D, would involve summing the individual terms of this progression.

So, for any number of seconds, n, the total distance fallen, D, would simply be the sum of this arithmetic progression.

$$D = S_A = \frac{n}{2}[2a + (n-1)d]$$

$$= \frac{n}{2}[32 + (n-1)32]$$

$$= \frac{n}{2}[32n]$$

$$= 16n^2$$

Seconds	Distance Fallen In That Second	Total Distance Fallen (D)
1	16	16
2	48	64
3	80	144
4	112	256
⋮	⋮	⋮
n	$16 + (n-1)32$	S_A with $a = 16$, $d = 32$, n = seconds in free fall

---

[1] This assumes that there is no friction (as in a vacuum), and that the body starts from rest. In actual free fall, the drop per second is less that the frictionless case; also, a terminal or maximum velocity exists.

If Mr. McCulloch counted a number each second, this formula gives his total free fall distance as $16 \cdot 70^2 = 78{,}400$ feet. Since he only fell 12,000 feet, he must have been doing some really fast counting (wouldn't you?).

Exercise 7

1. How far will a sky diver drop in 10 seconds of free fall?
2. A sky diver, who jumps out of the airplane at 10,000 feet wants to pull the cord at 3,600 feet. How many seconds should he wait?

NBA Draft

Arithmetic progression plays a part in the NBA draft of college basketball players. All the NBA teams that did not make the playoffs in the previous year (there were 11 in 1990) are eligible for the top three picks in the draft. The selections are determined randomly by drawing balls from a cylinder. The team with the worst record the previous year gets 11 balls in the cylinder; the team with the second-worst record gets 10 balls in the cylinder; the next worst team gets 9 balls, etc., until the best team of these non-playoff teams gets only 1 ball in the cylinder. The total number of balls in the cylinder is thus the sum of the arithmetic progression

$$S = 11 + 10 + 9 + \cdots + 1$$
$$= \frac{11}{2}(11 + 1) = 66$$

A team's chance of getting the prized first pick is then the number of balls they have in the cylinder divided by 66.

PROGRESSIONS

In 1990, the New Jersey Nets had the worst record, and so they got the most balls, or 11. Their chance of getting first pick was then $\frac{11}{66} = \frac{1}{6}$. As it turned out, the Nets did get the first pick and chose Derrick Coleman of Syracuse University. Derrick was just a student at my university; but I would gladly change bank accounts with him!

Note: A much more complicated draft lottery system is in place now.

Exercise 8

As a result of expansion, the NBA had 13 non-playoff teams in 1995. What was the chance that the worst team got the prized first pick?

Promotion Systems

Promotion systems in the military or government are based on ranks or levels, with promotions becoming more difficult to obtain as one goes up the ladder of success. Arithmetic progressions can serve as models for such systems. Let's illustrate this by considering the case of a gung ho recruit in the U.S. Army. The Army has nine levels of enlisted people, starting with private (E-1) and progressing through sergeant major (E-9). Our recruit becomes and E-1 as soon as he enters the Army. Then he must complete basic training and other goodies before promotion to E-2 (also called private) about three months later. Then, if he shines his boots properly, he get promoted to private first class (E-3) about nine months later. Once he is able to disassemble and assemble his rifle (not his gun), he will get promoted to corporal (E-4) about 15 months later. Thereafter, each promotion takes about six months more than the previous one.

Summarizing the time-in-rank picture for our recruit, we have

Promotion to	E-2	E-3	E-4	E-5	$\cdots$
Time required (months)	3	9	15	21	$\cdots$

These times follow an arithmetic progression with $a = 3$, $d = 6$, and $n = 8$. (There are only eight promotions if we disregard the first one, which is automatic with entry into the Army.) The sum of this progression, which represents the total time needed by a "good" person to reach the top enlisted rank is

$$S_A = \frac{8}{2}[2 \cdot 3 + (8 - 1)6] = 192 \ months, \quad or \quad 16 \ years$$

Exercise 9

Devise a new promotion system for the Army containing the same three-month wait for promotion to E-2 and a time-in-rank pattern that follows an arithmetic progression, but that cuts the total time required to reach the highest rank to 9 years or 108 months.

3.4 GEOMETRIC PROGRESSIONS

The general expression for the terms in a geometric progression is

Term Number	1	2	3	...	x	...	n
Value of Term	a	ar	ar^2	...	ar^{x-1}	...	ar^{n-1}

where

a = numerical value of first term
r = common multiple
x = term number
n = number of terms

Unique to geometric progressions is that each term (other than the first) is r times its predecessor. Thus, the third term is *r times the second term:* $r(ar) = ar^2$ (which axioms?). The value of r can be negative so that successive terms have alternating signs.

Some examples are:

Examples

						a	r	n
1. 3	6	12	24			3	2	4
2. 100	10	1	$\frac{1}{10}$	$\frac{1}{100}$	...	100	$\frac{1}{10}$	∞
3. 2	-4	8	-16	32		2	-2	5
4. -1	-3	-9	...	-729		-1	3	7
5. 8	-4	2	-1	$\frac{1}{2}$	$-\frac{1}{4}$	8	$-\frac{1}{10}$	6

Exercise 10

Provide the missing information

						a	r	n
a. 5	$\frac{25}{4}$	$\frac{125}{16}$						
b. -1	10	-100	1000	...				
c. __	__	__	__	__		7	.3	5
d. __	__	__	__	__	__	-4	$-\frac{1}{2}$	7
e. $\frac{1}{16}$	$\frac{3}{16}$	$\frac{9}{16}$	$\frac{27}{16}$					

Notice that the value of the xth term in a geometric progression, a_x, is

$$a_x = ar^{x-1}$$

For example, if a geometric progression has $a = 5$, $r = 3$, then the fourth term ($x = 4$) is

$$ar^{x-1} = 5 \cdot 3^{4-1} = 5 \cdot 3^3 = 135$$

Exercise 11

a What is the value of the 5th term in a geometric progression having $a = 6$ and $r = 2$?
b r can assume negative values. Consider the geometric progression with $r = -3$ and $a = 1$. Find the values of the first 5 terms.

PROGRESSIONS

c. r can assume values that are not whole numbers. Consider the geometric progression with $r = 1.1$ and $a = 1000$. Determine the value of the 8th term.

Summing A Geometric Progression

Many applications depend on knowing the sum (let's call it S_G) of a geometric progression.

$$S_G = a + ar + ar^2 + \cdots + ar^{n-2} + ar^{n-1}$$

A cute mathematical move (multiply both sides of the above formula by r) enables one to obtain a nice simple expression for S_G.

$$rS_G = ar + ar^2 + \cdots + ar^{n-1} + ar^n$$

Now, if we subtract the first equation from the second ...

$$rS_G = ar + ar^2 + \cdots + ar^{n-1} + ar^n$$
$$- [S_G = a + ar + ar^2 + \cdots + ar^{n-1}]$$
$$rS_G - S_G = -a + (ar - ar) + (ar^2 - ar^2) + \cdots + (ar^{n-1} - ar^{n-1}) + ar^n$$

Amazingly, all the terms except the first and the last cancel out. So,

$$rS_G - S_G = ar^n - a$$
$$S_G(r - 1) = a(r^n - 1)$$

Thus, the sum of the terms in a geometric progression is

$$S_G = \frac{a(r^n - 1)}{r - 1}$$

Exercise 12

Verify that the above geometric sum formula is algebraically equivalent to

$$S_G = \frac{a(1 - r^n)}{1 - r}$$

For you non-believers, let's check out these formulas on the geometric progression 3, 6, 12, 24, the sum of which we know is 45.

$$S_G = \frac{a(r^n - 1)}{r - 1} = \frac{3(2^4 - 1)}{2 - 1} = \frac{3(16 - 1)}{1} = 45$$

$$S_G = \frac{a(1 - r^n)}{1 - r} = \frac{3(1 - 2^4)}{1 - 2} = \frac{3(1 - 16)}{1 - 2} = \frac{3(-15)}{-1} = 45$$

They work!

Exercise 13

a. Using both versions of the geometric sum formula, sum the following geometric progressions. Check the results by old-fashioned addition.

    1   3   9   27   81
    20 10 5

b. Notice how in each case above, one version of the sum formula involves negative numbers while the other version does not. Does this difference depend on the value of r? Try to develop a rule to tell you which version of the sum formula to use if you wish to avoid negative numbers.

3.5 APPLICATIONS OF GEOMETRIC PROGRESSIONS

Now let's consider several applications of these geometric progressions.

Worker's Paradise

A classic problem involves the worker, Rich Quick, who offers his employer, McMaths, 40 years of faithful service for only 30 days of meager pay according to the scheme: 1¢ the first day, 2¢ the second day, 4¢ the third day, 8¢ the fourth day, etc. Rich could not even get a cup of coffee on his first week's pay; but beware of the growing power of geometric progressions.

Because Rich's pay on any day is double that of the previous day, we have a geometric progression with $r = 2$ and $a = 1$. So the sum of x days of pay is

$$S_G = \frac{a(r^x - 1)}{r - 1} = \frac{1(2^x - 1)}{2 - 1} = 2^x - 1$$

The total pay after 10, 20, and 30 days are (in cents)

Day	Total Pay	
10	$2^{10} - 1 = 1023$¢	Welfare, here I come!
20	$2^{20} - 1 = 1048576$¢	Over $10,000. Getting better.
30	$2^{30} - 1 = 1073741823$¢	Almost $11 million. I'll flip Big Maths for that anytime!

This illustration shows you the growth power, even if it is somewhat delayed, of a geometric progression. Judge Sands in the Yonker's NY desegregation case, discussed next, must have understood that.

Exercise 14

Unfortunately for Rich, Matt Maddox, the president of McMaths knows his math too. Matt offers Rich the payment scheme as long as $r = 1.5$. Under those conditions, show that Rich's total pay for all those years of flipping equations will be 383,500¢, or $3,835. This amounts to less than $100 for each of those 40 years. That calls for a strike, or at least some more bargaining. What value of r would give Rich a total of $1,000,000?

Yonkers, NY Doubling Fines

Yonkers, NY in 1988 was found in contempt of court for failure to meet a federal court-ordered desegregation housing plan. Judge Sands fined Yonkers for each day that they remained in contempt. The fines were such that they started with $100, and doubled each day thereafter. Thus, they were: $100, $200, $400, $800, etc. These fines form a geometric progression with $a = \$100$, $r = 2$, whose x^{th} day fine (F) and the geometric progression sum of all fines (S) are:

$$F = (100)2^{x-1}$$

$$S = \frac{a(r^x - 1)}{r - 1} = \frac{100(2^x - 1)}{2 - 1} = 100(2^x - 1) = (100)2^x - 100$$

The annual Yonker's budget was 337 million dollars. If the city didn't reverse itself and comply with the federal housing desegregation plan, we can solve for the time elapsed when the whole budget was kaput!

$$(100)2^x - 100 = 337,000,000$$

$$2^x = \frac{337,000,000}{10} - 100 = 3,369,000$$

$$x = \frac{\log_{10} 3,369,000}{\log_{10} 2} = \frac{6.527617}{.30103} = 21.7 \ days$$

Fortunately for Yonkers, an Appeals Court restricted the daily fines to a maximum of one million dollars. That would be reached in

$$(100)2^{x-1} = 1,000,000$$

$$x^{x-1} = 10,000$$

$$x - 1 = \frac{\log_{10} 10,000}{\log_2 2} = \frac{4}{.30103} = 13.3$$

$$x = 14.3 \quad or, \quad on \ the \ 15^{th} \ day$$

Even more fortunate for Yonkers is the fact that they did reverse themselves and obeyed the desegregation plan before the fines became onerous.

Exercise 15

a. What would the cumulative total of all the fines through the 14th day be?
b. If the $1 million per day fine began on the 15th day, how long would it take for the entire Yonker's budget to be used up?

Sure-Fire Gambling Scheme

The relationship between a term value and the sum of a geometric progression is the basis for success for a Lost Wages gambler. Let's see how he can "guarantee" a profit of $1 on a roulette wheel that pays double the bet. If he begins with a $1 bet and doubles his bet until he eventually wins, then

Bet Number	1	2	3	4	...	x
Bet Amount	1	2	4	8	...	2^{x-1}
Winning Payoff	2	4	8	16	...	2^x

To see the pattern, suppose that he lost the first two bets, and then won on the third bet. He would have bet a total of $1 + 2 + 4 = \$7$. However, the winning payoff on the third bet was $8. So, his net gain was $1.

We now show that the Lost Wages gambler is guaranteed $1 profit regardless of when he finally wins. Suppose that it takes x bets before he finally wins. His net profit at that point is

$$NetProfit = (\text{Winning Payoff on } x^{th} \text{ Bet}) - (\text{Total Bets to that Point})$$

$$= 2^x - (1 + 2 + 4 + \cdots + 2^{x-1})$$

$$= 2^x - \frac{2^x - 1}{2 - 1}$$

$$= 2^x - (2^x - 1) = \$1$$

Easy Street, here I come. But wait before you book that flight to Lost Wages. What if the gambler in our example only had $3 to bet? He would have gone broke before getting that "guaranteed" $1. And therein lies the catch. You need an infinite amount of money to be guaranteed that $1. And do you know anyone with an infinite amount of money who would risk it for only one more buck?

With a finite amount of money, we can determine the extent of a losing streak the gambler can withstand before going broke. Suppose he begins with $1023.

The total bet in x spins of the wheel is $2^x - 1$. If this exceeds $1023, he will be bankrupt. Solving for x,

$1023 = 2^x - 1$

$2^x = 1024$

so ... $x = \log_2 1024 = 10$

Thus, ten losses mean bankruptcy, since he would not have a penny left for the eleventh bet. He can withstand a nine-bet losing streak.

PROGRESSIONS

Exercise 16

a. Show that the betting sequence 3, 6, 12, 24, 48, ⋯ "guarantees" a $3 win.
b. Find the betting sequence that "guarantees" a $5 win.

Moore's Law of Computing Power

Gordon Moore, co-founder of Intel, observed that the number of transistors that chip makers could fit on a given silicon chip was doubling every 18 months. So, if some initial time, there were a transistors per silicon chip, then the sequence of number of transistors per silicon chip will be:

Initial	18 months	3 years	$4\frac{1}{2}$ years	6 years
a	$2a$	$4a$	$8a$	$16a$

The formula for the xth value in this sequence is:

$a2^{x-1}$ where x is the number of 18 month periods in the future.

For example, for 9 years into the future, which is 9/1.5 = 6 18-month periods, the formula gives:

$a2^5 = 32a$

or 32 times the initial number of transistors.

One can solve for the time when a certain multiple will be achieved. For example, when will there be 128 times as many?

$128a = a2^{x-1}$, or $128 = 2^{x-1}$ Solving this using logarithms gives $x = 8$, or 8 18 month periods, or 12 years.

Exercises 17

1. How many times the original value would occur in 9 years?
2. How long would it take for the number of transistors to multiply by 256?

Sales Forecasting

Sales of corporations often increase at approximately the same percentage every year. This is a so-called compound growth pattern. Each years sales are $(1 + g)$ times the previous years sales. Such is the nature of a geometric progression. So the year by year sales would be:

S_0 $S_0(1 + g)$ $S_0(1 + g)^2$ ⋯ $S_0(1 + g)^x$

where

S_0 is the current sales level
g is the growth rate (decimal)
x is time in years from now.

We can illustrate by considering the intense sales competition between Gangland Industries and C.A.R.T.E.L. International. Because of aggressive marketing, Gangland Industries (their salesmen say, "We'll make you a deal you *better not* refuse") has been winning a lot of "contracts" away from its larger competitor. Consequently, industry experts predict a 20 percent annual sales growth for Gangland, as opposed to a 10 percent figure for C.A.R.T.E.L. With current sales levels of $10 and $50 million, let's find out when Gangland will catch up in sales.

The two sales equations in the xth year are

$$S_G = 10(1.2)^x$$
$$S_C = 50(1.1)^x$$

By setting the sales level equal, we can find out when Gangland catches up.

$$S_G = S_C$$
$$10(1.2)^x = 50(1.1)^x$$
$$\frac{1.2^x}{1.1^x} = 5$$
$$1.091^x = 5$$
$$x = \log_{1.091} 5 = 18.5 \; years$$

Exercise 18

If Gangland, by even more aggressive marketing ("We'll make you a deal that your loved ones hope you won't refuse"), is able to increase its growth rate to 25 percent (everything else the same as stated above), when will they equal C.A.R.T.E.L. in sales?

Rule of 72

You saw the benefits of knowing the Rule of 72 back in Chapter 1. Now that we have an understanding of compound growth, we can develop the formula itself. Recall, the Rule of 72 states that the time in years for something to double (n) is approximated as

$$n = \frac{72}{i} \quad \text{where } i \text{ is the annual growth rate } (\%)$$

This rule is simply derived as follows: If some variable is growing at an annual compound rate if i, the formula to describe its value as of n years (P_0 is the initial value).

PROGRESSIONS

$P_n = P_0(1 + i)^n$

At the point when the prices have doubled from their initial value, we have

$P_n = 2P_0$

So

$P_0(1 + i)^n = 2P_0$

$2 = (1 + i)^n$

To solve for n, we need our friends, logarithms (you'll see why we use natural logs with base 2.718 later).

$ln\ 2 = ln\ (1 + i)^n$

$\qquad = n\ ln\ (1 + i)$

so

$n = \dfrac{ln\ 2}{ln\ (1 + i)} \cong \dfrac{.693}{ln\ (1 + i)}$

At this point, we must invoke a result from higher mathematics (like from the 100th story of the Empire State Building), which you are not responsible for on tests. (Does that mean you can skip this section? Absolutely not!!) This way-up result is that the denominator can be expressed as an infinite series of terms, as shown.

$n = \dfrac{.693}{i - \dfrac{i^2}{2} + \dfrac{i^3}{3} - \dfrac{i^4}{4} + -\cdots}$

As you can see, $ln\ (1 + i)$ can be represented as an infinite series of terms. On the surface, that seems to complicate matters. However, notice that each term is much less than its predecessor (since i is a decimal like .05, then i^2 would be much smaller, like $.05^2 = .0025$, and i^3 would even be much smaller yet). So Voila, we can drop all those terms with little consequence and get

$n \cong \dfrac{.693}{i} \cong \dfrac{.69}{i}$

Whoops! We derived the Rule of 69, and not the Rule of 72! But wait. Since 72 is divisible by more whole numbers and since it does better to accommodate the "dropped" terms, it turns out that we use

$n = \dfrac{.72}{i}$ (*i is decimal*)   or $n = \dfrac{72}{i}$ (*i is percentage*)

Exercise 19

a. How long would it take for money to double at a 6% annual rate?
b. If world population grows at an annual rate of 2%, how long would it take for population to double?
c. If a company predicts that its sales will double in 8 years, what annual rate of growth are they implying?

Chain Letters

Dear Friend:

Greetings. This is your chance of a lifetime. (And if you don't take it, you might have experienced your lifetime already!) This is how easy it works. Purchase this letter for $50 from your friend in the 10th position on the list of names below, and also send $50 to the first person on the list.

1. Godfather
2. Godmother
3. Godbrother
4. Godsister
5. Meyer Lansky estate
6. Al Capone Jr. estate
7. Jimmy Hoffa estate
8. Gambino family
9. John Gotti estate
10. U. Ben Had

Then redraft this letter, rub-out, I mean remove the first name, move all other names up one position, and place your name in the tenth position, then make 3 copies of this letter and sell them to 3 friends for $50 each. Once you accomplish that, you will have already made a $50 profit. But that is just the beginning. If no-one breaks the chain (and if they think of their loved ones, they won't!), soon $50 \cdot 3^{10}$ will roll in. Whoopee!!

Even the Post Office will benefit, as a total of $88,572 letters will be mailed out to top position people in this process. This will wipe out their chronic deficit.

In summary, if you don't want the Post Office in the red or yourself in a pool of red, you know what to do.

What an opportunity to solve the Post Office's deficit problem while becoming a millionaire! Could this be right?

If you send $50 to Godfather, include your name in the 10th place, while moving up the other nine people, and make 3 copies of that letter. If all three of your friends who receive your letter do likewise, they will send out a total of $3 \cdot 3 = 3^2 = 9$ letters, each with your name in the 9th spot. If those 9 people continue the chain, they will send out a total of $9 \cdot 3 = 3^3 = 27$ letters with your name in the 8th spot. And on it goes according to the following:

Your place:	10	9	8	7	6	...	1	
Letters:	3	3^2	3^3	3^4	3^5	...	3^{10}	←You get $50 · 3^{10}

Notice how the place number plus the exponent of 3 of the letters adds up to 11. This is

PROGRESSIONS

how we know how many letters you "will" receive when you get to the first spot. Right? The mathematical logic is correct and unbroken. That is more than could be said about the chain.

The total number of letters sent "would have been"

$$3^1 + 3^2 + 3^3 + \cdots + 3^{10} = \frac{3(3^{10} - 1)}{3 - 1} = 88{,}572$$

Again the letter mathematics were right.

We can generalize our understanding of chain letters. If there are N names on the letter and one sends to F friends, then

Your place:	N	$N-1$	$N-2$	$N-3$	$\cdots$	1
Letters:	F	F^2	F^3	F^4	$\cdots$	F^N

Getting $\$M$ per letter means a profit of MF^N. But don't hold your breath. And don't even try such a thing, since they are ILLEGAL!

Exercise 20

Derive the formula for the total number of letters sent for the general case. If stamps are 37¢ each, how much revenue "would" the Post Office have received?

Population Explosion

Have you ever thought about how many people are depending on you to finish mathematics, graduate, get married, and have children? In fact, their very lives depend on it. Yes, your descendants want you to think about the mathematics of it all, and then go do what you have to do to insure their chance to struggle through mathematics too!

Suppose you have 2 children, and each of them has 2 children ($2 \cdot 2 = 4$ grandchildren to you), and each of them has 2 children ($2^3 = 8$ great-grandchildren), and so on. If a generation is 25 years, then in the nth generation from now the world will be blessed with $ar^{n-1} = 2 \cdot 2^{n-1} = 2^n$ little ones, thanks to you. Thus the procreation progression is

$2, 4, 8, 16, 32, \cdots 2^n$

Altogether for the next n generations

$$S_G = \frac{a(r^n - 1)}{r - 1} = \frac{2(2^n - 1)}{2 - 1} = 2^{n+1} - 2; \quad \text{where } a = 2 \text{ and } r = 2$$

people are waiting for you to get the show on the road!! For example, in the next 200 years ($n = 8$), $2^9 - 2 = 510$ people are rooting for you.

Exercise 21

a How many people in the next 500 years are rooting for you?
b. If each descendant has 3 children, how many people in the next 200 years are rooting for you?

March Madness

The NCAA Division 1 men's basketball tournament starts with 64 teams. In the first round, there are 32 games. Those winners go on to play 16 games in the second round. Etc.

The total number of games needed to crown a national championship (and wipe out your entry to the office pool) is

$$32 + 16 + 8 + 4 + 2 + 1$$

This geometric progression with $a = 32$, $r = \frac{1}{2}$, and $n = 6$ has a sum

$$S = \frac{a(1 - r^n)}{1 - r} = \frac{32(1 - (\frac{1}{2})^6)}{1 - \frac{1}{2}} = 63$$

PROBLEMS

TECHNICAL TRIUMPHS

1. Determine the type of progression and the value of the requested term for the following progressions
   a. 4 9 14 19 $\cdots$ 10th term
   b. 100 50 25 12.5 $\cdots$ 7th term
   c. 80 70 60 50 $\cdots$ 15th term
   d. 2 3 4.5 6.75 $\cdots$ 8th term

2. Determine all the values in the following progressions
   a. $a = 8$ $r = \frac{1}{2}$ $n = 5$
   b. $a = -25$ $d = 4$ $n = 6$
   c. $a = 40$ $d = -100$ $n = 7$
   d. $a = -5$ $r = -2$ $n = 4$

3. Determine the sum of the following progressions
   a. 1 8 15 22 $\cdots$ 57
   b. The whole numbers from 1 to a billion.
   c. 2 2^2 2^3 2^4 $\cdots$ 2^{100}
   d. 1 .1 .01 .001 $\cdots$ 10^{-10}

4. An arithmetic progression with $a = 10$, $d = 5$, has 20 terms. Determine the sum of that progression.

5. A geometric progression with 15 terms has $a = 3$ and $r = 1.1$. Determine its sum.

6. The formula for the sum of an arithmetic progression is

$$16[12 + 5(n - 1)]$$

What are a, d, and n for that arithmetic progression?

7. The formula for the sum of a geometric progression is

$$\frac{7(1 - .8^9)}{.2}$$

What are a, r, and n for that progression?

8. The sum of an arithmetic progression is 500, $n = 10$, and $a = 5$. Determine d.

PROGRESSIONS

9. The sum of a geometric progression is 315 and $r = 2$, $n = 6$. Determine a.

10. The last term in an arithmetic progression is 34. If its sum is 210 and $n = 12$ and $d = 3$, find the value of the first term.

11. The sum of a arithmetic progress is 500. If $a = 10$ and $d = 5$, how many terms are in the progression?

12. The sum of a geometric progression is 1022. If there are 9 terms and $a = 2$, what is r?

13. If a price index grows 5% per year, and the initial value is 100, what will the index be in 10 years?

CONFIDENCE BUILDERS

1. If consumer prices increase 4% per year, use the Rule of 72 to determine how long it will take for prices to double. Use the more accurate Rule of 69.

2. The television show "Jeopardy" had a question board with 25 questions arranged in 5 categories, each with 5 levels of difficulty. The dollar value of the questions was

   10 10 10 10 10
   20 20 20 20 20
   30 30 30 30 30
   40 40 40 40 40
   50 50 50 50 50

   a. Use the appropriate progression formula to find the total dollar value for a category (column). What is the total winnings possible from answering all the questions?
   b. A "double jeopardy" board, with all prizes double that of the above board, was also used. What was the total value for a category? For the entire board?

3. Celebration of the Jewish holiday of Chanukah requires lighting 2 candles on the first day, 3 candles on the second day, 4 candles on the third day, and so on for the 8 days of the holiday. As a producer of Chanukah candles, how many should you put in a box?

4. A game involves moving from a starting point and picking up an object, returning the object to the starting point, then picking up a second object, returning it, then picking up and returning successive objects. If the 10 objects are 2 feet apart, and the nearest object is 20 feet away from the starting point, how many feet must one travel in order to complete the game?

5. Melvin Whipple received money equal to his age on his birthday from Grandma Winnie Whipple each year for his first 14 birthdays. If Melvin put the money into his piggy bank, how much did he have when he was 14?

6. A man, taking a ten-year job assignment, is given two alternative compensation schedules:

   Plan A   Starting salary of $10,000 with a 7 percent annual increase
   Plan B   Starting salary of $10,000 with a $1000 increase each year

   a. Which plan gives him the greater total salary over the first ten years?
   b. Which plan gives him the highest total if he remains on the job 20 years?

7. Ten years ago, a company established a Planning Department and staffed it with a college graduate, an assistant, and a secretary. The initial salaries and annual increments were as follows:

	Initial salary	Annual increment
College grad	$9000	$500
Assistant	$5000	$300
Secretary	$3000	$200

   Determine what the company paid for salaries to the Planning Department over this period.

8. Infinity Company offers volume discounts on their calculator as follows: The first calculator costs

$200, the second costs $198, the third cost $196, ..., the 51st costs $100. Restricting ourselves to the case of 51 or fewer calculators, where x is the number purchased,
   a. What is the cost of the xth calculator?
   b. What is the total cost of x calculators?

   If the volume discount is changed so that each calculator costs 98 percent of the previous one,
   c. What is the cost of the xth calculator?
   d. What is the total cost of x calculators?

9. As a train reaches the top of a long slope, the coupling of the last car breaks. The car begins to descend, passing over 2 feet the first second, 5 feet the second second, 8 feet the third second, and so on. If it takes 1 minute for the car to reach the bottom of the slope,
   a. How many feet will the train descend in the last second before it reaches bottom?
   b. What is the length of the entire slope?

10. Note: This is a trick question!

    MOTHER GOOSE
    As I was going to St. Ives,
    I met a man with 7 wives,
    Each wife had 7 sacks,
    Each sack had 7 cats,
    Each cat had 7 kits,
    Kits, cats, sacks, wives,
    How many were going to St. Ives?

11. Flunk-Out U has a donation plan for its newly graduated (if there are any) alumni. The alumni are asked to pledge $25 the first year after graduation and to increase their pledge 10% every year thereafter. See if you understand why no courses in progressions are offered at F.O.U.!

12. Colonel South offered Uran the following deal on TOW Missiles: For the first 300 TOWs: $15,000 for the first one, and for each additional TOW, the cost would decrease by $10

    For those above 300: The cost of each missile would be 1% less than that of the previous missile.
    a. How much would 300 TOW's cost?
    b. How much would 500 TOW's cost?

13. *The New York Times* (June 2, 1976) reported that on Memorial Day, after the Atlanta Braves suffered their fourth straight defeat (earlier they had lost 13 in a row), Ted Turner, the owner, took to the public address microphone and announced that all 3000 paid fans (actually it was 2994 for you trivia fans) could come to the next game free.

    Suppose Turner allowed the fans attending each successive game to come to the next game free until the Braves broke their losing streak. If 3000 newly paid fans at each successive game plus all the previous fans came to each game, how many straight Braves' losses (starting with the Memorial Day game) would it take to completely fill Atlanta Stadium (51,556 seats)?

14. The half-life of a certain radioactive material is 20 days. (Note: half-life is the time needed for the material to emit half its radiation). How long will it take for 99.9% of the radiation to be emitted?

MIND STRETCHERS

1. The eighteenth century economist Malthus was concerned with what he envisioned as a geometric growth in population, while fertile land and thus food grow only in arithmetic fashion. Discuss why Malthus was so concerned.

2. You and your spouse decide to have 3 children, as do all of your descendants thereafter. (Assume your descendants do not marry within the family.)
   a. After 5 generations, how many more people will inhabit the earth than if everyone decides to have 2 children?
   b. How many generations will it take for an additional 1000 people to be born?

PROGRESSIONS

3. A super ball bounces back to 80 percent of its previous peak height on each bounce. If it is initially dropped from a height of 5 feet, what is the total distance that it travels on its down-up-down movements? What is the total distance if the ball bounces back to 90 percent of its previous peak each time?

4. A little boy and his father were walking across the Green Gate Bridge when the boy asked how high the bridge was. The father, who felt silly in not knowing the answer, found a rock and decided to answer the boy's question and teach him some mathematics at the same time. When they got to the middle of the bridge, the boy dropped the rock. The father clocked the experiment from the time the rock left the boy's hand until impact into the water—it took exactly 5 seconds. How high is the bridge?

5. Suppose that friction causes the fall per second in earth's gravity to be half the frictionless case. Thus, in the first second, the fall will be 8 feet; in the second second, the fall will be 24 feet; in the third second, the fall will be 40 feet, and so on. Derive a formula that gives the total distance fallen depending on the number of seconds in free fall.

6. Use the formula you derived in Problem 5 (along with the knowledge that velocity at any time is half the frictionless case) to find out the Arizona sky diver's time in free fall and impact velocity (see page 65).

7. There was once a king who granted a favor to the subject who saved his life by agreeing to the subject's "meager" request of wheat grains on the 64 squares of a chess board as follows: 1 grain on the first square, 2 grains on the second square, 4 on the third, and so on, doubling the number of grains each time. Explain why the king chopped the subject's head off.

8. Show that the sum of the whole numbers from 1 to 10^k is the $2k$ digit number 5 followed by $(k-1)$ zeros followed by a repeat of the first k digits.

9. Derive the Rule that determines how long it takes for money in the bank to triple if the annual rate of interest is i.

10. Use the Rule you derived in problem 9 to provide understanding in the following real world situation.
    U.S. Assistant Secretary of Education, Finn, on the May 19, 1987 Donahue TV show stated, "we spend 3 times as much on education per kid as we did 25 years ago." His remark was intended to show that education expenses have grown tremendously, and thus to minimize the importance of proposed federal education budget cuts. Discuss.

11. Suppose you are asked to estimate a person's lifetime earnings, assuming that she begins work at a salary of $\$M$ per year. For the first n years, she gets a constant yearly increase of $\$d$ per year. For the next m years, her salary is expected to increase at a g percent rate per year. After this period, until retirement, her salary remains constant (this is a period of p years). Find the person's estimated lifetime earnings.

12. A man pledges money to his favorite charity as follows:

    In year 1, he pledges $\$d$;

    For years 2 through 20, he pledges D percent more each year than the previous year;

    For years 21 through 30, he pledges $\$\delta$ more each year than the previous year.
    a. How much will his pledge be in year 22?
    b. Develop the formula that tells how much money he will give to the charity over the 30-year period.
    c. If $d = 20$, $D = 10$, and $\delta = 50$, how much will his total donation be?

13. Congressman Wayne Pays hired two secretaries, Elizabeth Play and Prudence Workethic, on January 1, 1971, at initial salaries of $5000. Elizabeth's salary increased 20 percent each year, while Prudence's salary increased only 5 percent each year.
    a. Elizabeth typed 10 words-per-minute the first year and each year thereafter typed 1 word-per-minute less. Assuming she spent 100 days per year in the office and 5 minutes per day typing, how much was Elizabeth paid per word typed over the 5-year period of her employment?
    b. Prudence typed 60 words-per-minute the first year, and each year thereafter she typed an addition-

al 2 words-per-minute. Assuming she worked 250 days per year and spent 300 minutes per work day at typing, how much was she paid per word typed over the 5-year period of her employment?

*Note: Congressman Wayne Hayes was forced to resign in 1971 after a sex scandal involving a "no-work" secretary named Elizabeth Ray

14. Suppose that you have leaky plumbing such that half the water going down the drain drops out. When you empty the bucket, you immediately place it back under the drain and get back half of the original water. Assuming you start with a full bucket.
    a. On successive dumpings (starting with a full bucket), indicate the fractions of the bucket that contains water.
    b. How much water will you dump in the drain on 5 successive dumpings?
    c. How many dumpings would be necessary to get at least 99% of the water down the drain.

15. A cannoneer stacks 150 rounds of ammunition in the firing position so that the top layer contains three rounds and each lower layer contains one more round than the layer above. How many rounds are in the lowest layer?

16. Devise a chain letter plan whereby one letter (assuming there are no breaks in the chain) will return you $10,000.

17. Congress cost $25,000 to operate in its first year. Today, congress costs about 2 billion to operate. Before you go to Washington D.C. to stage a protest march, determine the annual percentage growth rate of expenses to run Congress.

CHAPTER 4

Mathematics of Finance

4.1 INTRODUCTION

Mathematics of finance is the generic name for mathematical applications involving flows of money in different time periods. These applications exist in finance, of course, but also in the related areas of investments, banking, accounting, and insurance.

This chapter is based on two old adages, "time is money" and "money doesn't grow on trees." If you don't believe the first adage, please send me $1000. I will give it back in five years. As for the second adage, if you are unsure as to how money does grow, consider this startling fact. Suppose your great great great . . . great grandparents deposited 1¢ in a bank that pays 6% interest at the time of Jesus's birth. Now, you, the caring philanthropist, would like to share the wealth with all 6 billion inhabitants of the world. Get some big gift boxes, as each person would get over 300,000 Earths filled with gold!!! (Note: We will show this astonishing calculation later. Stay tuned to this same channel.)

Learning Plan

Mathematics of finance applications generally involve arithmetic and/or geometric progressions, with the latter most widely used. We developed the progression theory necessary for these applications in Chapter 3. You will be using these theories often; so don't be afraid to go back and review them if you get stymied.

Here, we study first the mathematics of compound growth and compound interest. Then we develop formulas to evaluate future flows of money on a current basis. With this in hand, plus the understanding of progressions from Chapter 3, you will be ready to understand the important applications involving annuities, capital budgeting, depreciation methods, money theory, loan payments, and common stock valuation, and much more.

Again, you will need a good calculator to work the problems in this chapter.

4.2 COMPOUNDING

Many economic variables grow percentage-wise. For example, suppose that Endrun hires you at $50,000 starting salary and promises to give you a 10% raise every year. The raise at the end of the first year is found by multiplying your current base salary ($50,000) by the decimal equivalent of the raise (.10), or

Raise = 50,000(.10) = 5,000

Thus, your new base salary in the second year would be $55,000. It is this higher salary that would become the basis for your next raise, which would be:

Raise = 55,000(.10) = 5,500

That would make your third year salary 60,500. Notice how this raise is higher than the first raise. Such is the nature of compounding.

Exercise 1

Show that the next year's raise would be $6,050.

Let's now generalize this compound growth process. We do this in two stages. First, we generalize the starting point with a given specific constant percentage change. Then, we fully generalize with everything being variable.

Compound Growth

Consider a consumer price index which grows 5% per year. If the initial value of the index is P_0, then during the first year it would increase by 5% of P_0, or $.05P_0$. This increase, if added to the initial value gives the value of the index at the end of the first year, P_1.

$$P_1 = P_0 + .05P_0 = P_0(1 + .05) = P_0(1.05)$$

In the second year, growth is based on the previous year's level of P_1. So the increase in the index is $.05P_1$, which, if added to P_1, gives the value of the index at the end of the second year, P_2.

$$P_2 = P_1 + .05P_1 = P_1(1 + .05) = P_1(1.05)$$

Relating back to the initial value of the index gives

$$P_2 = P_1(1.05) = [P_0(1.05)](1.05) = P_0(1.05)^2$$

Continuing this process to get the index at the end of the third year

$$P_3 = P_2 + .05P_2 = P_2(1.05) = [P_0(1.05)^2](1.05) = P_0(1.05)^3$$

MATHEMATICS OF FINANCE

Exercise 2

a. If $P_0 = 100$, find the values of the price index for the first three years.
b. Show that $P_4 = P_0(1.05)^4$.

By now, you can see the pattern and appreciate that at this 5% growth rate, the consumer price level in any year x, symbolized by P_x, would be

$$P_x = P_0(1.05)^x$$

To generalize this formula, you can see that if the index went up 4% per year, .04 would replace .05 in the formula above. So why not think of a general growth rate of g. Obviously g would replace .05 in the formula to get the general compound growth formula

$$P_x = P_0(1 + g)^x$$

where

P_0 = initial value
g = annual growth rate (decimal)
x = number of years

Example

If the price of residential housing is forecasted to increase 7% per year, the price of a $100,000 house after x years will be

$$P_x = 100,000(1.07)^x$$

So, after 10 years it would be worth

$$P_{10} = 100,000(1.07)^{10} = \$196,715$$

Exercise 3

a. How much would the house be worth in 15 years?
b. If the real estate market turned a little sour and only increased 3% per year, determine the value of that house after 10 and 15 years.

Compound Interest

I hear tell that the River Bank is compounding interest every second. I reckon that if I put in a buck now, I could be one of those billionaires tomorrow.

Joe NoMath

Sorry Joe, but it doesn't work that way. But you won't understand until you see the mathematics behind it all.

In the previous section, we learned about compound growth in general. Now we will extend that knowledge to the compound growth of money in a bank.

In learning compound growth, we focused exclusively on yearly changes. For example, if you put $1000 in a savings bank that paid interest once a year at a rate of 6%, at the end of the year you would have

Beginning Balance	+	Interest	
1000	+	.06(1000)	
1000	+	60	= 1060

Alternatively, we could have expressed this as

$1000(1 + .06) = 1000(1.06) = 1060$

At the end of subsequent years, you would have

Year 2	Year 3	Year 4	
$100(1.06)^2$	$100(1.06)^3$	$100(1.06)^4$	etc.

This is essentially how we compute compound interest, EXCEPT for the little wrinkle that interest can be compounded more than once per year. Let's see how we handle this variation by tracing the growth of $1000 placed in a savings account that pays an annual rate of 6%, compounded quarterly.

The bank will credit interest every quarter, or three months. Now we might like them to give us 6% every three months, but bankers didn't get rich by doing dumb things like that. Rather, the bank would, since they credit interest 4 times per year, give us a quarter of the annual rate each time. Thus, in this case, we would get $\frac{6}{4} = 1.5\%$ interest each quarter. This is done by multiplying the current balance by the decimal equivalent of 1.5%, or .015. So after the first quarter, or three months, our balance will be

New Balance	=	Beginning Balance	+	Interest	
	=	1000	+	.015(1000)	
	=	1000	+	15	= 1015

Alternatively, we can get a different perspective on the process by factoring the 1000

New Balance = $1000(1 + .015) = 1000(1.015)$

If the money is left in the bank another quarter, the friendly tellers will credit you with another 1.5% interest payment on your new higher balance of $1015 as follows.

New Balance	=	Current Balance	+	Interest
	=	1015	+	.015(1015)
	=	1015	+	15.225

Notice how the second interest payment was $22\frac{1}{2}$¢ higher than the first. That is the beauty of compounding, whereby you get interest on your previous interest.

MATHEMATICS OF FINANCE

$$\text{Interest} = .015(1015) = .015(1000 + 15)$$
$$= \underset{\underset{\text{Interest on beginning balance}}{\uparrow}}{15} + \underset{\underset{\text{Interest on previous interest}}{\uparrow}}{.225}$$

We can relate the new balance back to the initial balance or $1000 as follows:

$$\begin{aligned}
\text{New Balance} &= 1015 + .015(1015) \\
&= 1015(1 + .015) \\
&= [1000(1.015)](1.015) \\
&= 1000(1.015)^2
\end{aligned}$$

Exercise 4

Show that the balance at the end of the third quarter will be $1045.6784, which is alternately written as $1000(1.015)^3$.

If the money is left for another quarter, the year-end balance will be

$$\begin{aligned}
\text{Year-end Balance} &= \text{Third quarter balance} + \text{Interest} \\
&= 1045.6784 + .015(1045.6784) = 1061.3636
\end{aligned}$$

Or, in terms of the beginning balance, since $1045.6784 = 1000(1.015)^3$

$$= 1045.6784(1 + .015)$$
$$= [1000(1.015)^3](1.015) = 1000(1.015)^4$$

Altogether $61.36 (to nearest penny) of interest was obtained. Thus the effective interest rate, defined as

$$\text{Effective interest rate} = \frac{\text{Actual year's interest}}{\text{Beginning balance}}$$

$$= \frac{61.36}{1000.00} = 6.136\%$$

was slightly higher than the stated or nominal 6 percent rate. This is always the case as a result of the compounding (or interest earned on interest) effect.

If the bank compounded semiannually, they would credit interest at a rate of 6/2 = 3% every 6 months. Monthly compounding would be done at the rate of 6/12 = .5% every month.

Exercise 5

a. Show how weekly compounding would be accomplished.
b. Verify that the balance at year-end for semiannual compounding would be $1000(1.03)^2$.

Now we understand the process of compound interest with some real-live numbers. But that specific case can't determine our bank-book balance for the myriad of other interest rate and compound period cases possible. So, we really need to generalize. And, of course, that means defining those dreaded "letters"

i (decimal) as the annual nominal interest rate
B_0 as the beginning balance
B_n as the balance at the end of the nth compound period
c as the number of compound periods per year.

Let's first understand the dynamics of any compound period. Suppose one starts that compound period with a balance of B. With an annual interest rate of i, credited c times per year, the interest rate for that particular compound period is $\frac{i}{c}$. Multiplying this by the beginning balance give $B \cdot \frac{i}{c}$, which is the interest for that compound period. Finally, adding this interest to the beginning balance gives the balance at the end of the compound period, or

$$B + B\frac{i}{c}$$

Using these ideas, the balance at the end of the first compound period (note that this could be at the end of 6 months, 3 months, 1 month, 1 week, 1 day, etc.) is

$$B_1 = B_0 + B_0\frac{i}{c}$$

At the end of the second compound period the balance is

$$B_2 = B_1 + B_1\frac{i}{c}$$
$$= \left(B_0 + B_0\frac{i}{c}\right) + \left(B_0 + B_0\frac{i}{c}\right)\frac{i}{c}$$
$$= B_0 + B_0\frac{i}{c} + B_0\frac{i}{c} + \left(B_0\frac{i}{c}\right)\frac{i}{c}$$

This representation of the balance at the end of the second compound period shows it to be the sum of the starting balance, B_0, plus two interest payments on the starting balance, $B_0(i/c) + B_0(i/c)$, plus interest on period 1 interest, $[B_0(i/c)]i/c$. The last term of interest on interest is the distinguishing feature of compound interest.

An alternative representation for B_2 is

$$B_2 = B_1 + B_1\frac{i}{c} = B_1\left(1 + \frac{i}{c}\right) \quad \text{but } B_1 = B_0\left(1 + \frac{i}{c}\right)$$
$$= \left[B_0\left(1 + \frac{i}{c}\right)\right]\left(1 + \frac{i}{c}\right)$$
$$= B_0\left(1 + \frac{i}{c}\right)^2$$

MATHEMATICS OF FINANCE

The ending balance in the third period is

$$B_3 = B_2 + B_2 \frac{i}{c} = B_2\left(1 + \frac{i}{c}\right)$$

$$= B_0\left(1 + \frac{i}{c}\right)^2\left(1 + \frac{i}{c}\right)$$

$$= B_0\left(1 + \frac{i}{c}\right)^3$$

Exercise 6

1. Show that the balance at the end of the fourth compound period is

$$B_4 = B_0\left(1 + \frac{i}{c}\right)^4$$

2. How would you calculate the balance for the end of the 15th compound period?

Basic Compound Interest Formula

The pattern of compound growth of the balance should be evident now. At the end of the nth compound period; the balance, B_n is

$$B_n = B_0\left(1 + \frac{i}{c}\right)^n$$

This is the basic compound interest formula. We will have occasion to use it many times.

It is important to realize that n in our compound interest formula equals the number of compound periods, not the number of years. If the time period of a problem is stated in terms of the number of years (y), then to find the number of compound periods (n), the conversion formula

$$n = cy$$

must be employed.

4.3 COMPOUND INTEREST APPLICATIONS

Now that we have a grasp of the compound interest process, as captured by the basic compound interest formula, we can apply our new knowledge to real world situations ranging from computing savings account balances to determining those effective interest rates.

Computing Bank Account Balances

$100 in an account paying 6 percent annual interest, compounded monthly, has the following balance at the end of eight months:

$$B_8 = 100\left(1 + \frac{.06}{12}\right)^8 = 100(1.005)^8 = \$104.07$$

$500 in an account paying 5 percent annual interest, compounded semiannually, has the following balance at the end of three years:

$$B_6 = 500\left(1 + \frac{.05}{2}\right)^6 = 500(1.025)^6 = \$579.85$$

$100,000 at 8 percent quarterly compounding grows in 3 years to the following amount.

$$n = c \cdot y = 4 \cdot 3 = 12$$

$$B_{12} = 100,000\left(1 + \frac{.08}{4}\right)^{12} = \$126,820$$

$5,000 at 9 percent compounded daily has the following balance at the end of 30 days (note: the banks use a 360 day year; thus $c = 360$)

$$5000\left(1 + \frac{.09}{360}\right)^{30} = \$5,037.63$$

Exercise 7

How much will $700 grow to in five years if it is compounded quarterly at an annual rate of 6 percent?

The Indians Weren't So Dumb After All

In 1624, the Indians sold Manhattan for $24. Their leader, Big Apple, promptly deposited the money in the Chase (White Man) Bank at 6 percent compounded quarterly. As of 2003, the balance was (Note 379 years means 4(379) = 1516 compound periods.)

$$B_{1516} = 24\left(1 + \frac{.06}{4}\right)^{1516} = 24(1.015)^{1516} = \$152,311,007,732$$

That is 152$^+$ billions!!!!

They are planning to buy all the Big Mac bonds, all the real estate on Manhattan, and still have some left over to buy a movie company so that they could win those thrillers at the pass.

Exercise 8

How much would the Indians have as of 2003 if they only received an annual rate of 5 percent?

Solving the World's Poverty Problem

You think the Indians are rich? They would pale in comparison to you, had only your great, great, great, great ... ancestors followed Ben Franklin's advice that a penny saved is a penny earned. Suppose that your wise ancestor put a penny into the Bethlehem Bank at the time of Jesus's birth (2003 years ago). Assuming a 6 percent interest rate compounded annually and no withdrawals ever, you would now have*

$$1(1.06)^{2003} = 4.8713 \cdot 10^{50} \text{¢} = \$4.8713 \cdot 10^{48}$$

That isn't quite a google of dollars; but man, that's a lot of bread. It is so much dough that if you shared that money in gold equally among the 6 billion inhabitants of the world, you would give each person over 300,000 Earth's completely filled with gold!!!!

Exercise 9

How much money would you have if the Bethlehem Bank only paid 4 percent?

Where to Stash Your Fortune

The Last National Bank gives 4.1 percent compounded annually, while the Shady Deal Bank gives an annual rate of 4 percent compounded quarterly. Where should you put your money?

If you put $\$B_0$ in each bank, the year-end balance would be

Last Bank $\quad B_0(1.041)^1 = 1.0410 B_0$
Shady Bank $\quad B_0(1.01)^4 \ = 1.0406 B_0$

So deal with Last Bank as long as Shady Bank isn't giving free steak knives with the opening of a new account.

Exercise 10

Could Shady Bank get your business if they compounded monthly?

More Compounds Syndrome

How often have you seen blaring advertisements stressing how "Our bank gives you more compounds than you-know-who bank"? From the sound of it, you might even think that doubling the number of compounds doubles the interest. Well, that's not how it works, as we will see now.

Let's consider the annual interest rate of 5 percent and increase the number of compounds per year and see what happens to the effective rate, defined as

---

*This problem was presented by Bernard Meltzer on his WOR talk show in the 1980's.

$$\text{Effective rate} = \frac{\text{Interest received in year}}{\text{Starting balance}}$$

$$= \frac{\text{Final balance} - \text{starting balance}}{\text{Starting balance}}$$

$$= \frac{B_c - B_0}{B_0}$$

$$= \frac{B_0\left(1 + \frac{i}{c}\right)^c - B_0}{B_0}$$

$$= \left(1 + \frac{i}{c}\right)^c - 1 \quad \text{or} \quad 100\left[\left(1 + \frac{i}{c}\right)^c - 1\right] \text{ expressed as a percent}$$

The effective interest rates for compounding 4 times (quarterly), 12 times (monthly), and 365 times (daily) are

Quarterly $\left(1 + \frac{.05}{4}\right)^4 - 1 = 1.05095 - 1 = 5.095\%$

Monthly $\left(1 + \frac{.05}{12}\right)^{12} - 1 = 1.05116 - 1 = 5.116\%$

Daily[1] $\left(1 + \frac{.05}{365}\right)^{365} - 1 = 1.05127 - 1 = 5.127\%$

More and more compoundings per year do not add materially to one's pocketbook. In fact, for 5 percent interest with compounding every instant (infinite number per year), the effective rate is only 5.12711 percent—not a heck of a lot more than daily compounding!

Exercise 11

a. Find the effective interest rate (still using 5 percent as the annual rate) for 365 · 24 (hourly) compounds per year.
b. Compute the quarterly, monthly, and daily effective rates assuming a 6 percent annual rate.

Multiple Investments

In a period of rising interest rates, it is wise to invest in short-term certificates of deposit (CD's) and then reinvest at higher rates when they expire. For example, consider the case of Manny Bucks, who invested $50,000 in a 6-month CD which pays at an annual rate of 9 percent compounded quarterly. When that expired, he reinvested in a 9-month CD paying at an annual 10

---

[1] Banks give you a bonus for daily compounding by dividing the annual interest rate by 360 so the effective interest rate really is

$\left(1 + \frac{.05}{360}\right)^{365} - 1 = 1.051998 - 1 = 5.1998\%$

MATHEMATICS OF FINANCE

percent rate compounded quarterly. We use the basic compound interest formula to find the ending balance of the first CD.

$$B_2 = 50{,}000\left(1 + \frac{.09}{4}\right)^2$$

Note that $B_0 = 50{,}000$, $c = 4$, and $n = 2$, since the CD expires after two quarters. This balance becomes the B_0 or beginning balance for the second CD, which has $i = .10$, $c = 4$, and $n = 3$. Consequently, his final balance is

$$\text{Final Bal} = B_0\left(1 + \frac{.10}{4}\right)^3$$

$$= \left[50{,}000\left(1 + \frac{.09}{4}\right)^2\right]\left(1 + \frac{.10}{4}\right)^3$$

Exercise 12

If Manny reinvests these proceeds in a 12-month CD which pays 10.5 percent compounded quarterly, what will his final balance be?

4.4 PRESENT VALUES

Since the mere passage of time allows money to grow, one should not be willing to evaluate money coming at some future date at its face value. For example, $105 due in one year is worth only $100 now since this latter amount invested risklessly at 5 percent compounded annually would grow to $105 in a year. In this section we resolve this problem by developing a method to relate money flows at different time periods.

The emphasis, until now, has been on how much one will have at some future date if one invests a certain amount now. However, for many important applications, the emphasis is reversed and the relevant question is, "How much should one be willing to pay now in order to receive a certain amount in the future?" In this light, let's solve for B_0 in terms of B_n in our basic compound interest equation (page ***).

$$B_0 = \frac{B_n}{\left(1 + \frac{i}{c}\right)^n}$$

To bring the time element more into focus, let's change the symbols so that F represents future amounts and P represents present amounts. Thus,

$$P = \frac{F}{\left(1 + \frac{i}{c}\right)^n}$$

for c compounds per year and

$$P = \frac{F_y}{(1 + i)^y}$$

for annual compounding, where y is the number of years.

P, called the *present value,* is the amount which, if invested now with compound interest or compound return, would grow to be F, the *future value.* As such, P can be thought of as the amount that F is worth now.

This may be a little hard to understand at first, so let's try an example. We already know how $100 in a savings account with 6 percent simple interest will grow in three years to a future value (F_3).

$$F_3 = 100(1 + .06)^3 = 100(1.06)^3 = \$119.10$$

If, however, we only knew that the future value at the end of three years at 6 percent annual compound interest was going to be $119.10, we could have solved for the initial deposit or present value (P).

$$\$119.10 = P(1 + .06)^3$$

$$P = \frac{\$119.10}{(1 + .06)^3}$$

$$P = \$100$$

This illustrates the logic of our present value formula.

$$P = \frac{F_y}{(1 + i)^y}$$

Exercise 13

a. Find how much of a deposit would be necessary for an ending balance of $1500 two years from now in an account paying 8 percent annual interest.
b. Find the initial deposit in an account which pays 6 percent annual interest if the balance at the end of four years will be $1000.

Note that the different interest rate[2] possibilities would result in different present values for the same future payment. For example, a $105 payment one year from now has a present value of $100 if $i = .05$, but only $87.50 if $i = .20$.

$$100 = \frac{105}{(1 + .05)^1} \qquad 87.50 = \frac{105}{(1 + .20)^1}$$

Now let's apply these present and future value concepts to some real problems.

---

[2] In many present value situations the interest rate being used is referred to as the "discount rate."

Savings Goals

Sarah and X. Cuse are proud parents of a baby boy. Hoping to send Junior to college, they want to know how much money to place in the savings bank now so that they will have $60,000 in 17 years. If their bank pays a 4 percent annual rate, and compounds quarterly, the formula

$$P = \frac{F_n}{\left(1 + \frac{i}{c}\right)^n}$$

with $i = .04$, $c = 4$, $n = cy = 4 \cdot 17 = 68$, and $F_{68} = 60{,}000$ tells the story.

$$P = \frac{60{,}000}{1.01^{68}} = 30{,}499.86$$

Ouch! It isn't a happy story. The Cuses don't have that kind of money. But don't despair. We will return later to show a more practical way to meet that savings goal.

Exercise 14

a. How much would the Cuses have to put into the bank now if the bank compounded monthly?
b. How much would it be if the bank paid 5 percent per annum and compounded quarterly?

Summing a Stream of Present Values

Many applications require us to sum a stream of present values. Here we illustrate this process in general and interesting applications involving lotteries.

Consider cash payments coming annually into the future. Suppose the payments come at the end of each year and we discount them back to the present at an i cost of capital. If the annual payments are $F_1, F_2, F_3 \cdots, F_n$, then their overall present value is

$$PV = \frac{F_1}{(1 + i)} + \frac{F_2}{(1 + i)^2} + \frac{F_3}{(1 + i)^3} + \cdots + \frac{F_n}{(1 + i)^n}$$

Alternatively, if the first payment comes immediately, the formula is slightly altered

$$PV = F_1 + \frac{F_2}{(1 + i)} + \frac{F_3}{(1 + i)^2} \cdots + \frac{F_n}{(1 + i)^{n-1}}$$

Often these series of terms are not "well-behaved" and therefore are difficult to sum with a neat simple formula. However, if the future cash payments are all equal, or even if they are a constant times the previous payment, then the series is a lovely geometric progression, which we indeed know how to sum with a neat formula. We will be doing lots of this in the rest of the chapter. To get a feel for this process, consider the following applications.

State Lottery

The states and the media play up the enormous grand prizes of state lotteries. However, winners do not receive those grand prizes at the time of winning. Rather, they get annual payments stretched out over many years. So you can see that present value ideas tell us that the "effective" prize is less than the stated amount. Let's see how this works by example.

Suppose you win the $50 million grand prize of a lottery. The money is likely to be paid off in 20 annual installments of $1.5 million each. (Note that your total payments are "only" $30 million. Guess who gets the rest?) Supposing an 8 percent cost of capital, the overall present value of these payments is:

$$PV = 1.5 + \frac{1.5}{1.08^1} + \frac{1.5}{1.08^2} + \cdots + \frac{1.5}{1.08^{19}}$$

This is a geometric progression with $a = 1.5$, $r = \frac{1}{1.08}$ and $n = 20$. So, its sum is:

$$PV = \frac{1.5\left[1 - \left(\frac{1}{1.08}\right)^{20}\right]}{1 - \frac{1}{1.08}} = \$15.91 \text{ million}$$

I doubt that you would turn down this $15.91 million present value. However, it is only about 30 percent of the stated grand prize. Such a reduction in value will be an important factor in deciding whether to play.

Exercise 15

What would the present value be if the cost of capital were 7 percent? 10 percent?

To illustrate the case where future amount in each year is a constant times the amount in the previous year, let's consider this lottery problem. Suppose the lottery has a payment plan to adjust for inflation. Instead of giving you the same amount each year, they start you off with a million and then increase the payment by 4 percent each year. For example, the payments are 1.5, 1.5 (1.04), 1.5 (1.04)^2, etc. So the net present value of this is

$$1.5 + \frac{1.5(1.04)}{1.08} + \frac{1.5(1.04)^2}{1.08^2} + \cdots + \frac{1.5(1.04)^{19}}{1.08^{19}}$$

Here r is a little more complicated. But note that each term is $\frac{1.04}{1.08}$ times its predecessor. So that is r

$$PV = \frac{a[1 - r^n]}{1 - r}$$

$$= \frac{1.5\left[1 - \left(\frac{1.04}{1.08}\right)^{20}\right]}{1 - \frac{1.04}{1.08}} = 21.465 \text{ million}$$

MATHEMATICS OF FINANCE

With a higher PV, this plan is better than the constant one.

We have now mastered the basics of compound growth, compound interest, present and future values, etc. With our knowledge of geometric and arithmetic progression, we are ready to tackle many applications in the so-called mathematics of finance. Such applications, involving annuities, capital budgeting, depreciation methods, money theory, loan payments, and common stock valuation, now follow in subsequent sections.

4.5 ANNUITIES

An annuity is a series of payments or collections at regular time intervals. Examples include (1) a person who puts an amount in the bank every three months in order to make some future expenditure, (2) a corporation that puts aside a certain amount of cash every year in a sinking fund to pay off a debt, and (3) a person who puts up an amount of money now so as to receive a fixed monthly income until death.

Annuity for Junior's College Education

Recall the plight of Sarah and X. Cuse who need a lump sum of $30,499.86 (see page 000) to finance Junior's college education. They don't have that much money now, but there is another way—the annuity method—to get him to college.

This is an example of a more general problem, which we can now investigate.

Consider the regular payment of p at the beginning of each compound period into a savings account paying an annual rate of i compounded c times per year. What is the total of the account after n compound periods?

Each payment gets compounded according to our basic compound interest formula and grows as follows:

Payment number	Compound period	Ending balance
	1 2 3 $\cdots$ $n-1$ n	
1	$p \longrightarrow$	$p\left(1 + \dfrac{i}{c}\right)^n$
2	$\quad p \longrightarrow$	$p\left(1 + \dfrac{i}{c}\right)^{n-1}$
3	$\quad\quad p \longrightarrow$	$p\left(1 + \dfrac{i}{c}\right)^{n-2}$
$\vdots$		$\vdots$
$n-1$	$\quad\quad\quad p \longrightarrow$	$p\left(1 + \dfrac{i}{c}\right)^{2}$
n	$\quad\quad\quad\quad p \rightarrow$	$p\left(1 + \dfrac{i}{c}\right)^{1}$

The total value of the annuity (A) is the sum of future values of the individual payments. (Summing from the bottom to the top.)

$$A = p\left(1 + \frac{i}{c}\right)^1 + p\left(1 + \frac{i}{c}\right)^2 + \cdots + p\left(1 + \frac{i}{c}\right)^{n-1} + p\left(1 + \frac{i}{c}\right)^n$$

This is a geometric progression with $r = [1 + (i/c)]$, $a = p[1 + (i/c)]$, and n = the number of payments, so the sum is

$$A = S_G = \frac{a(r^n - 1)}{r - 1}$$

$$= \frac{p\left(1 + \frac{i}{c}\right)\left[\left(1 + \frac{i}{c}\right)^n - 1\right]}{\left(1 + \frac{i}{c}\right) - 1} = \frac{p\left(1 + \frac{i}{c}\right)\left[\left(1 + \frac{i}{c}\right)^n - 1\right]}{\frac{i}{c}}$$

For example, if the Cuses put $300 in the savings bank every quarter, with $i = .04$, $c = 4$, and $n = 68$, they would have

$$A = \frac{300(1.01)(1.01^{68} - 1)}{\frac{.04}{4}} = \$29{,}306.83$$

after 17 years. Sorry, Junior, that's still not enough for you. We could try different values for the periodic payment, p, until we get one that results in about $60,000. Or, we could get smart and solve for p in terms of A, the known future need.
From above

known   unknown
  ↓   ↓

$$A = \frac{p\left(1 + \frac{i}{c}\right)\left[\left(1 + \frac{i}{c}\right)^n - 1\right]}{\frac{i}{c}}$$

So solving for p in terms of A, i, c, and n, we get

$$p = \frac{A\frac{i}{c}}{\left(1 + \frac{i}{c}\right)\left[\left(1 + \frac{i}{c}\right)^n - 1\right]}$$

In Junior's case, with $A = 60{,}000$, the quarterly payment must be

$$p = \frac{60{,}000(.01)}{1.01(1.01^{68} - 1)} = \$614.19$$

Note that the sum of the 68 annuity payments equals $41,764.92, which exceeds the lump sum payment of $30,499.86. Can you explain why?

Exercise 16

It Junior's bank raises its annual interest rate to 5 percent (everything else constant), how much will the quarterly payment be?

MATHEMATICS OF FINANCE

Sinking Fund for Fly Now Pay Later Airlines

Sinking funds, which are established to accumulate funds that will eventually pay off a bond debt, are widely used. To illustrate, consider the case of Fly Now Pay Later Airlines. They must pay off a $10 million bond in ten years. To do this, they established a sinking fund in which they put $\$p$ at the beginning of each year. If they get 5 percent interest compounded annually, what should p be?

Since $A = \$10,000,000$, $i = .05$, $c = 1$, and $n = 10$

$$10{,}000{,}000 = p(1.05)^{10} + p(1.05)^9 + \cdots + p(1.05)^1$$

1st year	2nd year	10th year
payment	payment	payment
grows to	grows to	grows in
in 10 years	in 9 years	1 year

$$= p(1.05) + p(1.05)^2 + \cdots + p(1.05)^9 + p(1.05)^{10}$$

You can sum this geometric progression on the right side with $a = p(1.05)$, $r = 1.05$, and $n = 10$. (Note: We summed the terms from right to left.) Then you can solve for p as

$$10{,}000{,}000 = \frac{a(r^n - 1)}{r - 1} = \frac{p(1.05)(1.05^{10} - 1)}{1.05 - 1}$$

$$p = \frac{10{,}000{,}000(.05)}{1.05(1.05^{10} - 1)} = \$757{,}186.43$$

Exercise 17

Fly Now Pay Later Airlines has the option of waiting five years before establishing a sinking fund to pay off the $10 million debt. Determine their yearly payment over the last five years if they do so.

More on Junior's College Fund

A word of caution is in order. The above annuity formulas were derived and thus apply only for a so-called regular annuity—one where equal payments are made at the beginning of every compounding period. For example, the formula we derived for p would correctly give Junior's equal beginning of the month payments for monthly compounding, as long as we recall that $c = 12$ and thus $n = 12 \cdot 17 = 204$.

$$p = \frac{60{,}000\left(\frac{.04}{12}\right)}{\left(1 + \frac{.04}{12}\right)\left[\left(1 + \frac{.04}{12}\right)^{204} - 1\right]} = \$205.15$$

But those formulas can't resolve any nonregular annuity plan, that is, yearly payments with quarterly compounding, unequal payments, or payments at the end of the compound period. For such variations, new formulas must be derived. But this won't be difficult for experts on progressions such as you. Let's illustrate by developing irregular annuity formulas for Junior and the go-go dancer, Ann Nudity.

If Sarah and X. Cuse decide to put money in the bank at the beginning of every year, then the 17 payments of $\$p$ will grow as follows:

Year   Value at end of year 17

```
1  2  3  ···  16  17
p ─────────────────→  p(1.01)^68  = future value of first payment
   p ──────────────→  p(1.01)^64  = future value of second payment
      p ───────────→  p(1.01)^60  = future value of third payment
         ⋮                            ⋮
               p ──→  p(1.01)^8   = future value of sixteenth payment
                  p→  p(1.01)^4   = future value of seventeenth payment
                      A = 60,000  = Balance at the end of year 17
```

These future values form a geometric progression, with $a = p(1.01)^4$, if we sum from bottom to top, $r = 1.01^4$, and $n = 17$. Its sum is

$$S_G = A = \frac{p(1.01)^4[(1.01^4)^{17} - 1]}{1.01^4 - 1}$$

$$60{,}000 = \frac{p(1.01)^4(1.01^{68} - 1)}{1.01^4 - 1}$$

Now solving for p

$$p = \frac{60{,}000(1.01^4 - 1)}{1.01^4(1.01^{68} - 1)} = \$2{,}420.52$$

Notice the differences between this formula for p and the one for a regular annuity. The net effect is that yearly payments are slightly less than four times the quarterly payments because of the longer exposure to interest.

Ann Nudity Saves for Acting School

Ann Nudity projects a $30,000 acting school need (she yearns for an illustrious career in the movies) when she retires from her career at the Go-Go Lounge 5 years from now. She plans to put an amount, call it p, in the savings bank, which pays a 6 percent rate compounded annually, and make additional payments at the beginning of every year. But since she forecasts that her salary will increase as her clothing decreases, Ann plans to make each payment 50 percent more than the previous one. Thus, the payment at the beginning of the second year will be

$$p + .5p = p(1 + .5) = 1.5p$$

And her payment at the beginning of the third year is 50 percent more, or

$$1.5p + .5(1.5p) = 1.5p(1 + .5) = 1.5^2 p$$

Exercise 18

a. Show what the fourth and fifth-year payments will be.
b. Show what the second-year payment of $1.5p$ will grow to five years from now.

The sum of the future values (five years hence) of the five yearly payments must equal $30,000.

$$30{,}000 = p(1.06^5) + 1.5p(1.06^4) + 1.5^2 p(1.06^3) + 1.5^3 p(1.06^2) + 1.5^4 p(1.06^1)$$

This is a geometric progression with $a = p(1.06^5)$, $r = 1.5/1.06 = 1.4151$, and $n = 5$, so the sum of the terms is

$$30{,}000 = S_G = \frac{a(r^n - 1)}{r - 1}$$

$$= \frac{p(1.06)^5 (1.4151^5 - 1)}{1.4151 - 1}$$

Solving for p

$$p = \frac{30{,}000(1.4151 - 1)}{1.06^5 (1.4151^5 - 1)} = \$1{,}990.67$$

Exercise 19

If the bank compounds quarterly, determine the value of p. *Hint:* This requires deriving a formula for p with the new conditions.

Retirement Annuities

Insurance companies and pension funds offer a type of "reverse annuity", whereby a retiree can pay an amount so as to receive a fixed yearly monthly or yearly income for life. These companies either take your fixed payment or they can use your accumulated IRA or pension fund amount as the payment.

Since the payments to the retiree are for life, these companies must have reliable data on life expectancies. They pay actuaries big bucks for such information.

The amount paid annually to a retiree depends on how many years he/she is expected to live and varies by individual. Fortunately, since the insurance company or pension fund deals with a large number of people, they can base their payments on life expectancies for a large class of people. For example, a man aged 65 would have one set of annual payments; but a woman of the same age would have a larger set of payments as women tend to live longer.

Now, let's get to the math of it all.

Since these companies invest the proceeds that the retiree gives them, their annual payments to the retiree are based on the annual rate of return, i, which they can earn on the money. Overall, the fixed initial payment (A) is then the present value of the annual set of payments.

Let's assume that the payments come at the end of the year.

The first payment (p) coming at the end of the first year has a present value of $p/(1 + i)$; the second payment coming at the end of the second year has a present value of $p/(1 + i)^2$, and so on. The total present value for a life expectancy of n years is

$$A = \frac{p}{(1 + i)} + \frac{p}{(1 + i)^2} + \cdots + \frac{p}{(1 + i)^n}$$

This is a geometric progression with $a = p/(1 + i)$, $r = 1/(1 + i)$, and $n = n$, the sum of which is

$$A = S_G = \frac{a[1 - r^n]}{1 - r}$$

$$= \frac{\frac{p}{1+i}\left[1 - \left(\frac{1}{1+i}\right)^n\right]}{1 - \left(\frac{1}{1+i}\right)} = \frac{p[1 - (1 + i)^{-n}]}{i}$$

Example

If an insurance company uses an i of 8 percent, and bases its charge on average life expectancy, how much should it charge 70-year-old people (life expectancy of 78) who want to receive $5000 per year until death?

$$A = \frac{5000(1 - 1.08^{-8})}{.08} = \$28{,}733.19$$

Normally, a retiree has a set pension amount. So, A is then a constant. Also, they generally prefer monthly payments. So in the above formulation, we can think of i as the month rate of return for the insurance company (annual divided by 12) and n as the number of months of life expectancy. Making these adjustments and solving for the unknown monthly payment, p:

$$p = \frac{Ai}{1 - (1 + i)^{-n}} = \frac{A\dfrac{i}{12}}{1 - \left(1 + \dfrac{i}{12}\right)^{-n}}$$

Now consider the specific case of Professor Ann Tique, who finally retired from Haavaard University with a million dollar pension. If Ann retired at age 70 and had a 15 year life expectancy (180 months), and the insurance company had a 12% rate of return, then her monthly payments would be:

$$p = \frac{1{,}000{,}000\left(\dfrac{.12}{12}\right)}{1 - \left(1 + \dfrac{.12}{12}\right)^{-180}}$$

$$= \$12{,}001.68$$

MATHEMATICS OF FINANCE

Exercise 20

1. If the insurance company can only earn a 6% rate of return, determine the monthly payments to Ann Tique.
2. Old Geizer, Ann's husband, also retired with a million dollar pension. Being older and a man with lower life expectancy (10 years), how much would his monthly payments be (assume a 6% rate of return)?

Sports Annuities

When reading the sports pages, aren't you often wondering if you haven't turned to the financial section? Michael Jordan holds the one-year salary record at $33 in his last season (1997–98) with the Chicago Bulls (Shak is not far behind at $29 million). As for multi-year total deals, there have been $100 million plus deals recently. They include Ken Griffy Jrs. 9-year $116.5 million deal with the Cincinnati Reds and Kevin Garnett's 6-year $126 million deal with the Minnesota Timberwolfs. But the mother-of-all sports contracts belongs to Alex Rodrigues. ARod signed a 10-year $252 million contract with the Texas Rangers in 2000!! (Note: the Rangers have finished in last place since then!!!) By the time you read this, maybe others have topped these figures.

The question before us is, "How can the owners afford such extravagant contracts?" Let's illustrate with owner Stein Brenner who just gave slugger Manny Bucks a 30-year $190 million contract. (Note: Manny will be collecting long after he retires!)

Unfortunately for Manny, he doesn't get the $190 million "up-front." Rather, he gets a $10 million signing bonus (Not bad for playing games!) and $6 million per year for the next 30 years. To pay Manny, Stein can establish a bank account that will provide just enough interest each year to meet the annual salaries. We now illustrate that process.

Next year's $6 million salary can be thought of as the future value of some unknown present value. Assuming a 10 percent annual interest rate, we solve for that present value

$$B_0 = \frac{6,000,000}{1.1^2} = \$4,958,677.69$$

To determine the total amount (above the signing bonus) that Brenner needs to deposit now, we sum the present value of all the future salary payments (in millions) as

$$PV = \frac{6}{1.1} + \frac{6}{1.1^2} + \frac{6}{1.1^3} + \cdots + \frac{6}{1.1^{30}}$$

This is a geometric progression with $a = \frac{6}{1.1}$, $r = \frac{1}{1.1}$, and $n = 30$, whose sum is then

$$S_G = \frac{\frac{6}{1.1}\left[1 - \left(\frac{1}{1.1}\right)^{30}\right]}{1 - \frac{1}{1.1}} = \frac{6\left[1 - \left(\frac{1}{1.1}\right)^{30}\right]}{.1} = \$56,561,488$$

Adding the signing bonus to this means that Stein Brenner will have to shell out "just" $66,561,488 now. Quite a savings from $190,000,000 though!

Exercise 21

a. If Brenner can only get a 7 percent interest rate, how much must he put in the bank now?
b. If Brenner can stretch the contract out to 45 years (10 percent interest), how much must he put in the bank?

4.6 CAPITAL BUDGETING

Investment projects typically involve an initial outlay of money with returns coming in future years. When a company is faced with alternative investments, each with its unique pattern of cash outflows and inflows, it must choose the best projects. For example, consider the choice among three investment projects (with equal initial cash investment) with the following returns (in millions) in their two-year life.

	Project		
Year	A	B	C
1	10	15	10
2	10	5	11
	20	20	21

Project C is clearly better than A, since their only difference is C's larger return in the second year. Although A and B have the same total return, since B's funds arrive more quickly and thus are available for reinvestment, B is a better investment. The comparison of B and C is much more complicated. C gives a greater total return, but B gives a quicker return. Thus, it is unclear which project would give a higher overall return, considering reinvestment of the funds.

The only way that projects like these can be compared is to put all the returns on a common time, usually present value, basis. To do this, an interest rate (i), which is called the *cost of capital* in the capital budgeting literature, must be determined. Entire books are devoted to the definition and determination of cost of capital. For our limited purpose here, it represents the minimum rate of return the firm will accept on its investments.

Supposing a 10 percent cost of capital for our hypothetical firm, then the present values of Projects B and C are

$$B \quad \frac{15}{1.1^1} + \frac{5}{1.1^2} = 13.64 + 4.13 = 17.77$$

$$C \quad \frac{10}{1.1^1} + \frac{11}{1.1^2} = 9.09 + 9.09 = 18.18$$

In this case, C with the highest present value is the best project.

MATHEMATICS OF FINANCE

Exercise 22

Show that Project B is better if the cost of capital is 25 percent.

The final decision on investment suitability must await the comparison between the present value of the returns and the size of the investment. For example, if Project C requires an investment of $19 million, it should not be undertaken.

The procedure outlined above is called the *net present value* method. We now develop its mathematics.

If the annual estimated cash inflows (assume they come at year end) from an investment project are $R_1, R_2, R_3, \ldots, R_n$ for its n year life, the present value of the entire stream (PV) is

$$PV = \frac{R_1}{(1+i)} + \frac{R_2}{(1+i)^2} + \frac{R_3}{(1+i)^3} + \cdots + \frac{R_n}{(1+i)^n}$$

PV is compared to I, the initial investment. If

$PV \geq I$ project should be favorable considered, although other projects may still prove better

$PV < I$ project should be dropped from consideration.

The set of terms in the total present value formula generally follow a pattern that can only be summed by the "brute force" method. However, future returns that are (1) constant each year, or are (2) increasing or decreasing at a constant percentage result in easily summed geometric progressions. We leave case (2) to the problems at the end of the chapter. Here, we just consider case (1), which approximates many investment project situations.

Consider the case of a constant annual return: $F_1 = F_2 = F_3 = \cdots = F_n = F$. Finding the present value of each and summing gives

$$PV = \frac{F}{(1+i)} + \frac{F}{(1+i)^2} + \frac{F}{(1+i)^3} + \cdots + \frac{F}{(1+i)^n}$$

This is a geometric progression with $a = \dfrac{F}{1+i}$, $r = \dfrac{1}{1+i}$, and $n = n$ the number of returns. The sum of this progression is

$$PV = \frac{\dfrac{F}{(1+i)}\left[\left(\dfrac{1}{1+i}\right)^n - 1\right]}{\dfrac{1}{1+i} - 1}$$

which can be simplified to

$$PV = \frac{F\left[1 - \left(\dfrac{1}{1+i}\right)^n\right]}{i}$$

For a long-lived investment project, $\left(\dfrac{1}{1+i}\right)^n$ is close to zero since $\left(\dfrac{1}{1+i}\right)$ is less than 1, and a number less than 1 raised to a power gets smaller as the exponent gets larger; for example,

$$\left(\frac{1}{2}\right)^2 = \frac{1}{4}, \quad \left(\frac{1}{2}\right)^3 = \frac{1}{8}, \quad \left(\frac{1}{2}\right)^4 = \frac{1}{16}.$$

So if n is large, we can approximate the present value as

$$PV \cong \frac{F}{i}$$

What a neat simple formula! Aren't you amazed that it can give us the sum of say 50 terms in such a simple way? For example, if a project returns $1 million per year for 50 years, its overall present value for a 10 percent cost of capital is $\frac{1}{.10}$ = $10 million.

I bet you recognize this formula from your finance course. Now, as Paul Harvey says, you know the story behind the formula.

A still unanswered question is, "How large must n be in order to use this simple formula?" If you really want to know, try the Mind Stretcher problem #11. By comparison the exact and simple approximation formulas, we see the bracketed term is <1, and so the exact formula is smaller.

$$\underbrace{\frac{R}{i}\left[1 - \left(\frac{1}{1+i}\right)\right]}_{<1} \quad < \quad \frac{R}{i}$$

   Exact   *Approximation*

When the bracketed term is near to 1 (say .90 or better), the approximation will be good. Let's check out the value of the bracketed term for some example cases.

Example cases ($i = .10$)

$$n = 10 \quad 1 - \left(\frac{1}{1.1}\right)^{10} = .634$$

$$n = 20 \quad 1 - \left(\frac{1}{1.1}\right)^{20} = .8516$$

$$n = 30 \quad 1 - \left(\frac{1}{1.1}\right)^{30} = .943$$

Exercise 23

Show for $i = .15$ that these three cases give .753, .939, and .985, respectively.

Using the 90 percent criteria, we can see that $n = 30$ is fine for $i = .10$, while $n = 20$ is good enough for $i = .15$.

MATHEMATICS OF FINANCE

In general, we solve for the "large enough n" to meet the 90 percent rule by doing

$$1 - \left(\frac{1}{1+i}\right)^n = .90$$

$$\left(\frac{1}{1+i}\right)^n = .10$$

$$\log_{10}\left(\frac{1}{1+i}\right)^n = \log_{10} .10 = -1$$

$$n = \frac{-1}{\log_{10}\left(\frac{1}{1+i}\right)} = \frac{1}{\log_{10}(1+i)}$$

For example, if $i = .10$, then $n = \dfrac{1}{.0414} = 24.2$ (say 25) years.

Exercise 24

a. Find the sufficient n if $i = .15$.
b. Using a 95 percent rule, derive the formula for the sufficient n.

4.7 DEPRECIATION METHODS

Two accelerated methods of depreciation accounting—sum-of-years' digits and double-declining balance—make use of progressions, as illustrated here.

Sum-of-Years' Digits Method

For the sum-of-years' digits method, an asset depreciated over an n year period is depreciated in the yth year by the fraction of the initial cost, D_y

$$D_y = \left(\frac{n - y + 1}{S_n}\right)$$

where S_n is the sum of the years during the life of the project; that is, the sum of the arithmetic progression of natural numbers from 1 to n.

The annual depreciation fractions are

$$\{D_1, D_2, D_3, \cdots, D_n\} = \left\{\left(\frac{n-1+1}{S_n}\right), \left(\frac{n-2+1}{S_n}\right), \left(\frac{n-3+1}{S_n}\right), \cdots, \left(\frac{n-n+1}{S_n}\right)\right\}$$

$$= \left\{\frac{n}{S_n}, \frac{(n-1)}{S_n}, \frac{(n-2)}{S_n}, \cdots, \frac{1}{S_n}\right\}$$

Example

Consider a company which will depreciate an investment in personal computers over a five-year period. The sum of the years digits is

$S_n = 1 + 2 + 3 + 4 + 5$

$S_n = \dfrac{n}{2}[2a + (n-1)d]$

$S_n = \dfrac{5}{2}[2(1) + 4(1)]$

$S_n = 15$

So the annual depreciation fractions are

Year (y)	Depreciation Fraction (Fy)
1	$D_1 = 5/15 = .33$
2	$D_2 = 4/15 = .27$
3	$D_3 = 3/15 = .20$
4	$D_4 = 2/15 = .13$
5	$D_5 = 1/15 = .07$
Total	1.00

Notice that the yearly depreciation charges follow an arithmetic progression with $a = n/S_n$, $d = 1/S_n$, and $n = n$ (the number of yearly depreciations). The sum of these depreciation fractions over n years (D) is

$$D = \dfrac{1}{S_n}[n + (n-1) + (n-2) + \cdots + 1]$$

$$= \dfrac{1}{S_n}(1 + 2 + 3 + \cdots + n)$$

But $S_n = 1 + 2 + 3 + \cdots + n$, so

$$D = \dfrac{1}{S_n}(S_n) = 1$$

Since the sum of the depreciation fractions equals 1, the entire cost of the asset is depreciated after n years, as it should be.

Exercise 25

a. Find the fraction of the total cost depreciated after 8 years for an asset depreciated over 20 years.

b. Show that the fraction of total cost depreciation after 6 years for an asset depreciated over n years is equal to

$$\frac{6}{2S_n}(2n - 6 + 1)$$

Hint: Sum the terms:

$$\frac{1}{S_n}[n + (n - 1) + (n - 2) + \cdots + (n - 6 + 1)]$$

Double-Declining Balance Method

For the double-declining balance method, an asset depreciated over an n year period is depreciated by a fraction of the initial cost in the yth year, D_y,

$$D_y = \frac{2}{n}\left(1 - \frac{2}{n}\right)^{y-1}$$

In the first year ($y = 1$), the depreciation by this method is

$$D_1 = \frac{2}{n}\left(1 - \frac{2}{n}\right)^0 = \frac{2}{n}$$

This is double the amount of straight-line depreciation, which employs an equal depreciation fraction every year, or $1/n$. And that, folks, is how double-declining balance got its name.

By substituting the different values of y, we find the annual depreciation fractions as

$$\{D_1, D_2, D_3, \cdots, D_n\} = \left\{\frac{2}{n}, \frac{2}{n}\left(1 - \frac{2}{n}\right)^1, \frac{2}{n}\left(1 - \frac{2}{n}\right)^2, \cdots, \frac{2}{n}\left(1 - \frac{2}{n}\right)^{n-1}\right\}$$

Example

If that investment in personal computers were to be depreciated over five years using the double-declining balance method, this is what the yearly depreciation fractions, D_y, would be.

Year	Depreciation Fraction (D_y)	Fractional Balance Remaining
1	$D_1 = 2/5 = .4$	$3/5$
2	$D_2 = 2/5 \, (3/5) = .24$	$3/5 \, (3/5) = (3/5)^2$
3	$D_3 = 2/5 \, (3/5)^2 = .144$	$3/5 \, (3/5)^2 = (3/5)^3$
4	$D_4 = 2/5 \, (3/5)^3 = .086$	$3/5 \, (3/5)^3 = (3/5)^4$
5	$D_5 = 2/5 \, (3/5)^4 = .052*$	$3/5 \, (3/5)^4 = (3/5)^5$
Total	$.922*$	

*In order to fully depreciate the asset by the double-declining balance method, an additional fraction of .078 of the asset would be depreciated in the fifth year.

These yearly depreciation fractions follow a geometric progression with $a = \frac{2}{n}$, $r = \left(1 - \frac{2}{n}\right)$, and $n = n$ (the number of depreciation fractions). The total fraction of the asset depreciated by this method in n years, D, is

$$D = \frac{a(1 - r^n)}{1 - r} = \frac{\frac{2}{n}\left[1 - \left(1 - \frac{2}{n}\right)^n\right]}{1 - \left(1 - \frac{2}{n}\right)}$$

$$= 1 - \left(1 - \frac{2}{n}\right)^n$$

The value of D cannot be equal to 1; thus this method alone can't depreciate the entire asset.[3] However, at the final year $\left(1 - \frac{2}{n}\right)^n$ is close to zero, and so the total accumulated depreciation is only slightly less than the cost of the asset. At any rate, when this method is used, an adjustment is made in the final year's depreciation so that the entire asset can be written off.

Let's review these two accelerated depreciation methods (with comparison to straight-line or equal yearly amounts) by developing a depreciation schedule for a five-year asset in our example.

Fraction of Asset Depreciated (D_y)

Year	Straight line	Sum-of-years' digits	Double-declining balance
1	.2	$\frac{5}{15} = .33$	$\frac{2}{5} = .4$
2	.2	$\frac{4}{15} = .27$	$\frac{2}{5} \cdot \frac{3}{5} = .24$
3	.2	$\frac{3}{15} = .20$	$\frac{2}{5} \left(\frac{3}{5}\right)^2 = .144$
4	.2	$\frac{2}{15} = .13$	$\frac{2}{5} \left(\frac{3}{5}\right)^3 = .086$
5	.2	$\frac{1}{15} = .07$	$\frac{2}{5} \left(\frac{3}{5}\right)^4 = .052$
Total	1.0	1.0	.922

Notice how the two accelerated methods take large depreciations in the early years. This is typical of them.

Exercise 26

Calculate the depreciation fractions for each method for a ten-year asset.

---

[3] If $n = 2$, the formula gives $D = 1$. However, this is an unrealistic case, since $a = 1$ and $r = 0$ means that the entire asset would be depreciated in the first year.

4.8 MONEY THEORY

The Federal Reserve, with power to adjust reserve requirements for commercial banks, controls the money supply of the country. When funds are placed in a commercial bank checking account, the bank can then lend up to a certain fraction of those funds. This maximum fraction has been about .8, but the Federal Reserve can change it up or down to meet the needs of overall monetary policy.

Let's now trace the theoretical effects of an introduction of funds into the banking system. I. Ben Frugal hasn't trusted the banks since the Depression; however, the high interest rates of late have finally persuaded him to switch his M, hidden under the mattress for 30 years, to the Last Bank and Truss Co. (a conglomerate). Calling L the maximum fraction of deposits they can lend, the Last Bank and Truss Co. can provide new loans of up to ML dollars. Suppose that it does just this for Rufus Botched, a man who wants to use the proceeds to pay for a roof job. Rufus immediately writes a check to pay the roofer, S. Bestos & Co., who happens to deal with another bank, the Chase Podunk Bank. With the inflow of ML dollars, Chase Podunk has available $(ML)L = ML^2$ dollars to lend. The bank then lends this amount of money to Joe Bananas, a local fruit and vegetable market owner. Joe writes a check made out to Will Hittman, who banks with the Water-Mellon National Bank. In turn, Water-Mellon Bank can lend $(ML^2)L = ML^3$ dollars, and the cycle goes on and on.

Frugal's original M, which, when under the mattress was not really in the money supply, has now set off a chain of events that, given k loan rounds, will result in total new money of

$$M + ML + ML^2 + ML^3 + ML^4 + \cdots + ML^k$$

These terms follow a geometric progression with $a = M$, $r = L$, and $n = k + 1$. So the total new money in circulation, T, is

$$T = \frac{a(1 - r^n)}{1 - r} = \frac{M(1 - L^{k+1})}{1 - L}$$

The value of T, of course, depends on the number of loan rounds. Theoretically, the highest value of T occurs for the case of an infinite number of loan rounds. For this case, $L^\infty = 0$ since $L < 1$, so

$$T = \frac{M}{1 - L}$$

Thus the money supply could increase by a multiple, $1/(1 - L)$, times the original input of M. Typically, $L = .8$ so that every dollar into the system can theoretically increase the money supply by $5. In practice it doesn't because there are time lags in the system and because the banks may decide to keep more than $(1 - L)$ fraction in reserve. Note that a withdrawal of a dollar could cause up to a $5 contraction in money supply.

Exercise 27

If the Federal Reserve by its open market operations introduces $1 billion dollars into the banking system, how much can the money supply eventually increase by if the maximum loan fraction is .9? .75? .5?

4.9 OH THOSE PAYMENTS

Many people have trouble meeting car and/or mortgage payments—a problem in which progressions play a part. Let's see how banks determine the amounts of those seemingly impossible payments.

Banks charge interest on the unpaid balance of a loan each month. Of course, you make a payment each month. The new balance of your loan is then the old balance plus interest minus your payment. For example, suppose you borrow $10,000 to buy a new car. The loan is at an annual rate of 12 percent for 4 years. In the first month the bank charge the monthly interest rate of 12/12 = 1 percent on the balance of $10,000.

Interest = .01(10,000) = $100

With your first monthly payment of $263.34 (We will see how one determines the monthly payment later.), the net balance of your loan is

Net Balance = 10,000 + 100 − 263.34 = $9,836.66

In the second month the bank charges 1 percent interest on this new balance (1 percent of 9836.66 is $98.37) and collects your second payment of $263.34 to arrive at the new net balance of

Net Balance = 9836.66 + 98.37 − 263.34 = $9,671.69

So on and on it goes for 48 months until you finally own an **OLD** car!!

Now let's generalize this process so that you can easily handle any loan situation.

Banks charge interest on the unpaid balance of a loan. Consider a M loan at an i interest rate per period. The first period's interest is Mi. If p is the payment for each period, then the balance at the end of the first period (B_1) is

$B_1 = M + Mi - p = M(1 + i) - p$

During the second period, the bank charges interest on this unpaid balance; so the second period's interest is $[M(1 + i) - p]i$. This, added to B_1 minus the second payment of p, gives the balance at the end of the second period.

$B_2 = [M(1 + i) - p] + [M(1 + i) - p]i - p$
$ = [M(1 + i) - p](1 + i) - p$
$ = M(1 + i)^2 - p(1 + i) - p$

The third period's interest is $B_2 i$, so the balance at the end of the third period is

$B_3 = B_2 + B_2 i - p$
$ = B_2(1 + i) - p$
$ = [M(1 + i)^2 - p(1 + i) - p](1 + i) - p$
$ = M(1 + i)^3 - p(1 + i)^2 - p(1 + i) - p$

MATHEMATICS OF FINANCE

One more period makes the pattern obvious. During the fourth period, interest is $B_3 i$, so the ending balance is

$$B_4 = B_3 + B_3 i - p$$
$$= B_3(1 + i) - p$$
$$= [M(1 + i)^3 - p(1 + i)^2 - p(1 + i) - p](1 + i) - p$$
$$= M(1 + i)^4 - p(1 + i)^3 - p(1 + i)^2 - p(1 + i) - p$$

Can you see that the balance at the end of the 20th period will be

$$B_{20} = M(1 + i)^{20} - p(1 + i)^{19} - p(1 + i)^{18} - \cdots - p(1 + i) - p$$

At the end of the loan (n periods), the balance is zero. Thus,

$$B_n = 0 = M(1 + i)^n - p(1 + i)^{n-1} - p(1 + i)^{n-2} - \cdots - p(1 + i) - p$$

Thus

$$M(1 + i)^n = p(1 + i)^{n-1} + p(1 + i)^{n-2} + \cdots + p(1 + i)^2 + p(1 + i) + p$$

A few simple mathematical manipulations will give us an interesting banking perspective about mortgages. Watch!

First, divide both sides by $(1 + i)^n$

$$\frac{M(1 + i)^n}{(1 + i)^n} = \frac{p(1 + i)^{n-1} + p(1 + i)^{n-2} + \cdots + p(1 + i)^2 + p(1 + i) + p}{(1 + i)^n}$$

Recalling that $\frac{a + b}{c} = \frac{a}{c} + \frac{b}{c}$, or in general a sum of numbers in the numerator divided by a single number in the denominator can be expressed as the sum of each numerator number divided by the denominator. So

$$M = \frac{p(1 + i)^{n-1}}{(1 + i)^n} + \frac{p(1 + i)^{n-2}}{(1 + i)^n} + \cdots + \frac{p(1 + i)^2}{(1 + i)^n} + \frac{p(1 + i)}{(1 + i)^n} + \frac{p}{(1 + i)^n}$$

$$= \frac{p}{(1 + i)} + \frac{p}{(1 + i)^2} + \cdots + \frac{p}{(1 + i)^{n-2}} + \frac{p}{(1 + i)^{n-1}} + \frac{p}{(1 + i)^n}$$

Revealed here is that the mortgage equals the sum of the present values of the payments. Bankers, deciding on whether to make a loan, look at mortgages or loans in this way.

To solve for p, we sum the right side, which is a geometric equation with $a = p/(1 + i)$, $r = 1/(1 + i)$, and $n =$ the *number of terms*.

$$M = \frac{\frac{p}{1 + i}\left[1 - \left(\frac{1}{1 + i}\right)^n\right]}{1 - \left(\frac{1}{1 + i}\right)} = \frac{p\left[1 - \left(\frac{1}{1 + i}\right)^n\right]}{i}$$

Now solving explicitly for p, we get

$$p = \frac{Mi}{1 - \left(\frac{1}{1+i}\right)^n}, \quad \text{or alternatively} \quad p = \frac{M(1+i)^n \cdot i}{(1+i)^n - 1}$$

Amazingly, this is the same formula as for a retirement annuity!

Examples

1. The Homeowners took out a 20-year, $60,000 mortgage at an annual interest rate of 12 percent. Thus the monthly payments on their home were

$$p = \frac{60,000(1.01)^{240} \cdot .01}{1.01^{240} - 1} = \$660.62 \quad \text{Note: } i = .12/12 = .01$$

The total of their 240 payments is $158,556 or over two times the mortgage. This is not at all uncommon.

2. If you take out a $10,000 car loan to be repaid over 24 months at a 12 percent annual (1 percent monthly) interest rate, your monthly payments will be

$$p = \frac{10,000(1.01)^{24} \cdot .01}{1.01^{24} - 1} = \$470.73$$

The total of all 24 payments will be $11,297.52

Exercise 28

1. Determine the Homeowners' monthly payment and the total amount of money paid to the bank if their annual interest rate is 11 percent on their 20-year, $60,000 mortgage loan.
2. Find the Homeowners' mortgage loan balance after a) one month, b) five months, c) one year, d) ten years.
3. If your car loan (see Example 2 above) is repaid over three years, what will the monthly payments be? What will the total of all payments be?
4. Silent Sam refuses to tell how much he owes the bank on his new house. However, his wife leaks the word that they pay $1,000 per month for payments on their 25-year, 14 percent mortgage. Now let the world in on his secret.

We now introduce an extension of the loan problem but complicated by different financing possibilities.

4.10 CASH BACK VERSUS LOW FINANCING

In recent times, sagging car sales lead to a new phenomena in the car-bargaining ritual. Car companies got smart and focused our attention on the decision to take a cash back or get low financing. With this stroke of genius, car companies took our minds off the age-old phenomena of hard

MATHEMATICS OF FINANCE

bargaining, which is still better than either option. But let's focus on the BIG DECISION. Since the cash back plan involves higher interest rates and the low financing option involves a shorter payment period, the decision is usually close, but complicated without a mathematical model. However, mathematical models are our thing, so let's see how we can make the best decision.

The decision hinges on bringing together the following two ideas from our earlier studies:

1. determining the monthly payment of the loan for each option.
2. summing the present values of these constant monthly payments to get an overall present value for each option.

Let's illustrate with an example. Lemon Motors offers $1000 off the sticker price of their $10,000 "Juicer" model with 12 percent annual financing over 4 years, or nothing off with 2.9 percent annual financing over 2 years. The comparison follows: (assume a cost of capital of 9 percent annually, or $.09/12 = .0075$ monthly.)

	$1000 Cash Back	**2.9% Financing**
Cost of car (loan amount)	9,000	10,000
Monthly payment	$\dfrac{9{,}000\left(\dfrac{.12}{12}\right)}{1-\left(\dfrac{1}{1+.01}\right)^{48}} = \$237.00$	$\dfrac{10{,}000\left(\dfrac{.029}{12}\right)}{1-\left(\dfrac{1}{1+\dfrac{.029}{12}}\right)^{24}} = \$429.37$
Present value	$\dfrac{237\left[1-\left(\dfrac{1}{1+.0075}\right)^{48}\right]}{.0075}$	$\dfrac{429.37\left[1-\left(\dfrac{1}{1+.0075}\right)^{24}\right]}{.0075}$
	$9523.79	$9398.54 ↑ Lower, so take the low financing rate

Actually, you should try some of that old-fashioned hard bargaining and get the price down to $8,500. That would be the best decision!

Exercise 29

The best alternative can change, depending on your cost of capital, or the best rate of return that you can get on your money. For example, the best you can do with your money might be to invest in a saving bank at an annual rate of 6 percent. Show that for that case, you should take the cash back.

4.11 COMMON STOCK VALUATION

Theory for common stock valuation states that the intrinsic value of a share of stock equals the present value of its future stream of dividends. So, if dividends are paid annually (at year end) in amounts $D_1, D_2, D_3, \cdots$, then the present value of this stream discounted at the rate of i (i can be thought of as the rate of return you wish to make in the market) is

$$PV = \frac{D_1}{(1+i)} + \frac{D_2}{(1+i)^2} + \frac{D_3}{(1+i)^3} + \cdots$$

These terms follow a geometric progression when either (1) The dividends remain constant or (2) The dividends grow or decline by a constant percentage each year. Let's investigate these cases.

For the case of constant dividends $D = D_1 = D_2 = D_3 = \cdots$, the total present value is

$$PV = \frac{D}{(1+i)} + \frac{D}{(1+i)^2} + \frac{D}{(1+i)^3} + \cdots$$

This is a geometric progression with

$$a = \frac{D}{1+i} \quad r = \frac{1}{1+i} \quad n = \infty$$

(theoretically, a common stock has an infinite life). So

$$PV = \frac{\left(\frac{D}{1+i}\right)\left[1 - \left(\frac{1}{1+i}\right)^\infty\right]}{1 - \left(\frac{1}{1+i}\right)}$$

$$= \frac{D}{i} \quad \text{since} \quad \left(\frac{1}{1+i}\right)^\infty = 0$$

Supply the missing algebra. Note the similarity of this result to the capital budgeting result (see page 106), when returns are assumed to be constant forever.

If dividends grow at a constant rate, g (for the case of a constant *declining* rate, g is negative) every year, then with $D_1 = D$, the stream of dividends will be

$$D, \quad D(1+g)^1, \quad D(1+g)^2, \cdots$$

Each of these dividends must be discounted to present value at the discount rate i, so the total present value is

$$PV = \frac{D}{1+i} + \frac{D(1+g)}{(1+i)^2} + \frac{D(1+g)^2}{(1+i)^3} + \cdots$$

This is a geometric progression with

$$a = \frac{D}{1+i} \quad r = \frac{1+g}{1+i} \quad n = \infty$$

so

MATHEMATICS OF FINANCE

$$PV = \frac{\left(\dfrac{D}{1+i}\right)\left[1 - \left(\dfrac{1+g}{1+i}\right)^{\infty}\right]}{1 - \left(\dfrac{1+g}{1+i}\right)}$$

If $g < i$, then

$$\left(\frac{1+g}{1+i}\right)^{\infty} = 0$$

so after a little algebra (which you should try to supply), we get

$$PV = \frac{D}{i-g} \quad (\text{when } g < i)$$

Let's observe the use of these formulas by listening in on some Wall Street investors.

Bernard Broke thinks that the dividends of Computer Ecological Systems, Inc., now $2 per share, will grow 5 percent per year indefinitely. His broker, I. Ben Bullish, thinks that they will remain constant. If Bernard wants to make 10 percent on his money, how much should he be willing to pay for the stock? How much would his broker advise him to pay?

Bernard's forecast for dividends translates into a present value of

$$PV = \frac{D}{i-g} = \frac{2}{.10 - .05} = \frac{2}{.05} = \$40$$

The broker's forecast translates into a present value of

$$PV = \frac{D}{i} = \frac{2}{.10} = \$20$$

Exercise 30

1. Find the numerical values for the present values of the first ten dividends from Computer Ecological Systems, Inc.
   a) using Bernard's forecast for dividends
   b) using his broker's forecast for dividends.
2. Explain how two people with identical forecasts for a common stock can be taking opposite actions—that is, one is selling and the other is buying.
3. For a 25 percent rate of return, find the price you should be willing to pay for Computer Ecological Systems, Inc., assuming dividend growth rates of a) 10 percent, b) 15 percent, c) 20 percent.
4. Suppose the price of Computer Ecological Systems, Inc. is $30 per share. Assuming Bernard's forecast is correct, what rate of return (i) will the stock provide if purchased at that price? If the broker's forecast is correct, what return will the stock provide?

PROBLEMS

TECHNICAL TRIUMPHS

1. An initial salary of $50,000 growing at a 5% compound annual rate will grow to

   $50,000 \, (1.05)^{10}$ in 10 years

   $50,000 \, (1.05)^{20}$ in 20 years

   Determine those future salary levels

2. Determine the effective interest rate ($i = .04$) for quarterly, monthly, and daily compounding.

3. Determine the following present values of $1000 coming in the future with annual compounding and $i = .07$
   a. $\dfrac{1000}{1.07^5}$
   b. $\dfrac{1000}{1.07^{15}}$
   c. $\dfrac{1000}{1.07^{30}}$

4. Solve for the future value (FV)

   $$PV = \$500 = \dfrac{FV}{1.08^6}$$

5. Determine the present value of the following sum

   $$\dfrac{1000}{1.1} + \dfrac{1000}{1.1^2} + \dfrac{1000}{1.1^3} + \cdots + \dfrac{1000}{1.1^{20}}$$

6. A $100,000 mortgage has a 7% interest rate. How much would the monthly payments be for
   a. 30 year mortgage
   b. 20 year mortgage
   c. 15 year mortgage

7. A car buyer has a chance of 2 loan plans. Find the monthly payments for each if the car costs $20,000 (no down payment).
   a. 8% for 24 months
   b. 6% for 48 months

8. A $200,000 asset is to be depreciated over 20 years. Determine the depreciation for the first year using
   a. straight line
   b. sum of years
   c. double declining balance

9. A new parent wishes to save for a child's college education. If she puts $100 into the bank every month and the bank pays 4% interest, compounded monthly, how much money will accumulate in 17 years?

10. Consider the following returns for 3 investments. (Each one involves the same initial investment)

MATHEMATICS OF FINANCE

Year	A	B	C
1	5	6	7
2	5	5	6
3	5	4	1

For a 10% cost of capital, which investment is the best.

CONFIDENCE BUILDERS

1. An Endrun executive profited from some off-the-books partnerships, and received 30 million dollars. If he invested that money in an off-shore bank that paid 6% compounded quarterly, how much money would he have when he got out of jail 20 years later.

2. If the price of gold is (currently $400 per ounce) forecasted to increase 10% per year.
   a. What will its price be in 5 years?
   b. How long will it take for the price to reach $1000 per ounce?

3. At age 14, Melvin decided to put his $400 lawn mowing business savings in the Solid Rock Bank to get the 4 percent compounded quarterly interest. The bank, which was not a member of FDIC, went bankrupt when Melvin was 22. How much did he lose?

4. A firm buys a machine with useful life of ten years at a cost of $10,000. Determine the depreciation schedule assuming
   a. straight-line depreciation,
   b. sum-of-years'-digits depreciation,
   c. double-declining balance depreciation.

5. Hans Zup deposited the one million dollars that he stole from Blinks in the Bank of Switzerland. If the bank pays 4 percent compounded quarterly, how much will he have when he gets out 20 years later?

6. Using the double-declining balance depreciation method, $2/n$ times the remaining book value is depreciated in each year, where n is the period over which the asset is depreciated. So for the first year, $1(2/n)$ fraction of the asset is depreciated, leaving $1 - (2/n)$ as the book value fraction at the end of the first year. The fraction $2/n$ times this, or $[1 - (2/n)](2/n)$, is the depreciation fraction taken in the second year.
   a. Show that the book value fraction at the end of the second year is $\left(1 - \frac{2}{n}\right)^2$.
   b. What depreciation fraction is taken for the third year?
   c. What is the book value fraction at the end of the third year?
   d. What depreciation fraction is taken during the fourth year?
   e. Are your results consistent with the actual depreciation fraction formula given in the chapter?

7. Triple-declining balance has never been used (mainly because the I.R.S. won't allow it). However, it would provide an even more accelerated method. Using it, $3/n$ of the remaining book value fraction is depreciated in each year.
   a. What fraction of the asset would be depreciated in the second year? in the third year?
   b. Derive the formula that gives the fraction of the asset depreciated in the yth year.
   c. What fraction of the asset is depreciated after n years?

8. Otto Mobil, who purchased a $20,000 car, was given two payment options
   Option 1: $2,000 cash back (still finance $20,000) and a 7.9% interest rate over 48 months
   Option 2: A 2.9% interest rate over 24 months
   Given that Otto has a 1% per month cost of capital, which option should he choose?

9. Al Zeimer just retired with his $500,000 pension. If the pension administrator bases their monthly payments on a 6% interest rate, how much will Al's monthly payment be? Note: Al is 65 and has a 80 year life expectancy.

10. Babe Ruth made $80,000 in 1930. When asked how he could make more than President Hoover, he said, "I had a better year!". Alex Rodriquez made 25.2 million in 2001.
    a. What annual compound salary growth rate would equate those two salaries?
    b. Considering that consumer prices have only gone up about 3% over that period, comment about baseball salaries today. (Hint: OVERPAID!!)

MIND STRETCHERS

1. This is your life: A Downhill Progression of Ken Lay (Enron Chairman) (the setting is 2002 at the time of congressional hearings on Enron's bankruptcy)
   a. As you testify before congress, you suddenly realized that you would soon be put in charge of a project to determine the nature of the new barber shop at Sing-Sing.
      One proposal calls for the inclusion of equipment for hair rinsing, tinting, curling, and so on. You think it's a little ahead of its time, but you feel the "new-look" prison will eventually need such a barber shop. For the n year project, you forecast no returns for K years; in year $(K + 1)$ the return should be $R; thereafter for the rest of the n years, returns should grow at the annual rate of g. If future returns are discounted at the rate of i, derive the formula for the present value of this project.
   b. Needless to say, your "barber shop" idea wasn't too well received by your superiors. In fact, they felt that a reassignment might cure your "revisionist" thinkings. So you were made janitorial officer.
      On the first day of the job, you realized the need for some new brooms. The supplier gives a volume discount. He charges $2.00 for the first broom ordered, $1.96 for the second, $1.92 for the third, and so on until you reach $1.00 for the twenty-sixth broom ordered. Thereafter, all brooms cost $1.00. You calculate that you will need to order 50 brooms. How much will they cost?
   c. Upon release from jail, you still had $1 million, down from the $700 million of Enron stock sales which was wiped out by stockholder lawsuits. You need $100 thousand per year to live on. If you take that account of money out of the bank (which pays 5% compounded annually) at the beginning of each year, how long can your money hold out?

2. *This Is Your Life: An Uphill Progression* After your release from jail you finally land a job displaying the "special of the week" merchandise for the Super Dooper Food Store.
   a. How many cans can you display if you use the triangular form as shown in the figure?

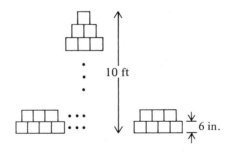

   b. Your work is so good that you are able to get a better job with Progression Rentals, Inc., as the "mathematician in charge of determining rental charge formulas." They have no fixed fee, but rather, charge $1.00 for the first mile driven, 98¢ for the second mile, 96¢ for the third mile, and so on. A customer inquires as to the charge for a 40-mile rental. What would it be?
   c. Your mathematical work is so good for Progression Rentals that sales each month are g percent more than the previous month. If sales the month you arrived were $S, how much were sales for the entire year of your employment?
   d. With such success, you were promoted to Vice President of Finance for the company. Your first assignment is to evaluate the investment in a capital asset. The asset returns $R per year for the first k years, after which its returns decrease by ρ percent per year until the nth year, at which time it

MATHEMATICS OF FINANCE

"dies" and is scrapped for $S. Find the present value of this asset if future returns are discounted at the rate of i.

3. Mr. Frugal has figured that he will need $15,000 for his son's college education ten years from now. He plans to invest a fixed sum at the beginning of each year for the first five years, then leave this sum to accumulate for the next five years. Assuming his investments can earn 6 percent compounded annually, how much must his yearly investment be?

4. Bernard Broke believes that Get-Rich-Quick stock, currently paying a $3 dividend per share, can maintain a 10 percent compound dividend growth rate for Y years; thereafter, the dividends will remain constant. Derive the formula for the present value of this assumed dividend stream if $i = .08$. How much should he be willing to pay for the stock if $Y = 5$?

5. A man puts $D into the savings bank initially and at the beginning of every second year. If the bank gives i percent interest compounded annually, how much money will he have in his account at the end of k (an odd number) years?

6. Zap Inc., after years of extensive research, developed an automatic zipper. Since it will take a large advertising effort to introduce the product, zero profits are expected for the first three years. In the fourth year, the company feels that a one million dollar profit can be realized. Profits for the subsequent ten years should grow so that the profit in any year is 20 percent greater than the profit in the previous year. Thereafter, profits should remain constant (until infinity). If Zap discounts future profits at a 10 percent discount rate ($i = .1$), determine the present value of profits from marketing its automatic zippers.

7. The Get-Rich-Quick Company was just taken over by a group of Nomath University graduates. Because of their lack of mathematical prowess, analysts project that dividends, currently $D per share (paid at year-end), will decline at an annual compound rate of d. For example, the dividend payment at the end of the second year will be $D - dD = D(1 - d)$. Ten years from now, you take over and reverse the trend so that dividends thereafter increase at an annual compound rate of g (for example, if the last dividend of the Nomath group is L, then your first dividend will be $L + gL = L(1 + g)$).
   a. Derive the formula for present value of the entire stream of dividends (until infinity) if future dividends are discounted at an annual rate of i.
   b. If $D = \$2$, $d = .3$, $g = .08$, and $i = .1$, what is the specific present value of the future dividends?

8. Ann Nudity puts all her savings into the bank at the end of the year. Initially she is able to save $1000 per year, but with growing fame, she saves 20 percent more each year up through the tenth year. Then with lessening fortunes, her savings decline 10 percent each year for the next ten years. If her bank pays 5 percent interest compounded annually, how much will she have at the end of 20 years?

9. Matt Maddix started a savings account with a $1000 deposit on January 1, 2004. Each January 1 thereafter, he plans to double his previous deposit. If the bank pays 5 percent interest compounded annually (payable at year-end), what will Matt's balance be as of December 31, 2008?

10. Stew Dent took out a $3000 loan at 10 percent annual interest to finance his freshman year. He must repay the loan in three equal annual installments (p) after graduation. If the first payment is made five years after taking out the loan, find p.

11. If a capital asset returns $1 million every year indefinitely, we showed in the chapter that the present value of these returns at a 10 percent discount rate would be $1/.1 = \$10$ million. In actuality, investments stop giving returns at some point. How many years of constant $1 million annual return are required before you should use the infinite returns formula, assuming that you wish to be no more than 5 percent away from the exact present value?

MODULE THREE

Modeling Variable Relationships

Understanding the relationships among variables is important in many aspects of life. The concept of a function, which we develop in Chapter 5, provides a way to record systematically such relationships in mathematical equation form. Classes of functions, which have important applications for management, are introduced in Chapter 6. Their characteristic equations, graphical "picture," and applications are investigated. In Chapter 7, we try to bring the ideas of Chapters 5 and 6 together so as to use functions as practical models for real world relationships.

CHAPTER 5

Functions

5.1 INTRODUCTION

> The word 'function' probably expresses the most important idea in the whole history of mathematics. Yet, most people hearing it would think of a 'function' as meaning an evening social affair, while others, less socially minded, would think of their livers.
>
> Kasner and Newman, *Mathematics and the Imagination* (New York: Simon & Schuster, 1940)

Relating and associating things are essential ingredients for learning and understanding our world. This process begins with the infant, who associates specific behavior on his or her part with certain responses by the parent, and progresses to the student, who (we hope) learns that good grades in mathematics are associated with diligent study and problem solving. The process continues to expand to the businessperson, who must have an understanding of the way sales are related to price, advertising, product quality, and other factors.

With our world of complex interrelationships, analysis of the environment is certainly enhanced by ways of systematically isolating and recording these relationships. The mathematical concept of a function, which is the subject of this chapter, is a step in this direction.

We begin our study with a visit to the locker room. The next voice you hear is that of Coach Hardnose of the Poison Ivy League Champions of Brawn University.

Football Mathematics

"OK you guys, this is a must game. So, we need a better way of getting our line blocking straight. And this function stuff is just the thing. Let me explain how we're going to use functions to win the game."

"But coach, we never learned this function stuff! You made us take basketweaving instead of math."

"OK, meatheads, I'll explain it to you. On first down, we take the ball round the right end, so everybody hits a man, like you see on the board here."[1]

---

[1] For those of you who are uneducated in football, the symbols in the football plays stand for the following: RE, right end; RB, right linebacker; RT, right tackle; MG, middle guard; LT, left tackle; LB, left linebacker; LE, left end.

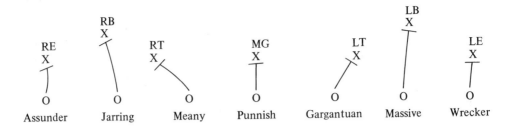

"Now, this is a function, because when each of us hits only one of them, we're functioning, as they say in higher math circles."

"On second down, we go around our left end, as you see here."

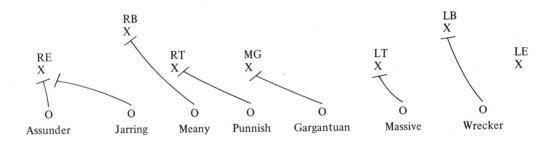

"Both Assunder and Jarring are going to hit their right end. You other guys hit like you see there. Nobody hits their left end! But you know what? We still have a function, because none of us is hitting more than one of them."

"On third down, we are going to surprise them by going up the middle. Look here!"

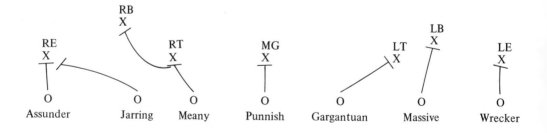

"Meany, you're the key to this play because you take out two men. You spoil the function by doing that, but things aren't that bad because we still have a relation—even if it isn't the kind of relation you're thinking about. Relations are a little different—one of us can hit more than one of them. Got that?"

"On fourth down, I'm planning a kick, because you guys probably won't remember this function stuff."

FUNCTIONS

Now let's take a look at the "instant replay". The offensive line for Brawn is a set of seven players, which we'll call the *domain*. The defensive line for the opposition, Slippery Rock, is also a set of seven players, called the *range*.

On first down, each player in the Brawn line (domain set) has a single man in the Slippery Rock line (range set) to block. This is pictorially represented as shown.

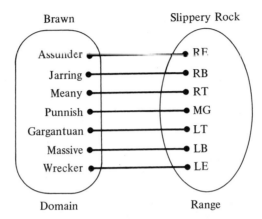

Notice the one-to-one correspondence of elements from the two sets. A rule (in this case a play) that assigns an element of the domain to a single element of the range is an example of a *function*.

Second down can be pictorially represented as follows.

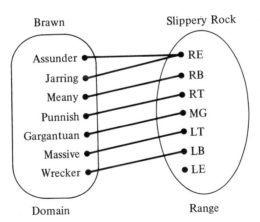

Here, notice that two elements of the domain are assigned to the same element of the range. But this two-to-one correspondence does not violate the function concept. As long as individual elements of the domain are assigned only to one element of the range (even if some range elements have two or more elements assigned to them), we have a function.

The assignment on third down, illustrated below, is such that it violates the requirements of a function.

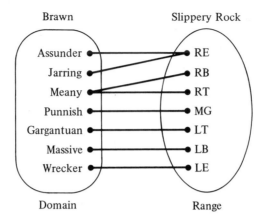

This results from the fact that one element of the domain (Meany) is assigned to block two different elements of the range. This one-to-two correspondence is not consistent with the definition of a function.

Any correspondence whatsoever between domain and range elements would be an example of what is called a *relation*. Thus, all three plays are examples of relations.

When a relation contains assignments of the one-to-one or many-to-one type, we call it a *function*. A function, then, is a special kind of relation.

Exercises 1

1. Are the following plays a function or a relation?

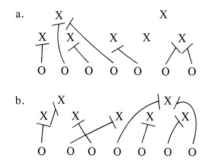

2. Construct two different function and two different relation plays.
3. Fred is married to Sarah, but has a mistress Mary. Mary is married to John, who has a lover, Jane. Ed, who isn't married, has had meaningful relationships with Mary, Sarah, and Jane. If the women constitute the domain and the men constitute the range, describe whether this represents a function, a relation, many relations, or a scandal.

Learning Plan

Now that you have in mind the general notion of a function, you can direct your attention to those types of functions that have applications in management. Functions involving assignments

FUNCTIONS

between two number sets are discussed first. Then you will see how and when a numerical function can be captured by a formula. In such cases, the function acts like a manufacturing process, whereby a raw material (the number) is mathematically operated on and changed until it exits from the manufacturing process as a single-valued final product. We will introduce the process concept here and expand upon it in the next chapter. Finally, graphs and how they are used to portray functions (and relations) will be explored. We make extensive use of graphical methods throughout this book.

5.2 NUMERICAL FUNCTIONS

Numerical functions and relations exist when both the domain and the range are sets of numbers. For example, in Figure 5.1 parts (a) and (c) represent functions, while part (b) doesn't. Do you know why?

Figure 5.1

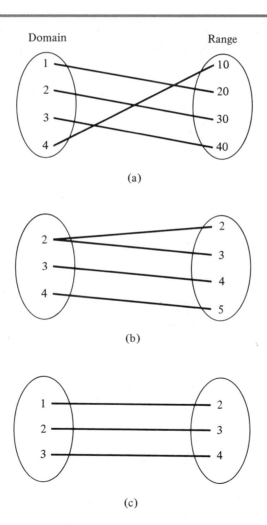

Focusing on the first case, let's use y and x to symbolize the range and domain elements respectively. We then can say that y is a function of x, symbolically written as

$y = f(x)$

Note that the symbol $f(x)$ is read "function of x" and not f times x or anything else. This notation is very convenient since that single value of y paired off with $x = k$ can be written as $f(k)$. For example,

$f(1) = 20$

$f(2) = 30$

$f(3) = 40$

$f(4) = 10$

Equations of Functions

Let's focus on the third case. It is a function, but more than that, there is a distinct regularity to its correspondence. Each domain element is assigned to a range element, whose numerical value is one larger than the domain element. One could symbolize this by the rule or equation

$y = f(x) = x + 1$

Using this equation,

$f(1) = 1 + 1 = 2$

$f(2) = 2 + 1 = 3$

$f(3) = 3 + 1 = 4$

We do indeed get the proper correspondences. Our equation really describes a process by which any domain element (x value) can be processed to yield the correct range element (y value). Pictorially, this is demonstrated in Figure 5.2.

Figure 5.2

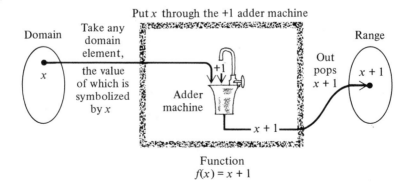

FUNCTIONS

Exercises 2

1. Consider part (a) of Figure 5.1, and show that a single equation is not sufficient to describe all the correspondences.
2. Construct a numerical function and a numerical relation using the domain: 0, 1, 2, 3, and the range: 5, 6, 7, 8.
3. Make up a domain and a range, and construct a function with a "regularity." Determine the equation that represents that regularity.

Let's develop these ideas further with a practical situation that illustrates the nature of univariate and multivariate functions.

Univariate Functions

Seymour Bottoms, owner of the Sinema Theater, needs a better understanding of the relationships among certain variables. To begin, let's assist him by relating ticket revenue and theater attendance.

If he charges $6 per sex fien ... er person, and if the theater has 100 seats, we can make the correspondence between all the possible attendance levels (the domain) and all the possible revenues (the range).

Notice that in Figure 5.3 each element of the domain is assigned one element of the range—so we can say that revenue is a function of attendance. Furthermore, where there is only one domain set, as we have here, we call it a function of one variable or *univariate* function.

The regularity of this function can be expressed by $y = f(x) = 6x$ (see Figure 5.4). For example, $f(10) = 6 \cdot 10 = 60$, $f(50) = 6 \cdot 50 = 300$.

We must be careful to use this equation only for domain elements—all the integers from zero through 100. Using it for numbers not in the domain, for example, $x = -7$, $x = 500$, and so on is a no-no. Figure 5.5 shows this pictorially. This emphasizes the need to document the do-

Figure 5.3

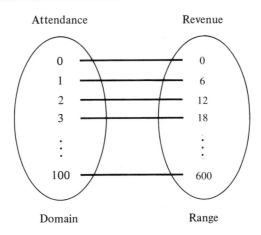

Figure 5.4

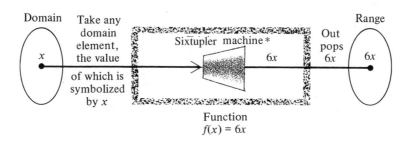

Figure 5.5

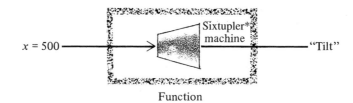

main set whenever one is specifying a functional rule. So, a complete statement of the revenue function is

$y = f(x) = 6x \quad \{x = 0, 1, 2, \ldots, 100\}$

Exercise 3

Had Seymour charged $7 per person, revenue would still be a function (although a different one) of attendance. Can you set up the correspondence and the formula to capture the regularity of that function?

Multivariate Functions

If, on ladies' day, Seymour charged $6 for men and only $3 for women, a given attendance level could mean more than one revenue level. For example, an attendance of two could give rise to $6 revenue (two women), $9 revenue (one man and one woman), or $12 revenue (two men). Because of the one to more-than-one correspondence, revenue is not a function of the single variable, attendance. Rather, revenue in this case is a function of two variables—M, the number of men in attendance, and W, the number of women in attendance. We symbolize and pictorially represent this process in Figure 5.6 as

$R = f(M, W) = 6M + 3W$

---

*In Seymour's case, this is a sextupler machine.

FUNCTIONS

Figure 5.6

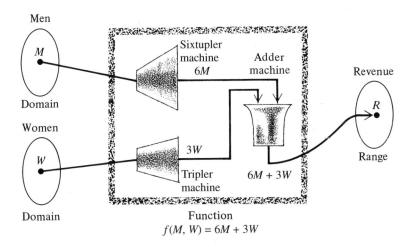

The uniqueness of the range element here preserves the functional relationship, despite the fact that two separate domains are involved.

Exercise 4

If the above entrance fees hold, but senior citizens (S) or people 62 and above are only charged $2, determine revenue as a function of the three variables M, W, and S. *(Note: M and W are now defined as people under 62 years old.)*

Functions of more than one variable, such as the example above, are called *multivariate* functions. They have many applications in the real world, some of which we shall cover in this book. However, since the theory for univariate functions, once mastered, can be extended to multivariate functions, we shall concentrate on functions of one variable for the balance of this chapter.

5.3 NOTATION

We now develop a simple, but effective, notation to help us master functions. Earlier we were introduced to the univariate functional notation:

$y = f(x)$

Again, recall that this is *not f* times *x*. Rather, it just indicates that "*y* is a function of *x*."

We now extend the notations idea to handle more than one function. Let's do this while we help Seymour Bottoms reach new depths of depravity with his other functions.

On the cost side of the picture, Seymour has both fixed costs (don't vary with attendance), such as film rental, projectionist, building rental, and so on, and variable costs (vary with attendance), such as cleaning fees, air conditioning, and so on. For the current x-rated attraction, *Underwear, with Selected Shorts*. Seymour estimates fixed costs as $100 per showing and variable costs as $.50 per person. The correspondence between attendance levels (domain) and total costs (range) is thus given in Figure 5.7. The following mathematical equation gives us that single range element (y) associated with each domain element (x):

$$y = f(x) = 100 + .5x \quad \{x = 0, 1, 2, \ldots, 100\}$$

Or, in process form, the cost function is as shown in Figure 5.8.

Since revenue and cost are different functions of attendance, it would be advantageous if we "name" them differently. Since we previously used $f(x)$ for the revenue function, we could then go down the alphabet from there to denote other functions. Thus, $g(x)$, $h(x)$, etc. could be used. So,

$f(x) = 6x$ Revenue function

$g(x) = 100 + .5x$ Cost function

If we had other functions of x, we could have gone down the alphabet further. For example,

$h(x) = \sqrt{x}$ For positive square root values only

Figure 5.7

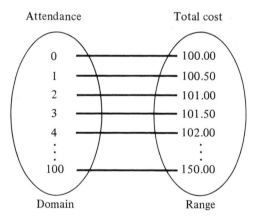

Figure 5.8

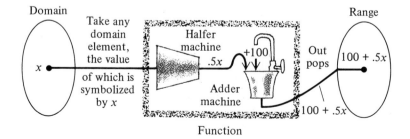

FUNCTIONS

might represent refreshment stand sales revenue as a function of x, and

$i(x) = 70 + \log_{10} x$

might represent the temperature in the theater as a function of attendance.

If we gave every function encountered a different letter name, we would soon run out of alphabet. We could resort to double, triple, and so on letter names, as for example, *ff(x)* and *ghc(x)*. But this would be more confusing than helpful. So, normally for any problem we start by naming the first function $f(x)$, and then going down the alphabet $g(x)$, $h(x)$, and so on. When we encounter another problem situation, we begin again with $f(x)$.

Nothing is sacred about the use of x and y as variable symbols. Using letters more reflective of the nature of the factors themselves is often advantageous. For example, our awareness and understanding of the variables would probably be greater if we used R for revenue, A for attendance, C for cost, S for sales at the refreshment stand, and T for temperature.

With these mnemonic variable names, the functions above would be written as

$R = f(A) = 6A$ Domain for all four cases is

$C = g(A) = 100 + .5A$ $\{A = 0, 1, 2, \ldots, 100\}$

$S = h(A) = \sqrt{A}$

$T = i(A) = 70 + \log_{10} A$

The value of a function can be determined by specifying the domain element. For example, the values of the four functions above for $x = 25$ are:

$f(25) = 6(25) = 150$

$g(25) = 100 + .5(25) = 112.5$

$h(25) = \sqrt{25} = 5$

$i(25) = 70 + \log_{10} 25 = 70 + 1.398 = 71.398$

Exercise 5

1. Find the values of all four functions above for $x = 100$.
2. Suppose that Seymour also had another function of x. What letter of the alphabet should be used to denote that function?
3. The cost function, $g(x) = 100 + .5x$ has a fixed and variable component. Which component would change if Seymour featured a film that was cheaper to rent (Cheaper in moral value too!!)? Could this new function be called $g(x)$?

5.4 GRAPHICAL REPRESENTATION

Someone once said, "A picture is worth a thousand words." If so, aren't two pictures worth two thousand words? We could go off the deep end and study the functional relationship between the domain (number of pictures) and the range (number of words) (see Figure 5.9). We would even see that the equation that captures this correspondence would be

Figure 5.9

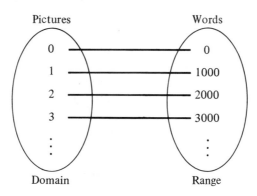

$y = f(x) = 1000x$

where

- x is the number of pictures
- y is the number of words.

But rather, let's see how graphical representation can assist in our understanding of functions. Suppose two real number lines, called *axes,* are drawn perpendicular to one another, as shown in Figure 5.10.

The horizontal axis is called the *abscissa,* or x-axis. The vertical axis is called the *ordinate,* or y-axis. The scales on the two axes need not be the same; for example, the y-axis can have a 10, 20, 30, and so on scale, while the x-axis has a 1, 2, 3, and so on scale. The intersection of the axes is called the *origin.*

Each point in this two-dimensional space has a unique position relative to the origin. That is, each point can be reached by starting at the origin, moving a certain distance horizontally along the x-axis, then moving a certain distance vertically along the y-axis. For example, point A is one unit to the right and one unit above the origin. Point B is two units to the left and three units above the origin. Point C is three units to the left and one unit below the origin. Point D is two units to the right and two units below the origin.

By convention, we attach positive values to distances to the right of and above the origin, and negative values to distances to the left of and below the origin. Two numbers—the first being the horizontal distance or x-coordinate, the second being the vertical distance or y-coordinate—uniquely describe the location of any point.

This coordinate system is called the *Cartesian coordinate system* after René Descartes, the seventeenth-century Frenchman, who pioneered in its development. Using this notation, the locations or x- and y-coordinates of the four points are shown in tabular form.

Point	(x, y)
A	(1, 1)
B	(–2, 3)
C	(–3, –1)
D	(2, –2)

FUNCTIONS

Figure 5.10

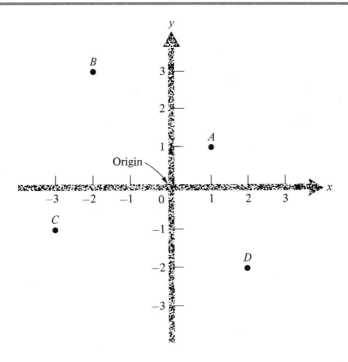

Exercise 6

Locate the following points in the Cartesian coordinate system: (4, 0), (–1, –1), (2.5, 3), and (0, 1).

The concept of a function can now be viewed in graphical terms. For this purpose, let's return to the three relations on page 129. If we let the domain elements be horizontal distances and the range elements be vertical distances, then each corresponding pair of elements describes a point in Cartesian space, which we can plot for the three cases as shown in Figure 5.11.

Recall that (a) and (c) are functions. The thing that rules out (b) as a function is that a single value of x ($x = 2$) is assigned to two different values of y. Graphically, this is two points located above a single x value.

The above cases contained a finite number of corresponding pairs. When we have a case with an infinite number of corresponding pairs, plotting results in something similar to that shown in Figure 5.12. In this figure only (a) represents a functional assignment, since only in (a) do we have each x value assigned to only one value of y. For (b) and (c), there are x values that are assigned to more than one value of y. For such cases, where the curve "bends back on itself," y is not a function of x.

You may have gotten the impression that a correspondence that isn't of the functional type is useless, bad, undesirable, nefarious, or even treasonous. Not so. To the contrary, some of the most desirable, sought-out correspondences are not functions. See Figure 5.13.

Sometimes y is not a function of x, but x is a function of y. This means that if the domain

Figure 5.11

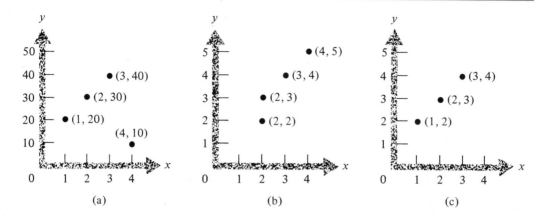

Figure 5.12

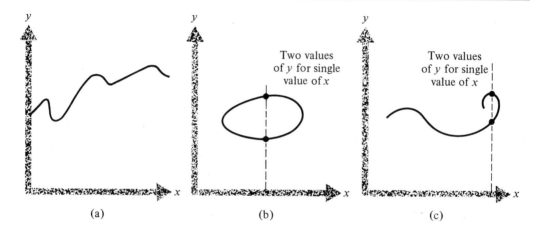

Figure 5.13

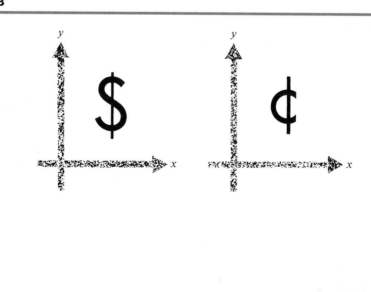

FUNCTIONS

Figure 5.14

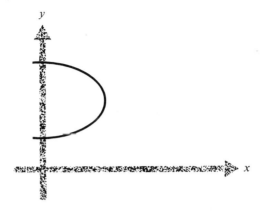

and range were interchanged, a functional correspondence would exist (see Figure 5.11b). Graphically, this case is illustrated in Figure 5.14. Symbolically, we can say for such cases that

$y \neq f(x)$ but $x = g(y)$

5.5 CLASSIFICATION OF FUNCTIONS

If one studies many functions, one discovers similarities, which are especially obvious when graphing. Also, certain graphical shapes have similar equations. Thus, one could learn about a lot of apparently different functions by classifying functions that result in certain "graphs" and studying their general equations.

In Chapter 6, we get close and friendly with four categories of functions (linear, quadratic, exponential and logarithmic) that have the greatest relevance for management and life applications. For no extra charge, we explore some other functions too. For each category, we learn how they graph and how those graphs change with the parameters of the equation. Technically, all the above is referred to in mathematical circles as "Analytic Geometry". Could that be why the chapter is so named? But, we go much beyond such mathematical technicalities by presenting several real world applications for each function type.

Chapter 7 is all about using our Chapter 6 knowledge to select mathematical equations that can best model real situations. We do this by learning how patterns of data and their changes lead to specific functions that best model those realities. We end the chapter with a "Curvy Road to Profit" section that is sure to bring glee to all present and future profit-seeking entrepreneurs.

PROBLEMS

TECHNICAL TRIUMPHS

1. Are the following assignments from the domain (x) to the range (y) functions?

a.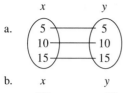

b.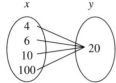

2. Graph the assignments in problem #1 on cartesian coordinates.

3. In a given application, there are 7 different functions. If the first function was called $f(x)$, what would the other sex functions be called.

4. If $y = f(x) = x^2$ for the x domain: $\{1, 2, 3, 4\}$
   a. Show the four assignments from domain to range.
   b. Graph these assignments on Cartesian coordinates.
   c. Is it valid to connect the 4 points in the graph?
   d. If one connected the 4 points, what would that imply about the domain.

5. Are the following univariate or multivariate functions?
   a. $y = x + x^2$
   b. $y = x + \log_{10} x$
   c. $y = 2^x$
   d. $y = xy$

6. Consider the equation

   $y = \sqrt{x}$ {All positive real numbers}

   a. Is y a function of x?
   b. Is x a function of y?

CONFIDENCE BUILDERS

1. A firm's sales are expected to be from $10 or $11 million if they spend $1 million on advertising; however, if they spend $2 million on advertising, they expect sales to be from $15 or $16 million. Are sales a function of advertising?

2. Your bathroom scale always reads 2 pounds more than the scale at the gym.
   a. Is bathroom scale weight a function of gym scale weight?
   b. Is gym scale weight a function of bathroom scale weight?

3. Consider an office with 3 secretaries (domain) and 15 managers (range). Suppose each secretary is assigned to work for 5 managers. Is this assignment a function or a relation?

4. The profits (π) for the Maxi Taxi Company are given by the equation

   $\pi = -2.5Q^2 = 200Q - 3500$

   where Q is the number of riders. Q in terms of profit is

   $Q = 40 \pm \sqrt{200 - .4\pi}$

   a. Are profits a function of riders?
   b. Are riders a function of profits?

FUNCTIONS

5. On page 106, we derived the following formula for the present value (PV) of a long-lived asset that returns $R each year if the discount rate is i:

$$PV = \frac{R}{i}$$

   a. Is this an example of a univariate or a multivariate function?
   b. If R were specified to be $1 million, what type of function would result?

6. In 1976 the state of New York introduced a plan to have the renewal fee for a driver's license based on the number of moving violations (for example, speeding) incurred by the driver. The new plan has a basic license fee for those with no moving violations; however, those with one or two moving violations pay double, and those with three or more moving violations pay triple the basic fee.
   a. Is the cost of a license a function of number of moving violations under this plan?
   b. What are the range and domain?
   c. If the basic fee is $10, set up the function that assigns moving violations to license fees.

7. A student in Professor S. O. Terrick's Mismanagement class believes that the exams are so confusing that her grade could be A, B, or C for a given high level of study and C, D, or F for a given low level of study. Is her grade a function of study? Explain.

8. A company does the grocery shopping for elderly people who are unable to drive to the store. The company charges a fee that equals 15% of the cost of the order.
   a. Illustrate the process that results in fee (F) as a function of the cost of the order (C).
   b. Is this a regular function that can be captured by a mathematical equation? If so, do it.

9. Three children, Lauren, Jeffrey, and Marlund, are assigned household chores. Lauren does dishes and vacuuming. Jeffrey does setting the table and dishes. Marlund does dusting. Is that assignment a function?

10. Dizzy Land charges $20 per person except for those under 12, who are charged $10 admission. Determine the function that relates total revenue (Z) to the number of people 12 or over (x) and the number of people under 12 (y).

11. Coach Hardnose wants to know the functional rule that assigns a scrimmage line to a field goal kicking distance if the ball is kicked from 7 yards back of the line and the goal posts are 10 yards behind the goal line.

12. Bea Ribe is taking the course "Accounting Procedures to Hide Illegal Foreign Payments." Her grade on the final (G) is the following function of hours of study (S) for the domain $S = \{1, 5, 25\}$.

    $$G = f(S) = 50 + 20 \log_5 S$$

    a. Determine the range.
    b. Can we find the grade for 50 hours of study?
    c. Graph the function.

13. Certain legal steps must be performed in buying a house. For these services, attorney Lee Gall charges $100 plus 1 percent of the sales price for a conventional mortgage and $50 plus $\frac{1}{2}$ percent of the sales price for an assumable mortgage.
    a. For a conventional mortgage, show the functional process (call it $f(x)$) to determine Lee Gall's fee for the sales price (in thousands) domain: $\{20, 25, 30, 35, 40\}$.
    b. Do the same for assumable mortgages.

14. Citibank sets their prime rate $1\frac{1}{2}$ percentage points above the three-week average for the 90-day rate for commercial paper.
    a. Is Citibank's prime rate a function of 90-day commercial paper rate?
    b. What is the equation that currently relates the two rates?

c. Considering that prime rates only are changed in increments of $\frac{1}{4}$ point (a 90-day paper rate of 5 percent would mean a $6\frac{1}{2}$ percent prime rate, but this wouldn't change until the 90-day paper rate went up to $5\frac{1}{4}$ percent or down to $4\frac{3}{4}$ percent), is prime rate a function of 90-day paper rate?

15. A construction flag-person works varying amounts due to weather and other factors. For any regular (R) hours up to 40 per week, he gets $8 per hours. Overtime hours (O), or those in excess of 40, are paid at the double rate of $16 per hour. If he is limited to 20 overtime hours per week:
    a. What is the domain for R?
    b. What is the domain for O?
    c. Illustrate the multivariate function that gives his total pay in process form.
    d. Find the mathematical equation for this multivariate function.
    e. If he pays 29% of his total pay in taxes, find his after-tax pay function.

MIND STRETCHERS

1. (For baseball fans only.) Consider a bunt play with a runner on first and second. Describe a function that assigns a fielder to each base yet allows for fielding a bunt.

2. Sylvia Porter was a syndicated columnist on economic subjects in the 1970's. In one of her articles she addressed the question, "How much refrigerator storage capacity do you need?" "A good rule of thumb is to have 9 cubic feet for a family of two, with an additional cubic foot for each additional family member and an additional two more feet if you entertain frequently."
    a. Set up a table to show storage capacity recommended for families of two through eight who do not entertain frequently.
    b. Repeat part (a) for families who entertain frequently.
    c. Is storage capacity a multivariate function of family size and the frequency of entertaining? Does Porter's statement enable you to define this function precisely?
    d. If the entertaining variable is designated as zero if you entertain no more than once per week, and 1 if you entertain more than once per week, can storage capacity be defined precisely as a multivariate function?
    e. Which of the following multivariate functions represents the case described in part (d)?

    $C = 9 + 2M + 2E$ C = capacity

    $C = 9 + 1(M - 2) + E$ M = family members

    $C = 9 + 1(M - 2) + 2E$ E = 0 or 1, depending on entertainment

    Regarding freezer space, Porter recommends 2 cubic feet for each family member.
    f. Develop the equation that describes this function.

3. The normal probability distribution that describes IQ scores is:

$$y = f(u) = 266(2.718)^u \quad \text{where } u = \left(\frac{x - 100}{15}\right)^2$$

Explain why this is referred to as "a function of a function."

4. There are many different defenses in the game of basketball. Three widely-used defenses are: zone, man-to-man (or is it person-to-person), and double team a superstar. Illustrate these defenses with assignments from a domain (our players) to a range (their players). Which defenses are functions? Which are relations? Is it necessary to use a function in order to play good defense?

CHAPTER

Analytic Geometry

6.1 INTRODUCTION

Analytic geometry is the study of the graphical properties of functions. This branch of mathematics had its inception in the seventeenth century when the French mathematicians Descartes and Fermat* recognized the importance of graphical methods.

Every numerical function can be graphed. These "pictures" of functions reveal a dimension of the variable relationship that the equation alone cannot provide. Consequently, graphing allows a deeper understanding of the function itself, along with its relationship to other functions. For example, understanding the graphical picture of a family of functions helps identify and apply such functions effectively.

Click There we are on our wedding day.

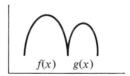

Click Oh $f(x)$, why did you have to show that picture? I joined Obese Observers about then.

---

*See, he did more than leave us his mysterious "Last Theorem".

143

Click And there is our family on $h(x)$'s first birthday. He wasn't walking yet, but his shape gives away his Quadratic ancestry.

Click There we are at Dizzy Land with the Linears. Our family doesn't socialize with them anymore since we could never agree on common interests to apply ourselves to. Besides, they never wanted to see our beautiful slides.

z-z-z-z-z-z-z
Dear, show the Exponentials those exciting slides from our trip to Mathachusetts.
z-z-z-z-z-z-z-z

Learning Plan

With a little knowledge of analytic geometry, you can use family "pictures" and graphical effects of the terms in an equation to move quickly from a specific function to its graph. In doing so, you can gain a deeper understanding of the relationship of the variables. The development of these skills is the purpose of this chapter, where the analytic geometry of the following families of functions is presented.

1. Linear
2. Quadratic
3. Exponential
4. Logarithmic
5. Power
6. Polynomial
7. Mixed

These functions have been chosen for study because they are applicable to important everyday phenomena. Examples of such applications are presented for each family of functions.
We devote more time and energy to the first four categories as they have more real world applications.
For each family of functions

1. the general equation is specified
2. the graphical effects of the various terms in the equation are investigated

ANALYTIC GEOMETRY

3. the overall graph of the functions is developed
4. applications to daily life are presented.

With the graphical perspective of mathematical relationships in hand, one can more effectively model the real world. For example, conceiving of some practical phenomenon in graphical terms can lead us to formulate an equation to represent it. This is the subject of Chapter 7, so be patient!

For the time being, let's concentrate on developing the graphical "new dimension" to our understanding of equations.

6.2 LINEAR FUNCTIONS

Linear functions are the simplest and easiest to work with. That alone should make us feel kindly toward them. However, their simplicity belies their usefullness and wide applicability. They serve to model many real world relationships. Also, they are essential to our understanding of calculus.

Most likely, you have encountered linear functions before in your mathematics background. Typically, they have been represented in one of the following four notational ways:

$y = mx + b$

$y = ax + b$

$y = a + bx$

$y = b_0 + b_1 x$

There is nothing wrong with any of these ways of expressing linear functions. Maybe you even got comfy with one of them, and now like a close friend do not want to abandone it. Unfortunately, these formulai are not informative about the nature of the constants. For example, why should m stand for slope (and not mud) as it does in the first formula.

My fellow Americans, it is time for a change! We are adopting this heretofore unheard of notation.

$y = f(x) = sx + i$

where

s stands for slope
i stands for y-intercept

We will have more to say about s and i later.

The simplicity of linear functions and their ability to model many real world relationships make them practical for many kinds of applications. We have already seen them in action in the car-rental decision of Chapter 1, the Sinema Theater case of Chapter 5, and the relation between pictures and words.

$G = f(M) = .1M + 10$ Gertz cost

$B = f(M) = .05M + 20$ Bavis cost

$R = f(A) = 6A$ Sinema revenue

$C = f(A) = .5A + 100$ Sinema cost

$y = f(x) = 1000x$ Picture is worth 1000 words

Linear functions always graph as straight lines.[1] Let's illustrate with the following examples, which are graphed in Figure 6.1.

$y = f(x) = 2x + 3$

$y = g(x) = -x + 6$

x	$f(x)$	$g(x)$
0	3	6
1	5	5
2	7	4
3	9	3
4	11	2

Figure 6.1

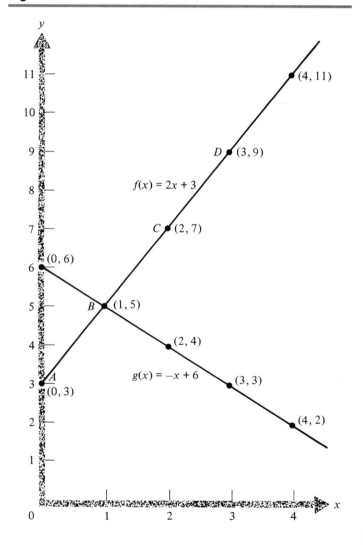

---

[1] Do you realize that by drawing continuous straight lines, we are assuming the domain to be the set of real numbers? This will be our assumption throughout unless otherwise stated.

ANALYTIC GEOMETRY

Exercise 1

Graph the following linear functions:

$y = f(x) = 3x + 2$

$y = g(x) = -2x + 1$

The orientation of the above lines could have been predicted merely by noting the values of the constants in their equations. We will illustrate this by developing the role and meaning of the constants, i and s, in the general linear equation. (See Figure 6.2)

i is for Intercept

If $x = 0$, then $sx = 0$ and

$y = f(x) = sx + i$

$= f(0) = 0 + i = i$

Thus, i is the range element paired off with $x = 0$. So the coordinate $(0, i)$, shown in Figure 6.2,

Figure 6.2

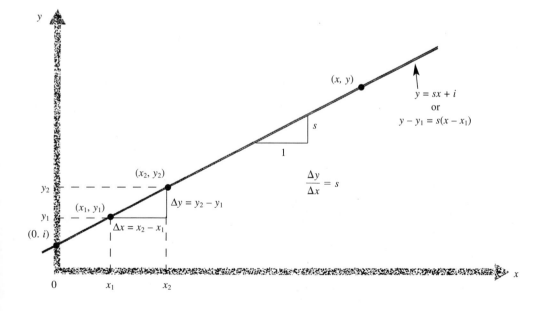

is a point on the line. Because this point lies on the y-axis, i is called the *y-intercept,* or just *intercept.* Look at the two linear functions we graphed earlier. In the first case, $y = 2x + 3$, the line crossed the y-axis at $y = 3$. In the second case, $y = -x + 6$, the line crossed the y-axis at $y = 6$. These values of y were, of course, equal to i in the respective equations.

Exercise 2

Return to the two linear functions in exercise 1. Determine the y-intercepts by merely looking at their equations?

s is for Slope

Understanding the meaning of the constant, s, in the general linear function is important because it serves to differentiate linear functions from all others. Consider any two points, (x_1, y_1) and (x_2, y_2), on the general linear function. Since y_1 and y_2 are the process outputs for x_1 and x_2 inputs respectively, we can say (See Figure 6.2)

$$y_1 = f(x_1) = sx_1 + i$$
$$y_2 = f(x_2) = sx_2 + i$$

Subtracting the first equation from the second,

$$y_2 - y_1 = (sx_2 + i) - (sx_1 + i)$$
$$= s(x_2 - x_1)$$

and defining

$\Delta y = y_2 - y_1$ as the change in y

$\Delta x = x_2 - x_1$ as the change in x

we have

$$\Delta y = s \cdot \Delta x$$

so

$$s = \frac{\Delta y}{\Delta x}$$

It should be emphasized that x_1 and x_2 were any values of x. Thus, Δx could have been very small, very large, or anywhere in between. Regardless, the ratio, $\Delta y/\Delta x$, is always constant and equal to s. This is a very important result that will be used time and again, especially when we get to calculus.

As a special case of this result, when the change in x is one unit ($\Delta x = 1$), $\Delta y = s \cdot 1 = s$. And this again occurs regardless of the position along the x-axis.

ANALYTIC GEOMETRY

To illustrate these ideas, let's compute $\Delta y/\Delta x$ between several points along the specific linear function $y = f(x) = 2x + 3$. Notice that $\Delta y/\Delta x = 2$ always, which is the value of s in the equation of this line. (See the tables below and Figure 6.1 on page 146.)

Point	(x, y)	Point to point	Δx	Δy	$\dfrac{\Delta y}{\Delta x}$
A	(0, 3)	A to B	1	2	2
B	(1, 5)	A to C	2	4	2
C	(2, 7)	A to D	3	6	2
D	(3, 9)	B to C	1	2	2
		B to D	2	4	2
		C to D	1	2	2

Exercise 3

Consider this same function for point $E(100, 203)$ and point $F(-1000, -1997)$. Show that $\Delta y/\Delta x$ is still 2 when these points are related to points A, B, C, and D.

The symbol s is the ratio of the "change in y" to the "change in x," or the "rise" divided by the "run." It is called the *slope* of the line. (We have mnemonically called it s to remind you of slope.) Imagine piloting a plane as it ascends along the linear function. Then the plane would increase altitude by s units for every 1 unit of ground distance.

The constant slope property of the linear function is the key to understanding when such functions are applicable in the everyday world. For example, if you estimate that *for every* dollar increase in rent, a large apartment complex will suffer two vacant apartments, *blink* goes the "linear light": $\Delta y = -2$ apartments when $\Delta x = \$1$, *regardless* of the value of x itself. So if rent is $300, the owner will lose two occupants if he or she raises the rent to $301; if rent is $350, two occupants will be lost if rent is raised to $351. No matter what the rent level is, increasing it $1 will result in two lost renters. Keep this idea in mind, as we will refer to it many times!

Now that we understand the meaning of the constants s and i, a special caution is in order. A function is still linear if only one of these two constants is explicitly present in the equation. For example:

$y = 5x$ so $s = 5$ and $i = 0$ $y = 10$ so $i = 10$ and $s = 0$

Notice that when $i = 0$, the line goes through the origin; when $s = 0$, the line has zero slope and thus is horizontal.

Point-Slope Formula

Hopefully, you are now very comfortable with the explicit formulation

$y = sx + i$

of the linear function. Not to destroy that comfort level, but rather to be complete, I must inform you that there is an alternative, but equivalent formulation, called the point-slope formula, which can be expressed as

$y - y_1 = s(x - x_1)$

It is often easier to use when fitting linear functions to real world data.

We derive the point-slope formula from a situation whereby a specific point (x_1, y_1) on the line is known. And with the slope (s) being constant between any two points, we can denote the general point as (x, y) and say (See Figure 6.2 on page 147).

$$s = \frac{\Delta y}{\Delta x} = \frac{y - y_1}{x - x_1}$$

So,

$y - y_1 = s(x - x_1)$ Point-Slope Formula

For example, if we know that one point on a straight line is $(x_1, y_1) = (4, 7)$ and the slope is -2, then

$y - 7 = -2(x - 4)$

Of course, we could isolate y as

$y = 7 - 2(x - 4)$

Or even put the equation in explicit form

$y = 7 - 2x + 8 = 15 - 2x$

Exercise 4

Determine the point-slope formula if a straight line has a slope of 3 and goes through the point $(x_1, y_1) = (1, 2)$. Then, put the equation in explicit form.

We summarize all our knowledge about linear functions, then slopes, intercepts, equation forms, etc. in Figure 6.2.

Graphing Linear Functions

Graphing linear functions is about as simple a task as you can have in mathematics. Since they graph as straight lines, all you need to do is locate two points, connect those points with a straight line, and extend this line to include the entire domain.

Both the explicit and point-slope versions of the linear function give the slope and one point along the line directly in their formula. They differ in that the explicit version gives the y-intercept point, while the point-slope version can give any point.

ANALYTIC GEOMETRY

Figure 6.3

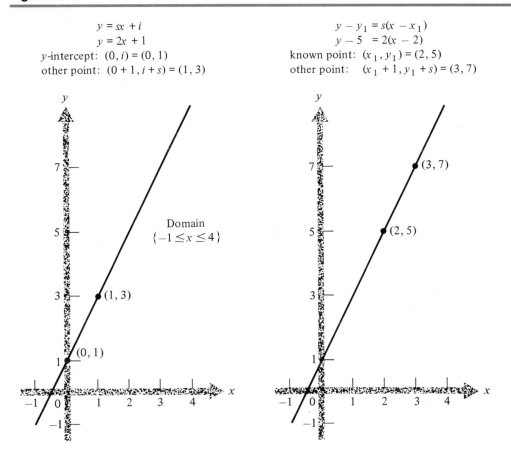

With one point and the slope, we can always find a second point. One way to find a second point is to use the fact that y increases by s units when x increases by one unit. Thus, for the explicit form with the known point $(0, i)$, a second point is $(0 + 1, i + s)$ or $(1, i + s)$. For the point-slope version with the known point (x_1, y_1), a second point is $(x_1 + 1, y_1 + s)$. Figure 6.3 illustrates finding a second point from the point given directly in the explicit and point-slope formulas for the same linear function.

Exercises 5

Plot the following linear functions:
1. $y = 3x + 2$
2. $y - 2 = 1(x - 3)$
3. $y = \frac{1}{2}x$
4. $y = 6$
5. $y + 1 = -2(x - 4)$
6. $y = 4x - 1$

Construct several other linear functions and plot them. Don't stop until you can plot a linear function in less than 10 seconds!

Multivariate Linear Functions

Until now, we have only studied univariate linear functions, or those in which one variable is a function of a single other variable. But alas, the world is often more complicated than that. For example, consider Sinema Theatre, a microcosm of the real world. If admission price were the same for everyone (say $6), then total ticket revenue (R) would be the following univariate function of total attendance (Q).

$$R = f(Q) = 6Q$$

But reality tells us that children under 18 (What are they doing there? Do their mothers know?) have a lower admission fee (say $3). So the actual total revenue is a function of two variables, the number of adults (A) and the number of children (C).

$$R = g(A,C) = 6A + 3C$$

This is a multivariate linear function. Notice how the basic "linear term" of a constant times the variable is repeated more than once.

Such functions exhibit a conditional-constant slope property, which we illustrate now. Suppose 50 children were in attendance. Substituting $C = 50$ in the multivariate formula gives

$$R = 6A + 3(50) = 6A + 150$$

which we easily recognize as a univariate linear function with a slope of 6. So holding $C = 50$, the revenue increases $6 per each additional adult. Symbolically

$$\frac{\Delta R}{\Delta A} = 6 \quad \text{if} \quad C = 50$$

Regardless of the specific value of C, as long as it is held constant at say C^*, the revenue function would be univariate with a slope of 6.

$$R = 6A + 3C^* = 6A + \text{constant}$$

Graphically speaking, univariate linear functions are straight lines (one dimensional figures with length but no width) on two-dimensional graph paper. For multivariate functions, their dimensions are always one less than the space in which they are described. For example, our Sinema revenue function would be a two-dimensional flat surface (plane) in three dimensions, as shown in Figure 6.4.

If senior citizens (S) get in for $1, we can amend our function further

$$R = h(A,C,S) = 6A + 3C + 1S$$

Don't ask me to graph that one. I'll only say that it is a three-dimensional surface (called a hyperplane) in four dimensions. We would need Einstein to make sense out of that.

There is no end to the number of terms that a multivariate linear function can have. The only restriction is that each additional term must be a constant times a variable. So in general, multivariate linear functions can be symbolized as

ANALYTIC GEOMETRY

Figure 6.4

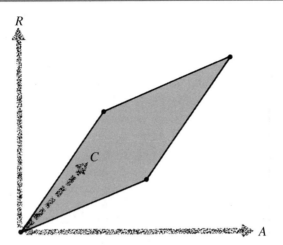

$$y = i + s_1x_1 + s_2x_2 + s_3x_3 + \ldots + s_nx_n$$

where i is the constant term, the s's stand for the respective conditional slopes and the x's stand for the n different variables.

Let's see two more of these functions.

Sam Dunk, star basketball player for the Oshkosh Gosh's, can score points in three ways: three point shots (x), regular two-point field goals (y), and 1 point foul shots (z). Thus, his total points is given by the following function

$$T = f(x, y, z) = 3x + 2y + 1z$$

A manufacturer makes 4 different products, A, B, C, and D, with variable costs of $8, $20, $55, and $250, respectively. If the fixed cost is $500,000, then the total cost of producing x units of A, y units of B, z units of C (whoops, we ran out of alphabet!), so let's use w units of D, is

$$T = f(x, y, z, w) = 500{,}000 + 8x + 20y + 55z + 250w$$

Exercise 6

If the manufacturer had a fifth product, of which it produced v units at a variable cost of $800 per unit, revise the total cost function to include it.

6.3 APPLICATIONS OF LINEAR FUNCTIONS

Demand Curves

Linear functions are often used to approximate demand curves. (A demand curve is the graph of a function relating Q, the number of units sold, to P, price per unit.) Of course, a linear function is exact only if for *every* dollar increase in price sales suffer by the *same* amount. For example, S. Lumlord, with a 600-apartment complex, is faced with the demand function

$Q = 1000 - 2P$ $\{200 \leq P \leq 500\}$ (See Figure 6.5)

If he charges $200 per apartment, all 600 apartments will be rented. But for every dollar increase in price, two apartments will go vacant. For example, the same loss in tenants occurs in raising the rent from 200 to 201 as for 400 to 401.

Exercises 7

1. Can you attach any meaning to the intercept value of 1000?
2. Suppose the demand curve was $Q = 900 - 4P$. How many tenants would be lost for each dollar increase in rent? What would be the highest rent that still attracts at least one tenant?

Revenue Functions

For the case of constant price, total revenue (R) is a linear function of sales (Q) of the form

$R = PQ$

where P is price (see Figure 6.6). For example, a product with a $6 sales price has the revenue function, $R = 6Q$. This was the case for Sinema Theater in Chapter 5.

Notice that the intercept point is the origin (no sales means no revenue). The price is the slope of the line, because the change in revenue (ΔP) for a change in sales of one unit is $P.

Figure 6.5

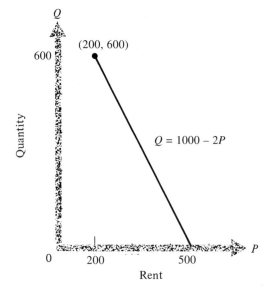

ANALYTIC GEOMETRY

Figure 6.6

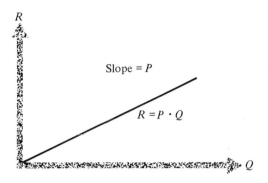

$$\frac{\Delta R}{\Delta Q} = P$$

In other words, if you sell one more item (for $P), you will add $P to your total revenue.

Exercises 8

1. Plot the revenue function for a product with a $2 selling price.
2. Consider a constant-price product whose total revenue is $10,000 if 2000 units are sold. Plot its revenue function.

Total Cost Function

Total costs for a manufacturing firm are usually divided up between fixed and variable costs. Fixed costs (F) are constant and don't vary with the level of production. Included here are non-changing costs such as the president's salary, interest on debt, and so on. Variable costs are those that vary directly with the production level (Q). Examples are labor and materials costs. If, furthermore, each unit of production has the same variable cost (v), then total cost (T) is the following linear function of production.

$T = F + vQ$ (See Figure 6.7)

Exercise 9

Hardog Co. manufactures pet rocks. If their fixed costs are $50,000, and each pet rock costs $2 to make, determine and graph their total cost function.

Straight-Line Depreciation

It makes sense that straight-line depreciation formulas graph as straight lines. In fact, both book value (B) and accumulated depreciation (A) of an asset are linear functions of the time (t) since the purchase of the asset, as given below. (See Figure 6.8)

Figure 6.7

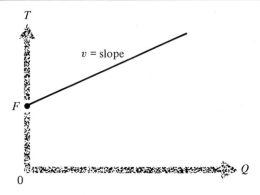

Figure 6.8

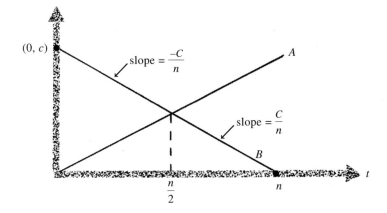

$$B = C - \left(\frac{C}{n}\right)t$$

$$A = \left(\frac{C}{n}\right)t$$

where

    C is the cost of the asset
    n is the accounting life
    t is time in years

For example, a \$10,000 asset depreciated over 5 years would have $C = 10{,}000$ and $n = 5$. So the two functions are

$$B = 10{,}000 - \left(\frac{10{,}000}{5}\right)t = 10{,}000 - 2{,}000t$$

$$A = \left(\frac{10{,}000}{5}\right)t = 2000t$$

ANALYTIC GEOMETRY

Exercises 10

1. Consider an asset costing $50,000 that is depreciated in straight-line fashion over ten years. Graph the book value and accumulated depreciation functions.
2. Show that in general $B = A$ when $t = n/2$.

Temperature Conversion

Scientific work involves use of the metric system, which employs the Centigrade (C) temperature scale. Americans have been brought up on the Fahrenheit (F) temperature scale. The following linear conversion formula relates the two temperature scales.

$$C = \frac{5}{9}F - \frac{160}{9} \quad \text{(See Figure 6.9)}$$

Notice that 32 degrees Fahrenheit (freezing point) corresponds to zero degrees Centigrade, and 212 degrees Fahrenheit (boiling point) corresponds to 100 degrees Centigrade.

For every 1 degree increase in F, C only increases 5/9 degree.

Exercises 11

1. Give the Centigrade weather report if the day's high and low Fahrenheit temperatures are forecasted to be 80 and 50, respectively.
2. The *Guiness Book of World Records (1973)* reports the highest recorded temperature as 127.4 degrees Fahrenheit (Ouargla, Algeria) and the lowest recorded temperature as –126.9 degrees Fahrenheit (Vostok, Antarctica). Convert these temperatures to Centigrade readings.
3. What temperature has equal Fahrenheit and Centigrade readings?

Figure 6.9

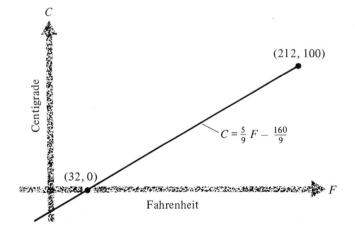

Figure 6.10

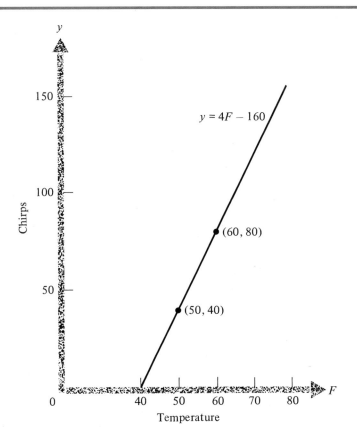

Cricket Chirps

Harold R. Jacobs in his interesting book, *Mathematics: A Human Endeavor,*[2] shows that the number of chirps (y) that a cricket makes per minute is the following linear function of the temperature (F) in Fahrenheit degrees.

$y = 4F - 160 \quad \{F \geq 40\} \quad$ (See Figure 6.10)

For every 1 degree increase in temperature, four additional chirps will ensue.

Exercises 12

1. Make a cricket thermometer by solving for $F = f(y)$. Suppose your pet cricket records 60 chirps per minute some evening. Without checking your TV weather report, determine the temperature.

---

[2] Harold R. Jacobs, *Mathematics: A Human Endeavor,* San Francisco: W. H. Freeman, 1970.

2. Although Bionic Cricket follows this function to supersonic chirp levels, everyday crickets can't do over 200 chirps per minute. What then is the upper limit for the temperature domain?
3. For you students who use the Centigrade (C) system, determine y (number of cricket chirps) in terms of C

Common Stock Betas

Common stock investors need to better understand the risks they are taking. To help them in this endeavor, they can become aware of the relationships between individual stock returns and those of the market as a whole. To quantify such a risk measure, researchers gather data on the annual rates of return (dividend yield plus percentage change in price) for individual stocks (R) and the overall market (M). They then use calculus to find the best-fitting straight line for the data (We will illustrate these methods when we study least-squares regression in Section 15.12 on page 573.). In summary, the researchers fit the following model:

$R = \alpha + \beta M$

where

α = riskless rate of return (something like a Certificate of Deposit return that has no risk.
β = slope of the line that relates individual stock return to market return. This is a measure of risk.

Relatively risk-free stocks, like utilities (don't tell this to Three-Mile Island stockholders) have β values under 1, while risky high growth-oriented stocks have β over 1. For example, consider the following lines for the utility Green Power Company (R_U) and the growth stock Artificial Stupidity Systems (R_G) (I better not give you its acronym!):

$R_U = 6 + .4M$

$R_G = 6 + 1.8M$

If in a given year we expect the market to have a 20% rate of return, we can forecast returns for these two companies as

$R_U = 6 + .4(20) = 14\%$

$R_G = 6 + 1.8(20) = 42\%$

So, why ever buy utilities? The answer, of course, lies with the fact that markets do go down sometimes. For example, if the market return were -10% one year, then the expected returns for these two companies would be

$R_U = 6 + .4(-10) = +2\%$

$R_G = 6 + 1.8(-10) = -12\%$

Low betas mean that the stock tends to move less than the market in both directions. High betas mean that the stock's movements in both directions exaggerate those of the market.

Exercises 13

1. Graph the above two functions over the domain.
2. Explain the movements relative to the market for a stock with a beta of 1.
3. Under what market conditions would the return of the utility stock equal that of the growth stock?

Postage Costs—Linear Step Function

The cost of mailing a letter first class allows us to learn a variation of linear functions. In 2003, a 37 cent stamp was needed for any letter up through one ounce. Although it isn't exactly so, let's for the time being assume that you need a second 37 cent stamp if the weight falls anywhere from just over 1 ounce to 2 ounces. Etc. Graphically, this stamp cost looks like steps, as shown in Figure 6.11.

If we only consider the domain of whole numbers, as indicated by the dots, we would have the function

$$y = f(x) = 37x \quad \{x = 1, 2, 3, \cdots\}$$

But, to consider all possible weights, we need to invoke the following new notation

Figure 6.11

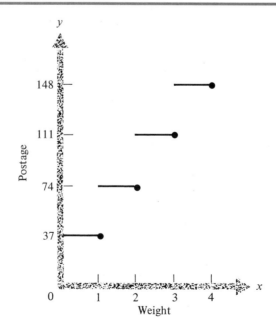

ANALYTIC GEOMETRY

[x] which equals x if x is an integer, and which equals the next higher integer if x is not an integer

For example

[2] = 2

[2.1] = 3

Using this notation, then the 2003 assumed postal cost function would be

$y = f(x) = 37[x]$

In reality, only the first ounce costs 37 cents in 2003. All further ounces, or part thereof, cost 23 cents. So the actual postage cost function in 2003 was

$y = f(x) = 37[x]$ for $x \leq 1$

$\quad\quad = 37 + 23[x - 1]$ for $x > 1$

For example, if you mail a letter weighing 1.7 ounces, it would cost

$37 + 23[1.7 - 1]$

$37 + 23[.7]$

$37 + 23(1) = 37 + 23 = 60$ cents

Exercises 14

1. Graph the actual 2003 postage-cost function.
2. Since the postage cost is probably higher as you sit there, unable to write home for money, graph the current postage-cost function.

Segmented Linear Functions

Sometimes two or more linear functions with different slopes and different domains are needed to describe actual relationships. These so-called segmented linear functions are best demonstrated with the following federal income tax application.

Income Tax Woes

Your author composed this section on tear-stained paper, as I prepared my 2001 federal taxes.

For taxable incomes under 100,000, taxpayers have detailed tables listing the tax due for every possibility. They also provide a formula for these cases. For taxable incomes above 100,000, which you will certainly achieve once you master the math in this book, they only provide a formula. The IRS has separate formulas for four different types of filers. Shown below is only Schedule X, the information for single taxpayers.

This schedule essentially gives us, in column three, a variation of the point-slope formula

2001 Tax Rate Schedules

Schedule X—Use if your filing status is **Single**

If the amount on Form 1040, line 39, is: Over—	But not over—	Enter on Form 1040, line 40	of the amount over—
$0	$27,050	.... 15%	$0
27,050	65,550	$4,057.50 + 27.5%	27,050
65,550	136,750	14,645.00 + 30.5%	65,550
136,750	297,350	36,361.00 + 35.5%	136,750
297,350	....	93,374.00 + 39.1%	297,350

LINE 39 AMOUNT	TAX DUE
0 – 27,050	$y = 0 + .15(x - 0)$
27,050 – 65,500	$y = 4057.50 + .275(x - 27,050)$
65,550 – 136,750	$y = 14,645 + .305(x - 65,550)$
136,750 – 297,350	$y = 36,361 + .355(x - 136,750)$
>297,350	$y = 93,374 + .391(x - 297,350)$

For TAX DUE, we use a variation of the point-slope formula, or

$$y = y_1 + s(x - x_1)$$

If you are having any trouble going from the schedule to the above formulas, don't fret. That process is "curve-fitting", which is the subject of the next chapter.

Exercise 15

Put these formulas is the linear explicit form, or $y = i + sx$

In Figure 6.12, we graph these 5 straight lines. Notice that they have successively higher slopes (marginal tax rates).

6.4 QUADRATIC EQUATIONS

Quadratic equations are characterized by squared* variables. The general equation of a quadratic in two variables, x and y, written in implicit form, is

---

*An equation with neither x nor y squared, but an xy term, is considered quadratic, since the sum of the exponents equals 2.

ANALYTIC GEOMETRY

Figure 6.12

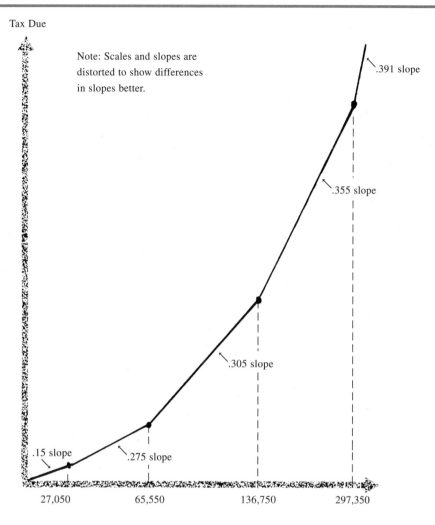

$$Ax^2 + Bxy + Cy^2 + Dx + Ey + F = 0$$

where A, B, C, D, E, and F are constants.

Notice how all terms in the implicit formulation are to the left of the equal sign. The independent and dependent variables of the explicit formulation are not indicated. We use this implicit formulation because it provides the bases for identifying the type of quadratic.

Some examples of quadratic equations, with the values of the constants are given below.

	A	B	C	D	E	F
$x^2 - xy + 2y^2 - 3x + 4y - 8 = 0$	1	−1	2	−3	4	−8
$3x^2 + 2y^2 + 6 = 0$	1	0	2	0	0	6
$y^2 - 3x + 10y - 5 = 0$	0	0	1	−3	10	−5
$8xy - x = 0$	0	8	0	−1	0	0

Notice how some terms in the general expression have zero constants, which effectively wipes them out in the equation. The quadratic nature is not changed by having a few terms wiped out, as long as there exists at least one of the x^2, xy, or y^2 terms. So, no more than 2 constants from A, B, and C can be zero. We can see why by setting $A = B = C = 0$. The only terms left are

$Dx + Ey + F = 0$

which is a linear function, as seen once we explicitly solve for y.

$Ey = -Dx - F$

$$y = \left(\frac{-D}{E}\right)x - \left(\frac{F}{E}\right)$$

Exercises 16

Determine the constants: A, B, C, D, E, and F for the following quadratic equations.
1. $6x^2 + 3xy - 5x + 10 = 0$
2. $xy - 2y^2 - 3y = 0$
3. $y^2 - 8x = 0$

It is important to note here that some quadratic equations do not represent functions. For example, the last exercise above can be solved for y

$y^2 = 8x$
$y = \pm\sqrt{8x}$

So there are 2 values of y for a specific value of x. Such a one to two assignment from domain (x) to range (y) violates the definition of a function.

Quadratic Roots Formula

Equations that only have x terms or only y terms are called quadratic equations in one variable. For example

$x^2 + 3x + 2 = 0$ $B = 0, C = 0, E = 0$
$2y^2 - 4y + 15 = 0$ $A = 0, B = 0, D = 0$

ANALYTIC GEOMETRY

Such equations have 2 so-called "roots" or solutions. Thus, 2 values of the unknown (either x or y) exist that satisfy such equations.

Consider the general formulation of a quadratic in one variable, x.

$$ax^2 + bx + c = 0 \quad \text{where } a, b, c \text{ are constant}$$

Think back to eighth grade and see if you don't remember the following "monster" formula that solves these equations

$$x = \frac{-b \pm \sqrt{b^2 - 4ac}}{2a}$$

Does this bring back bad memories?
Let's illustrate by finding the roots of the above examples

$$x^2 + 3x + 2 = 0$$
$$a = 1 \quad b = 3 \quad c = 2$$
$$x = \frac{-3 \pm \sqrt{3^2 - 4(1)(2)}}{2(1)}$$
$$= \frac{-3 \pm \sqrt{1}}{2} = \frac{-3 \pm 1}{2}$$

$x = -1, -2$ are the two roots

If you are good at factoring, you probably could have gotten these 2 roots that way

$$x^2 + 3x + 2 = (x + 1)(x + 2) = 0$$
$$\downarrow \qquad \downarrow$$
$$x = -1 \quad x = -2$$

The second example above could not be factored in your head no matter how good you are at factoring. Don't worry that we changed the variable to y. The quadratic roots formula works no matter what the variable name is.

$$2y^2 - 4y - 15 = 0$$
$$a = 2 \quad b = -4 \quad c = -15$$
$$y = \frac{-(-4) \pm \sqrt{(-4)^2 - 4(2)(-15)}}{2(2)}$$
$$= \frac{4 \pm \sqrt{136}}{4}$$
$$= \frac{4 + 11.66}{4}; \quad \frac{4 - 11.66}{4}$$
$$= 3.915; \quad -1.915$$

Exercises 17

Find the roots of the following quadratic equations
1. $x^2 - 9x + 3 = 0$
2. $2y^2 + 4y - 6 = 0$

Identification

The graph of a quadratic equation in two variables always turns out to be one of the following four forms: circle, ellipse, parabola, or hyperbola, as shown in Figure 6.13. It is easy to identify the form directly from the implicit equation by the following rules, which depend on the value of B (coefficient of the xy term).

Quadratic identification rules	
If an xy term exists ($B \neq 0$), calculate $\delta = B^2 - 4AC$. Then when $\delta < 0$, it's an ellipse $\delta = 0$, it's a parabola $\delta > 0$, it's a hyperbola	If no xy term exists ($B = 0$), when $A = C$, it's a circle $A \neq C$, but A and C have same sign, it's an ellipse $A = 0$ or $C = 0$, but not both, it's a parabola A and C have opposite signs, it's a hyperbola

Let's identify and plot several quadratic equations.

Example 1

$x^2 - y + 4 = 0$

The correspondence of constants to the general equation is $A = 1$, $B = 0$, $C = 0$, $D = 0$, $E = -1$, and $F = 4$. Since $B = 0$ (no xy term present), and $C = 0$ but $A \neq 0$, it's a parabola.

To plot, it helps to put the equation in explicit form.

$y = f(x) = x^2 + 4$

Setting x at various values gives values of $f(x)$. Noting that $f(x)$ is the same as $f(-x)$ for all

Figure 6.13

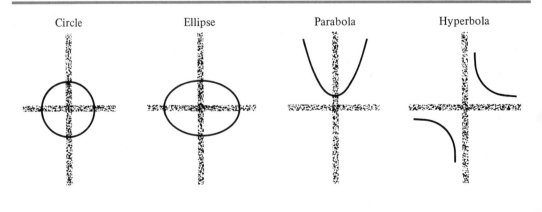

Circle　　Ellipse　　Parabola　　Hyperbola

ANALYTIC GEOMETRY

x, along with the fact that we have an identified parabola, makes plotting easier. See the table and Figure 6.14.

x	$y = x^2 + 4$
0	4
1	5
2	8
−1	5
−2	8

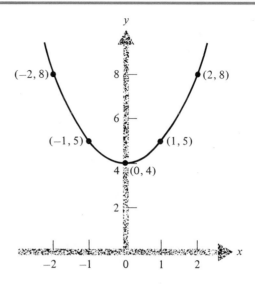

Figure 6.14

x	$y = \dfrac{2}{x+1}$
0	2
1	1
2	$\frac{2}{3}$
10	$\frac{2}{11}$
↓	↓
∞	0
−1	∞
−2	−2
−3	−1
↓	↓
−∞	0

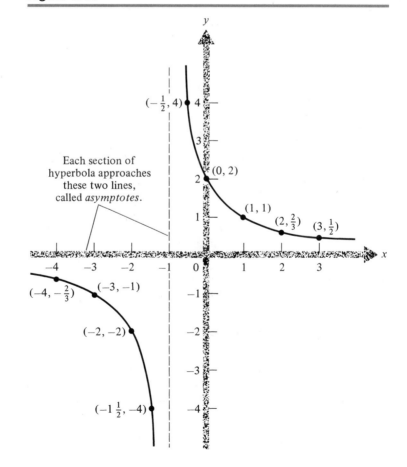

Figure 6.15

Example 2

$xy + y - 2 = 0$

The correspondence of constants to the general equation is $A = 0$, $B = 1$, $C = 0$, $D = 0$, $E = 1$, and $F = -2$. An xy term is present, so $\delta = B^2 - 4AC$ must be determined and compared to zero. Since $\delta = 1^2 - (4 \cdot 0 \cdot 0) = 1$, it is a hyperbola. Again, the equation in explicit form eases the plotting chore.

$y(x + 1) = 2$

$$y = g(x) = \frac{2}{x + 1}$$

Some well-chosen values of x, and our knowledge that we have a hyperbola, allow for quick plotting. See Figure 6.15 and the accompanying table.

Example 3

$x^2 + y^2 - 9 = 0$

x	$y = \pm\sqrt{9 - x^2}$
0	± 3
3	0
−3	0

Figure 6.16

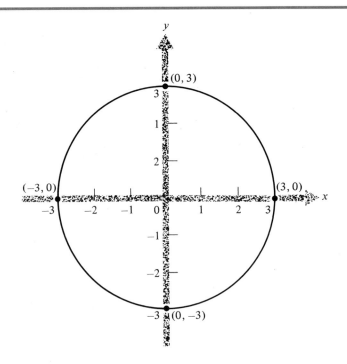

ANALYTIC GEOMETRY

The correspondence of constants to those in the general equation is $A = 1$, $B = 0$, $C = 1$, $D = 0$, $E = 0$, and $F = -9$. Since $B = 0$, and both A and C equal 1, it is a circle.

Solving explicitly for y gives

$y^2 = 9 - x^2$

$y = \pm\sqrt{9 - x^2}$

This circle is symmetric about the vertical axis, as $x = +k$ gives the same value of y as $x = -k$. When $x > 3$ or $x < -3$, x^2 is larger than 9, and we have a negative number under the square root sign. So our circle, on the real plane, is limited to the $-3 \leq x \leq 3$ domain.

Our knowledge of a circle, its symmetry, and domain allows for plotting with only a few points, as shown in Figure 6.16 and the accompanying table.

Example 4

$x^2 + xy + y^2 - 13 = 0$

The correspondence of constants to the general equation is $A = 1$, $B = 1$, $C = 1$, $D = 0$, $E = 0$, and $F = -13$. Because $B \neq 0$, we must find $\delta = B^2 - 4AC = 1 - 4 = -3$. Since δ is negative, the identification rules tell us that it is an ellipse.

Continuing as before, we try to solve explicitly for y. However, this time we get nothing but aggravation.

$xy + y^2 = 13 - x^2$

$y(x + y) = 13 - x^2$

No matter what we do, the y refuses to be isolated. So, let's take a new approach by setting a value of x in the original implicit equation and solving for y. Setting $x = 0$ gives

$0 + 0y + y^2 - 13 = 0$

or

$y = \pm\sqrt{13}$

Setting $y = 0$ gives $x = \pm\sqrt{13}$. The four points found thus far are all of equal distance from the origin (see Figure 6.17). Without having identified it as an ellipse, it would be tempting to connect the points in circle fashion as the dotted line indicates.

Since we were successful in determining points by arbitrarily substituting in values of x or y, let's continue by setting $x = 1$.

$1^2 + 1y + y^2 - 13 = 0$

or

$y^2 + y - 12 = 0$

This can be factored as

Figure 6.17

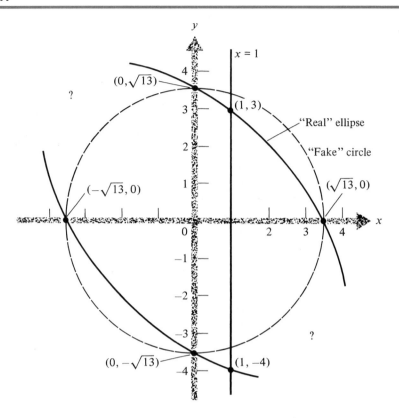

$(y + 4)(y - 3) = 0$

so, $y = -4$ and $y = +3$ are paired off with $x = 1$. These points confirm graphically that we don't have a circle and also establish the "tilt" of the ellipse. We still don't know how far out the ellipse extends, as indicated in Figure 6.17.

Before we proceed, it would be helpful to see exactly what we have been doing. For example, when we set $x = 1$ above, we had two equations: the ellipse and the vertical line at $x = 1$.

$x = 1$

$x^2 + xy + y^2 - 13 = 0$

The intersection of these two equations is two points: (1, 3) and (1, −4).

So, by setting one of the variables equal to a constant, we establish a vertical (setting x) or a horizontal (setting y) line whose intersection with the ellipse determines points on the ellipse. For example, if we set $x = 2$ in the ellipse equation, we are looking for the two points that are common to the ellipse and the line $x = 2$ (see Figure 6.18):

$2^2 + 2y + y^2 - 13 = 0$

$y^2 + 2y - 9 = 0$

Now, how does one isolate y? Factoring doesn't work with this one. Manipulating, in conventional ways such as

ANALYTIC GEOMETRY

Figure 6.18

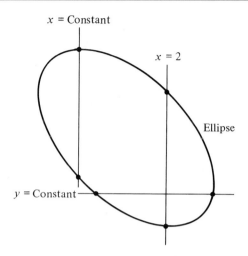

$y^2 + 2y = 9 \quad y^2 = 9 - 2y$

$y(y + 2) = 9 \quad y = \pm\sqrt{9 - 2y}$

$y = \dfrac{9}{y + 2}$ Rats!

leaves one with y in terms of itself, certainly not a joyous state of affairs.

We need a rescue. Wait, look off to the horizon. The quadratic roots formula cavalry is coming!!

$y^2 + 2y - 9 = 0$

$a = 1 \quad b = 2 \quad c = -9$

$y = \dfrac{-2 \pm \sqrt{2^2 - 4(1)(-9)}}{2(1)}$

$= \dfrac{-2 \pm \sqrt{40}}{2} = \dfrac{-2 \pm 6.32}{2} = 2.16; \, -4.16$

Now we can complete the graph of the elipse armed with this wonderful quadratic roots formula.

Setting of one variable	Quadratic equation in other variable	Solution or roots	Coordinates
$y = 2$	$x^2 + 2x - 9 = 0$	$x = \dfrac{-2 \pm \sqrt{4 - (4 \cdot 1)(-9)}}{2}$	$(-1 + \sqrt{10}, 2)$
		$x = \dfrac{-2 + \sqrt{40}}{2}; \dfrac{-2 - \sqrt{40}}{2}$	$(-1 - \sqrt{10}, 2)$
		$x = -1 + \sqrt{10}; \, -1 - \sqrt{10}$	

$x = -3$	$y^2 - 3y - 4 = 0$	$y = \dfrac{-(-3) \pm \sqrt{9 + 16}}{2}$	$(-3, 4)$
		$y = 4; -1$	$(-3, -1)$
$x = 5$	$y^2 + 5y + 12 = 0$	$y = \dfrac{-5 \pm \sqrt{25 - (4 \cdot 1 \cdot 12)}}{2}$	Imaginary coordinates*
		$y = \dfrac{-5 \pm \sqrt{-23}}{2}$	

*Whenever we get a negative number under the square root sign, it means that the curve does not exist on the real plane for that value of the variable.

Exercises 18

1. Set $x = -4$, $y = -2$, and $x = 3$ and find the roots of the resulting quadratic equations. Use these points along with the others to plot the ellipse.
2. Identify and plot the following quadratic equations:
   a. $x^2 + y^2 + 2x - 3y - 4 = 0$
   b. $2x^2 - y^2 + 3x + 5 = 0$
   c. $x^2 + xy + 2y^2 - 10 = 0$
   d. $3x^2 + x + y + 1 = 0$

6.5 APPLICATIONS OF QUADRATIC EQUATIONS

Revenue Functions

Whenever the demand function for a commodity is linear and downward sloping, the revenue function is quadratic. Let's see why.

The total revenue (R) for a product is the price per unit (P) times the number of units sold (Q).

$R = PQ$

A linear, downward-sloping demand curve is represented by

$Q = sP + i$

where $s < 0$. So

$R = P(sP + i)$
$ = sP^2 + iP$

Consider the specific case of S. Lumlord's demand curve on page 154

$Q = 1000 - 2P$

ANALYTIC GEOMETRY

So revenue is

$$R = P \cdot Q = P(1000 - 2P) = 1000P - 2P^2$$

Putting this in implicit form and making the correspondence of constants in the general quadratic equation

$$-2P^2 + 1000P - 1R = 0$$

$A = -2 \quad D = 1000 \quad E = -1$

We see that $B = 0$, and that $A \neq 0$, but $C = 0$; thus we have a parabola on our hands.

Exercise 19

Graph the above revenue function for S. Lumlord.

Unit Cost Function

Whenever total cost is a linear function of production level (see page 155), unit cost (U) or total cost (T) divided by the number of units produced (Q) is a hyperbolic function of production. Let's see why.

$$T = F + vQ$$

$$U = \frac{T}{Q} = \frac{F + vQ}{Q} = \frac{F}{Q} + v$$

Writing this equation in implicit form gives

$$UQ - vQ - F = 0$$

Since U and Q are variables, and F and v are constants, we have a quadratic equation with a product-of-variables term (UQ). Thus, to identify it, we must compute the value of $\delta = B^2 - 4AC = 1^2 - (4 \cdot 0 \cdot 0) = 1$. Since δ is greater than 0, our identification rules tell us we have a hyperbola on our hands. The unit cost function graphs as shown in Figure 6.19.

Exercise 20

Determine and plot the unit cost function for the Hardog Company as defined in the exercise on page 155.

Sum of Arithmetic Progression

Recall that we derived the formula for the sum of an arithmetic progression in section 3.2 as

$$S_A = \frac{n}{2}[2a + (n-1)d]$$

Figure 6.19

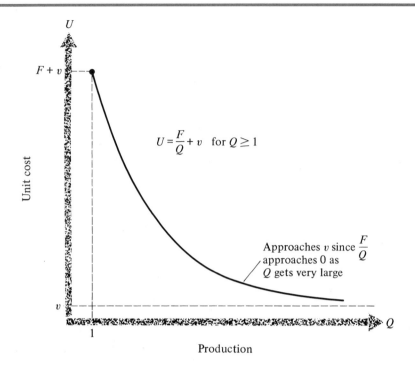

With a little algebraic manipulation, this can be written as

$$S_A = \left(\frac{d}{2}\right)n^2 + \left(a - \frac{d}{2}\right)n$$

With a and d as constants, we have a quadratic equation in the variables S_A and n.

Exercises 21

1. Use the identification rules to show that the arithmetic sum formula is a parabola.
2. Recall the Melody Tent situation in section 3.3. There we determined that $a = 64$ and $d = 8$, so the particular arithmetic sum formula is

$$S_A = 4n^2 + 60n$$

Plot S_A (the total number of seats) against n (the number of rows of seats). Use the quadratic roots formula to determine how many rows of seats are needed to have a total of 8800 seats.

Bivariate Normal Distribution

An important probability distribution in statistical theory is the *bivariate normal*. Picture taking a horizontal slice out of such a distribution, as shown in Figure 6.20. The equation of the points along that intersection is

ANALYTIC GEOMETRY

Figure 6.20

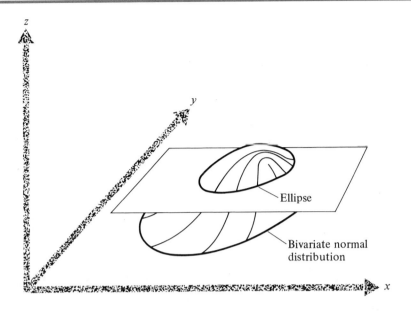

$$x^2 + y^2 - 2\rho xy = k$$

where ρ, k are constants and $\rho < 1$. It so happens that this intersection line is an ellipse (remarkable!).

Exercises 22

1. Use the quadratic identification rules to verify that the above equation represents an ellipse.
2. For $\rho = .5$ and $k = 2$, plot the ellipse.

Unitary Elasticity Demand Curve

A classic demand curve—one that no good economics book would dare be without—is the so-called *unitary elasticity demand curve*. We will have more to say about it when we cover calculus. Here we shall merely illustrate its equation

$$Q = \frac{k}{P}$$

where

    k is a constant
    Q is quantity demanded
    P is price per unit

Believe it or not, this equation represents a hyperbola. Let's see why by putting it in implicit form.

$PQ = k$

$PQ - k = 0$

The variables here are P and Q. With a PQ product term present, $B \neq 0$, so identification requires calculation of $\delta = B^2 - 4AC = 1$. Since δ is positive, it is a hyperbola.

Exercise 23

If $k = 1000$, plot the unitary elasticity demand curve.

Random Walk

You are a sleepwalker and you are standing $\frac{1}{2}$ foot from the edge of a cliff. Your steps are 1-foot long and you are equally likely to walk in any direction. On your next step, then, you will end up at some point along a 1-foot radius circle which centers at your present position. The equation of this circle—using your initial point as $(x, y) = (\frac{1}{2}, 0)$—is

$x^2 + y^2 - x - \frac{3}{4} = 0$ (See Figure 6.21)

Exercises 24

1. Use the quadratic identification rules to verify that the above equation is a circle.
2. Suppose a friend can reach you and save you, but only after you have taken one step. What is the probability that you'll never have to work another quadratic problem again? (*Hint:* Find the points where the cliff intersects the circle. The fraction $\theta°/360°$ is the answer.)

Figure 6.21

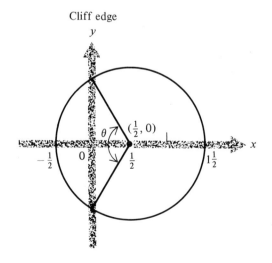

6.6 EXPONENTIAL FUNCTIONS

Exponential functions have many applications, some of which we encountered in the module on finite mathematics. They are especially important in depicting the possible outcomes of uncertain events. The most illustrious member of this family of functions is the normal (bell-shaped) curve.

Mathematically, exponential functions are characterized by a constant that has an exponent, which itself is a function of x. Their general representation is

$$y = f(x) = b^{g(x)}$$

where b is a positive constant and $g(x)$ is a function of x.

Examples

$y = 2^x \qquad y = .5^{2+x}$

$y = 10^{-6x} \qquad y = 4^{1/x}$

$y = 3^{-x^2} \qquad y = 2.3^{2x^2+3x-6}$

Beware, for it is common for students to confuse these functions with other types. For example, $y = 2^x$ is an exponential function, but $y = x^2$ is a parabola.

Exponential functions have varied graphs. Let's get a feeling for these differences by investigating the nature of important types of exponential functions. First, consider the exponential function with $g(x) = cx$. Thus

$$y = f(x) = b^{cx}$$

where, b and c are real numbers ($b > 0$).

We can get a feel for this type of function by setting c equal to 1 and considering $b = 2$ and $b = 3$. These two cases are tabled below.

x	$y = 2^x$	$y = 3^x$
−3	$\frac{1}{8}$	$\frac{1}{27}$
−2	$\frac{1}{4}$	$\frac{1}{9}$
−1	$\frac{1}{2}$	$\frac{1}{3}$
0	1	1
1	2	3
2	4	9
3	8	27

Does it surprise you that the range elements for integer values of x form a geometric progression? Of course, with the real number domain, all the intermediate values are generated, as indicated by the continuous graph in Figure 6.22.

All such functions go through the point (0, 1) since $b^0 = 1$. The range is always positive, since a positive number raised to any real power is positive.

Figure 6.22

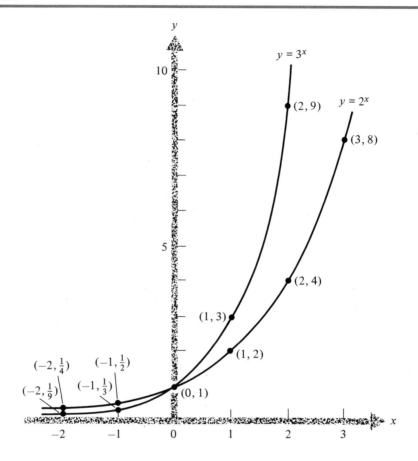

For the same b values, let's consider the case of $c = -1$. The resulting functions are the reciprocals of the previous ones, since

$$b^{-x} = \frac{1}{b^x} = \left(\frac{1}{b}\right)^x$$

Figure 6.23

x	$y = 2^{-x}$	$y = 3^{-x}$
-3	8	27
-2	4	9
-1	2	3
0	1	1
1	$\frac{1}{2}$	$\frac{1}{3}$
2	$\frac{1}{4}$	$\frac{1}{9}$
3	$\frac{1}{8}$	$\frac{1}{27}$

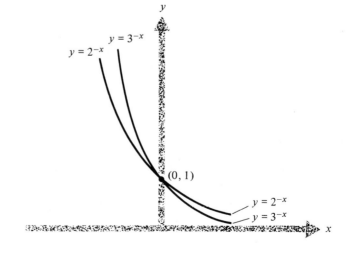

ANALYTIC GEOMETRY

Figure 6.24

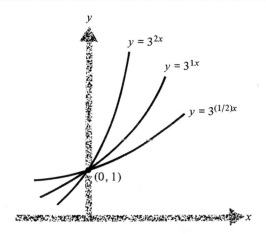

Therefore, when b^x is large, b^{-x} is small, and vice versa, as illustrated by the graphs in Figure 6.23. Had we rotated 3^x and 2^x around the vertical axis, we would have obtained 3^{-x} and 2^{-x}. Again, the entire range is positive, and all curves pass through the point (0, 1).

We can see the effect of c's value by graphing the cases of $c = \frac{1}{2}$, 1, and 2 for the same base, $b = 3$ (see Figure 6.24).

The value of c greatly affects the slope of the curve, with higher values of c causing higher slopes. However, all curves still go through the point (0, 1).

Exercises 25

1. Plot $y = (\frac{1}{2})^x$. Explain why it has the same graph as $y = 2^{-x}$.
2. Plot $y = (\frac{1}{3})^{-x}$. Explain why it has the same graph as $y = 3^x$.
3. Plot and compare $y = 2^x$ with $y = 2^{2x}$.
4. Plot and compare $y = 4^{-x}$ with $y = 5 \cdot (4^{-x})$.

Now consider another class of exponential functions with $g(x) = cx^2$. Thus, its formula would be

$$y = f(x) = b^{cx^2} \quad \text{where } b > 0, c \text{ is a real number}$$

To get a feel for the graph of such functions, let's set $b = 2$ and consider the two cases of $c = +1$ and $c = -1$. See Figure 6.25 and the table below.

x	$y = 2^{x^2}$	$y = 2^{-x^2}$
0	1	1
1	2	$\frac{1}{2}$
-1	2	$\frac{1}{2}$
2	16	$\frac{1}{16}$
-2	16	$\frac{1}{16}$
$\pm k$	2^{k^2}	$\frac{1}{2^{k^2}}$

Figure 6.25

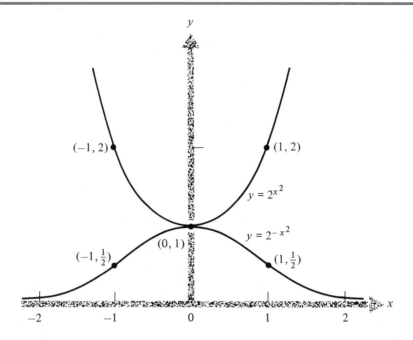

Symmetry, meaning $f(-x) = f(x)$, exists in both cases. However, with $c = +1$, a parabola-like curve emerges, while with $c = -1$, a bell-shaped curve results. The latter is none other than a cousin to the famous normal curve. In general, a parabola-like curve results if $c > 0$, while a bell-shaped curve results if $c < 0$. In all cases, such curves go through the point $(0, 1)$.

Exercise 26

Graph $y = 2^{cx^2}$ for $c = +2$ and $c = -2$. Compare these curves to the cases with $c = +1$ and -1.

6.7 APPLICATIONS OF EXPONENTIAL FUNCTIONS

Worker's Paradise

Recall the worker in section 3.5 who offered his services for a meager 1¢ on day 1, 2¢ on day 2, 4¢ on day 3, and so on through 30 days. Back then, you were given that his salary, y, on the xth day is $y = 2^{x-1}$. But this is merely $2^x \cdot 2^{-1} = \frac{1}{2}(2^x)$. (This manipulation was brought to you by exponent law no. 1. Lo and behold, this is an exponential function of the form $y = ab^{cx}$, where $a = \frac{1}{2}$, $b = 2$, and $c = 1$. Recall also that the domain is limited to the natural numbers up through 30. So,

$$y = \tfrac{1}{2}(2^x) \quad \{x = 1, 2, 3, \ldots, 30\}$$

ANALYTIC GEOMETRY

The sum of the 30 "paychecks" (S_G) is given by the geometric progression sum formula

$$S_G = \frac{1(2^{30} - 1)}{2 - 1} = 2^{30} - 1$$

Now suppose we wish to know the pay for some other number of paychecks. For x days of work, the geometric sum formula gives

$$y = 2^x - 1$$

which is an exponential function. Notice the relationship between the single day pay, $\frac{1}{2}(2^x)$, and the total pay to that time, $(2^x - 1)$. The latter is one penny less than twice the former. For example, the pay on day 4 is 8¢, while the pay up through day 4 is 15¢. (This property was very important to the success of the Lost Wages gambler in section 3.5. Why?)

Exercises 27

1. Plot the single day and the total pay functions.
2. Consider the case for a tripling of pay each day, that is, 1, 3, 9, and so on. How many days are necessary to become a millionaire?

Future and Present Values

The future value (F) of $1 invested now at an i rate compounded annually is an exponential function of time (n) in years.

$$F = (1 + i)^n$$

The present value (P) of $1 n years hence, if money is discounted at an i annual rate, is

$$P = \frac{1}{(1 + i)^n} = (1 + i)^{-n}$$

These are exponential functions of the first type, with $b = (1 + i)$, $c = 1$ and -1, respectively. The domain is the set of natural numbers.

Exercise 28

Sketch the above functions for $i = .10$.

Sales Forecasting

For the case of a constant percentage increase in sales every year (which is reasonably close to the truth for many corporations), sales follow an exponential pattern (analogous to the bank balance that grows by the same percentage every year). So, if the annual sales growth rate is g (expressed as a decimal), then sales t years from now (S_t) are the following exponential function of t:

$$S_t = f(t) = S_0(1 + g)^t$$

where S_0 is current sales.

We can illustrate by considering the intense sales competition between Gangland Industries and C.A.R.T.E.L. International. Because of aggressive marketing, Gangland Industries (their salesmen say, "We'll make you a deal you *better not* refuse") has been winning a lot of "contracts" away from its larger competitor. Consequently, industry experts predict a 20 percent annual sales growth for Gangland, as opposed to a 10 percent figure for C.A.R.T.E.L. With current sales levels of $10 and $50 million, let's find out when Gangland will catch up in sales.

The two sales functions are

$$S^G = 10(1.2)^t$$
$$S^C = 50(1.1)^t \quad \text{(See Figure 6.26)}$$

By setting the sales levels equal, we can find out when Gangland catches up.

$$S^G = S^C$$
$$10(1.2)^t = 50(1.1)^t$$
$$\frac{1.2^t}{1.1^t} = 5$$
$$1.091^t = 5$$
$$t = \log_{1.091} 5 = 18.5 \text{ years}$$

Figure 6.26

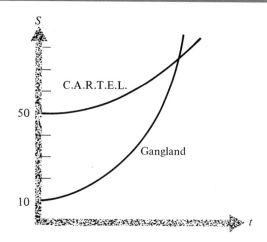

Exercise 29

If Gangland, by even more aggressive marketing ("We'll make you a deal that your loved ones hope you won't refuse"), is able to increase its growth rate to 25 percent (everything else the same as stated above), when will they equal C.A.R.T.E.L. in sales?

Exponential Probability Distribution

The exponential probability distribution serves to approximate the failure pattern of a product. Its formula is

$$y = \lambda e^{-\lambda x}$$

where

> x is age of product
> λ is a constant reflecting product wear-out
> e is a mysterious (but important) number equaling 2.718 . . .
> y is failure rate (fraction of total product failing per time period)

This function, reflecting a "rate," cannot be fully appreciated until we study calculus. Neither can you appreciate the constant, $e = 2.718. \ldots$ For the time being, let's approximate e by 3 (a less ominous number) and plot the function to see its shape (see table and Figure 6.27). Assuming $\lambda = .5$, then

$$y = .5(3^{-.5x})$$

x	y
0	.5
1	$.5(3^{-.5}) = .29$
2	$.5(3^{-1}) = .17$
4	$.5(3^{-2}) = .06$

Figure 6.27

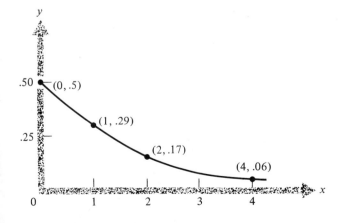

x	$T = 1 - 3^{-.5x}$
0	0
2	$\frac{2}{3}$
4	$\frac{8}{9}$

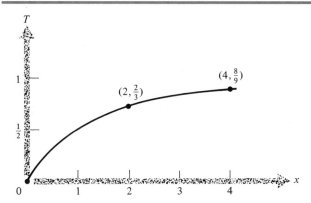

Figure 6.28

The total fraction of failures (T) from a production run after x years can be shown to be

$$T = 1 - e^{-\lambda x}$$

For $\lambda = .5$, the graph (approximate e by 3) is shown in Figure 6.28.

Exercise 30

A durable product is one with a small λ, and a "flimsy" product is one with a large λ. Sketch the cases of $\lambda = .1$ and $\lambda = 1$. For each case, how long would it take for half the production run to fail?

Normal Probability Distribution

The normal probability distribution describes many everyday phenomena and is important in the theory of statistics. As an example of its use, consider the following normal curve, which represents the distribution of IQ scores (with an average of 100 and a standard deviation of 15) in a population of 10,000 people:

$$y = 266 \cdot e^{-[(x-100)/15]^2}$$

where x is IQ score and y is frequency of occurrence (see table and Figure 6.29).

x	y
100	266
85 or 115	$\frac{266}{e} \cong 98$
70 or 130	$\frac{266}{e^4} \cong 5$
55 or 145	$\frac{266}{e^9} \cong 0$

Note: Powers of e can be found by using Table 2 in the Appendix.

ANALYTIC GEOMETRY

Figure 6.29

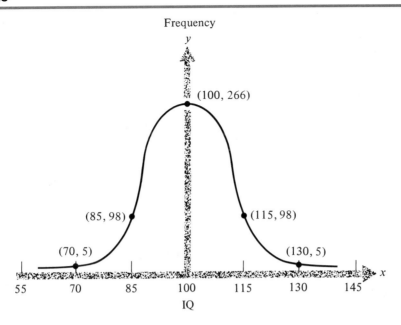

Thus, about 2.7 percent (266/10,000) of the people will score 100, the average. For scores further and further from the average, there are fewer and fewer occurrences; about 1 percent of the people score 85 or 115, while only about .05 percent of the people score 70 or 130.

Exercise 31

If the IQ test described above really had a standard deviation of 20, the following normal distribution of scores would hold:

$$y = 200 \cdot e^{-[(x-100)/20]^2}$$

Plot and compare this distribution to the previous one.

6.8 LOGARITHMIC FUNCTIONS

In section 2.9 logarithms were introduced. It might not be a bad idea to return and review their properties. Here, we shall consider some examples where a variable in the real world is a logarithmic function of another variable. Such relationships are especially important in learning situations.

In general, all logarithmic functions are of the form

$$y = f(x) = \log_b[g(x)]$$

where $g(x)$ is a function of x and b is the base of the logarithmic system.

Examples

1. $y = \log_{10} x$
2. $y = \log_2 x^3 = 3 \log_2 x$ (This result is brought to you by the third law of logarithms)
3. $y = \log_3(7x) = \log_3 7 + \log_3 x$ (Which law allows this?)
4. $y = \log_e(x^2 + x)$

We will concentrate on the case of $g(x) = kx^c$, where k and c are constants. A little algebraic manipulation reveals

$$y = \log_b(kx^c) = \log_b k + \log_b x^c$$
$$= \log_b k + c \log_b x$$
$$= a + c \log_b x$$

where the constant a is substituted for the constant $\log_b k$.

The basic shape of a logarithmic function can be discovered by plotting the case of $a = 0$ and $c = 1$, for the bases, b, of 2 and 4, as illustrated in the table and Figure 6.30.

x	$y = \log_2 x$	$y = \log_4 x$
1	0	0
2	1	$\frac{1}{2}$
4	2	1
8	3	$\frac{3}{2}$
$\frac{1}{2}$	−1	$-\frac{1}{2}$
$\frac{1}{4}$	−2	−1

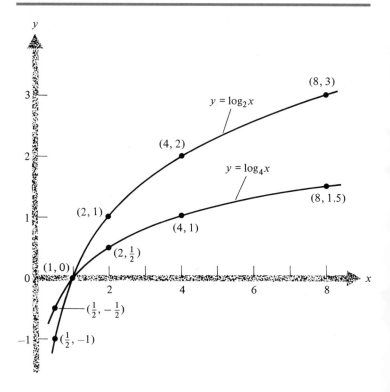

Figure 6.30

Changing the base doesn't change the general shape of the curve. However, the "slope" of the curve declines as the base gets larger.

ANALYTIC GEOMETRY

Exercises 32

1. Plot $y = \log_{10} x$ and compare it to the above functions.
2. Can you use the graph in Figure 6.30 as a basis to quickly sketch $y = \log_3 x$?

In the general equation, a serves to raise or lower the entire curve by a constant amount. It doesn't change the slope one iota. The value of c does affect the slope of the curve. These ideas are illustrated in Figure 6.31. Note that all such functions go through the point $(1, a)$.

Figure 6.31

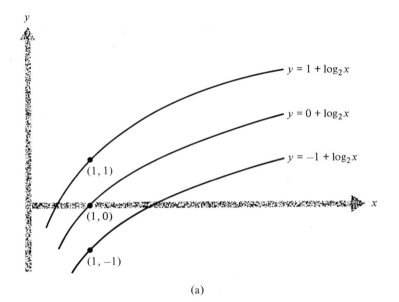

(a)

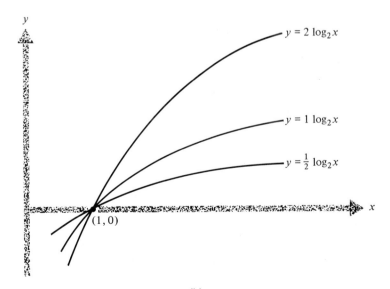

(b)

Exercises 33

1. Plot $y = a + \log_3 x$ for the cases of $a = -2$ and $+2$.
2. Plot $y = c \log_4 x$ for the cases of $c = -2$ and $+2$.
3. Plot $y = 10 + 2 \log_5 x$.

6.9 APPLICATIONS OF LOGARITHMIC FUNCTIONS

Learning Curves

Learning theory implies that the more one studies a subject, for example, this °#≠÷! mathematics, the more one will learn. However, more and more studying does lead to diminishing marginal returns. Logarithmic curves depict such a learning model. For example, consider one's grade (y) on the material in this chapter as a function of time (x) spent in studying. A function such as

$$y = f(x) = 20 + 40 \log_{10} x \quad \{1 \le x \le 100\}$$

might be reasonable to describe this relationship (see Figure 6.32).

x	$y = 20 + 40 \log_{10} x$
1	20
10	60
100	100

Figure 6.32

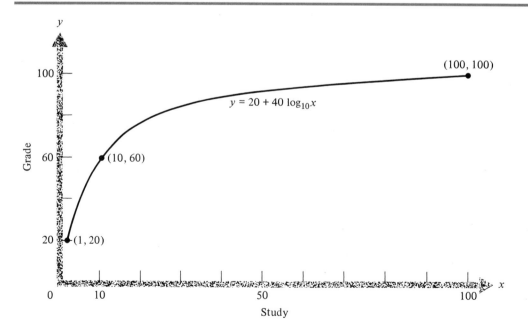

Notice that a study increase from 1 to 10 hours has the same effect (+40 points) as a study increase from 10 to 100 hours. This effect generalizes, as the following development shows. If

$$y = a + c \log_b x$$

then for any two points, (x_1, y_1) and (x_2, y_2), $y_2 - y_1$ will be

$$y_2 - y_1 = (a + c \log_b x_2) - (a + c \log_b x_1)$$
$$= c \log_b x_2 - c \log_b x_1$$
$$= c(\log_b x_2 - \log_b x_1)$$
$$= c \log_b \left(\frac{x_2}{x_1}\right)$$

The last step was brought to you by the second law of logarithms. Notice that for $(y_2 - y_1)$ to be constant (say 40 points), since c is a constant, x_2/x_1 must be a constant. In the previous case, x_2/x_1 was 10 when $(y_2 - y_1)$ was 40 points. So a study increase from 1 to 10 hours, 10 to 100 hours, 5 to 50 hours, and so on all result in a grade gain of 40 points. Likewise, studying 5 hours instead of 1 hour would have the same net grade gain as studying 25 rather than 5 hours.

Exercises 34

1. Graph the following four learning functions for $\{1 \leq x \leq 25\}$.
   a. $y = 30 + 20 \log_5 x$
   b. $y = 50 + 20 \log_5 x$
   c. $y = 10 + 40 \log_5 x$
   d. $y = 20 + 10 \log_5 x$
2. If the first equation above represents the learning curve of the average student for the work of this chapter, discuss the likely aptitude for mathematics and the prior exposure to these topics of the students whose learning curves are given by the other three equations.

Walking Pace

Marc and Helen Bornstein reported in the article, "The Pace of Life," *Nature*, February 19, 1976, the results of their research on walking pace in 15 cities of varying sizes [from Psychro, Greece (365 population) to Brooklyn, New York (2,602,000 population)]. They found the following logarithmic model fits the data best:

$$V = f(P) = .05 + .86 \log_{10} P$$

where V is walking velocity in feet per second, and P is population.

For example, the model predicts that the average walking pace in a city of 100,000 people is

$$V = f(10^5) = .05 + .86 \log_{10} 10^5 = .05 + .86(5)$$
$$= 4.35 \text{ feet/second}$$

which is equivalent to 3 miles per hour.

Considering the population domain from 100 to 10,000,000, the functional values and resulting graph in Figure 6.33 are given.

P	V (feet/second)
10^2	1.77
10^3	2.63
10^4	3.49
10^5	4.35
10^6	5.21
10^7	6.07

Figure 6.33

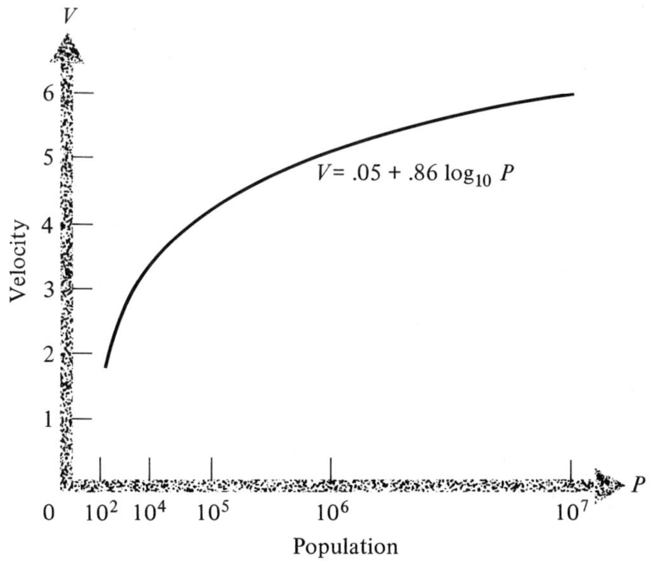

$V = .05 + .86 \log_{10} P$

Exercises 35

1. In mile-per-hour units, the walking pace function is

   $V = f(P) = .034 + .59 \log_{10} P$

   For the same set of domain points, find mile-per-hour walking velocity.
2. The P scale was distorted in Figure 6.33 in order to show all the points. The result was a typical log-shaped curve. If we plot V versus the base 10 exponent of P (e.g., if $P = 10^5$, use 5 on the P-axis), we get the so-called semilog scale.
   a. Graph the walking pace function (feet/second) in semilog form.
   b. What is the shape of your semilog graph?
   (*Note:* All logarithmic functions of the form $y = a + c \log_b x$ graph as straight lines when using semilog scales, as above.)

6.10 RELATIONSHIP OF EXPONENTIAL AND LOGARITHMIC FUNCTIONS

Exponential and logarithmic functions are closely linked mathematically, as we will now show.

If y is an exponential function of x, with base b, of the form

ANALYTIC GEOMETRY

$$y = f(x) = b^x$$

then by taking the logarithm of both sides, base b

$$\log_b y = \log_b b^x = x$$

we see that x is a logarithmic function of y.
   We can generalize further by considering the more general exponential function

$$y = f(x) = b^{g(x)}$$

Taking the logarithm of both sides reveals

$$\log_b y = \log_b b^{g(x)} = g(x)$$

so $g(x)$ is a logarithmic function of y.

Examples

1. If $y = 2^x$ then $x = \log_2 y$
2. If $y = 3^{x^2}$ then $x^2 = \log_3 y$

Exercise 36

Show that if $y = 5^{-x}$ then $x = -\log_5 y$.

6.11 MISCELLANEOUS FUNCTIONS

In the previous sections, we introduced the most important functions for management application. However, there are other functions that have importance for management. These functions—power, polynomial, and mixed—are introduced here along with some sample applications.

Power Functions

Functions of the form

$$y = x^n$$

where n is a real number are called *power functions*. We have already had occasion to know and love some special cases of them; for example, the linear function $y = x^1$, the parabola $y = x^2$, and the hyperbola $y = x^{-1}$.
   Other examples are

Figure 6.34

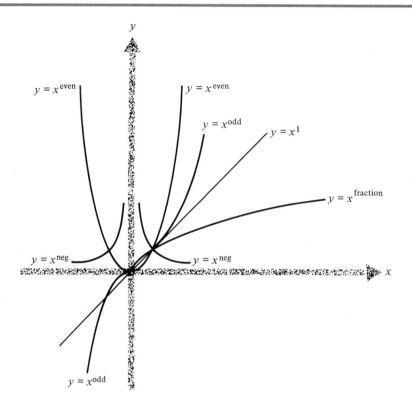

$y = x^{2/3}$ $y = x^{100}$

$y = x^5$ $y = x^{-1/2}$

$y = x^{-4.2}$ $y = x^7$

The shapes of power functions vary widely depending on the magnitude of the exponent, as illustrated in Figure 6.34.

When n is a fraction ($0 < n < 1$), the power function has a logarithmic-like shape. As the fraction approaches 1, the shape approaches that of the straight line, $y = x^1$. For even exponents ($n = 2, 4, 6, \ldots$), power functions have a parabola-like appearance. For odd exponents (3, 5, 7, $\ldots$), power functions are parabola-like, except the left side has been twisted around to be under the x-axis. For negative exponents, power functions are hyperbola-like.

Exercises 37

1. Plot power functions for the cases of $n = 4, 5, \frac{1}{2}$, and -2.
2. Plot and compare the power function $y = x^{1/2}$ to the logarithmic function $y = \log_2 x$.

ANALYTIC GEOMETRY

Polynomials

Polynomials are the sums (or differences) of power functions having natural number exponents. Examples are

$$y = f(x) = \frac{1}{3}x^3 - \frac{11}{4}x^2 + \frac{9}{2}x + 6$$

$$y = g(x) = \frac{1}{4}x^4 - x^3 + x^2$$

We will employ the first example above later in our study of calculus, so let's get to know it better (see Figure 6.35).

This polynomial has a "roller-coaster" portion between $x = 0$ and $x = 6$. To the left of this portion, it is all downhill; to the right, it is all uphill. Notice that two turning points (change from up to down or vice versa) exist. Although it is true only for certain polynomials, these "regular" types have one less turning point than the highest exponent in the equation. For example, the polynomial here has 3 as its highest exponent, so it has two turning points.

x	$f(x)$
0	6
1	$8\frac{1}{12}$
2	$6\frac{2}{3}$
3	$3\frac{3}{4}$
4	$1\frac{1}{3}$
5	$1\frac{5}{12}$
6	6

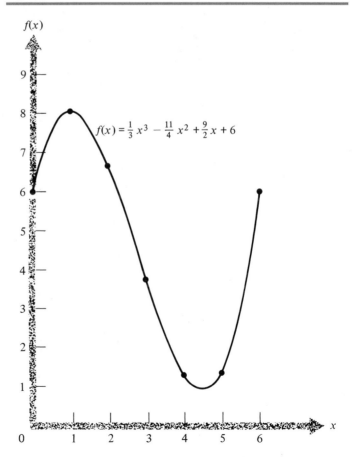

Figure 6.35

Exercise 38

Graph the polynomial $y = g(x) = \frac{1}{4}x^4 - x^3 + x^2$ for x between -1 and $+3$. Show that it has three turning points at $x = 0$, $x = 1$, and $x = 2$.

Mixed Functions

Some situations in our everyday world are so complex that a single functional type is not sufficient to represent them. In such cases a mixed functional relationship may be relevant.

Mixed functions are sums, products, or quotients of the "pure" functions; for example,

$y = f(x) = x + 8(2^{-x})$ Linear plus exponential

$y = g(x) = x^2 \log_2 x$ Parabolic times logarithmic

$y = h(x) = x^{1/2}(x^3 + x^2 + x)$ Power times polynomial

$y = i(x) = \dfrac{5 \log_{10} x}{8 - 3x}$ Logarithmic divided by linear

Mixed functions can have just about any graphical picture, depending on the component functions and the arithmetic operations involved. Knowledge of the graph of the component functions is of great help. We will illustrate in Figure 6.36 with $f(x)$, defined above.

By adding the ordinates for the two component functions, we get the graph of $f(x)$. Since $8(2^{-x})$ approaches zero as x gets large, the sum of x and $8(2^{-x})$ approaches just x as x gets large.

Exercise 39

Use your knowledge of the shape of x^2 and $\log_2 x$ to predict the graph of $g(x)$ in the above examples. For x of 1, 2, 4, 8, 16, and 32, find $g(x)$ and graph the results. How good was your prediction?

x	$8(2^{-x})$	$f(x)$
1	4	5
2	2	4
3	1	4
4	.5	4.5
5	.25	5.25

Figure 6.36

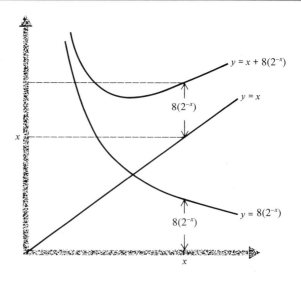

6.12 APPLICATIONS OF MISCELLANEOUS FUNCTIONS

Economic Order Quantity

In the calculus module, we will develop the EOQ (economic order quantity) model. This model shows that under certain assumptions, the order size (Q) for stocking a product should be the following power function of annual sales (S):

$$Q = f(S) = aS^{1/2}$$

where a is a constant.

Exercise 40

Consider a firm using this order size formula with $a = 4$. Graph the order size for annual sales between 100 and 10,000 units.

Volume

The volume of an object is the product of its length times its width times its height. For the special case of a cube, which has equal length, width, and height (let each dimension have magnitude of x), the volume (V) is the following power function of x (See table and Figure 6.37):

$$V = f(x) = x^3$$

Figure 6.37

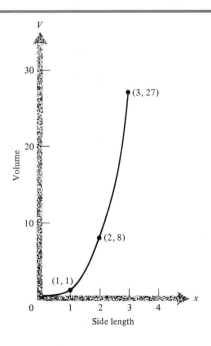

x	$V = x^3$
0	0
1	1
2	8
3	27

Exercises 41

1. What is the volume of a die whose edge dimensions are all $\frac{1}{2}$ inch?
2. Show that a die with twice the edge dimension of another has eight times the volume.
3. A box with a square base with dimensions of x inches has a height that is $\frac{1}{2}x$. What is its volume?

Revenue with Parabolic Demand

Consider a parabolic demand curve of the form

$$Q = f(P) = aP^2 + bP + c$$

where

Q is quantity sold
P is price
a, b, c are constants

Then, total revenue (R) is the following polynomial (notice that parabolas are polynomials too) function of price:

$$R = g(P) = PQ$$
$$= P(aP^2 + bP + c)$$
$$= aP^3 + bP^2 + cP$$

For example, the parabolic demand curve

$$Q = P^2 - 20P + 100$$

has a total revenue curve

$$R = PQ$$
$$= P(P^2 - 20P + 100)$$
$$= P^3 - 20P^2 + 100P$$

Exercise 42

Consider the above demand curve. The domain is $\{1 \leq P \leq K\}$. If K is the price where demand drops to zero, find K.

Government Spending

In 1973 Governor Reagan of California proposed a plan to limit state government spending and taxes. As reported in the October 16, 1973, *Wall Street Journal,* the plan would "limit state government expenditures to a declining percentage of California's personal income, starting at the current rate of 8.75% and dropping gradually [in linear fashion—*Ed.*] to 7.25% in 1989."

If we assume an annual 8 percent growth in California personal income from the $100 billion level of 1973, this plan would mean that the maximum state budget t years from 1973 (B) would be the following function of t:

$$B = f(t) = (8.75 - .1t)1.08^t$$

where B is in billions.

This mixed function (linear times exponential) graphs as shown in Figure 6.38. Note that we have just graphed the early years of such a plan. At some point, if the plan is continued indefinitely, B will reach a peak and thereafter decline. We will consider this problem more generally and with the help of calculus on page 514.

Exercises 43

1. The *Wall Street Journal* article reported "the plan would still enable the state to triple its spending over 15 years, according to the sponsors." Is this correct under the assumption of an 8 percent annual growth of personal income?
2. To see that the state budget will eventually decline, show that $f(80)$ is less than $f(70)$.

Figure 6.38

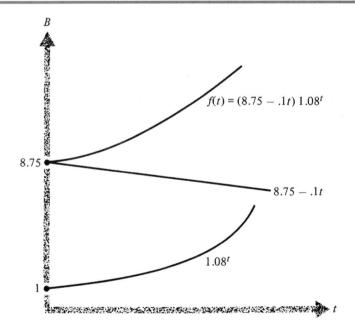

6.13 SUMMARY

In this chapter we have considered the types of equations that are most important for management applications. In each case we have developed insights into how they look graphically, and in doing so we have gained added insight into the equations themselves. What we really didn't do is to consider an everyday, practical problem or situation and try to fit an equation or equations to the problem. This is the subject of the next chapter.

PROBLEMS

TECHNICAL TRIUMPHS

1. Determine the slope (s) and intercept (i) of the following linear functions. Graph each for the domain $\{0 \leq x \leq 10\}$.
   a. $y = 9x + 7$
   b. $y = 10 - 4x$
   c. $y = 6x$
   d. $y = 300$

2. Write the equations of the linear functions which have
   a. $s = 2 \quad i = 8$
   b. $s = -5 \quad i = 20$

3. Consider the following two quadratic equations. For each, find the values of the six constants (A, B, C, D, E, F) in the general quadratic equation.
   a. $3x^2 - 2xy - 7x + 10y = 0$
   b. $6y^2 + 3x^2 + 9xy - 19 = 0$

4. Identify and graph the two quadratic equations in question #3.

5. Determine the roots of the following quadratic equations:
   a. $3x^2 + 4x - 7 = 0$
   b. $-y^2 + 5x - 9 = 0$

6. Graph the following exponential function:

   $y = f(x) = 3^{x^2} \quad \{-2 \leq x \leq +2\}$

7. Graph the following logarithmic function:

   $y = f(x) = 20 - 6 \log_{10} x \quad \{1 \leq x \leq 5\}$

8. The following compound interest formula gives one's bank balance after x years if one begins with $1000 and the bank pays 6% annually.

   $B = 1000(1.06)^x$

   a. What type of function is this?
   b. Graph the function for $\{x \leq 5\}$.

9. The following mixed function can be described as a sum of what types of functions.

ANALYTIC GEOMETRY

$$y = 5 \log_{10} x^2 + \frac{10}{x}$$

Graph it for the domain $\{1 \leq x \leq 10\}$.

10. What type of function is the following?

$$y = f(x) = 3x^3 - 8x^2 + 5x + 10$$

Graph it for the domain $\{0 \leq x \leq 3\}$.

CONFIDENCE BUILDERS

1. Fahrenheit (F) temperature is related to Centigrade (C) temperature according to function

$$F = f(c) = \frac{9}{5}C + 32$$

   a. If the lowest possible value of C is absolute zero or -273, state the domain for this function.
   b. What are the slope and intercept values?
   c. Graph the function.

2. Kelvin (K) temperature is related to Centigrade (C) temperature as follows:

$$K = g(C) = C - 273$$

   a. Is this a linear function? Why?
   b. Relate Fahrenheit (F) temperature to Kelvin (K) temperature. (You'll need help from the formula in problem #1 to do this.)

3. Snowfall (S) in inches is approximated by the following linear function of precipitation (P) in inches.

$$S = f(P) = 10P \quad \{P \geq 0\}$$

   a. How much snow would result from one inch of precipitation?
   b. If 9 inches of snow fell in a storm, how much precipitation occurred?
   c. What does the slope of 10 represent?

4. A car has a gas tank with a 10-gallon capacity. It gets 20 mile per gallon on regular unleaded gasoline. Consequently, the range (R) of miles that it can travel before needing gas is the following linear function of gasoline (G) in the tank at the beginning of a trip.

$$R = f(G) = 20G$$

   a. What is the domain?
   b. How far can the car go before needing gas if it starts with 5 gallons?
   c. If grandma's house is 75 miles away, how much gas is needed to go back and forth?
   d. If the car would get 25 miles per gallon on super unleaded gas, determine the revised range function, $R = g(G)$.

5. Okun's law on the 1975 unemployment rate (U) was described in a *New York Time's* article ("Okun's Law on Jobless," July 2, 1975) as the following function of gross national product percentage increase (G).

$$U = f(G) = 9 - \tfrac{1}{3}(G - 4)$$

a. What type of function is this?
b. What unemployment rate would result from a 10 percent increase in gross national product?
c. Graph the function.

6. An advertising executive believes that the following function represents total sales (S) as a function of the dollar expenditure in radio advertising (R) and newspaper advertising (N).

$$S = 6{,}000{,}000 + 3R + 4N$$

a. What does 6,000,000 represent?
b. Which advertising media produces a better sales return? Why?
c. How much additional sales will result from $1000 additional radio advertising?
d. If radio and newspaper advertising is budgeted at 1 and 2 million respectively, what would total sales be according to the executive's model?

7. Suppose that you are in the "Lula Loop" business. Since your product is of the fad type, all its sales will occur within the first 12 months of introduction. You forecast that the sales rate (S) in thousands of units per month over the first four months of growing sales and subsequent period of declining sales is the following function of time (t) in months since introduction.

$$S = f(t) = 4t \quad \{t \leq 4\}$$
$$S = g(t) = 24 - 2t \quad \{4 < t \leq 12\}$$

a. What type of functions are these?
b. What will the sales rate be as of $t = 1, 4, 10$ months?
c. At what time(s) will the sales rate be 6?
d. Graph the functions.

8. The total annual production cost (C) at a manufacturing plant is given by the following linear function of output (Q).

$$C = f(Q) = 2{,}500{,}000 + 1000Q$$

a. Graph this function.
b. What is the value of the intercept? What economic interpretation can you give it?
c. What is the value of the slope? What economic interpretation does it have?
d. If total output is forecasted to be 10,000 units, what would the forecast for total cost be?
e. If total cost last year was $11,000,000, how much was total output?

9. The average height of females (y) in feet is approximated by the following function of age (x) in years.

$$y = f(x) = 2 + 1.25 \log_3(x + 1) \quad \{x \leq 20\}$$

a. What type of function is this?
b. What is the average height of an eight-year-old girl?
c. At what age is the average height five feet?
d. Graph this function.

10. A mother estimated that the number of diapers per week (D) required for her baby was the following function of the baby's age (x) in years.

$$D = f(x) = \frac{100}{2^x} \quad \{x \leq 2\}$$

a. What type of function is this?
b. How many diapers per week are needed when the baby is just born? One year old? Two years old?
c. Graph the function.

ANALYTIC GEOMETRY

11. The probability distribution (y) for the number of defective light bulbs (x) in a batch of 1000 is

    $y = f(x) = .1e^{-.3(x-50)^2}$ $\{x = 0, 1, 2, \ldots, 1000\}$

    a. What type of function is this?
    b. Find $f(x)$ for $x = 46, 50, 54,$ and 58.
    c. Graph the function.

12. Sturgis' rule, which assists in the classification of statistical data, states that the number of classes (C) should be the following function of the number of observations (n).

    $C = f(n) = 1 + 3.3 \log_{10} n$

    a. What type of function is this?
    b. How many classes does Sturgis' rule suggest for 100 observations? for 1000 observations?
    c. Graph Sturgis' rule.

13. A researcher found that the average weight (W) of children in their first eight years of life was given by the following function of age (t) in years.

    $W = f(t) = 8(2^{-t}) + 30 \log_3(t + 1)$

    a. Determine the average weight of 2-year-old children.
    b. Graph the function.

14. The shelf life of milk (L) in days is approximately the following function of temperature (T) in Fahrenheit degrees at which it is stored.

    $L = f(T) = \dfrac{50}{T - 32}$ $\{T \geq 32\}$

    a. What type of function is this?
    b. What is the shelf life of milk stored at 40°? at 50°?
    c. What temperature would result in a 10-day shelf life?
    d. Graph the function.

15. Sound travels a distance (D) in feet equal to the following function of time (t) in seconds.

    $D = f(t) = 1100t$

    a. What type of function is this?
    b. If you see a lightning bolt (light travels so fast that you practically see it when it happens) and hear the thunder 5 seconds later, how far away was the bolt?
    c. If a lightning bolt hits 500 feet from you, how long will it take to hear the thunder?
    d. Graph $f(t)$.

16. The cost of a one-way domestic airplane ticket (C) in 2003 was approximately the following function of distance (x) in hundreds of miles.

    $C = f(x) = 100 + 35 \log_2 x$ $\{x \geq 2\}$

    a. What type of function is this?
    b. How much would a one-way ticket from Washington, D.C., to Chicago (approximately 800 miles) cost?
    c. How far could you travel (back and forth) for $300?
    d. Graph the function.

17. The average cost of a funeral (F) is forecasted to be the following function of time (t) in years after 2004.

    $F = f(t) = 6000(1.05)^t$

    a. What was the average funeral cost in 2004?
    b. What type of function is this?
    c. If you have a 50-year life expectancy in 2004, what would your funeral cost?
    d. When would the average funeral cost $15,000?

18. An individual, attempting to add to her vocabulary, noticed that during the first part of a session she learned quickly, but as the session wore on she learned less quickly. After a statistical study of the number of new words learned (W) and the time of the learning session (t) in minutes, the following function was found to relate the variables.

    $W = f(t) = 10 \log_{10} 2t \quad \{1 \leq t \leq 100\}$

    a. What type of function is this?
    b. How many new words could be learned in a five-minute session? in a 50-minute session?
    c. What length of session would be required to learn 15 new words?
    d. Graph the function.

19. The volume of a sphere (V) is the following function of the radius (r):

    $V = f(r) = \frac{4}{3}\pi r^3$ where π is 3.1417

    a. What type of function is this?
    b. What is the volume of a sphere with a radius of 2 feet?
    c. Graph the function.

20. Since the interest on municipal bonds is not taxable, their effective yields are higher. Effective yield (y) is computed by dividing the municipal bond interest rate by a factor that depends on an individual's tax bracket (x). For the case of a 6 percent municipal bond, the effective yield (in percent) is the following function of x (expressed in decimal form):

    $y = f(x) = \dfrac{6}{1-x}$

    a. What type of function is this?
    b. What is the effective yield for a person in the 40 percent tax bracket? in the 20 percent bracket?
    c. What income tax bracket would a person have to be in to have a 9 percent effective yield?
    d. Graph the function.

MIND STRETCHERS

1. The Big Bend Pretzel Company produces two types of pretzels (A and B). Production per day is given by the equation

   $2x^2 + y^2 + 4x + 3y - 70 = 0$

   where x is units of A produced, and y is units of B produced.
   a. Identify the type of equation.
   b. How many units of A can be produced if no B is produced?

ANALYTIC GEOMETRY

c. How many units of B can be produced if no A is produced?
d. If three units of A are produced, how many units of B can be produced?
e. Plot the portion of this equation achievable in the real world.

2. You must display T cans of beer in triangular display using n layers as shown in the figure.

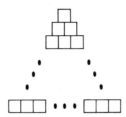

a. Find $T = f(n)$. (*Hint:* Arithmetic progression)
b. What type of function is $f(n)$?
c. Solve for n in terms of T.
d. If $T = 210$, how many layers are needed?

3. Slick Industries and C.A.R.T.E.L are battling it out for control of the grease and palm oil business (they grease the palms of politicians). C.A.R.T.E.L. currently has 80 percent of the market, but Slick's tough new advertising campaign ("We can be persuaded to forget that secretary who has been on your payroll for five years without ever setting foot in the office.") promises to change that. Thus, industry experts forecast the sales for the two companies to be the following functions of time (t) in years from now.

$S_s = f(t) = 20(1.3)^t$

$S_c = g(t) = 80 - 10 \log_{10}(t + 1)$

a. Graph the two functions. What type of function is each?
b. From the graph approximate the time it will take for Slick's sales to catch up to C.A.R.T.E.L.'s.

4. After two games for the School of Management "Capitalists" softball team, the author was batting 3 for 6, or .500. At that point, I vowed to go on a hitting streak and get x consecutive hits. If this happens, my batting average (B) would be the following function of x.

$B = f(x) = \dfrac{3 + x}{6 + x}$

a. What will the batting average be if $x = 4$?
b. What type of function is $f(x)$?
c. Graph $f(x)$ for $\{1 \leq x \leq 9\}$.
The other players on the team were convinced that the author would go on an out streak and thus make x straight outs.
d. Determine my batting average function of x, or $g(x)$ if that very very unlikely event happens.
e. What type of function is $g(x)$?
f. Graph $g(x)$ for $\{1 \leq x \leq 9\}$.

5. Having failed mathematics, Joe NoMath decided to join the army and was sent to the front to take Hill 367, a very strategic spot. Joe's Intelligence Officer told him that he had the following two alternatives, both of which would enable the capture of Hill 367.
Commit half the troops
Commit all the troops

Joe wanted to pick the alternative that would minimize deaths among his men. After a great deal of thought, however, Joe was able to estimate the number of deaths (D) of his men for the two alternatives as the following functions of enemy soldiers (E) in thousands thrown into battle.

Commit half: $D = f(E) = .1 + .05E + .01E^2$

Commit all: $D = g(E) = .2 + .04E + .005E^2$

  a. What types of functions are $f(E)$ and $g(E)$?
  b. Graph the functions (on a single graph) for $\{1 \leq E \leq 10\}$.
  c. Develop a rule to decide which alternative to choose once the value of E is estimated.

6. The cost (C) in dollars of heating a certain home on a day with an average temperature of 20 degrees Fahrenheit is given by the following function of desired inside temperature (T).

$$C = f(T) = .00002(T - 20)^3 \quad \{T > 20\}$$

  a. What type of function is this?
  b. How much more would it cost to keep a 70 degrees rather than a 65 degrees inside temperature?
  c. Graph the function.

7. A real estate investor thinks that the apartment building just purchased for $200,000 will increase in value $40,000 each year. If this is true, and the investor sells the building x years from now, the present value (PV) of the sales price for a 10 percent discount rate is given by the function

$$PV = f(x) = (200{,}000 + 40{,}000x)(1.1)^{-x}$$

  a. What type of function is this?
  b. Find $f(1)$ and $f(5)$.
  c. Graph the function for $x \leq 10$. Does the graph indicate that there is a best time to sell?

8. The world population (approximately 6 billion in 2003) is growing at an annual rate of 1.9 percent. If this rate continues, the world population (P) as a function of time (t) in years after 2003 will be

$$P = f(t) = 6(1.019)^t$$

  a. What type of function is this?
  b. How long will it take for the population to double the 2003 level?
  c. When will the population reach 20 billion?
     If after ten years the population growth rate drops to 1 percent per year so that the function

$$P = g(t) = f(10) \cdot (1.01)^{t-10}$$

holds thereafter,
  d. When will the population double the 2003 level?
  e. When will the population reach 20 billion?

9. A company with two large discount stores in a region (A and B in the diagram) wishes to locate the district office at a point equally distant from each. The set of such points, illustrated by point P in the diagram, can be found by employing the Pythagorean theorem as follows:

$d_1 = d_2$

so

$d_1^2 = d_2^2$

ANALYTIC GEOMETRY

thus

$$(x - 2)^2 + (y - 5)^2 = (x - 11)^2 + (y - 10)^2$$

a. Simplify the above equation to find the equation for the set of equally distant points.
b. What type of equation did you find?
c. If the company wished to be 10 miles from each store, where should they locate?
d. If they want to locate the district office on a major highway on the line $x = 7$, what point should they choose?

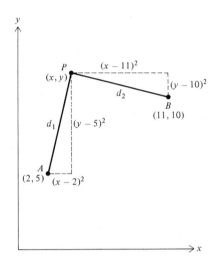

10. A beer distributor who innovated with discount prices and self-service experienced big sales gains. Sales receded from their peak when competition appeared. However, when the innovator brought his son into the business, new sales gains were recorded. This pattern of sales (S) in thousands of cases per year over the ten years of the business is captured by the following functions of time (t) in years since the innovation.

$S = f(t) = 10(1.5)^t$ $\{t = 1, 2, 3, 4\}$

$S = g(t) = 70 - 5t$ $\{t = 5, 6, 7\}$

$S = h(t) = t^3 - 21t^2 + 147t - 308$ $\{t = 8, 9, 10\}$

a. Describe the nature of the three functions.
b. Graph the sales functions.

11. A student's cumulative grade point average (y) is a weighted average of the prior semester's cumulative average (A) and the current semester's average (x), as defined by the formula,

$$y = f(x) = \frac{CA + cx}{C + c}$$

where C is the prior semester's credits, and c is the current semester's credits.
a. For the case of a beginning sophomore with $C = 30$, $A = 2.0$, and $c = 15$, find $y = f(x)$.
b. What type of function is it?

c. If the student must have a 2.5 cumulative average at the end of the semester in order to transfer to a different school, what is the minimum semester average that he or she must achieve?
d. Graph $f(x)$.
e. Consider a second-semester senior who has completed 105 hours with a 1.9 cumulative average. Determine his or her final cumulative average as a function of the final semester's average.
f. What must his or her final semester's average be in order to graduate (2.0 cumulative average required for graduation)?
g. Consider your own case. Determine your own new cumulative average as a function of your current semester's average.
h. How well would you have to do this semester to raise your cumulative average by .2?

12. Phil Z. Calfitness is in training to break the world's record for the most push-ups in 10 minutes. Typically he starts strong, tires, then gets his second wind and winds up in a flourish. Mathematically, his push-up rate (y) is the following function of time (t) in minutes.

$y = 60(3^{-2t})$ \qquad $\{0 \leq t \leq 5\}$

$y = 20 + 30 \log_2(t - 4)$ \qquad $\{5 < t \leq 10\}$

a. Graph the function.
b. Will his second wind ever allow him to match his initial ($t = 0$) rate? If so, when?

13. After 100 times at bat, a baseball player's slugging performance (S) can be determined from the following function of his home runs (H), triples (T), doubles (D), and singles (S):

$$S = f(H, T, D, S) = \frac{4H + 3T + 2D + 1S}{100}$$

a. What type of function is this?
b. If Darryl Rasberry had 5 home runs, 2 triples, 8 doubles, and 15 singles, what would his slugging percentage be?
c. Holding everything else constant, by how much does S change for every home run? Does that make baseball sense?
d. (Sports fans only) Describe what S is really measuring.

14. George Blanda was a versatile football player. He played quarterback and also did all the kicking for Oakland. Thus, he could score 6-point touchdowns (T), 3-point field goals (F), and 1-point extra points (E). Determine his total points (P) as a multivariate function of those variables.

CHAPTER 7

Curve Fitting

7.1 INTRODUCTION

Mathematical
Model Agency
Executive: I lined up an outstanding set of curves for your selection of "Playmath of the Month."
Hugh Heffer: Each is a picture in itself. What a range of beauty.
Exec: I knew full well at the origin that it would not be a straightforward decision.
Hugh: You are right. This will require more than a sketchy look. I must make extensive observations to shape this decision.
Exec: It figures that you will have to coordinate all your power to choose.
Hugh: Do you think that I could plot a relation with the winner in my domain?
Exec: No! Intercept such thoughts. They pointed out that they consider it radical to serve such a function.
Hugh: Rats!

Real world curve fitting situations can't compare to Hugh's in certain respects. Nevertheless, our little introduction does illustrate the stages of curve fitting as a management tool. Hugh began with an understanding of what are ideal centerfold dimensions. He then sorted different models until finding the one with the right kind of curves. By setting the parameters and the domain, he obtained the ideal specific model, which he can employ to good advantage.

In the usual (admittedly duller) curve-fitting situation, we begin with a general understanding of the relationship between variables. Many general types of equations could reflect this relationship, with each type working best under a given set of practical conditions. Then, with empirical data, this model's parameters and domain can be specified and employed.

We will illustrate this process with the case of S. Lumlord. He has 600 identical apartments for rent. At $200 rent per apartment all 600 were rented. Recently, he raised the rent to $220 and observed a lot of grumbling and finally an occupancy of 560 apartments. In sum, he observed two paired observations—(rent, apartments) = (200, 600) and (220, 560) (see Figure 7.1)—from an unknown functional relationship. Fortunately, the relationship is not comple-

Figure 7.1

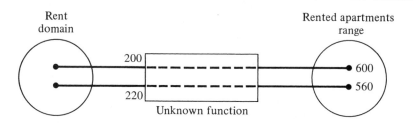

Figure 7.2

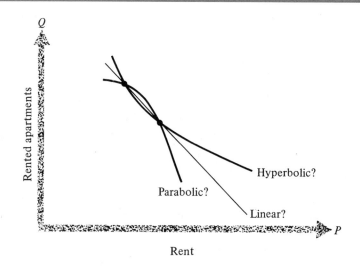

ly unknown. Lumlord does know that increases in rent result in decreases in occupancy (other things being equal). Yet, many mathematical functions are consistent with this tendency and the two observed points (see Figure 7.2). Only an understanding of actual rental market conditions, as they translate into slopes over the domain, allows for choosing a realistic mathematical model.

Let's pause to investigate this slope concept.

Basics of Slope

In chapter 6, we were introduced to the concept of slope. There we saw that straight lines (linear functions) have constant slope and curving lines have varying slopes.

The concept of slope of a curve will be studied extensively when we reach calculus. However, a rudimentary understanding of slope is useful here for curve fitting purposes. Mathematical slope is basically the same concept that anyone would use while hiking through the countryside (see Figure 7.3).

CURVE FITTING

Figure 7.3

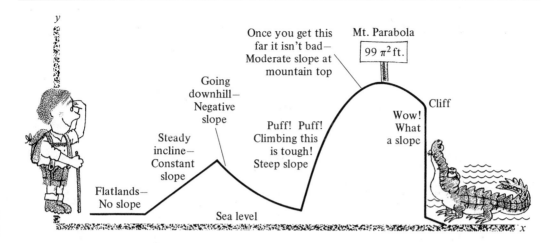

The slope s between two points on a curve is measured by the ratio

$$s = \frac{\text{change in } y}{\text{change in } x} = \frac{\Delta y}{\Delta x}$$

Let's see how an understanding of slope can aid in curve fitting. If S. Lumlord believes that for every $10 increase in rent the occupancy loss will remain the same (constant slope), then he thinks a linear model is most appropriate. If he believes that an initial $10 increase in rent will result in a large occupancy loss (high negative slope) with subsequent $10 rent increases resulting in smaller occupancy losses (low negative slope), then he thinks he has a hyperbolic curve. If Lumlord thinks that an initial $10 increase in rent would have little effect because of tenant inertia (low negative slope), but that subsequent increases would cause a back-up of moving vans (high negative slope), then he sees a parabolic curve. Figure 7.4 demonstrates these cases.

Figure 7.4

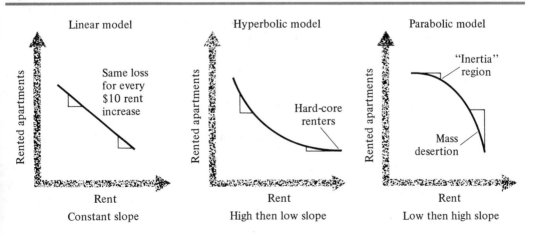

Learning Plan

In this chapter we will develop methods to employ mathematical functions (models) to capture relationships between actual variables.

This chapter presents a contrast to the previous chapter upon which it is built. Chapter 6 presented the concept, "Here is a function. See how it graphs for different parameter settings." Chapter 7 says, "Here is the real world. Since I want to understand it better by relating certain variables, I had better know quite a lot about the individual variables and their joint tendencies, and I probably need a graphical picture of their movements. Also, I must know the various models of the previous chapter so that I can capture my understanding of the real world with a good mathematical model." The following stages of learning are employed to meet our goals.

1. Define major mathematical functions useful in curve fitting, and review the graphical shapes and slopes of each.
2. Define the role of parameters, and show how their values uniquely specify a function.
3. Show how observations of the world allow parameter and domain specification.
4. Develop applications of curve fitting from various areas of life, with special attention given to demand, revenue, cost, and profit curves.

We will be slowed a bit in this chapter because we have not yet learned calculus or efficient methods to solve systems of equations. But the extra effort required here should provide a better appreciation for the advanced topics when they appear later. (Isn't it a fact that we can't appreciate something unless we recall how difficult life was without it?)

7.2 CURVE-FITTING MODELS

In this section, we define five functions that are useful in curve-fitting applications. These five functions—linear, hyperbola, parabola, exponential, and logarithmic—are mathematically specified and graphed in Figure 7.5 These do not comprise all the useful curve-fitting models; however, they are the simplest, and they do reflect the most important joint relationships among variables. More complicated models can sometimes improve on the fit of a variable relationship. In such cases, the methods employed here for the basic models can be similarly applied with the more complicated models.

Parameters

The constants a, b, c, s and i in the curve-fitting models in Figure 7.5 are called *parameters*. Their values don't affect the general shape of the curve—it is the operations performed on x (squaring, taking the logarithm, and so on) that do. Rather, the parameters serve to differentiate curves of the same type. For example, the values of s and i in the linear function don't change its basic shape (straight as an arrow), but serve to locate that arrow and its direction. We need as many observation points as parameters to fit a model. For example, two points determine a unique linear function; however, we need three points to determine a unique three-parameter parabola.

CURVE FITTING

Figure 7.5 Basic curve-fitting models.

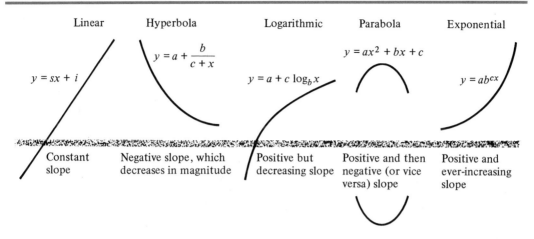

We'll illustrate with the S. Lumlord problem on page 207. Recall that a rent of $200 resulted in all 600 apartments being rented, while a rent of $220 resulted in 560 apartments being rented.

First, assume a linear model. Each point provides an equation to solve for the linear function parameters, s and i.

$$Q = f(P) = sP + i$$

where

Q is number of apartments rented
P is rent per apartment

Thus

$$f(200) = 600 = s \cdot 200 + i$$

$$f(220) = 560 = s \cdot 220 + i$$

With two equations and two unknowns, we can solve simultaneously . . . Whoa!

You may be asking, "How does one solve simultaneously?" I suggest you go to section 2.8 where simultaneous equations are solved using the substitution method. But to solidify your understanding, let's illustrate the substitution method with this case.

The method involves solving for one variable in terms of the other variable in one equation. So

$$i = 600 - 200s \quad \text{from equation 1}$$

Then the method involves substituting this equation for i into the other equation, or

$$560 = 220s + (600 - 200s)$$

After some simple algebra, we find

$s = -2$

Then, substituting this value of s in the equation for i gives

$i = 600 - 200(-2) = 1000$

And finally, we have the fitted function

$Q = f(P) = 1000 - 2P$

There are other (even better) methods to solve simultaneous equations. We will develop them in Chapter 9. But for now the problems in this chapter can be handled very nicely by the substitution method.

If we assume that a hyperbolic model of the form

$$Q = f(P) = \frac{a}{b + P}$$

best fits the situation, we can use the two points of information to specify its parameters, a and b, as follows:

$$f(200) = 600 = \frac{a}{b + 200}$$

$$f(220) = 560 = \frac{a}{b + 220}$$

After some algebraic manipulation, we arrive at the system of equations

$a = 600b + 120{,}000$

$a = 560b + 123{,}000$

which, if solved simultaneously, gives $a = 168{,}000$ and $b = 80$. So the specified hyperbolic function is

$$Q = f(P) = \frac{168{,}000}{80 + P}$$

The linear and hyperbolic models going in different directions through the two known points are shown in Figure 7.6.

Exercise 1

a. Determine the parameter values for the parabola, of the form $Q = f(P) = aP^2 + c$, which goes through the two points (200, 600) and (220, 560).
b. Graph this parabolic model and the above linear and hyperbolic models. Compare.

CURVE FITTING

Figure 7.6

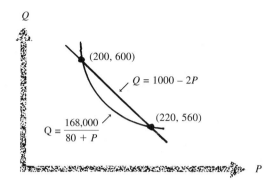

We have just witnessed that two observation points allow for the specification of a unique two-parameter linear function, or a unique two-parameter hyperbola, or a unique two-parameter parabola. This was possible because, in each case, we were able to generate two equations with two unknowns. Now let's see what happens if we try to fit the three-parameter parabola

$$Q = f(P) = aP^2 + bP + c$$

to the observations. Each observation provides one equation:

$$600 = f(200) = a \cdot 200^2 + b \cdot 200 + c$$
$$560 = f(220) = a \cdot 220^2 + b \cdot 220 + c$$

Solving for c in the first equation and substituting in the second equation give

$$-40 = 8{,}400a + 20b$$

Now b can be solved for in terms of a,

$$b = -420a - 2$$

but it can't be determined uniquely. And, consequently, the value of c can't be determined uniquely. Thus, there are many sets of parameters (a, b, c) that "work" for the two points. In other words, there are many three-parameter parabolas that go through the two points, as illustrated in Figure 7.7.

Let's see how a third point will allow us to identify the real parabola. Suppose that a rent of $300 is forecast to result in 200 rented apartments.

(P, Q)	Equation
(300, 200)	$200 = 90{,}000a + 300b + c$
(200, 600)	$600 = 40{,}000a + 200b + c$
(220, 560)	$560 = 48{,}400a + 220b + c$

Figure 7.7

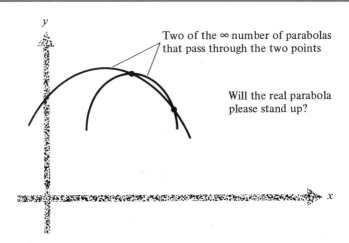

This system of equations[1] can be solved best by methods developed in Chapter 9. However, the method of substitution, which we already know, will work. In the first equation solving for c in terms of b and a gives

$c = 200 - 90{,}000a - 300b$

Substituting this for c in the other two equations yields the system of two equations

$400 = -50{,}000a - 100b$

$360 = -41{,}600a - 80b$

that has the solution $a = .004167$ and $b = -6.0835$. Substituting these values in the first equation and solving for c give $c = 1{,}650$. Thus, the fitted parabola is

$Q = .004167P^2 - 6.0835P + 1{,}650$

In general, k distinct observation points allow for determination of a unique k parameter curve. But what happens if we have more than k points?

Figure 7.8 shows that three points determine not one, but three different two-parameter straight lines. They would determine three different two-parameter hyperbolic functions too. (Why would they determine only one three-parameter parabola?)

So, in fitting a k parameter curve, if we have more than k points, we can do the following:

1. give up!
2. use only k representative points
3. fit the "best-fitting" line

The first alternative is abhorrent in light of your zeal for learning math. Isn't it? Oh well, suppose it is anyway and let's go on to the second alternative.

---

[1]Notice that these equations are linear in a, b, c despite the fact that the original function is quadratic.

CURVE FITTING

Figure 7.8

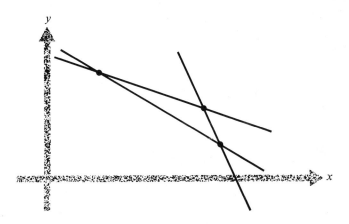

Figure 7.9

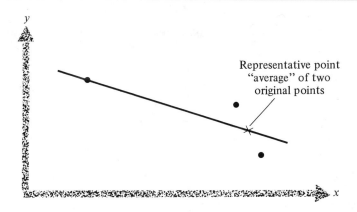

Representative point "average" of two original points

In the graphical example of Figure 7.8, a point midway between the two points farthest to the right could be taken as representative of them. Then a unique straight line could be fitted to the two representative points, as shown in Figure 7.9.

To illustrate this approach with an actual problem consider a baby who is used to being rocked to sleep. Suddenly, the parents say, "This is not the way it should be!" and the baby is left to cry itself to sleep. The time required to fall asleep without rocking for the next several days is given and graphed in Figure 7.10.

Using the two parameter hyperbolic model

$$y = \frac{a}{b + x}$$

we need two representative points to fit the line. If days 1 and 7 are considered representative, then substituting $x = 1$ and $x = 7$ into the hyperbolic model and simplifying give the following system of two equations:

Figure 7.10

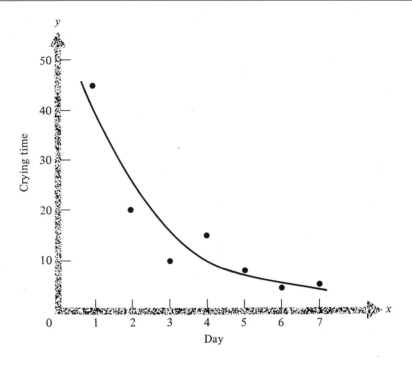

$$45 = \frac{a}{b+1}$$

$$5 = \frac{a}{b+7}$$

Solving these equations simultaneously gives $b = 33.75$ and $c = -.25$, so the fitted model is

$$y = f(x) = \frac{33.75}{-.25 + x}$$

Exercises 2

1. If the fitted hyperbolic model continues to hold after the first week, how long will it take before the baby falls asleep in 2 minutes?
2. Fit an exponential function of the form $y = a \cdot 2^{bx}$ through the two representative points $(1, 45)$ and $(7, 5)$.

As for the third alternative, we are not ready for a full understanding. Finding the "best-fitting" line for a set of data with more (sometimes many more) points than parameters requires calculus. In addition, matrix algebra is useful in setting up the equations needed for solution. You can take an advanced peak at Chapters 8, 9, 10, and 15 where these methods are fully developed. But in the meanwhile, to give you a bare understanding here, consider the following

CURVE FITTING

data set which describes x, the amount of study of 8 students and their resulting grade, y, out of 25 points on a quiz.

	Observations		Supplementary calculations	
Student	x Hours Study	y Grade	x^2	xy
1	2	16	4	$2 \cdot 16 = 32$
2	1	9	1	$1 \cdot 9 = 9$
3	3	17	9	$3 \cdot 17 = 51$
4	1	12	1	$1 \cdot 12 = 12$
5	4	22	16	$4 \cdot 22 = 88$
6	2	13	4	$2 \cdot 13 = 26$
7	1	8	1	$1 \cdot 8 = 8$
8	2	15	4	$2 \cdot 15 = 30$
	$\Sigma x = 16$	$\Sigma y = 112$	$\Sigma x^2 = 40$	$\Sigma xy = 256$

Let's find the best linear function for this data.

Use of calculus gives the following two equations to solve for the best fitting linear function

$$ni + (\Sigma x)s = \Sigma y$$

$$(\Sigma x)i + (\Sigma x^2)s = \Sigma xy$$

where

n is the number of observations
Σx is the sum of the x values
Σy is the sum of the y values
Σx^2 is the sum of the squares of the x values
Σxy is the sum of the products of x times y

Notice how various sums are needed. Carrying out these sums and substituting in the above equations yields

$$8i + 16s = 112$$
$$16i + 40s = 256$$

which if solved simultaneously gives us the best-fitting, or so-called "least-squares" line (See Figure 7.11).

$$y = f(x) = 6 + 4x$$

This line is such that the sum of all the deviations (distance from the point to the line) squared is a minimum—and thus the name "least squares."

We will be back to this problem in later chapters once we get the necessary foundation.

Figure 7.11

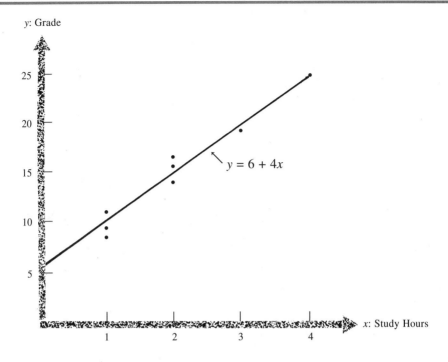

Exercise 3

Calculate the sum of the squared deviations for all the points. For example, the deviation shown in the figure is $12 - (4 + 6(1)) = 2$. So that squared deviation is $2^2 = 4$. Note that calculus insures that the sum of the squared deviations for this line is smaller than for any other possible line.

Choosing the Best Model

We have seen that several models will go through the same set of observation points. But which model is best? The answer to this question depends on how well the model fits the entire set of points inherent in the real relationship.

To choose the best overall curve, we must have an understanding of the actual relationship over the entire domain and match that pattern to the various patterns embodied in the curve-fitting models.

In the next several sections, we will see how real world slope and change patterns among variables lead to specific models. The slope patterns and the appropriate model that captures those patterns that are discussed here are:

CURVE FITTING

1. constant slope—linear model
2. constant percentage change—exponential model
3. slope increasing at decreasing rate—logarithmic model
4. negative slope with lower bound—hyperbolic model
5. reversing slope—parabolic model

7.3 CONSTANT SLOPE—LINEAR MODEL

The hallmark of linear functions is their constant slope. No other function can make that claim.

First, let's review by showing that the assumption of constant slope leads to the linear model

$$\frac{\Delta y}{\Delta x} = \text{constant} = s$$

$$\Delta y = s(\Delta x)$$

$$y - y_1 = s(x - x_1)$$

If $x_1 = 0_1$ then $y_1 = i$. This leads to the explicit formulation

$$y - i = s(x - 0)$$

$$y = sx + i$$

The linear function is simplest to fit and understand. Also, it can serve as a compromise fitting function, since it is easy to assume constant slope when little is known about the real slope pattern. In some cases, several linear relationships over limited domains can be employed, even when one overall linear function isn't justified.

The ease and advantages of fitting a linear function cause it to be used more than is warranted in our actual world. One must be cautious that its single assumption—constant slope, holds.

Whenever you encounter a situation where the change in y is *always* the same *for any* change in x, the linear light should blink in your brain. Statements such as

1. Insurance pays the same amount *for every* dollar of medical cost
2. I pay the same additional sales tax *for every* dollar of my restaurant bill.
3. *Each dollar increase* in the price of a barrel of crude oil results in the *same decrease* in demand.

all indicate constant slope.

Exercises 4

Consider the following real world situations. Discuss why or why not a linear function can model it.
a. A salesperson makes 10% commission on all sales.

b. It takes more heat to raise the home temperature from 75 to 76 degrees than it does from 65 to 66 degrees.
c. The amount depreciated on an asset is the same every year.
d. Interest in a bank is compounded quarterly at a 5% annual rate.
e. A runner moves quickly at first, but tires near the finish.
d. One's ideal weight increases the same amount for every additional inch of height.

Now let's fit some linear functions

Medical Bills

A health insurance plan has a $50 deductible clause, and pays 80 percent ($.80 per dollar) of the bill above $50. If we define

y = insurance company payment

x = medical bill

then the $.80 payment *for each* dollar above $50 translates into constant slope of

$$\frac{\Delta y}{\Delta x} = .80$$

Since $y_1 = 0$ when $x_1 = 50$, we can use the point-slope formula to get

$y - 0 = .80(x - 50)$

$\quad y = .80x - 40 \quad \{x > 50\}$

Exercises 5

1. If your allergist, Dr. Paul Linn, just sent you a bill for $200, how much will the insurance company pay?
2. Develop the linear function that gives your payment for an $x medical bill ($x > 50$).

U.S. Crude Oil Production

U.S. Crude oil production is greatly affected by the prices set by OPEC. Since U.S. oil production is fairly costly, as compared to OPEC, some U.S. wells will become economically viable as world prices, dictated by OPEC, increase.

Recent studies have shown that U.S. producers will pump 8 million barrels per day if world crude oil prices are $15 per barrel. However, for every dollar increase in price, .3 million additional barrels will be produced in the U.S.

The phrase "for every dollar increase" means that slope is constant, and thus a linear function with slope.

CURVE FITTING

$$s = \frac{\Delta Q}{\Delta P} = \frac{.3}{1} = .3$$

where

Q = millions of barrels per day
P = price per barrel

holds. Since we know the point $(Q, P) = (8, 15)$, we start by using the point-slope formula

$Q - Q_1 = s(P - P_1)$

$Q - 8 = .3(P - 15)$

or $Q = 3.5 + .3P$

Exercises 6

a. What would U.S. production be if the price rose to $30 per barrel.
b. If for every dollar increase in price, an additional .5 million barrels were produced, determine the revised linear function.

Internal Revenue Service Blues

Each year, the Internal Revenue Service lays a segmented linear function on us for determining our federal tax. Oh happy days! In 2001 single tax-payers were confronted with following 5 segment linear function shown in Figure 7.12.

Let's focus on the lowest two income categories. (It is too depressing to go any further.) Notice that for a given segment, you pay the same amount of additional tax for every addition-

Figure 7.12 2001 Tax Rate Schedules

2001 Tax Rate Schedules

Schedule X—Use if your filing status is **Single**

If the amount on Form 1040, line 39, is: Over—	But not over—	Enter on Form 1040, line 40	of the amount over—
$0	$27,050	.... 15%	$0
27,050	65,550	$4,057.50 + 27.5%	27,050
65,550	136,750	14,645.00 + 30.5%	65,550
136,750	297,350	36,361.00 + 35.5%	136,750
297,350	....	93,374.00 + 39.1%	297,350

al dollar of income. For example, in the lowest interval, you pay $.15 *for every* additional dollar of taxable income.
Defining

y as the tax due

x as the taxable income

In the first segment, the line goes through the point $(x, y) = (0, 0)$ with slope of .15. So with the intercept, $i = 0$

$$y = sx + i = .15x$$

In the second taxable income category, the line goes through the point (27,050, 4,057.50) with a slope of .275. Using the point-slope formula

$$y - y_1 = s(x - x_1)$$
$$y - 4{,}057.50 = .275(x - 27{,}050)$$

which translates to

$$y = .275x - 3{,}381.25 \quad \{27{,}050 \leq x \leq 65{,}550\}$$

Exercise 7

Not successful? Want to get back at all those rich people? Compute the segmented linear function for the highest income bracket.

7.4 CONSTANT PERCENTAGE CHANGE— EXPONENTIAL MODEL

If the y variable changes by a constant percentage of its latest value for a unit change in x (as is the case for many applications including compound interest, sales for a growth company, and so on), then, as we show here, the exponential model is the appropriate one.

Mathematically, a constant percentage change means that

$$\frac{\Delta y}{\Delta x} = ky \quad \text{becomes} \quad \Delta y = ky$$

where k is a constant and $\Delta x = 1$. If y_0 is the initial value of the variable, then

$$y_1 - y_0 = ky_0$$

where y_1 is the value of y when x is 1. Thus

CURVE FITTING

$$y_1 = y_0 + ky_0$$
$$= y_0(1 + k)$$

Now letting x increase by 1

$$y_2 - y_1 = ky_1$$
$$y_2 = y_1 + ky_1$$
$$= y_1(1 + k)$$
$$= y_0(1 + k)(1 + k)$$
$$= y_0(1 + k)^2$$

During the next period

$$y_3 - y_2 = ky_2$$
$$y_3 = y_2 + ky_2$$
$$= y_2(1 + k)$$
$$= y_0(1 + k)^2(1 + k)$$
$$= y_0(1 + k)^3$$

Exercise 8

Show that $y_4 = y_0(1 + k)^4$

Can you see the pattern? If not, review the development of the compound interest formula in Chapter 4—this is the same pattern, and it generalizes for any period, x, as

$$y_x = y_0(1 + k)^x$$

This is an exponential function of the form described in Chapter 6, with the correspondence of symbols being as shown:

Chapter 6 version: $y = ab^{g(x)}$

Current version: $y = y_0(1 + k)^x$

Since in our applications of the exponential model the x variable is often time, we frequently replace it with the letter t.

To fit an exponential function of this type, we only need to know the initial value of $y(y_0)$ and the constant percentage change (k) expressed as a decimal. Let's illustrate with a few applications.

Growth Companies

Stock market investors are perpetually searching for "growth companies," or those whose sales and earnings grow at a high percentage rate per year.

If C.A.R.T.E.L.'s sales (currently $15 billion) are expected to increase 15 percent per year, and their earnings (currently $2 billion) are expected to increase 10 percent per year, then the respective sales (S) and earnings (E) functions are

$S = 15(1 + .15)^t = 15(1.15)^t$
$E = 2(1 + .10)^t = 2(1.1)^t$ t is time in years from now

Exercises 9

1. If Gangland Industries' sales (currently $5 billion) are expected to increase 20 percent per year, and earnings (currently $1 billion) are expected to increase 17 percent per year, fit the respective sales and earnings functions.
2. Because of their convincing sales staff, Gangland Industries' sales and earnings growth rates are expected to lead C.A.R.T.E.L.'s indefinitely. Using the growth rates determined above, find out when Gangland's sales will catch up to C.A.R.T.E.L.'s. Do the same for earnings.

Population Growth

The world population was approximately 6 billion in 2003 and growing at about 2 percent per year. Assuming a continuation of this growth rate, the future population will be

$P = 6(1 + .02)^t$ where t is time in years after 2003

At this rate, the population will double in

$12 = 6(1.02)^t$

$2 = 1.02^t$

$t = \log_{1.02} 2 = 35$ years

Exercise 10

If the population growth rate drops to 1 percent per year, fit the population function and determine how long it would take for the population to double.

Car Depreciation

The resale value of a car declines about 25 percent per year. So, the first year's loss of value (Δy) is

$\Delta y = .25C$ where C is the initial cost

CURVE FITTING

For example, a $20,000 car depreciates .25(20,000), or $4,000, during the first year.

Conversely, resale value after one year (V_1) is

$V_1 = .75(20,000)$

Assuming the same yearly percentage decline rate thereafter, we can generalize to find the value in the tth year for a $20,000 car.

$V_t = 20,000(.75)^t$

In general, for a car with initial cost of C, the future value function is

$V_t = C(.75)^t$

Exercises 11

1. Graph the value function for the $20,000 car.
2. If a $20,000 car depreciates 30 percent per year, determine its value function.

7.5 INCREASING AT DECREASING RATE— LOGARITHMIC MODEL

For some practical applications, the dependent variable continues to increase, but at a decreasing rate (slope is always positive, but decreasing). This is the basic shape of the logarithmic function, which can be used to model such cases. Here we will illustrate fitting the logarithmic model for learning and interest-yield curves.

Learning Curves

Learning situations are typified by early quick accumulation of knowledge and later slow accumulation of the fine points. For example, think about learning the material covered in this chapter. An hour or two of studying the chapter could provide a great deal of understanding on curve fitting—especially since it is so well written! With succeeding hours of study and problem solving, additional understanding would accrue, but at a lower rate. After an extended period of study (say 15–20 hours), the basics would be well learned, and additional study would help improve one's grade on an exam, but only to a limited degree. Logarithmic functions are good models for such learning situations, as the following episode illustrates.

Corporal Klod, put in charge of typing the morning reports for all the companies in the battalion, was sent to Morning Report School. Sgt. Foulmouth warned him, "You < ÷ ≠ × ! you better learn it right!" At first, Klod had to type each report several times to get it correct. Gradually, he learned the tricks of the trade and typed the reports faster and faster.

Lt. Foureyes, a recent graduate of East Point, suggested the following logarithmic curve to forecast Klod's speed in future days:

$y = f(x) = c \log_5 x$

where x is days on the job, and y is number of correct reports per day. At time $x = 0$, y realistically should be zero. Since the above model gives $y = -\infty$ when $x = 0$ ($\log_b 0 = -\infty$), an amendment of the model is in order. Fortunately, a simple translation of the axis can overcome this problem.

$y = f(x) = c \log_5(x + 1)$

The revised model still suffers from "initialization" problems. If Klod is able to do any reports right away, this model, which gives $y = 0$ when $x = 0$, can't reflect it. Thus, the final model,

$y = a + c \log_5(x + 1)$

where a represents the initial production level, was settled upon.

Given the same basic learning model, Sgt. Foulmouth and Lt. Foureyes still have a basic disagreement on how to improve Klod's performance. The sergeant believes the best approach is force: "You ⊀=Δ*! you better do a !+){><! good job!" The lieutenant believes that he should slowly build Klod's confidence in himself.

Under Foulmouth's plan, Klod would be scared and produce immediate results (three per day initially), although his progress would be slowed by the mental anguish (he could do only six per day as of day 124). Under the lieutenant's plan, he would initially do only one per day, but progress nicely to be able to do seven per day as of day 124.

	Sergeant's plan	*Lieutenant's plan*
Curve fitting equations	$3 = a + c \log_5 1$	$1 = a + c \log_5 1$
	$6 = a + c \log_5 125$	$7 = a + c \log_5 125$
Solving for a	Since $\log 1 = 0$, the first equation in each case gives a directly.	
	$a = 3$	$a = 1$
Solving for c	Substituting the values of a in the second equation.	
	$6 = 3 + c \log_5 125$	$7 = 1 + c \log_5 125$
	$c = 1$	$c = 2$
Fitted model	$y = 3 + 1 \log_5(x + 1)$	$y = 1 + 2 \log_5(x + 1)$

Exercises 12

1. How many reports per day will Klod be able to do under each plan on day 624?
2. When will Klod be able to do eight reports per day under each plan?
3. After what time will the lieutenant's plan result in more correct reports?
4. Graph the daily report production for each plan.

Yield Curves

Yield curves relate the interest rate for bonds to the time until redemption. Interest rates on government and corporate bonds generally increase, rapidly at first but then more slowly, as the

CURVE FITTING

time to redemption increases. Consequently, logarithmic functions serve as excellent models for yield curves. Let's illustrate by fitting a logarithmic yield curve of the following form to data for government bonds:

$$y = a + c \log_4 x$$

where y is interest rate (percent) and x is years until redemption.

Suppose short-term (3 months, or $\frac{1}{4}$ year) Treasury bills yield an average interest rate of 5 percent, while long-term (16 years) government bonds yield an average interest rate of 8 percent. Substituting this information into the yield model and solving

$$5 = a + c \log_4 \tfrac{1}{4} = a + c(-1)$$
$$8 = a + c \log_4 16 = a + c(2)$$

gives $a = 6$ and $c = 1$. So the fitted model is

$$y = 6 + 1 \log_4 x$$

Exercises 13

1. Graph the yield curve.
2. Look in a current financial publication to find the interest rates for various maturities of government bonds. Graph the points. Does the curve seem to be logarithmic? If so, take the points corresponding to 3-month and 16-year bonds and fit a yield curve (use the form $y = a + c \log_4 x$).

7.6 NEGATIVE SLOPE WITH LOWER BOUND—HYPERBOLIC MODEL

Functions that decrease (negative slope) in such a way as to approach a lower bound are nicely modeled by the three-parameter hyperbolic function,

$$y = f(x) = a + \frac{b}{c + x}$$

The parameter, a, represents the lower bound since $y = a$ when $x = \infty$. Such functions result from dividing a linear function by a linear function, as we saw upon deriving the unit cost function on page 173. In that case, the parameter, c, was zero. In other cases, we can derive the functions from the basic information of the application. Let's illustrate with the liquidity trap idea.

Liquidity Trap

Monetary theory tells us about the so-called liquidity trap—that people will maintain some reserve cash position, regardless of how high interest rates go. The idea of a minimum cash level

Figure 7.13

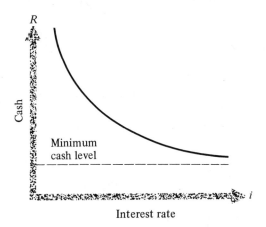

despite the level of interest rates suggests the graph of Figure 7.13 which is modeled by the three-parameter hyperbolic function

$$R = a + \frac{b}{c + i}$$

where R is cash reserve, and i is interest rate.

If people will maintain cash reserves of $50 billion despite the level of interest rates ($a = 50$), and reserves of $200 billion and $100 billion for interest rates of 3 percent and 10 percent, respectively, then substituting in our model gives

$$200 = 50 + \frac{b}{c + 3}$$
$$100 = 50 + \frac{b}{c + 10}$$

Solving for b in each equation,

$b = 150(c + 3)$

$b = 50(c + 10)$

and setting these equations equal

$150c + 450 = 50c + 500$

$$c = \tfrac{1}{2}$$

so

$b = 150(3.5) = 525$

CURVE FITTING

and thus the cash reserve function is

$$R = 50 + \frac{525}{\frac{1}{2} + i}$$

Exercises 14

1. How large a cash reserve does the model forecast for a 5.5 percent interest rate?
2. If people will maintain cash reserves of $75 billion despite the interest rate, fit the revised model.

7.7 SLOPE REVERSAL—PARABOLIC MODEL

Whenever a function increases and then decreases, or vice versa, the parabolic model that incorporates such a slope reversal can serve as a good model.

As we will see many times, the parabolic model results from multiplying a variable by a linear function of that variable. For example, if y is a linear function of x, then yx is

$$y = f(x) = sx + i$$
$$z = yx = (sx + i)x$$
$$= sx^2 + ix$$

You can see from the quadratic identification rules on page 166 that this is indeed a parabola.

This model will hold if, for example, demand (Q) is a linear function of price (P) and revenue is defined as PQ.

$$\text{Revenue} = R = PQ = P(sP + i)$$
$$= sP^2 + iP$$

In other cases, the particulars of the situation will point to the parabolic model, which will then have to be fitted. Let's illustrate with Miles Runner's bid for the four-minute mile.

Four-Minute Mile

Miles Runner begins his mile run at a velocity of .3 miles per minute. As the race wears on, Miles wears out so that by the time three minutes are up, he is going at a rate of .2 miles per minute. Then, miraculously, he catches his second wind and puts on a last-minute kick so that his velocity at four minutes is .25 miles per minute.

The situation suggests a velocity function that declines, then increases. This is ideal for the parabola. Let's fit one of the form

$$v = f(t) = at^2 + bt + c$$

where

> a, b, c are constants
> t is time in minutes
> v is velocity in miles per minute

Since $v = .3$ when $t = 0$, we immediately know that $c = .3$. Then the other two points of information result in the following two equations:

$.2 = 9a + 3b + .3$

$.25 = 16a + 4b + .3$

which, if solved simultaneously, give

$a = \frac{1}{48}$

$b = -.096$

So the fitted parabola is

$v = f(t) = \frac{1}{48}t^2 - .096t + .3$

Exercise 15

Roger Rannister, Miles Runner's chief competitor, also has a parabolic velocity function. He starts a little faster with an initial velocity of .35 miles per minute. However, he tires more easily, and at the three-minute point his velocity is .18 miles per minute. His "stretch kick" brings his velocity up to .25 miles per minute at the four-minute point. Determine Roger's velocity function.

Saving Gas

Ever since the oil boycott of the United States in 1974, Americans have been trying to find ways to conserve on gasoline. One way is to drive at the velocity at which one's engine runs most efficiently. Velocities lower or higher than that cause engines to waste gasoline. Such a situation suggests a parabolic (increasing then decreasing) curve to model miles per gallon (m) as a function of velocity (v) (see Figure 7.14). Assuming the parabolic model:

$m = f(v) = av^2 + bv + c$

where a, b, and c are parameters (constants), we can fit this model for a particular engine by using three points of data. For example, if

$m = 20$ when $v = 60$

$m = 18$ when $v = 70$

$m = 10$ when $v = 30$

CURVE FITTING

Figure 7.14

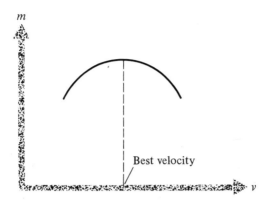

then we get the following three equations:

$20 = a \cdot 60^2 + b \cdot 60 + c = 3600a + 60b + c$

$18 = a \cdot 70^2 + b \cdot 70 + c = 4900a + 70b + c$

$10 = a \cdot 30^2 + b \cdot 30 + c = 900a + 30b + c$

This is a system of three linear equations. Rather than solve it now by brute-force methods, let's wait for the more rewarding methods described in the linear systems module.

7.8 THE "CURVY" ROAD TO PROFITS

In this section we develop important curve-fitting applications along the road to profits.

We begin our trip by fitting demand curves to various market conditions. A demand curve relates quantity demanded, Q, to price, P.

$Q = f(P)$

Revenue, R, is defined as price times quantity,

$R = PQ$

but since Q is a function of P, we can find R as a function of P:

$R = g(P) = P \cdot f(P)$

Costs are normally related to quantity produced. But since quantity is a function of price, we can find costs as a function of price:

$C = h(Q) = h(f(P)) = i(P)$

Since profits, π, are defined as revenue minus cost,

$$\pi = R - C$$

we can find profits as a function of price

$$\pi = j(P) = g(P) - i(P)$$

Let's see how it all works in practice.

Demand Curves

Demand curves relate the quantity that consumers are willing to purchase of a product at various prices. Since consumers will purchase less of a product if prices are increased (with the possible exception of certain items, such as wines and perfumes, which acquire more prestige at a higher price), demand curves have negative slope.

S. Lumlord earlier provided us with examples of demand curves in the apartment market. Now let's consider three other cases dealing with No-Extras air fare, warm underwear, and kinked televisions.

No-Extras

Creaky Airlines has just introduced a No-Extras service (customers say that their advertising has finally caught up with the actual service) on the Des Moines—Nashville route. Their Boing 706.5 (it would have been a 707 if it wasn't for the missing one-half wing) has 100 regular coach seats ($95 per seat) and 80 No-Extras seats. They need our demand curve fitting prowess before they can determine the price of those No-Extras seats.

Market research has shown that the regular coach seats would be sold out and the No-Extras seats would be empty if their price were $90 or more. But for every dollar under $90, two No-Extras tickets would be purchased—one would be a switch from regular seats, and the other would be a new customer.

We will now determine the quantity of No-Extras passengers (Q_{NE}) and the quantity of regular passengers (Q_R) as functions of No-Extras price (P). Because the stated phenomena happen *for every* dollar decrease in No-Extras fare, both functions are linear.

First, let's consider the No-Extras demand. The slope of this linear demand curve is

$$\frac{\Delta Q_{NE}}{\Delta P} = \frac{+2}{-1} = -2$$

Since the point $(P, Q_{NE}) = (90, 0)$ is on the line, we can use the point-slope formula as

$$Q_{NE} - Q_1 = s(P - P_1)$$
$$Q_{NE} - 0 = -2(P - 90)$$
$$Q_{NE} = -2P + 180$$

The slope of the regular seat demand curve is

CURVE FITTING

$$\frac{\Delta Q_R}{\Delta P} = \frac{-1}{-1} = +1$$

Since the point $(P, Q_R) = (90, 100)$ is on the line, we can use the point-slope formula to fit the demand curve as

$$Q_R - Q_1 = s(P - P_1)$$
$$Q_R - 100 = 1(P - 90)$$
$$Q_R = 1P + 10$$

(*Note:* This is an example of a cross-demand curve, since it relates the quantity of one item to the price of another competing item.)

If Creaky Airlines wanted to know the consequences of a No-Extras fare of, say, $65 per seat, we are now in position to help. Our demand curves give the respective ticket sales as

$$Q_{NE} = -2 \cdot 65 + 180 = 50$$
$$Q_R = 1 \cdot 65 + 10 = 75$$

Exercises 16

1. Fit the respective demand curves if three No-Extras (two from regular fare and one new fare) are purchased for every dollar reduction in No-Extras price.
2. For the demand curves found in exercise 1, what are the ticket sales consequences of a $65 No-Extras price?

Warm Underwear

Consider the situation for a consumer product—men's electric undershorts. This novel product for outdoor types has a customer mix of the jet-set crowd and the merely curious. Price increases above some nominal figure will scare away the curious; however, the jet-set customers would pay just about any price (it is a prestige item for the man who has everything). This analysis of the market results in a demand curve that can be represented by a hyperbolic model of the form

$$Q = f(P) = \frac{b}{c + P} \quad \text{(See Figure 7.15)}$$

Note: $a = 0$ in the general model, since $Q = 0$ when $P = \infty$.

Knowledge of two points on the demand curve is necessary to determine the two parameters, b and c. So, if consumers would buy 2 million units at a price of $3 and 1 million units at a price of $10, then

$$2 = \frac{b}{c + 3}$$
$$1 = \frac{b}{c + 10}$$

Figure 7.15

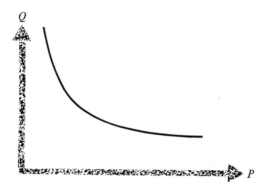

Solving these equations simultaneously gives $b = 14$ and $c = 4$. Thus, the fitted demand curve is

$$Q = f(P) = \frac{14}{4 + P}$$

where Q is sales in millions of units, and P is price per unit.

This model can now be used to forecast the demand for other price levels. For example, the forecasted demand for a price of $6 would be

$$Q = f(6) = \frac{14}{4 + 6} = 1.4 \text{ million}$$

Exercises 17

1. If one million shorts could be sold only if the price were dropped to $8 (everything else the same as the above example), fit the revised hyperbolic demand curve.
2. Use the revised demand curve to forecast sales at $6.

Kinked Television Demand

Kinked demand curves, which comprise two different-sloping linear functions, can serve to model demand in an oligopoly market.

Blurry Television Co. is in an oligopoly market dominated by a few large producers. The industry's normal price for a portable 21-inch color set is $300. Blurry and the rest of the industry all fall within a few dollars of this normal price. If Blurry reduces its price, the competition will generally follow, and Blurry's sales will rise by only 40,000 units per $10 decrease. If Blurry increases its price, the competition won't follow, and Blurry will suffer a loss of 100,000 units per $10 increase. Blurry's current sales are 1 million units per year.

The kinked demand curve hinges on the point

CURVE FITTING

$(P_1, Q_1) = (300, 1{,}000{,}000)$

For $P < 300$, it has the constant slope

$$\frac{\Delta Q}{\Delta P} = \frac{+40{,}000}{-10} = -4000$$

For $P > 300$, it has the constant slope

$$\frac{\Delta Q}{\Delta P} = \frac{-100{,}000}{+10} = -10{,}000$$

Using the point-slope formula for each section gives

$P < 300$

$$Q - Q_1 = s(P - P_1)$$
$$Q - 1{,}000{,}000 = -4000(P - 300)$$
$$Q = -4000P + 2{,}200{,}000$$

$P > 300$

$$Q - 1{,}000{,}000 = -10{,}000(P - 300)$$
$$Q = -10{,}000P + 4{,}000{,}000$$

Now we can help Blurry forecast sales under various pricing situations. For example, for prices of $250 and $350, our kinked demand curve forecasts sales as

$P = 250$

$$Q = -4000(250) + 2{,}200{,}000 = 1{,}200{,}000 \text{ units}$$

$P = 350$

$$Q = -10{,}000(350) + 4{,}000{,}000 = 500{,}000 \text{ units}$$

Exercises 18

1. If Blurry gains 50,000 sales for every $20 price decrease and loses 150,000 sales for every $20 price increase, fit the revised kinked demand curve.
2. Using this revised demand curve, forecast Blurry sales for $250 and $350 prices.

Revenue Functions

At the beginning of this section we showed how total revenue (R) defined as price (P) times quantity (Q)—where $Q = f(P)$ is the demand function—is itself a function of price. Now let's consider the nature of revenue functions for constant, linear, and hyperbolic demand functions. Also, we will find the specific revenue functions for the demand curves fitted earlier.

Constant Demand

Constant demand,

$$Q = c$$

where c is a constant, could not hold over wide price domains. However, for cases of limited supply, such as for a rare life-saving medical device, it could hold.

For constant demand, the revenue function

$$R = QP = cP$$

is a linear function, which goes through the origin, with slope of c, as shown in Figure 7.16.

For example, if a rare life-saving medical device (only 100 in existence) would be completely sold out to hospitals for any price less than $25,000, the revenue function would be

$$R = 100P \quad \{P < 25{,}000\}$$

Exercise 19

If S. Lumlord can rent all his apartments for any price less than $300, determine the revenue function for $P < 300$.

Figure 7.16

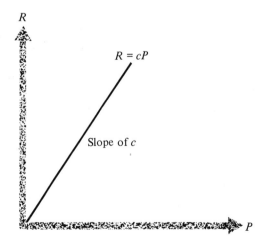

CURVE FITTING

Linear Demand

Linear demand results in a parabolic revenue function, as illustrated now.

$$R = PQ = P(sP + i)$$
$$= sP^2 + iP$$

The constants, s and i, in the revenue function are the slope and intercept values of the demand curve. Since s is negative, such parabolas graph as shown in Figure 7.17. (*Note:* We will need calculus to show why.)

I'll bet you wonder where the revenue is highest. You could determine it now by the exhaustion method: try all values of P for a specific function and calculate each R. But if you are patient, you'll be able to do it with calculus and I guarantee you'll be able to do it within 10 seconds or double your money back.[2]

Now let's find the total revenue curve for each linear demand curve fitted earlier in the chapter.

S. Lumlord $\quad Q = -2P + 1000$

$\qquad R = PQ = P(-2P + 1000) = -2P^2 + 1000P$

Creaky Airlines $\quad Q_{NE} = -2P + 180$

$\qquad R_{NE} = PQ_{NE} = P(-2P + 180) = -2P^2 + 180P$

$\qquad Q_R = 1P + 10$

---

[2]The author reserves the right to restrict this offer to Phi Beta Kappa students living in Alaska with IQ's over 200.

Figure 7.17

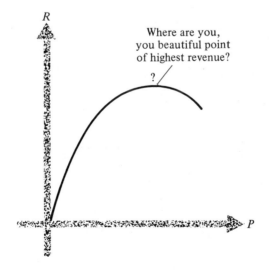

Kinked TV's
$$R_R = 95Q_R = 95(1P + 10) = 95P + 950$$
$$Q = -4{,}000P + 2{,}200{,}000 \quad \{P < 300\}$$
$$R = PQ = P(-4{,}000P + 2{,}200{,}000)$$
$$= -4{,}000P^2 + 2{,}200{,}000P$$

Exercise 20

Determine the kinked TV revenue curve for $\{P > 300\}$.

Notice that the revenue curve is parabolic in all cases.

Hyperbolic Demand

In this section, we will consider three types of hyperbolic demand and determine the nature of the revenue function for each.

The hyperbolic demand function of the form

$$Q = f(P) = \frac{b}{P} \quad a = c = 0 \text{ in the general hyperbolic model}$$

results in a constant revenue

$$R = PQ = P\left(\frac{b}{P}\right) = b$$

This is the classical case in economics of unitary elasticity. (We will develop the elasticity concept in section 14.2.)

The slightly different hyperbolic demand curve

$$Q = f(P) = a + \frac{b}{P} \quad c = 0 \text{ in the general hyperbolic model}$$

results in a linear revenue curve

$$R = PQ = P\left(a + \frac{b}{P}\right) = aP + b$$

This is every manager's wish—revenue rising endlessly in linear fashion as price increases. Sorry, but it will never happen.

The hyperbolic demand model

$$Q = f(P) = \frac{b}{c + P} \quad a = 0 \text{ in the general hyperbolic model}$$

results in a hyperbolic revenue function

CURVE FITTING

$$R = PQ = P\left(\frac{b}{c+P}\right) = \frac{bP}{c+P}$$

Exercise 21

Using the quadratic identification rules, verify that this is indeed a hyperbola.

The electric underwear demand curve was of this particular hyperbolic form.

$$Q = f(P) = \frac{14}{4+P}$$

Thus, its revenue curve is

$$R = PQ = \frac{14P}{4+P}$$

which graphs as shown in Figure 7.18.

Notice that the revenue increases, but at a decreasing rate, and approaches the limit of $14 million. This results from the fact that the number 4 in the denominator is insignificant as P gets very large. So revenue approaches

$$\frac{14P}{P} = 14$$

The demand function

$$Q = \frac{a}{P^2}$$

which isn't hyperbolic, but graphs as if it were, results in the hyperbolic revenue curve

Figure 7.18

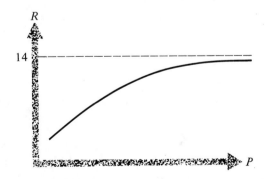

$$R = PQ = P\left(\frac{a}{P^2}\right) = \frac{a}{P}$$

So R goes down as P goes up. This could well be the truth for railroad and inner-city transit systems.

Cost Curves

The total cost (C) of producing Q units of product is the sum of two parts, fixed cost (F) and variable cost (V):

$$C = F + V$$

Fixed costs are those that don't change over realistic domains of production (for example, costs related to plant, buildings, and administration); thus F is a constant. Variable costs are those that vary with production (for example, raw materials and direct labor). V can be expressed as

$$V = vQ$$

where v is variable cost per unit. Furthermore, v itself is a function of Q. Let's investigate several possibilities of this function. (1) v could be relatively constant over wide ranges of production; (2) v could increase or decrease with Q—the increase would be true if high production incurs extra costs of inefficient workers—the decrease would be true if economies of scale are achieved.

We can now substitute the two variable cost functions in the total cost equation and observe the nature of the resulting total cost curve.

Variable cost case	Total cost curve, $C = F + vQ$	Total cost curve type
(1) constant $v = c$	$C = F + cQ$	linear
(2) Linear $v = aQ + b$	$C = F + (aQ + b)Q = F + bQ + aQ^2$	parabola

Exercise 22

Suppose variable cost is a hyperbolic function given by

$$v = a + \frac{b}{Q}$$

where a and b are constants. Find the total cost curve and identify its type.

Now let's fit a few cost curves.

The company that produces electric undershorts has fixed costs totaling $500,000. Each pair of electric undershorts produced results in an additional cost of $3.

CURVE FITTING

The constant variable cost ($v = c = 3$), as given above, results in a linear total cost curve. In this case, it is

$$C = F + vQ = 500{,}000 + 3Q$$

S. Lumlord has a monthly fixed cost of $40,000 relating to his 600 apartment complex. In addition, each rented apartment costs him $40 per month (heat, upkeep, etc.), and each unrented apartment costs him $20 per month (needs minimum heat so the pipes don't freeze).

Since the variable costs are constant for each rented or unrented apartment, we have

$$C = F + vQ + v_U Q_U$$

where

> F (fixed cost) is $40,000
> v (variable cost for rented apartment) is $40
> Q is number of rented apartments
> v_U (variable cost of unrented apartment) is $20
> Q_U (number of unrented apartments) is $600 - Q$

Substituting these values in the cost equation gives

$$C = 40{,}000 + 40Q + 20(600 - Q)$$
$$= 52{,}000 + 20Q$$

Exercise 23

If both the rented and unrented variable costs increase by $10 over those in the above example, determine S. Lumlord's total cost function.

Creaky Airlines has a $5000 fixed cost (crew, gasoline, terminal facilities, overhead, etc.) for each flight on the Des Moans to De Troit route. In addition, there is a $20 cost for each regular passenger and a $5 cost for each No-Extras passenger.

With constant variable costs for the two different classes of passengers, we have

$$C = F + v_R Q_R + v_{NE} Q_{NE}$$

where

> F (fixed cost) is $5000
> v_R (variable cost for regular passenger) is $20
> Q_R (number of regular passengers) is $P + 10$
> v_{NE} (variable cost of No-Extras) is $5
> Q_{NE} (number of No-Extras passengers) is $-2P + 180$
> P is price of No-Extras seat

Substituting these values in the total cost curve gives

$C = 5000 + 20(P + 10) + 5(-2P + 180)$

$ = 6100 + 10P$

Exercise 24

If the price of jet fuel increases and Creaky's fixed costs thus increase to $5500 per one-way trip, determine the revised total cost curve (assuming all other costs remain the same).

Blurry Television has fixed costs of $5,000,000. By buying raw materials in volume, it is able to reduce variable costs. Their engineers have estimated that variable costs per unit produced are $200 initially, but then they decline $.00001 = 10^{-5}$ dollars per unit. This constant-slope property means that a linear variable cost curve is applicable. This curve has intercept of 200 and slope of $-.00001$.

$v = 200 - .00001Q$

Consequently, Blurry's total cost curve is

$C = F + vQ = 5,000,000 + (200 - .00001Q)Q$

$ = 5,000,000 + 200Q - .00001Q^2$

This is a parabolic function, which we should have anticipated, given the linear variable cost curve.

Exercise 25

Suppose Blurry actually incurred inefficiencies as a result of increased production such that the slope of the variable cost function was $+.00001$ (all other things the same).
1. Determine this revised total cost function.
2. Graph and compare the revised and original total cost functions.

7.9 PROFIT CURVES—BREAKEVEN ANALYSIS

Profit (π) is defined as total revenue (R) minus total cost (C).

$\pi = R - C$

In turn, total revenue and total cost are defined as

$R = PQ$

$C = F + vQ$

so profits are equal to

CURVE FITTING

$$\pi = PQ - F - vQ$$

where

> P is price per unit
> Q is number of units sold
> F is fixed cost
> v is variable cost per unit

The nature of the profit function depends on each of these terms. Furthermore, as we saw earlier, Q is a function of P(demand curve) and v is a function of Q.

We have now arrived at the most important function for management. Of course, as greedy capitalists, we want achieve the maximum profit. Unfortunately, we don't have the necessary mathematical tools yet. We need calculus to achieve that goal. So, be patient. The calculus rescue is only a few chapters away!

In the meanwhile, we will now investigate profit functions for various realistic revenue and cost curves. Here as a consolation prize, we can determine the useful, but not optimal, break-even points, or where profits are zero.

Linear Revenue and Cost Curves

Linear revenue and cost functions imply constant price and variable cost per unit. We'll illustrate this case with an example with $F = 2000$, $P = 10$, and $v = 5$. Profits are then

$$\pi = 10Q - 2000 - 5Q = 5Q - 2000 \quad \text{(See Figure 7.19)}$$

For low levels of output, $R < C$, so profits are negative. At some output level, called the breakeven point ($Q = 400$ here), profits equal zero. Beyond the breakeven point, profits are positive.

Before going on to other revenue and cost situations, we should generalize this case. Assuming constant price and variable cost, the profit function becomes

Figure 7.19

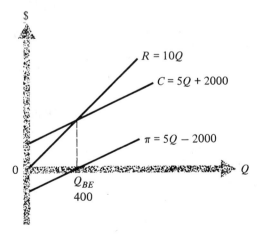

$$\pi = PQ - F - vQ$$
$$\pi = (P - v)Q - F$$

At the breakeven point $\pi = 0$, so

$$0 = (P - v)Q - F$$

Solving for Q, which is the breakeven output (call it Q_{BE}),

$$Q_{BE} = \frac{F}{P - v}$$

For our particular values of $F = 2000$, $P = 10$, and $v = 5$ (note $P > v$)

$$Q_{BE} = \frac{2000}{10 - 5} = 400$$

Exercises 26

1. The assumption of constant P implies what demand curve? For what market type is such an assumption justified?
2. Graphically show the consequences of $P = v$ and $P < v$.

Parabolic Revenue and Linear Cost

The assumption of constant P is an unrealistic one. Unfortunately, we are faced with downward-sloping demand curves—we can sell more only at lower prices. Maxi Taxi Co. can serve to illustrate the profit consequences of the more realistic linear downward-sloping demand curve, with resulting parabolic revenue.

The Maxi Taxi Co. operates in Hicktown with one car driven by—you guessed it—Max himself. Max charges a flat fee for all rides within the town. Currently, his flat fee is $1, for which he averages 50 customers per day. For every 25¢ increase in fixed fee, he thinks he'd lose 10 customers.

The phrase *for every* denotes constant slope and thus a linear demand function. The slope is

$$s = \frac{\Delta Q}{\Delta P} = \frac{-10}{25} = -.4$$

Since the point $(P, Q) = (100, 50)$ is on the line, we can use the point-slope formula to get its equation (P is in cents):

$$Q - 50 = -.4(P - 100)$$
$$Q = 90 - .4P \quad \text{or} \quad P = 225 - 2.5Q$$

Consequently, revenue as a function of Q is

CURVE FITTING

$$R = f(Q) = PQ = (225 - 2.5Q)Q = 225Q - 2.5Q^2$$

If Max's daily fixed costs are $35, plus an average variable cost of $.25 per customer (mainly gasoline), then the linear function with intercept of 35 and slope of .25 models his cost curve.

$$C = \underset{\text{(in dollars)}}{.25Q + 35} = \underset{\text{(in cents)}}{25Q + 3500}$$

Max's profit in cents, as a function of number of riders, is

$$\pi = R - C = (225Q - 2.5Q^2) - (25Q + 3500)$$
$$= -2.5Q^2 + 200Q - 3500$$

This is a parabolic function, rising then declining as Figure 7.20 shows. The point of maximum profit, which we so desperately want to find, will elude us until we learn the calculus needed for its solution. However, in the meanwhile, we can find the breakeven point. Small consolation!

At breakeven, $\pi = 0$:

$$0 = -2.5Q^2 + 200Q - 3500$$

This is a quadratic equation in one variable, which can be solved by the quadratic roots formula

$$Q_{BE} = \frac{-200 \pm \sqrt{200^2 - 4(-2.5)(-3500)}}{2(-2.5)} \cong 26 \text{ and } 54 \text{ (not in domain)}$$

Figure 7.20

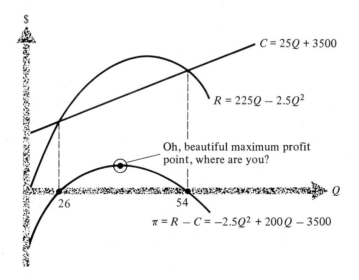

which translates into a break even price of

$$P_{BE} = 225 - 2.5(26) = \$1.60$$

Exercise 27

Verify that the above profit curve is a parabola.

Exercises 28

Determine the profit curve for the following cases, for which we earlier developed revenue and cost curves.

1. S. Lumlord
2. Creaky Airlines
3. Blurry Television

PROBLEMS

TECHNICAL TRIUMPHS

1. Suppose your normal weekly salary is $200, and you get $10 for every hour of overtime. Determine your total income as a function of hours of overtime.

2. Determine your age as a function of the calendar year.

3. Suppose a wing of a hospital needs one nurse for every ten patients in addition to a head nurse. Determine the number of nurses needed as a function of the number of patients.

4. A salesperson receives compensation of $5000 base salary plus a commission of 1 percent of sales.
   a. Determine the total income (I) as a function of sales (S).
   b. How much must be sold in order to earn $60,000?

5. A firm believes that sales (S) in millions from advertising (A) in millions follows a logarithmic function of the form

$$S = f(A) = a + c \log_2 A \quad \begin{array}{l} \{A \geq 1\} \\ a, c \text{ are constants} \end{array}$$

   If $S = \$10$ million when $A = \$1$ million, and $S = \$20$ million when $A = \$4$ million.
   a. Determine the sales function.
   b. How much sales should $8 million of advertising yield?

6. The output in units per day (y) for someone learning a certain industrial job is thought to be the following function of time on the job (x): $y = a + c\log_{10}x$, where a and c are constants. On day 1, the person produces 6 units. On day 10, the person produces 8 units. How many units will be produced on the 100th day?

7. Gasoline mileage per gallon (m) for a certain car model is 20 miles per gallon at a velocity (v) of 40 miles per hour and 18 miles per gallon at a velocity of 60 miles per hour.

CURVE FITTING

a. Fit a parabolic function of the form $m = f(v) = av^2 + bv$ to the data.
b. What would gas mileage be at a velocity of 50 miles per hour?

8. Gypyou Co. has just introduced a new product. Market researchers forecast that the sales rate (S) will climb initially but then decline (as consumers realize that they were taken again) according to the parabolic function of time (t) in months from product introduction.

$$S = f(t) = at^2 + bt + c$$

If $S = 22.8$ when $t = 1$, $S = 34$ when $t = 5$, and $S = 30$ when $t = 10$, specify the three equations needed to solve for the parameters, a, b, and c.

9. A firm recorded that when it charged \$1 for its product, revenue (R) was \$1000; at \$2 per unit, revenue was \$1500. If R is assumed to be a parabolic function of price (P) of the form

$$R = aP^2 + bP \quad \text{where } a \text{ and } b \text{ are constants}$$

a. Determine the revenue function.
b. What will R be if $P = \$3$?
c. Plot the revenue function.

10. If the average height (H) of children during the first 4 years is an exponential function of age (t) of the following form:

$$H = g(t) = a \cdot 4^{ct}$$

where a, c are constants, $H = 20$ inches when $t = 0$, and $H = 40$ inches when $t = 4$. Determine the function $g(t)$.

CONFIDENCE BUILDERS

1. The Fly-Now-Pay-Later Airlines advertises group charter, all-expense-paid round trips to London. The charter plane seats 150 passengers, and the charge per person is \$300 plus an additional \$5 for each empty seat. Let x be the number of empty seats. Derive the equation that gives total revenue (R) as a function of x.

2. It is estimated that formal college education will increase one's average lifetime earnings according to the logarithmic function

$$L = 400{,}000[1 + .4\log_3(x + 1)]$$

where L is lifetime earnings and x is years of college education.
a. If the cost per year of a college education is approximately \$70,000 including lost income, determine the "net additional lifetime income" (N) as a function of x.
b. How much is a four-year college education worth?

3. A firm's revenue (currently \$20 million) is growing at a rate of 10 percent per year. Its total cost (currently \$25 million) is growing at a rate of 5 percent per year.
a. Determine revenue and total cost as functions of time (t) where $t = 0$ now.
b. Determine the profit function.
c. When will the firm break out of the red?

4. A certain type of light bulb has a 50 percent chance of being "alive" after 100 hours of use. If y is the chance of being alive and x is the hours of use, fit the failure curve $y = a \cdot 2^{cx}$ to the data.

5. A textbook publishing company has a fixed cost of \$300,000 per year and a variable cost of \$40 per book. The sales price per copy is \$100.

a. How many books does the company have to sell to reach the breakeven point?
b. What is the profit (loss) at 10,000 books sold?

6. A certain manufacturer of gnurds is able to sell all his production at $5 per unit. Past records show that two points on his linear total cost function are $(Q, C) = (100, 1000)$ and $(200, 1300)$.
   a. Determine Q_{BE}, the breakeven output level.
   b. How much would the manufacturer have to lower fixed costs so that Q_{BE} would be 300 units?

7. If every dollar of advertising (up to $1 million) brings in $2 worth of sales (sales are $5 million even with no advertising), and each advertising dollar over $1 million brings in $.50 in sales, determine sales (S) as a function of advertising budget (A).

8. In January 1973, 4000 people died in automobile accidents in the United States with the 65-mile-per-hour speed limit. In January 1974, after the speed limit was reduced to 55 miles per hour, 3000 people died in automobile accidents. If automobile accident deaths (D) is a parabolic function of speed limit (L) of the form

   $$D = f(L) = aL^2 + bL \quad \text{where } a \text{ and } b \text{ are constants}$$

   a. Fit the function.
   b. How many deaths does the function forecast for a 50-mile-per-hour speed limit?

9. A fruit stand operator buys bananas at wholesale for 15¢ per pound and sells them for 25¢ per pound. If he doesn't sell them in 3 days, they become overripe and worthless. If he buys 100 pounds from his wholesaler (and can't get any more for 3 days), determine banana profits (π) as a function of demand (D). (*Note:* D can be larger than 100.)

10. A Gangland Industries' subsidiary, Laundry Associates (they launder illegal campaign contributions), expects an annual 20 percent growth rate in sales for the next 5 years, followed by a 10 percent annual growth rate thereafter. If sales are currently $10 million, find sales ($S$) as a function of time (t) in years from now.

11. A study regarding the number of dinners served in a dining car of a passenger train showed that on average 60 percent of the passengers order a meal, provided that the number of passengers is 1000 or less. Because of the limited space in the dining car, it has been found that only 30 percent of the excess over 1000 passengers order meals in the dining car. Express the number of meals served (M) as a function of the number of passengers (n).

12. The WASHIT Laundromat has noticed that in response to increased business, its electric bill has been increasing. The last three months of bills have been as follows.

Month	Electric Bill
1	$50.50
2	51.00
3	51.50

   a. Assuming that a linear function is relevant, express the electric bill (B) as a function of time (t) in months.
   b. If the minimum electric bill (assuming no customers) is $10, and each complete washing cycle uses $.05 of electricity, express B as a function of W, the number of complete wash cycles.
   c. Using the results of parts a and b, how many wash cycles are expected in month number 10?

MIND STRETCHERS

1. The Syracuse Tiddlers of the Eastern Tiddly Wink League undertook a study of home game attendance. They found that when tickets were priced at the regular rate of $1 per seat, they averaged 2500 fans per game. However, for games in which they offered half-price tickets, they averaged 4000 fans

CURVE FITTING

per game. Tillie Winkle, the owner, thinks that average game attendance (A) is best related to price per ticket (P) as

$$A = f(P) = \frac{a}{b + P^2} \quad \text{where } a \text{ and } b \text{ are constants}$$

    a. Use the data to fit the attendance function $f(P)$.
    b. Graph the attendance function.
    c. Determine the revenue function.
    d. If total costs per game are constant at $2000, determine the breakeven attendance level.
Tedlee Winkler, the ticket manager, thinks that another function,

$$A = g(P) = \frac{a}{b + P}$$

best relates attendance and ticket price.
    e. What type of function is $g(P)$?
    f. Fit $g(P)$ to the data.
    g. Graph $g(P)$ and compare it to $f(P)$.
    h. Determine the revenue function for this case.
    i. Determine the breakeven attendance level for this case.

2. Al Lee has just signed to fight Willy Dive for the heavyweight championship in the Philadelphia Spectrum (Al Lee predicts a "dilly in Philly"). The promoters feel that the entire arena would be sold out (30,000 seats) at $20 per seat. However, for every $2 increase in seat price, 1000 seats would go empty.
    a. Determine fight attendance (Q) as a function of seat price (P).
    b. Determine total revenue (R) as a function of seat price.
    c. Identify the attendance and revenue functions.
    d. What seat price would result in $750,000 revenue?
    e. If total costs are $500,000, what is the breakeven attendance? the seat price?
    f. If in addition to the half-million cost in part e, Al Lee gets 10 percent of the total revenue, what is the breakeven attendance? the seat price?

3. President Akie Demia of Ivory Tower University wants to estimate student enrollment (E) and total university costs (C) as functions of tuition level (t). At $30,000 tuition per year, she believes that 10,000 students would enroll; however, for each $1,000 increase in tuition, 200 fewer students would enroll. On the cost side, there are fixed costs of $260 million and each student enrolled adds an additional $5,000 cost. Determine the following:
    a. $E = f(t)$.
    b. $C = g(t)$.
    c. Revenue as a function of t.
    d. Profit as a function of t.
    e. The breakeven number of students.

4. The number of pounds of tomatoes (T) in thousands that can be grown on a one-acre plot is approximated by the following function of the number of tomato plants (x) in thousands:

$$T = 50 \ln x \quad \{x \geq 2\}$$

If a one-acre plot has $100 fixed costs, each tomato plant costs $.50, and the farmer can get $.20 per pound for tomatoes, determine for a one-acre plot the following functions of x:
    a. Total costs.
    b. Total revenue.

c. Total profit.
d. Determine the breakeven level for x.

5. Alvin Pell operates a 1000 tree orchard. As of September 1, Al's trees average a yield (Q) of 300 pounds of apples, for which he can get $.20 per pound on the market. Four weeks later, the market price will be $.10 per pound. After September 1 (and until the middle of October), Al's apples grow such that the yield per tree increases 50 pounds each week. Price per pound of apples (P) is an exponential function of time in weeks after September 1 (x) of the form

$P = f(x) = a \cdot 2^{-cx}$ where a and c are constants

Al's total cost would be $30,000 if he picked the apples on September 1; however, for each week that he delays, he incurs an additional $1000 cost. Determine the following as functions of x:
a. Yield per tree.
b. Price per pound.
c. Revenue per tree.
d. Total orchard revenue.
e. Total orchard cost.
f. Total orchard profit.

6. A firm that can sell any amount of its product for $10 per unit has a linear total cost function. If no items are produced, its total cost is $1000. Its breakeven output level is 200 units.
a. Write the equations giving total revenue and total cost as functions of output level.
b. Illustrate the situation graphically.
c. By how much would fixed costs have to be reduced (holding other costs and prices constant) in order to have a breakeven output level of 150 units?

7. The racing time for Born Loser to run the mile in his last six races, along with the weight of the jockey, were

Race	Time	Weight
1	2:08	120
2	2:06	105
3	2:05	110
4	2:02	100
5	2:08	110
6	2:00	105

a. Graph the data.
b. Determine the best fitting linear function that relates time to weight.

8. A firm is engaged in breakeven analysis for its main product. If total revenue (R) and total cost (C) are both linear functions of total output (Q)—all variables in units of millions—and if $C = 10$ when $Q = 0$, and $C = 25$ when $Q = 5$,
a. Determine the cost function.
b. What price per unit (P) will result in a breakeven output level of 50 million units?
The following year the firm was able to reduce variable costs to $2.90 per unit. If, under the new conditions, $R = 60$ when $Q = 20$,
c. What would fixed costs have to be reduced to in order to keep the breakeven point at 50 million units?

9. Please cook the roast for 20 minutes at 500°, and then for 30 minutes for each pound at 315°. The turkey should be cooked for 30 minutes at 500° and then for 20 minutes at 315° for each pound of bird. Of course, to plan your cooking day effectively, you'll need to know total cooking time of the roast (r) and total cooking time of the turkey (t) as functions of the weight (w). Go to it, all you gourmet cooks, and fit those functions. For drinks, you will mix 1 part of lemonade syrup with 5 parts of

CURVE FITTING

water. Find the total amount of lemonade as a function of the syrup used. How much syrup should you use if you need a quart of lemonade? Cheers!

10. A cloth diaper service charges (weekly basis) a fixed fee of $5, $.40 per diaper for the first 25, $.25 per diaper for diapers 26 through 100, and $.10 per diaper for all diapers over 100.
    a. Express their total charge (C) as a function of number of diapers ordered (n) per week.
    b. Graph this function.

    Bamper's disposable diapers cost $.35 each.
    c. Express Bamper's total cost (B) as a function of n.
    d. Suppose parents wish to use the diapers with the lowest cost. Provide them with a rule showing which type of diaper to select depending on the weekly usage.

11. A recruiter for the Monkey Business School would like to develop a function relating number of applicants (A) and the time (t) in days spent in recruiting. Even if no recruiting is done, experience has shown that about 200 applications will be made; 10 days of recruiting would yield 50 additional applicants.
    a. If recruiting follows the laws of diminishing returns, suggest a 2 parameter curve to fit the data
    b. Fit that curve
    c. Recruiting costs $100 per day, net profit per enrollee is $200 per student, 75 percent of the applicants are accepted, and 50 percent of these enroll in the university. Determine net profit minus recruiting costs as a function of time spent in recruiting.

MODULE FOUR

Linear Systems

Sets of linear equations are called linear systems. Such systems arise in various real world situations. We learn the basics of linear systems in Chapter 8. Chapter 9 shows us how to solve small linear systems by algebraic methods. A new kind of algebra, matrix algebra, is presented in Chapter 10 and allows us to solve larger linear systems. Chapter 11 gives us the knowledge to optimize linear functions subject to constraints. A comprehensive overview and learning plan for this module can be found in Section 8.1.

CHAPTER

Linear Systems: Overview

8.1 INTRODUCTION

Please recall the car rental problem in Chapter 1 and the gas mileage curve-fitting problem in Chapter 7.

Car rental problem
Gertz: $C - .10M = 10$
Bavis: $C - .05M = 20$

Gas mileage problem
$3600a + 60b + c = 20$
$4900a + 70b + c = 18$
$900a + 30b + c = 10$

These are linear systems of two and three equations, respectively. Notice that all the equations are characterized by variables raised only to the first power. Squared variables, "logarithmed" variables, and so on are definitely no-no here. Also notice that the equations are written with all the variables on the left side and the constant term on the right side of the equal sign. As you'll see later, this implicit formulation for writing equations enhances the solution process.

Both of the applications above arose from the mathematical formulation of a practical problem. Other situations give rise to systems with 2, 3, 4, or any finite number of linear equations.

Now consider the following problem:

Maximize profit $5x + 10y$

Subject to $x + 2y \leq 60$

$4x + 3y \leq 70$

This is an example of a linear programming system. It is similar to the previous linear systems in that all variables are raised to the first power. It differs in that a linear function must be maximized subject to a set of linear *in*equalities.

The delightful outcomes (cheaper car rental, more efficient gas usage, maximum profits, etc.) accruing to those well versed in the topic of linear systems should be sufficient incentive to learn about these systems and their solutions. Such is our task in the next four chapters, which comprise the linear systems module.

Chapter and Module Learning Plan

Here in Chapter 8, we begin a four-chapter journey toward linear system fun and profit. We start with the basics by exploring the nature of linear equations of two variables (univariate) and more than two variables (multivariate). In doing so, we reluctantly give up the explicit formulation of linear equations, which we have mastered, for the implicit formulation. Believe me, you will grow to love the implicit formulation too, once you see that it helps us solve the systems. Speaking of "solving", we then strive to understand system composition, order, and the distinction between variables and unknowns before we can reach that happy solution state. Then, the best part begins. We develop several real world linear systems applications. Completing the applications by solving the systems and understanding the real world better must wait until Chapter 9.

Chapter 9 is all about solving linear systems. You got a taste for that endeavor in Section 2.8, where we solved linear systems simultaneously by the substitution method. But, we can do better. Recall your high school teacher, Miss OldMath, trying to show you how adding and subtracting equations mysteriously lead to solutions. Hopefully, you now will have a fighting chance to understand what that was all about. That so-called "elimination method" leads us to develop the algebraic solutions for simple systems. Once we study those algebraic solutions, we can see the patterns that a man named Cramer first saw. His so-called "Cramer's Rule", which employs determinants, is a beautiful and easy method to solve small linear systems. We try out this new toy on several applications. But, since using this method on large systems would wear out even the fastest computer, we will need help. Matrix algebra in Chapter 10 will come to our rescue.

Chapter 10 provides us with an efficient method to solve any linear system, no matter how large. But, this method is based on a new kind of algebra, called matrix algebra. Matrices are essentially tables of numbers. Matrix algebra is the set of rules that allows us to work with and manipulate matrices. Using matrix algebra, we can express a linear system in a compact matrix equation form. It then provides methods to manipulate matrices to systematically solve the linear system. At first, matrix algebra will be strange to you. But, with study and practice, you will soon master it and use it to solve small linear systems manually. Computers love working with matrices. So, we turn over the large linear systems to them.

In Chapter 11, we attack the linear programming problem. Basically, that is to maximize or minimize some objective function, such as profits or costs, subject to various constraints. First, we formulate some practical applications. Then, with the help of algebraic methods, both ordinary and matrix, we learn how to solve such problems. We develop the Simplex Method, which systematically captures the various algebraic steps needed to proceed to a solution. We finish that chapter by solving some practical linear programming problems, manually and by the computer.

8.2 BASIC CONCEPTS

In this section we introduce some basic concepts of linear system analysis, including the specification of the linear family in implicit equation form, the composition and order of linear systems, the differentiation between variables and unknowns, and the meaning of a solution to a linear system.

Linear Family of Equations in Implicit Form

In this section we take a close look at the family of linear equations, which comprise the basic building blocks of linear systems. You already met the baby of the family—the straight line—whose explicit form is

$$y = sx + i$$

You probably got to the point where you could quickly graph these babies, since the explicit form gives the slope (s) and the intercept (i) directly. Great! Now let's blow it all by a change of notation.

You may find this hard to believe, but it is often more fruitful to express linear equations in implicit form. For the straight line, the implicit form is

$$ax + by = k$$

where a, b, and k are constants.

With that maneuver, you can't graph now what was trivial before. For example, the graph of the straight line $1x + 2y = 3$ is a big mystery now—a giant step backwards in mathematical methods! But wait. We can relate a, b, and k to our old friends s and i by solving this new notation explicitly for y:

$$ax + by = k$$
$$by = k - ax$$
$$y = -\left(\frac{a}{b}\right)x + \left(\frac{k}{b}\right)$$

Comparing this to the explicit form, we see the following equivalence of symbols:

$$\text{slope} = s = -\frac{a}{b}$$
$$\text{intercept} = i = \frac{k}{b}$$

Thus, $1x + 2y = 3$, with $a = 1$, $b = 2$, and $k = 3$, has a slope of $-\frac{a}{b} = -\frac{1}{2}$ and an intercept of $\frac{k}{b} = \frac{3}{2}$. Whew! You can graph it after all!

Exercises 1

Graph the following straight lines by finding the slope and intercept directly from their implicit formulation.

1. $2x + 1y = 4$
2. $3x - 5y = -10$

3. $-4x + 1y = -6$
4. $\frac{1}{2}x + \frac{1}{4}y = 1$

Another member of the linear family, the *plane,* is obtained by adding a third variable, (z). The general expression for a plane in implicit form is

$ax + by + cz = k$

where a, b, c, and k are constants and x, y, and z are variables.

We met this member of the family back in Chapter 5 when we derived Seymour Bottoms' revenue function as

$R = 6M + 3W$

which, in implicit form, is

$R - 6M - 3W = 0$

where R, M, and W are the three variables.

Planes are flat surfaces in three-dimensional space (like a sheet of paper in a room). They can have many orientations, as illustrated in the graphs in Figure 8.1.

Three points (not lying on a straight line) determine the equation of a plane.

Figure 8.1

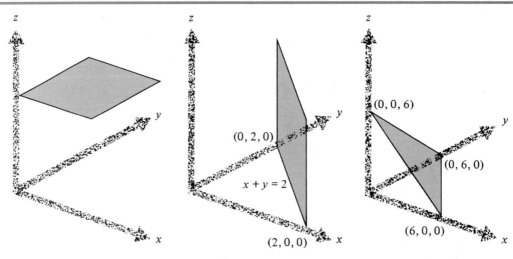

(a) The equation of a plane k units above the floor; that is, $z = 10$ might be the equation of the ceiling.

(b) The equation of a plane dividing the room diagonally. Plane extends up from the line $x + y = 2$ on the floor.

(c) The equation of a "tilted" plane that goes through points $(0, 0, 6)$, $(6, 0, 0)$, and $(0, 6, 0)$.

LINEAR SYSTEMS: OVERVIEW

By adding a fourth variable (w), we obtain another member of the linear family, generally described as

$$ax + by + cz + dw = k$$

where x, y, z, and w are variables, and a, b, c, d, and k are constants.

Examples

1. $x + y + z + w = 10$
2. $5x - 2y + 4z - \frac{1}{2}w = -3$
3. $-x + 10y + 3z - 6w = 0$
4. $7x + 0y - 8z + 2w = \frac{1}{2}$

This is a three-dimensional surface in four-dimensional coordinate space. Now, folks, I know you paid a lot of hard-earned money for this text and deserve many visual aids, but there is no way I'll graph that one for you.

Note that the two-variable linear equation has one dimension (length but not thickness) in two-dimensional space; the three-variable case (plane) has two dimensions (length and width but no thickness) in three-dimensional space; the four-variable equation is a three-dimensional surface in four-dimensional space. In general, linear equations with n variables are $(n - 1)$ dimensional surfaces in n dimensional space.

To keep the record straight, linear equations with four or more variables are called *hyperplanes*.

System Composition and Order

Linear systems are sets of linear equations, each containing a certain number of variables. We describe a linear system by its order, which is the number of equations followed by the number of variables. Some examples are as follows:

2×2 *System*

$2x - y = 3$
$x + 2y = 4$

3×3 *System*

$x - y + z = -2$
$2x + z = 1$
$x - 5y - 2z = 3$

2×3 *System*

$x + y + z = 10$
$2x - y - z = 5$

4×4 *System*

$4x - 2y + z + w = -13$
$-x - z - w = 2$
$ y + 3z + 2w = 0$
$2x + 5y - z = 20$

Solution

A solution to a linear system is a set of values for the unknowns that satisfy all the individual equations. Only square systems (with an equal number of equations and unknowns) result in a unique solution. For the three square cases above, the solutions are

2×2 $x = 2$ $y = 1$

3×3 $x = 2$ $y = 1$ $z = -3$

4×4 $x = -1$ $y = 4$ $z = -2$ $w = 1$

These solution values, when substituting in each equation, satisfy the equality. For example, substituting $x = 2$ and $y = 1$ in the 2×2 system gives

$2 \cdot 2 - 1 = 3$ or $3 = 3$

$2 + 2 \cdot 1 = 4$ or $4 = 4$

Nonsquare systems either have no solution, or they have an infinite number of solutions. The 2×3 system given above has an infinite number of solutions, two of which are ($x = 5$, $y = 10$, $z = -5$) and ($x = 5$, $y = 3$, $z = 2$).

Exercises 2

1. Substitute these two solutions in the 2×3 system to show that equality holds for each equation.
2. Substitute the given solution values in the 3×3 and the 4×4 linear systems to show that the equality holds for every equation.

Variable vs. Unknown

The words *variable* and *unknown* have been used almost interchangeably up until now. That may have caused you some confusion; so let's clarify and differentiate the two terms.

In the context of a specific equation, say, $2x - y = 3$, x and y are considered variables since they vary over the length of that straight line. In other words there exist an infinite number of (x, y) pairs that satisfy that equation. However, in the context of a linear system, say, the 2×2 given above, there is only one (x, y) pair that satisfies both equations. And that set of numbers happens to be unknown prior to solution. Hence in the system context, x and y are considered unknowns.

8.3 APPLICATIONS OF LINEAR SYSTEMS

Here we formulate some practical applications of linear systems. Our examples touch on investments, gridiron romance, demand and supply equilibrium, production, least-squares curve fitting, and input-output analysis. Once we develop methods to solve such systems in Chapters 9 and 10, we will return to each application for solution.

Investments

Bernard Broke has $100 thousand to invest between two stocks: Gangland Industries, which has a 4 percent dividend rate, and C.A.R.T.E.L., which has an 8 percent dividend rate. If Bernard

LINEAR SYSTEMS: OVERVIEW

wishes to have a $5 thousand annual dividend income, how much should he invest in each company?

System Formulation

Let

x = Gangland investment (thousands)

y = C.A.R.T.E.L. investment (thousands)

The fact that the entire $100 thousand must be divided between the two investments leads to the mathematical statement,

Gangland investment + C.A.R.T.E.L. investment = 100

$$x + y = 100$$

The fact that the $5 thousand annual total dividend is the sum of the individual company dividends leads to the second equation,

Gangland dividend + C.A.R.T.E.L. dividend = 5

$$.04x + .08y = 5$$

So, the complete two-equation linear system is

$$x + y = 100$$
$$.04x + .08y = 5$$

Exercise 3

Develop the linear system needed to tell Bernard how to achieve $12,000 annual income by investing $200,000 between a stock paying a 5 percent dividend and a bond yielding 10 percent interest.

Gridiron Romance

Bea Lock, the first female center in intercollegiate history, is in love with the head cheerleader, Ben Dover. She knows that she can catch him (aside from a cross-body block) only if she can slim down to a size 10 dress—and in the process lose her All-American rating.

Obese Observers has a new four-week diet that promises to do the trick for Bea. By following this plan, she should lose weight at the rates of

9 pounds per week as of the end of week 1 ($x = 1$)

2 pounds per week as of the end of week 3 ($x = 3$)

If the diet's weight loss rate, y, is captured by the following function,

$$y = f(x) = ax + \frac{b}{x+1}$$

where x is time in weeks from start of the diet, find the values of a and b that fit the two data points.

System Formulation

Since $f(1) = 9,\quad 9 = a \cdot 1 + \dfrac{b}{1+1}$

Since $f(3) = 2,\quad 2 = a \cdot 3 + \dfrac{b}{3+1}$

Upon simplification, we get the complete linear system needed to ring those wedding bells:

$1a + .5b = 9$

$3a + .25b = 2$

Exercise 4

If the weight loss rate of the diet was captured by the following function,

$y = f(x) = ax + bx^2$ (where a and b are constants)

set up the linear system needed to find the values of a and b.

Demand and Supply Equilibrium

Market research has shown that consumers would buy 1,500,000 Mickey Mouse digital watches at $100 per watch. However, for every $10 increase in price, they would purchase 50,000 fewer watches.

On the supply side, manufacturers would produce 2,500,000 of these watches if they could sell them for $200. However, for every $10 decrease in price, they would produce 100,000 fewer watches.

System Formulation

Because of their constant slope, both demand and supply curves are linear. Defining,

Q = quantity

P = price

and using the point-slope formula, we derive the demand and supply curves as

LINEAR SYSTEMS: OVERVIEW

Demand: $Q - 1{,}500{,}000 = \dfrac{-50{,}000}{+10}(P - 100)$

Supply: $Q - 2{,}500{,}000 = \dfrac{-100{,}000}{-10}(P - 200)$

Upon simplification, the 2 × 2 system in implicit form is

$Q + 5{,}000P = 2{,}000{,}000$

$Q - 10{,}000P = 500{,}000$

Exercise 5

If consumers would purchase 60,000 fewer watches and manufacturers would produce 80,000 more watches for every $10 increase in price (everything else the same as above), derive the revised linear system.

Production

A custom manufacturer is going out of business. It has an inventory of 12 gizmos, 10 doodads, and 7 thingamajigs, which it wants to use up *completely* in the manufacture of products A, B, and C.

The following table gives the parts requirements for the three products.

	A	B	C
Gizmos	1	4	2
Doodads	3	1	1
Thingamajigs	2	3	0

How many of each product should be manufactured so as to use up *all* the inventory?

System Formulation

Let

x = units of product A

y = units of product B

z = units of product C

The exhaustion of each raw material leads to one mathematical equation. For example, the 12 gizmos will be used up in A production (1 for each unit, or a total of $1x$), B production (4 for each unit, or a total of $4y$), and C production (2 for each unit, or a total of $2z$). This relationship leads to the equation

$1x + 4y + 2z = 12$

Similarly, equations can be written for the exhaustion of doodads and thingamajigs. See if you can develop these equations before looking below, where they are given along with the gizmo equation as the complete three-equation linear system.

$1x + 4y + 2z = 12$

$3x + 1y + 1z = 10$

$2x + 3y = 7$

Exercise 6

Upon search of his warehouse the manufacturer found 5 more gizmos, 4 more doodads, and 5 more thingamajigs. Set up the linear system that will tell how much of each product to produce in order to use up *all* the inventory.

Least-Squares Regression

Least-squares regression is a management tool for predicting one variable on the basis of another or others; for example, predicting sales as a function of advertising, or job performance as a function of years of experience and civil service examination score. Initial insights on the relationship among variables can be obtained by plotting the observations on a scatter diagram, as illustrated in Figure 8.2.

When, as shown, the points seem to align themselves around a straight line, we have the problem of determining the equation of the "best-fitting" line. This determination depends on the criteria of best fitting that we employ. A widely accepted one is the least-squares criteria. Using it, the deviations from the points to all possible straight lines are considered. The line

Figure 8.2

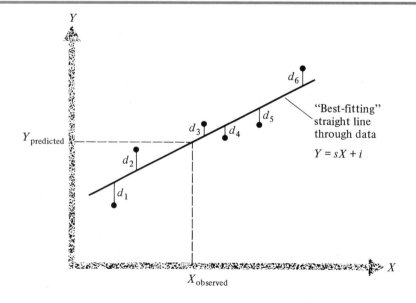

LINEAR SYSTEMS: OVERVIEW

with smallest sum of squared deviations ($\Sigma d^2 = d_1^2 + d_2^2 + \cdots + d_n^2$, where n is the number of points of data) is selected.

The seemingly endless search of all possible straight lines is done very quickly using calculus. We will see and understand the mechanics of this feat in Chapter 15. In the meanwhile, it is sufficient to know that the calculus derivation yields a set of so-called normal equations, which, if solved, give the parameters i and s of the least-squares regression line. For the simple regression case of one dependent and one independent variable, these normal equations, which, lo and behold, are linear and form a linear system, are

$ni + (\Sigma X)s = \Sigma Y$

$(\Sigma X)i + (\Sigma X^2)s = \Sigma XY$

where n is the number of observations

ΣX is the sum of the X values

ΣY is the sum of the Y values

ΣX^2 is the sum of the squares of the X values

ΣXY is the sum of the products of X times Y

To illustrate these ideas, consider the prediction of the number of correct answers on a mathematics quiz (Y) on the basis of the hours of study (X). Data for a sample of eight students are in the following table.

	Observations		Supplementary calculations	
Student	X	Y	X^2	XY
1	2	16	4	2 · 16 = 32
2	1	9	1	1 · 9 = 9
3	3	17	9	3 · 17 = 51
4	1	12	1	1 · 12 = 12
5	4	22	16	4 · 22 = 88
6	2	13	4	2 · 13 = 26
7	1	8	1	1 · 8 = 8
8	2	15	4	2 · 15 = 30
	$\Sigma X = 16$	$\Sigma Y = 112$	$\Sigma X^2 = 40$	$\Sigma XY = 256$

For this data the normal equations, which comprise a 2 × 2 linear system, are

$8i + 16s = 112$

$16i + 40s = 256$

Exercise 7

The nineteenth century geneticist Sir Francis Galton coined the word *regression* for this type of study. He did so after observing the regressive tendency for tall fathers to have shorter sons and short fathers to have taller sons, as illustrated in Figure 8.3.

This tendency doesn't hold today, as the current generation generally tends to be taller than

Figure 8.3

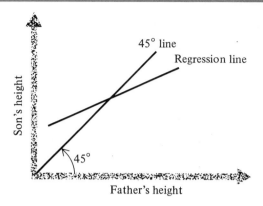

the previous generation. However, the following data illustrates the pattern Galton observed in the last century.

X Father's height	Y Son's height
75 inches	73 inches
70	72
65	68
63	65
72	71

1. Graph this data in scatter diagram form.
2. Develop the normal equations.

Since practical relations usually involve many variables, *multiple* (more than one independent variable) linear regression has found wide application. For example, by adding a mathematics aptitude predictor variable, we could use the following linear (plane) model to forecast grades:

$Y = a + bX + cZ$

where

Y is the math quiz score
X is the hours of study
Z is the mathematics aptitude.

The normal equations (which again require calculus to derive) needed to solve for the constants a, b, and c and fit this model are

$$na + (\Sigma X)b + (\Sigma Z)c = \Sigma Y$$
$$(\Sigma X)a + (\Sigma X^2)b + (\Sigma XZ)c = \Sigma XY$$
$$(\Sigma Z)a + (\Sigma XZ)b + (\Sigma Z^2)c = \Sigma YZ$$

LINEAR SYSTEMS: OVERVIEW

where the different required sums are illustrated below in the context of the data.

Observations			Supplementary calculations				
X	Z	Y	X^2	Z^2	XZ	XY	YZ
2	3	16	4	9	6	32	48
1	1	9	1	1	1	9	9
3	2	17	9	4	6	51	34
1	2	12	1	4	2	12	24
4	3	22	16	9	12	88	66
2	2	13	4	4	4	26	26
1	0	8	1	0	0	8	0
2	3	15	4	9	6	30	45
$\Sigma X = 16$	$\Sigma Z = 16$	$\Sigma Y = 112$	$\Sigma X^2 = 40$	$\Sigma Z^2 = 40$	$\Sigma XZ = 37$	$\Sigma XY = 256$	$\Sigma YZ = 252$

In this case, the normal equations form the following 3 × 3 linear system:

$8a + 16b + 16c = 112$

$16a + 40b + 37c = 256$

$16a + 37b + 40c = 252$

Exercise 8

If Galton had bothered to check the heights of the mothers, he might have found a different story. In order to determine the effect of this additional variable, we have gone to the historical archives to find the heights of those five mothers (no expense is spared here!). The complete set of data is

X	Z	Y
Father's height	Mother's height	Son's height
75 inches	63 inches	73 inches
70	70	72
65	66	68
63	65	65
72	64	71

Determine the set of three normal equations.

Input-Output Analysis

Leontief received the 1973 Nobel Prize in economics for his pioneering work in input-output analysis. The input-output model allows study of the flow of goods and services in an economy by breaking it up into its various industries and their interactions. Planned economies, such as in the former Soviet Union, made extensive use of such models. In our unplanned economy such models allow for the tracing of effects resulting from some autonomous change. For example, if consumers decide to purchase fewer automobiles, the demand for steel, glass, tires, and so on falls. In turn, the reduced demand for steel affects the railroad, coal, and other industries. And on and on the effects go. Without a model such as input-output, this complex interaction of effects could not be adequately understood.

An actual input-output model would have one equation for each industry in the economy. This would mean several hundred equations for a developed economy. Here, we will just take a simple case to get a feel for another application of linear systems. Consider a hypothetical economy with only four industries: materials, transportation, food, and construction.

The following table provides information on the interactions in the economy.

	Materials	*Transportation*	*Food*	*Construction*	*Final demand*
Materials	.3	.1	.2	.5	1
Transportation	.3	.2	.4	.1	3
Food	0	.1	.1	0	5
Construction	.2	.3	.1	.1	4

The numbers in the table reflect the dollar value of row industry products used in the manufacture of one dollar's worth of column industry product. For example, .5 dollar, or 50 cents of materials industry input, is required for each dollar of construction industry output.

By going down a column in the table, we get all the input requirements for a dollar's worth of output for that industry. For example, each dollar of transportation industry output requires .1 dollar, or 10 cents of materials; .2 dollar, or 20 cents of transportation;[1] .1 dollar, or 10 cents of food (coffee, tea, or milk on the plane); and .3 dollar, or 30 cents of construction input. The columns need not add up to one, since each industry adds value to its product.

System Formulation

Let

M = value of materials industry output

T = value of transportation industry output

F = value of food industry output

C = value of construction industry output

By using the numbers in a row, we can get information regarding the flow of an industry's product back to itself, to other industries, and to final demand. Let's illustrate with the materials industry. As seen in Figure 8.4, our equation is

Total materials output		*Materials for materials*		*Materials for transportation*		*Materials for food*		*Materials for construction*		*Final demand*
M	=	$.3M$	+	$.1T$	+	$.2F$	+	$.5C$	+	1

---

[1] It may seem funny at first, but note that each industry needs some of its own product in order to produce and sell new products. For example, the transportation industry needs railroads to move automobiles to the dealers.

LINEAR SYSTEMS: OVERVIEW

Figure 8.4

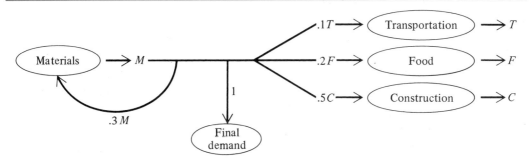

We can exemplify the logic of this equation by considering the .5C term. Since for every dollar of construction .5 dollar of materials is needed, then with total construction output of C, .5C dollars of materials are required by the construction industry. Final demand (which could be expressed in millions, billions, or whatever the situation calls for) is the value of product purchased by consumers for consumption.

Using the logic described above, see if you can understand why the following equations hold for the transportation, food, and construction industries, respectively.

$T = .3M + .2T + .4F + .1C + 3$

$F = 0M + .1T + .1F + 0C + 5$

$C = .2M + .3T + .1F + .1C + 4$

We now rearrange our equations to have only the final demand on the right of the equal sign. Illustrating with the materials industry equation,

$M - .3M - .1T - .2F - .5C = 1$

In this form total materials industry output (M) minus the materials industry output flowing to the various industries equals final demand.

One further simplification can be made by factoring M out of the first two terms to get $M - .3M = .7M$. Now, expressing all equations in this form gives the complete linear system.

$+.7M - .1T - .2F - .5C = 1$

$-.3M + .8T - .4F - .1C = 3$

$- 0M - .1T + .9F - 0C = 5$

$-.2M - .3T - .1F + .9C = 4$

Exercise 9

Describe the change to the input-output linear system if

1. New technology is developed in the construction industry such that each dollar of output requires only 40 cents of materials input.

2. Each dollar of transportation output requires more transportation input, say 25 cents.
3. Consumers increase their final demand for food to 6.

PROBLEMS

TECHNICAL TRIUMPHS

1. Consider the implicit linear function

   $2x - 5y = 4$

   Determine the explicit form and find its slope and intercept.

2. Given the explicit linear function

   $y = 7x + 10$

   Put it in implicit form.

3. What is the order of the following linear system?

   $2x - 3y + 4z = 9$

   $-6x + 5y - 2z = 8$

   Explain whether this system could have a unique solution. If not, what would be needed to obtain a unique solution?

4. Which of the following points $(x, y) = (2, 5)$ or $(x, y) = (4, -1)$ is a solution for the following linear system?

   $x - y = 5$

   $2x + 4y = 4$

5. If one has 4 variables, how many equations are needed to have a unique solution?

6. Consider the inequality

   $x + y \leq 3$

   How many of these points $(x, y) = (1, 3), (5, -3), (9, -6)$ satisfy this inequality?

7. Consider the equation:

   $x + y + z + w + v = 4$

   a. How would you describe such an equation?
   b. How many such equations would be needed to find a unique solution?

8. Describe in geometric terms the nature of the following linear equations:
   a. $6x - 5y + 4z = -10$
   b. $x + 4y = 6$
   c. $2x - 3y + 5z - 6w = 0$

LINEAR SYSTEMS: OVERVIEW

9. Are $y = 10$ and $x = 7$ linear equations? Explain using a graphical approach.

10. Find the slope and intercept of the following implicit linear equation without reverting to transforming it to an explicit linear equation.

$8x - 5y = 7$

CONFIDENCE BUILDERS

1. Zerox leases two models of photocopiers. Model A costs $500 per month, plus $.07 per copy. Model B costs $1000 per month, plus $.05 per copy. Formulate the linear system needed to determine the breakeven number of copies.

2. A driver is traveling along Route 2.718 when he sees a deer dart onto the road ahead of him. He immediately hits the brakes. If his velocity (v) in feet per second during braking is the following function of time in seconds (x) since hitting the brakes,

$v = a + bx^2 + c2^x$

where a, b, and c are constants, and if

v is 90 feet/second   when x is 0

v is 80 feet/second   when x is 1

v is 50 feet/second   when x is 2

set up the three-equation linear system needed to solve for the parameters a, b, and c.

3. "I, Howard Hoarde, being of sound mind and body, do bequeath my $2 billion fortune to my good friend S. Chottiner (x dollars), to the fellow who gave me a lift and 20 years ago lent me 25¢ (y dollars), and to a fund for medical research of gout (z dollars). I wish dear old Sherman to receive twice the sum of that fellow, but only one-half the sum of the medical fund."

   Set up the linear system required to allocate Howard's money.

4. The tolls for passage across the Pollution River Bridge are $.50 for cars, $1 for trucks, and $2 for buses.
   a. A total of 4000 vehicles (cars and trucks only) crossed the bridge on a certain day, yielding a total toll revenue of $2300. Develop the linear system required to determine how many cars (x) and trucks (y) crossed the bridge.
   b. A total of 5000 vehicles, including buses (z), crossed the bridge on another day, yielding a total toll revenue of $3100. If there were three times as many trucks as buses, develop the 3 × 3 linear system required to determine the number of each type of vehicle that crossed the bridge.

5. A firm is facing a demand curve such that it can sell 100 units at $10 each, but for each dollar increase in price, they will sell 5 less units. The firm is willing to supply 40 units at $5 and for every dollar increase in price they would supply and additional 10 units. Determine their demand and supply curves.

6. In 2003, the long distance telephone company, 10-10-987, offered service within USA at 3 cents per minute, plus a connect charge of 39 cents per call. AT&T offered long distance service at 7 cents per minute with no connect charge. Set up the two equations needed to solve for the breakeven calling time.

7. An investor wants to determine the market beta (β) of a stock. In 2002, the stock lost 30% while the overall market lost 20%. In 2003, the stock gained 40% while the overall market gained 25%. Set up the two equations of the form:

$y = \beta x + i$

where y is the stock's return and x is the overall market return.

MIND STRETCHERS

1. a. Little Penny Pincher has 34 coins—nickels (N) and dimes (D)—in her piggy bank, with a total value of $2.70. Set up the linear system needed to determine the number of each type of coin.
   b. Granny Buck gave little Penny a handful of change, which she promptly deposited in her piggy bank to add to the fortune described in part a. Granny's change, which included quarters (Q), brought the total value in the piggy bank to $6.30 for the 58 coins. Set up all the linear equations possible based on the information given. How many did you get? Do you have a square system? How many more equations do you need to have a square system?
   c. The following information enables you to write that additional equation to square the system in part b. Suppose that quarters, nickels, and dimes weigh 1, $\frac{1}{2}$, and $\frac{1}{4}$ ounce, respectively, and that the coins in total weigh 30 ounces. Now set up the 3 × 3 system needed to determine the number of each type of coin.
   d. After several visits by her other grandmother, Granny Gelt, Penny wound up with 110 coins (including half-dollars (H), which weigh $1\frac{1}{2}$ ounces) worth a total of $16.00 and weighing a total of 65 ounces. Set up all the linear equations possible. How many did you get? Do you have a square system? How many more equations do you need to get in order to have a square system?
   e. If Penny has twice as many quarters as half-dollars, determine the missing equation needed to square the system developed in part 1d.

2. Recall the Melody Tent situation as described on page 62. Suppose that on one sweaty summer afternoon, 1700 people, including adults (A) paying $3 each, children 6 to 16 years old (C) paying $2 each, and children under 6 or babies (B) who get in free, paid a total of $2600 to see "The Bugs Bunny Children's Show with Batman and Robin." When the crowd had to free Robin from the spell of Dr. Danger, a roar measured at 120 decibels was recorded (about equal to a Concorde takeoff). Assuming that each adult, child, and baby contributed .01, .08, and .10 decibels, respectively, to the din, set up the 3 × 3 linear system required to determine the number of adults, children, and babies in attendance.

3. a. In March 1976 the New York Telephone Company offered its nearly 150,000 home telephone customers in the Syracuse metropolitan area an alternative monthly telephone service charge plan. Until then the service charge was $9.55 regardless of the number of telephone calls. Now, the user could choose the alternative plan of $4.37 plus $.082 per local telephone call.
   Symbolizing the number of calls as N and the cost of the service as C, set up the linear system needed to solve for the breakeven number of calls.
   b. The New York Telephone Company has another service charge plan whereby the user is charged $6.71 plus $.082 for each call over 50. Set up the linear system needed to solve for the breakeven point between this plan and the constant charge of $9.55 plan.
   c. If the plans in part a. were updated in 2003 to have a $30 fixed service charge regardless of number of calls and an alternative plan of $10 plus 10 cents per local call, set up the linear system to solve for the breakeven number of calls.

4. Consider the hypothetical economy described earlier in the input-output section. Suppose a fifth industry, education and training, is included in the economy. This industry has a final demand of 2, and each dollar of its output requires 10 cents of materials, 15 cents of transportation, 5 cents of food, 10 cents of construction, and 20 cents of its own input. Also, for each dollar of materials, transportation, food, and construction output, 10 cents of education and training input is required.
   Set up the five-equation linear system for this economy.

5. The Sharp model of investments assumes that the returns on common stocks are linearly related to returns in the market as a whole. If returns of Fly-By-Night Airlines (y) and the overall market (x) over the past 6 years were

Year	x	y
1	4	10
2	0	3
3	-2	-2
4	2	4
5	7	6
6	4	9

determine the normal equations for the simple least-squares regression line.

6. Sy Kosis, a nervous investor, disputes the simplicity of the Sharp model, and claims that Fly-By-Night returns (y) should be related in multiple regression fashion to overall market returns (x) and airline industry returns (z). For that purpose, he gathered the following data:

Year	x	z	y
1	4	9	10
2	0	1	3
3	-2	-4	-2
4	2	2	4
5	7	3	6
6	4	8	9

Develop the normal equations for the multiple regression equation.

CHAPTER

Linear System Solution Methods

9.1 INTRODUCTION

Chapter 8 gave us a basic understanding of the nature of linear systems. But it left us without the punch line of solving them. So we now need Chapter 9 to give us the big prize of solving linear systems and applying the results to real world decision making.

Learning Plan

In this chapter we will explore methods to solve square linear systems. Quite possibly, you merely memorized these methods a long time ago in Miss Oldmath's class. Here, you will be encouraged to learn what really lies behind them. Once you begin to understand the fundamental principles, you will be able to see the solution patterns, which are beautiful to say the least. You will then see how these patterns can be captured by *Cramer's rule,* which, incidently, uses determinants.

While the methods developed in this chapter will work on any square linear system, they can efficiently handle only 2×2 or 3×3 systems. So in this chapter we limit ourselves to these cases.

At the conclusion of this chapter we will use Cramer's Rule to solve the applications developed in the previous chapter.

It is interesting to note that the methods developed in this chapter provide valuable insights into the matrix solution methods of the next chapter and the linear programming methods of Chapter 11. With so much riding on this development, it is crucial that you understand this chapter.

9.2 SOLUTION BASICS

In this section we will delve into and behind the methods you learned long ago to solve linear systems. In the process we will see the effects of adding and subtracting equations, develop the ideas of independent and dependent equations, and recognize the different possible solution cases. Since visual aids can do so much to illuminate these ideas, let's focus on the following graphable 2 × 2 linear system; which we call the "reference system":

REFERENCE SYSTEM

Implicit form	*Explicit form*
$3x + y = 5$	$y = 5 - 3x$
$-x + 2y = 3$	$y = \frac{3}{2} + \frac{1}{2}x$

Methods from the Past

Sometime in the past you were given three methods for solving linear systems: graphical, substitution, and elimination. Only the elimination method will be explored in depth in this chapter.

Graphical methods are excellent for providing insights into the solution process and we will take advantage of this. However, such methods are helpless for any system larger than 2 × 2. Even for the 2 × 2 system the accuracy of the graphical method depends on the care of graphing. In the case of our reference system careful graphing reveals that the two lines intersect at $x = 1$ and $y = 2$ in Figure 9.1.

The substitution method is described in detail in section 2.8. Briefly, it involves solving explicitly for one variable in one of the equations. For our reference system

$y = 5 - 3x$ from the first equation

Figure 9.1

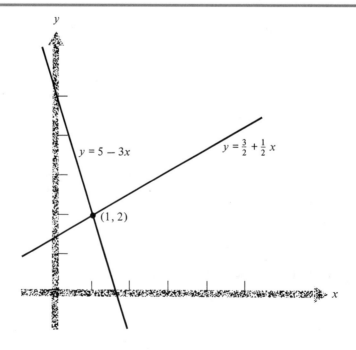

LINEAR SYSTEM SOLUTION METHODS

Then this expression is substituted (and thus the name) for that variable in the other equation. For our reference system

$-x + 2y = 3$ second equation

$-x + 2(5 - 3x) = 3$ substituting expression for y from first equation

We are then left with one equation having only one unknown. Upon simplification, we find

$-7x = -7$

$x = 1$

Then by substituting this value of x in either equation, we obtain the solution value for y. For example,

$y = 5 - 3x = 5 - 3 \cdot 1 = 2$

$y = \frac{3}{2} + \frac{1}{2}x = \frac{3}{2} + \frac{1}{2} \cdot 1 = 2$

The substitution method works fine for 2×2 systems. However, it has shortcomings for larger systems and can be effectively used only in conjunction with the elimination method.

Elimination Method

The elimination method to solve linear systems (as taught in high school by Mr. Oldmath) had something to do with multiplying equations by constants and adding or subtracting equations. And, although you probably got to the point of being able to do the manipulations to reach the solutions, why this process works probably remains a well-kept secret until this day. Now we hope to uncover this mystery, since these methods provide the key to understanding all the advanced solution methods presented in this text.

Many different systems can have the same solution. Such systems are called equivalent. Fortunately, we can always devise an equivalent system that is easier to solve than our original system. We do this by a series of revisions to the system's equations. The effects of these revisions, which are multiplication of an equation by a constant and addition and subtraction of equations, will now be explored.

Constant Times an Equation

To begin, let's multiply both sides of the second equation in our reference system by 2.

$2(-x + 2y) = 2 \cdot 3$

$-2x + 4y = 6$

$4y = 6 + 2x$

$y = \frac{3}{2} + \frac{1}{2}x$

Since we arrive at the explicit form of our original equation, multiplication by 2 has no effect on the points represented by our equation.

Exercises 1

1. Multiply both sides of the first equation in our reference system (in implicit form) by 3. Then explicitly solve for y. Does this operation have any effect on the set of points represented by the first equation?
2. Repeat exercise 1, except multiply by -3.

You have just seen a few cases where multiplication of an equation by a constant didn't change its set of points. But could these be coincidental and due to a fortuitous choice of constants? The only way we can answer this question is to introduce general conditions. So let's multiply the general two-variable linear equation, $ax + by = k$, by any nonzero constant, represented by λ.

		Slope	Intercept
General equation	$ax + by = k$	$-\dfrac{a}{b}$	$\dfrac{k}{b}$
Multiply by λ	$\lambda(ax + by) = \lambda k$		
"Revised" equation	$\lambda ax + \lambda by = \lambda k$	$-\dfrac{\lambda a}{\lambda b} = -\dfrac{a}{b}$	$\dfrac{\lambda k}{\lambda b} = \dfrac{k}{b}$

The "revised" equation has the same slope and intercept as the original equation. So it isn't revised at all, which just goes to show that multiplication of a linear equation by a constant doesn't change its set of points one iota.

Addition and Subtraction of Equations

Adding the two equations in our reference system gives

$$3x + y = 5$$
$$+(-x + 2y) = +3$$
$$\overline{(3x - x) + (y + 2y) = 5 + 3}$$
$$2x + 3y = 8 \quad \text{or } y = \tfrac{8}{3} - \tfrac{2}{3}x$$

Subtracting the second equation from the first gives

$$3x + y = 5$$
$$-(-x + 2y) = -3$$
$$\overline{(3x + x) + (y - 2y) = 5 - 3}$$
$$4x - y = 2 \quad \text{or } y = -2 + 4x$$

Combining multiplication and addition operations by adding 2 times the second equation to the first equation gives

LINEAR SYSTEM SOLUTION METHODS

$$3x + y = 5$$
$$+(-2x + 4y) = +6$$
$$\overline{x + 5y = 11} \quad \text{or } y = \tfrac{11}{5} - \tfrac{1}{5}x$$

Alas, this is all interesting, but we need some analysis of what's happening.

In the three examples, we ended up with different linear equations with no readily apparent relationship. Also, each equation involved two unknowns and didn't allow for solution of a one-unknown equation. But wait, perhaps graphing these equations, along with our reference system equations, will reveal something.

Figure 9.2 shows that they all go through the solution point (1, 2).

Figure 9.2

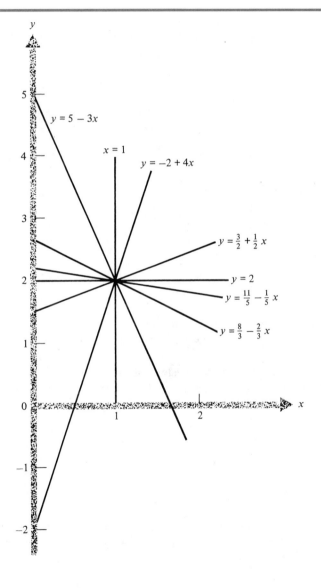

Exercises 2

Perform the following operations on our reference equations and graph the resulting equations.

1. Subtract the first equation from the second.
2. Add 3 times the first equation to the second.
3. Multiply both equations by 4 and add the resulting equations.
4. Multiply both equations by –5 and add the resulting equations.

Do all the resulting equations go through the point (1, 2)? They should, or you goofed!!

The lines $x = 1$ and $y = 2$ also go through the solution point. Both of these lines could have been generated by the appropriate multiplication and addition operations. For example, the line $y = 2$ is obtained by adding 3 times the second equation, $3(-x + 2y = 3)$ or $-3x + 6y = 9$, to the first equation.

$$\begin{array}{r} 3x + y = 5 \\ +(-3x + 6y) = +9 \\ \hline 0x + 7y = 14 \quad \text{or } y = 2 \end{array}$$

Notice that these operations were just right to make the x coefficient in the resulting equation zero, which left an equation with only one unknown, y.

At this point, we have an equivalent system in "triangular" form.

$3x + y = 5$

$\quad y = 2$

In this form, y's solution value is immediately known. If we wish, we could use the substitution method to solve for x. However, let's see how the elimination method does this job.

We need not start with the original system of equations. We can begin with the above equivalent system and generate the line $x = 1$ by subtracting the second equation from the first.

$$\begin{array}{r} 3x + y = 5 \\ -(+ y) = -2 \\ \hline 3x + 0y = 3 \quad \text{or } x = 1 \end{array}$$

Alternatively, we could have obtained the line $x = 1$ from the original system of equations. One way this can be accomplished is by multiplying the second equation by $\frac{1}{2}$ and subtracting that result from the first equation.

LINEAR SYSTEM SOLUTION METHODS

$$3x + y = 5$$
$$-(-\tfrac{1}{2}x + y) = -\tfrac{3}{2}$$
$$3\tfrac{1}{2}x + 0y = 3\tfrac{1}{2} \quad \text{or } x = 1$$

Exercise 3

Show that multiplying the first equation by 2 and subtracting it from the second equation in the reference system accomplishes the same result.

Our first few unsuccessful and later successful equation manipulations were characterized by multiplying the second equation by some nonzero constant, λ,

$$\lambda(-x + 2y) = \lambda \cdot 3$$
$$-\lambda x + 2\lambda y = 3\lambda$$

and adding the result to the first equation:

$$3x + y = 5$$
$$+(-\lambda x + 2\lambda y) = +3\lambda$$
$$\overline{(3x - \lambda x) + (y + 2\lambda y) = 5 + 3\lambda}$$
$$(3 - \lambda)x + (1 + 2\lambda)y = 5 + 3\lambda$$

The resulting general equation gives a different straight line through the solution point (1, 2) for each value of λ.

Exercise 4

The reader should verify that this equation always goes through the point (1, 2) regardless of the value of λ. Do this by substituting $x = 1$ and $y = 2$ in the equation and simplify to the point where it is obvious that equality holds.

When we added the two equations, λ was in effect equal to 1.

$$(3 - 1)x + (1 + 2 \cdot 1)y = 5 + 3 \cdot 1$$
$$2x + 3y = 8$$

Subtracting the second equation from the first is in effect using $\lambda = -1$, as illustrated.

$$[3 - (-1)]x + [1 + 2(-1)]y = 5 + 3(-1)$$
$$4x - y = 2$$

In general a negative λ means that λ times the second equation is subtracted from the first equation.

Of course, these λ values left us with equations still containing two unknowns. Only by setting λ in such a way as to "zero" a coefficient, as shown next, can we arrive at a solution value.

The x coefficient, $(3 - \lambda)$, equals zero when $\lambda = +3$, which yields the equation $0x + 7y = 14$, or $y = 2$.

The y coefficient becomes zero when $\lambda = -\frac{1}{2}$, which results in the equation $3\frac{1}{2}x + 0y = 3\frac{1}{2}$, or $x = 1$.

Example

Consider the system

$$2x - y = 7$$

$$x + 3y = 14$$

Taking λ times the second equation gives

$$\lambda(x + 3y = 14) = \lambda x + 3\lambda y = 14\lambda$$

This added to the first equation is

$$2x - y = 7$$
$$\lambda x + 3\lambda y = 14\lambda$$
$$\overline{(2x + \lambda x) + (-y + 3\lambda y) = 7 + 14\lambda}$$
$$(2 + \lambda)x + (3\lambda - 1)y = 7 + 14\lambda$$

The x coefficient is zero if $2 + \lambda = 0$, or $\lambda = -2$. The y coefficient is zero if $3\lambda - 1 = 0$, or $\lambda = \frac{1}{3}$. Using these two λ values yields the solution

$\lambda = -2$ then $-7y = -21$ or $y = 3$

$\lambda = \frac{1}{3}$ then $\frac{7}{3}x = 7 + \frac{14}{3}$ or $x = 5$

Exercises 5

Using the λ approach, solve the following linear systems.

1. $3x - 2y = 5$
   $x + 4y = 11$

2. $4x + 3y = -1$
   $2x + y = -1$

Later, we will use this method to solve general linear systems of two or three equations. But first we must investigate the difference between independent and dependent equations, since the solution of a system depends on the number of each type present.

Independent and Dependent Equations

Our efforts in the previous section showed that a linear combination of two-variable linear equations

$$k_1\varepsilon_1 + k_2\varepsilon_2$$

where k_1 and k_2 are constants, and ε_1 and ε_2 stand for equations 1 and 2, is another linear equation (we could call it ε_3) that goes through their intersection.

This statement generalizes to linear equations of any number of variables. For example, the linear sum of two planes is another plane that goes through the original line of intersection, as illustrated in Figure 9.3.

The following example system with $\varepsilon_3 = \varepsilon_1 + \varepsilon_2$ illustrates this phenomenon.

ε_1: $x + y + z = 6$

ε_2: $2x + 6y + 3z = 9$

ε_3: $3x + 7y + 4z = 15$

Because the third equation was derived from the other two, we can view it as being *dependent*. In general, if an equation in an n equation linear system can be derived by summing a linear combination of the other $(n - 1)$ equations as

$$\varepsilon_n = k_1\varepsilon_1 + k_2\varepsilon_2 + k_3\varepsilon_3 + \ldots + k_{n-1}\varepsilon_{n-1}$$

where the k's are constants, but not zero, it is said to be dependent. On the other hand, if an equation cannot be written as a linear combination of the other equations, it is said to be *independent*.

Entire sets of linear equations are considered dependent or independent, depending on whether or not they contain equations that can be written as linear combinations of other equations in the system. In the previous example we had a set of three dependent equations. Not only could the third equation be written as a linear combination of the first two, but also the first equation could be written as a linear combination of the other two

Figure 9.3

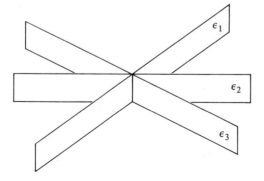

$\varepsilon_1 = 1\varepsilon_3 + (-1)\varepsilon_2$

$\quad = \varepsilon_3 - \varepsilon_2$

and the second equation could be written as a linear combination of the first and third.

Exercises 6

1. Find k_1 and k_3 such that $\varepsilon_2 = k_1\varepsilon_1 + k_3\varepsilon_3$.
2. Suppose that the third equation in our example system was ε_3: $4x + 8y + 5z = 21$. Can you find k_1 and k_2 such that $\varepsilon_3 = k_1\varepsilon_1 + k_2\varepsilon_2$? Is the set of equations dependent or independent?
3. Suppose that the third equation in our example system was ε_3: $4x + 8y - 2z = 21$. Is it possible to find nonzero values of k_1 and k_2 such that $\varepsilon_3 = k_1\varepsilon_1 + k_2\varepsilon_2$? Is this set of equations dependent or independent?

Normally it takes an equal number of independent equations and unknowns to uniquely solve for all the unknowns. In our example system on page 283 we have three unknowns but only two independent equations. Consequently, we don't have an unique solution and, in fact, have an infinite number of solutions along the line of intersection. The third equation, which is dependent in the example system, doesn't reduce this infinite set of solution points. However, a third independent equation would intersect this line of solution at a single point, leaving that point as the unique solution to the system.

Exercise 7

A student developed the following two equations to solve a linear system:

$x + y + z = 500$

$5x + 1y + .1z = 500$

She was unable to derive a third equation, but then she got a bright idea. Since the second equation expressed revenue in dollars for three classes of patrons, she derived the revenue equation expressed in cents, or $500x + 100y + 10z = 50{,}000$, to complete a 3×3 system.
   Explain the fallacy of this procedure.

Solution Possibilities

The nature of the solution to a linear system depends on the number of unknowns, number of independent equations, and the number of parallel equations. Table 1 (page 285) organizes these possibilities and gives an example of each.
   Notice that having the same number of independent equations as unknowns isn't sufficient to have a unique solution because of the parallel equation possibility. But aside from that case, underdetermined cases (number of independent equations is less than the number of unknowns) result in an infinite number of solutions, exactly determined cases (equal number of indepen-

LINEAR SYSTEM SOLUTION METHODS

Table 1 Systems of linear equations

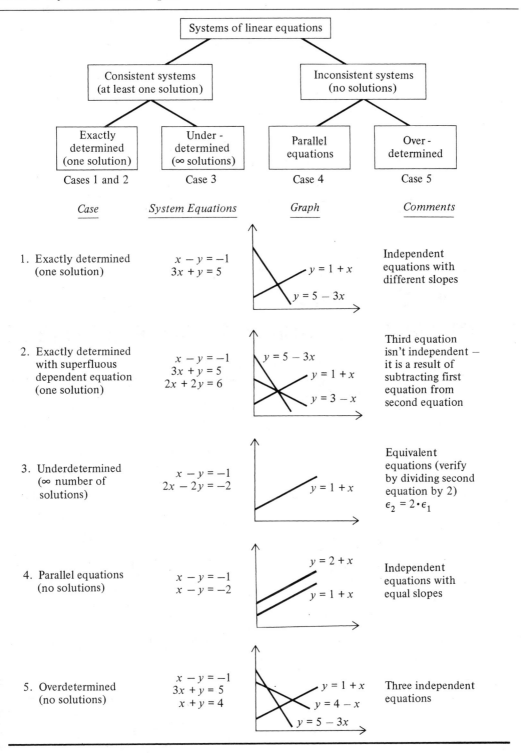

dent equations and unknowns) result in one solution, and overdetermined cases (number of independent equations exceeds the number of unknowns) result in no solution. This patterns holds in general for a linear system of any size.

Exercises 8

Graph the following systems and describe which of the cases (exactly, under- or overdetermined; or parallel equations) they fit.

1. $x + y = 5$
   $x + y = 6$

2. $x + y = 5$
   $-x + 2y = 4$

3. $x + y = 5$
   $2x - 3y = -5$
   $3x + 4y = 18$

4. $x + y = 5$
   $-2x - 2y = -10$

5. $x + y = 5$
   $2x - 3y = -5$
   $3x - 2y = 0$

9.3 GENERAL SOLUTION OF 2 × 2 LINEAR SYSTEMS

In this section we will use our accumulated theory and insights to solve the general case of a 2×2 linear system, which we represent as

$$a_1 x + b_1 y = k_1$$

$$a_2 x + b_2 y = k_2$$

where the a's, b's, and k's represent constants.

Multiplying the second equation by the constant, λ, and subtracting the result from the first equation gives

$$(a_1 x + b_1 y) - \lambda(a_2 x + b_2 y) = k_1 - \lambda k_2$$

$$(a_1 - \lambda a_2)x + (b_1 - \lambda b_2)y = k_1 - \lambda k_2$$

Solving by "zeroing" coefficients gives

x solution

"zero" y coefficient
$b_1 - \lambda b_2 = 0$

$$\lambda = \frac{b_1}{b_2}$$

Using this λ

$$\left(a_1 - \frac{b_1}{b_2}a_2\right)x + 0y = k_1 - \frac{b_1}{b_2}k_2$$

y solution

"zero" x coefficient
$a_1 - \lambda a_2 = 0$

$$\lambda = \frac{a_1}{a_2}$$

Using this λ

$$0x + \left(b_1 - \frac{a_1}{a_2}b_2\right)y = k_1 - \frac{a_1}{a_2}k_2$$

LINEAR SYSTEM SOLUTION METHODS

$$x = \frac{k_1 - \frac{b_1}{b_2}k_2}{a_1 - \frac{b_1}{b_2}a_2}$$

$$y = \frac{k_1 - \frac{a_1}{a_2}k_2}{b_1 - \frac{a_1}{a_2}b_2}$$

$$x = \frac{\frac{k_1 b_2 - b_1 k_2}{b_2}}{\frac{a_1 b_2 - b_1 a_2}{b_2}}$$

$$y = \frac{\frac{k_1 a_2 - a_1 k_2}{a_2}}{\frac{b_1 a_2 - a_1 b_2}{a_2}}$$

$$x = \frac{k_1 b_2 - k_2 b_1}{a_1 b_2 - a_2 b_1}$$

$$y = \frac{k_1 a_2 - k_2 a_1}{a_2 b_1 - a_1 b_2} = \frac{a_1 k_2 - a_2 k_1}{a_1 b_2 - a_2 b_1}$$

Let's see if these general formulas work on our reference system.

$3x + y = 5$
$-x + 2y = 3$

thus

$a_1 = 3 \quad b_1 = 1 \quad k_1 = 5$
$a_2 = -1 \quad b_2 = 2 \quad k_2 = 3$

$$x = \frac{5 \cdot 2 - 3 \cdot 1}{3 \cdot 2 - (-1)1}$$

$$y = \frac{3 \cdot 3 - (-1)5}{7}$$

$$= \frac{7}{7} = 1$$

$$= \frac{14}{7} = 2$$

They do!

Exercises 9

Using the general formulas, solve the systems

1. $2x - y = 13$
   $x + 3y = -4$

2. $x - 2y = 0$
   $3x + 4y = 20$

Patterns exist in these general solutions. They will be more obvious after we see their extension to 3×3 systems. For the present we can explore the significance of the common denominators.

The denominators, $a_1 b_2 - a_2 b_1$, which we will call D, in the solution for both x and y determine if a unique solution exists. If $D = 0$, since division by zero yields ∞, no unique solution exists. If $D \neq 0$, a unique solution exists.

We saw earlier that the underdetermined and parallel lines cases didn't have unique solutions. Here we illustrate that their D values are zero.

Underdetermined case

$x - y = -1$
$2x - 2y = -2$

$D = a_1 b_2 - a_2 b_1$
$\quad = 1(-2) - 2(-1)$
$\quad = 0$

Parallel lines case

$x - y = -1$
$x - y = -2$

$D = a_1 b_2 - a_2 b_1$
$\quad = 1(-1) - 1(-1)$
$\quad = 0$

Exercises 10

1. Show that the systems

   $x - 2y = 10 \qquad 4x + 12y = 8$
   $3x - 6y = 20 \qquad -3x - 9y = -6$

   have no solution by computing D. Verify by plotting.

2. Show that the general case of two unknowns but one independent equation

   $ax + by = k$
   $\lambda a x + \lambda b y = \lambda k$

   has no solution.

3. Show that the general case of parallel lines

   $ax + by = k_1$
   $ax + by = k_2$ $\quad k_1 \ne k_2$

   has no solution.

In summary we have shown that the "left side" constants (a's and b's here) alone determine whether a unique solution exists. This generalizes for any size of system.

9.4 GENERAL SOLUTION OF 3 × 3 LINEAR SYSTEMS

In this section we will develop methods for solving 3×3 linear systems.

If, by adding and subtracting equations by the elimination method, such a system can be reduced to an equivalent system in the following "triangular" form,

$x \quad y \quad z$
$\quad \ \ y \quad z$
$\quad \quad \ \ z$

LINEAR SYSTEM SOLUTION METHODS

then, by working backwards and using the substitution method, successive equations will only have one unknown. For example, the solution to the following triangular system,

$$x + y + z = 6$$
$$y - z = -1$$
$$z = 3$$

is

$$z = 3$$
$$y = -1 + z = -1 + 3 = 2$$
$$x = 6 - y - z = 6 - 2 - 3 = 1$$

This easy solution seems too good to be true. We had better go back to the original system,

$$x + y + z = 6$$
$$3x + 4y + 2z = 17$$
$$-x + 2y + z = 6$$

to see how the elimination method pulled it off.

First, x can be eliminated from the second equation by adding -3 times the first equation to it.

$$3x + 4y + 2z = 17$$
$$\underline{-3x - 3y - 3z = -18}$$
$$0x + 1y - 1z = -1$$

Then, x can be eliminated from the third equation by adding the first equation to it.

$$-x + 2y + z = 6$$
$$\underline{x + y + z = 6}$$
$$0x + 3y + 2z = 12$$

At this point we have the equivalent system

$$x + y + z = 6$$
$$y - z = -1$$
$$3y + 2z = 12$$

Finally, y can be eliminated from the revised third equation by adding -3 times the revised second equation to it.

$3y + 2z = 12$

$-3y + 3z = 3$

$0y + 5z = 15$ or $z = 3$

And so we arrived at the equivalent triangular system given earlier:

$x + y + z = 6$

$y - z = -1$

$z = 3$

Geometrically, the second and third equations in the triangular equivalent system (each a plane as shown in Figure 9.4)

$z = 3$ (horizontal plane)

$y - z = -1$ or $z = y + 1$

intersect in a line parallel to the x-axis. This line of intersection is described by

$z = 3$

$y = 2$

$x =$ any real number

The plane (first equation)

$x + y + z = 6$ or $z = 6 - x - y$

intersects this line at the single point where $x = 1$. Thus, the solution (point common to all three planes) is $(x, y, z) = (1, 2, 3)$.

Figure 9.4

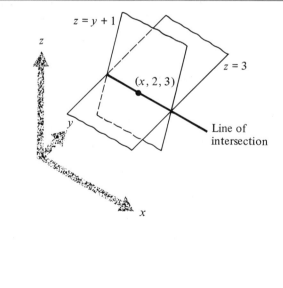

LINEAR SYSTEM SOLUTION METHODS

Exercises 11

Triangularize and solve the following systems.

1. $\quad x + y + z = 1$
   $\quad 2x - y + 4z = -2$
   $\quad -x + 3y - 3z = -7$

2. $\quad -x + 2y + 3z = 10$
   $\quad 3x - 2y + z = -16$
   $\quad -2x - y - 4z = 5$

The triangularization procedure can be applied to the general 3×3 linear system

$$a_1 x + b_1 y + c_1 z = k_1$$

$$a_2 x + b_2 y + c_2 z = k_2$$

$$a_3 x + b_3 y + c_3 z = k_3$$

where the a's, b's, c's, and k's are constants, to derive the general solution. But it would be one heck of a job. However, so that you can appreciate the "easier" methods developed later, let's outline the procedure.

First, x is removed in the second equation by adding $-(a_2/a_1)$ times the first equation to it. Then, x is removed from the third equation by adding $-(a_3/a_1)$ times the first equation to it.

Exercise 12

Carry out these operations and show that the revised second and third equations containing only y and z are

$$\left[b_2 - \left(\frac{a_2}{a_1}\right) b_1 \right] y + \left[c_2 - \left(\frac{a_2}{a_1}\right) c_1 \right] z = k_2 - \left(\frac{a_2}{a_1}\right) k_1$$

$$\left[b_3 - \left(\frac{a_3}{a_1}\right) b_1 \right] y + \left[c_3 - \left(\frac{a_3}{a_1}\right) c_1 \right] z = k_3 - \left(\frac{a_3}{a_1}\right) k_1$$

If we call these revised equations ε_2 and ε_3, respectively, and their y coefficients β_2 and β_3, respectively, then y can be eliminated from the third equation by the operation

$$\varepsilon_3 - \left(\frac{\beta_3}{\beta_2}\right) \varepsilon_2$$

The result would be a triangularized set of three "hairy" equations, with z as the only variable in the third equation, y and z as the variables in the second equation, and all three variables in the first equation.

At this point working backwards to solve for z, then y, then x would be possible, but not very probable! If you aren't frightened by monster mathematical expressions, try it. Most readers will be content to accept the final result, which is given in Table 2.

Table 2   Linear System Solution Patterns

System

$$\begin{array}{ll} 2\times 2 & 3\times 3 \\ a_1x + b_1y = k_1 & a_1x + b_1y + c_1z = k_1 \\ a_2x + b_2y = k_2 & a_2x + b_2y + c_2z = k_2 \\ & a_3x + b_3y + c_3z = k_3 \end{array}$$

Solution

$$x = \frac{k_1b_2 - k_2b_1}{a_1b_2 - a_2b_1} \qquad x = \frac{k_1b_2c_3 + k_3b_1c_2 + k_2b_3c_1 - k_3b_2c_1 - k_1b_3c_2 - k_2b_1c_3}{a_1b_2c_3 + a_3b_1c_2 + a_2b_3c_1 - a_3b_2c_1 - a_1b_3c_2 - a_2b_1c_3}$$

$$y = \frac{a_1k_2 - a_2k_1}{a_1b_2 - a_2b_1} \qquad y = \frac{a_1k_2c_3 + a_3k_1c_2 + a_2k_3c_1 - a_3k_2c_1 - a_1k_3c_2 - a_2k_1c_3}{\text{same denominator as for } x}$$

$$z = \frac{a_1b_2k_3 + a_3b_1k_2 + a_2b_3k_1 - a_3b_2k_1 - a_1b_3k_2 - a_2b_1k_3}{\text{same denominator as for } x}$$

Patterns

1. Identical denominators for a given system—same denominators for x and y in two-equation system; same denominators for x, y, and z in three-equation system.
2. Equal number of terms in numerator and denominator.
3. Equal number of plus and minus terms in numerator and denominator.
4. Each term is a product of two constants for the two-equation system, three constants for the three-equation system. (In general each term is a product of n (number of equations) constants.)
5. Two terms in numerator and denominator of two-equation system, but six (not three as you may have suspected) terms for three-equation system. It turns out that the number of terms is $n!$ ($2! = 2$, $3! = 6$).
6. Each term contains one and only one constant from each equation; for example, $a_1b_2c_3$, where the subscripts give the number of the source equation.
7. Each term in the denominator contains one coefficient from each variable; for example, one of the a's and one of the b's for the two-equation system.
8. The k's only appear in the numerators.
9. Coefficients belonging to the unknown to be solved for do not appear in the numerator.
10. The subscripts form all possible permutations of the numbers 1 to n, where n is the number of equations in the system.
11. Numerators are identical to denominators, *except* that the coefficients of the unknown to be solved for are replaced by the k's. For example, in solving for y in the two-equation system, the numerator is identical to the denominator except that k_1 replaces b_1 and k_2 replaces b_2.

Generalization

All these patterns observed for 2×2 and 3×3 systems generalize to any size of square linear system.

LINEAR SYSTEM SOLUTION METHODS 293

With this much grief for a "little" three-equation system, you can just imagine what it would be like solving a four-, five-, or ten-equation system by these methods. Oh, there must be a better way!! Cramer, save us!!

9.5 CRAMER'S RULE

The eighteenth century mathematician Cramer discovered a far less painful way to solve linear systems. In a nutshell Cramer began by observing and comparing the algebraic solution to the general two-equation (we did that) and three-equation (we almost did that) systems. He detected various solution patterns, which he showed would exist for any size system. We retraced this development in Table 2 by listing the two- and three- unknown solutions, observing the patterns, and seeing their generalization.

Cramer captured these patterns by determinants. In essence these are mathematical procedures that enable us to operate on arrangements of the system coefficients so that one arrives at the right coefficient in the right place with the right sign. It seems too good to be true! But wait—even better news is coming. For two- and three-equation systems, determinants have visual representations.

Cramer's Rule for 2 × 2 Systems

Two-equation systems require 2×2 determinants for solution. The four numbers in such a determinant are arranged as

$$\begin{vmatrix} a & b \\ c & d \end{vmatrix}$$

and are processed by multiplying the "down-to-the-right" diagonal elements (ad) and adding -1 times the product of the other elements (cb) as follows:

$$\begin{vmatrix} a & b \\ c & d \end{vmatrix} = ad - cb$$

Examples

1. $\begin{vmatrix} 3 & 2 \\ 1 & 4 \end{vmatrix} = 3 \cdot 4 - 1 \cdot 2 = 10$

2. $\begin{vmatrix} 5 & 3 \\ 4 & -2 \end{vmatrix} = 5 \cdot (-2) - 4 \cdot 3 = -22$

3. $\begin{vmatrix} 1 & 2 \\ 2 & 4 \end{vmatrix} = 1 \cdot 4 - 2 \cdot 2 = 0$

4. $\begin{vmatrix} 2 & 1 \\ -3 & 2 \end{vmatrix} = 2 \cdot 2 - (-3) \cdot 1 = 7$

Exercises 13

1. $\begin{vmatrix} 5 & 2 \\ 4 & 3 \end{vmatrix} =$

2. $\begin{vmatrix} 1 & 2 \\ 3 & 4 \end{vmatrix} =$

3. $\begin{vmatrix} 2 & 3 \\ 4 & 6 \end{vmatrix} =$

4. $\begin{vmatrix} 5 & -2 \\ 3 & 22 \end{vmatrix} =$

Using this scheme, then the solution to the two-equation linear system is

$$x = \frac{D_x}{D} = \frac{\begin{vmatrix} k_1 & b_1 \\ k_2 & b_2 \end{vmatrix}}{\begin{vmatrix} a_1 & b_1 \\ a_2 & b_2 \end{vmatrix}} = \frac{k_1 b_2 - k_2 b_1}{a_1 b_2 - a_2 b_1}$$

$$y = \frac{D_y}{D} = \frac{\begin{vmatrix} a_1 & k_1 \\ a_2 & k_2 \end{vmatrix}}{\begin{vmatrix} a_1 & b_1 \\ a_2 & b_2 \end{vmatrix}} = \frac{a_1 k_2 - a_2 k_1}{a_1 b_2 - a_2 b_1}$$

The reader should check back to Table 2 to verify that this scheme of setting up the determinants does indeed duplicate the algebraic solutions for x and y.

This method of setting up determinants is called *Cramer's rule.* Notice that in solving for x and y, the bottom determinant is composed of left-side coefficients—those to the left of the equal sign—in the arrangement that they appear in the system. The top determinant is the same as the bottom determinant *except* the right-side coefficients (k's) replace the coefficients associated with the variable to be solved for. So, when solving for x, k's replace the a's; when solving for y, k's replace the b's. What a snap!

Example (Reference System)

$3x + y = 5$

$-x + 2y = 3$

$$D = \begin{vmatrix} 3 & 1 \\ -1 & 2 \end{vmatrix} = +(6) - (-1) = 7$$

$$D_x = \begin{vmatrix} 5 & 1 \\ 3 & 2 \end{vmatrix} = +(10) - (3) = 7$$

LINEAR SYSTEM SOLUTION METHODS

$$D_y = \begin{vmatrix} 3 & 5 \\ -1 & 3 \end{vmatrix} = +(9) - (-5) = 14$$

(with diagonal products $-(-1 \cdot 5)$ and $+(3 \cdot 3)$)

So

$$x = \frac{D_x}{D} = \frac{7}{7} = 1$$

$$y = \frac{D_y}{D} = \frac{14}{7} = 2$$

which is the solution we found earlier.

Exercises 14

Solve the following systems by Cramer's rule:

1. $2x - y = 7$
   $-x + 2y = -5$

2. $x + 2y = 12$
   $4x - 3y = 4$

Now let's expand these ideas to three-equation systems.

Cramer's Rule for 3 × 3 Systems

Cramer's rule for setting up determinants to solve a three-equation system is

$$x = \frac{D_x}{D} \qquad y = \frac{D_y}{D} \qquad z = \frac{D_z}{D}$$

$$x = \frac{\begin{vmatrix} k_1 & b_1 & c_1 \\ k_2 & b_2 & c_2 \\ k_3 & b_3 & c_3 \end{vmatrix}}{\begin{vmatrix} a_1 & b_1 & c_1 \\ a_2 & b_2 & c_2 \\ a_3 & b_3 & c_3 \end{vmatrix}} \qquad y = \frac{\begin{vmatrix} a_1 & k_1 & c_1 \\ a_2 & k_2 & c_2 \\ a_3 & k_3 & c_3 \end{vmatrix}}{\begin{vmatrix} a_1 & b_1 & c_1 \\ a_2 & b_2 & c_2 \\ a_3 & b_3 & c_3 \end{vmatrix}} \qquad z = \frac{\begin{vmatrix} a_1 & b_1 & k_1 \\ a_2 & b_2 & k_2 \\ a_3 & b_3 & k_3 \end{vmatrix}}{\begin{vmatrix} a_1 & b_1 & c_1 \\ a_2 & b_2 & c_2 \\ a_3 & b_3 & c_3 \end{vmatrix}}$$

Notice how the bottom determinant for all three unknowns is the same and is made up with left-side coefficients in the order they appear in the system. The top determinants are the same as the bottom one *except* the k's replace the coefficients associated with the variable to be solved for. Thus, in solving for x, the k's replace the a's; in solving for y, the k's replace the b's; in solving for z, the k's replace the c's.

Now if we only knew a way to find the value of a 3 × 3 determinant, we'd be all set. Fortunately for us, here again a very neat visual device exists for this task. Starting with the base determinant (D) formed with the natural arrangement of left-side coefficients,

$$D = \begin{vmatrix} a_1 & b_1 & c_1 \\ a_2 & b_2 & c_2 \\ a_3 & b_3 & c_3 \end{vmatrix}$$

we annex the first two columns as follows:

$$\underbrace{\begin{vmatrix} a_1 & b_1 & c_1 \\ a_2 & b_2 & c_2 \\ a_3 & b_3 & c_3 \end{vmatrix}}_{\text{original array}} \quad \underbrace{\begin{matrix} a_1 & b_1 \\ a_2 & b_2 \\ a_3 & b_3 \end{matrix}}_{\substack{\text{annexed first} \\ \text{two columns}}}$$

Then we add all the down-to-the-right products and subtract from the total the sum of the down-to-the-left products. The result is the value of D.

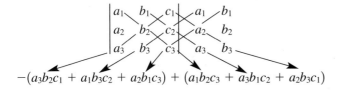

$$-(a_3b_2c_1 + a_1b_3c_2 + a_2b_1c_3) + (a_1b_2c_3 + a_3b_1c_2 + a_2b_3c_1)$$

To get the value of D_x, we need only substitute k's for a's in the base determinant, annex the first two columns, and apply the rules.

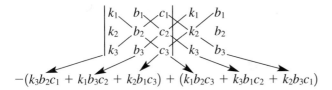

$$-(k_3b_2c_1 + k_1b_3c_2 + k_2b_1c_3) + (k_1b_2c_3 + k_3b_1c_2 + k_2b_3c_1)$$

We find the value of D_y by replacing the b's by k's, or

$$\begin{vmatrix} a_1 & k_1 & c_1 \\ a_2 & k_2 & c_2 \\ a_3 & k_3 & c_3 \end{vmatrix} \begin{matrix} a_1 & k_1 \\ a_2 & k_2 \\ a_3 & k_3 \end{matrix}$$

$$-(a_3k_2c_1 + a_1k_3c_2 + a_2k_1c_3) + (a_1k_2c_3 + a_3k_1c_2 + a_2k_3c_1)$$

Exercises 15

1. Show that D_x/D and D_y/D duplicate the solutions for x and y given in Table 2.
2. Set up the constant array to find D_z. Use the rules for a 3×3 determinant to find the six products.
3. Show that D_z/D duplicates the solution for z given in Table 2.

Now let's use Cramer's rule and determinants to quickly solve the following system, which we belabored earlier.

LINEAR SYSTEM SOLUTION METHODS

$$x + y + z = 6$$
$$3x + 4y + 2z = 17$$
$$-x + 2y + z = 6$$

$$D = \begin{vmatrix} 1 & 1 & 1 \\ 3 & 4 & 2 \\ -1 & 2 & 1 \end{vmatrix} \begin{matrix} 1 & 1 \\ 3 & 4 \\ -1 & 2 \end{matrix} = 8 - 3 = 5$$

$$-(-4 + 4 + 3) + (4 - 2 + 6)$$

$$D_x = \begin{vmatrix} 6 & 1 & 1 \\ 17 & 4 & 2 \\ 6 & 2 & 1 \end{vmatrix} \begin{matrix} 6 & 1 \\ 17 & 4 \\ 6 & 2 \end{matrix} = 70 - 65 = 5$$

$$-(24 + 24 + 17) + (24 + 12 + 34)$$

$$D_y = \begin{vmatrix} 1 & 6 & 1 \\ 3 & 17 & 2 \\ -1 & 6 & 1 \end{vmatrix} \begin{matrix} 1 & 6 \\ 3 & 17 \\ -1 & 6 \end{matrix} = 23 - 13 = 10$$

$$-(-17 + 12 + 18) + (17 - 12 + 18)$$

$$D_z = \begin{vmatrix} 1 & 1 & 6 \\ 3 & 4 & 17 \\ -1 & 2 & 6 \end{vmatrix} \begin{matrix} 1 & 1 \\ 3 & 4 \\ -1 & 2 \end{matrix} = 43 - 28 = 15$$

$$-(-24 + 34 + 18) + (24 - 17 + 36)$$

$$x = \frac{D_x}{D} = \frac{5}{5} = 1$$

$$y = \frac{D_y}{D} = \frac{10}{5} = 2$$

$$z = \frac{D_z}{D} = \frac{15}{5} = 3$$

Exercises 16

1. Show by Cramer's rule that $(x, y, z) = (4, 3, 2)$ is the solution to the linear system:

    $$x - y + z = 3$$
    $$2x + 3y - 4z = 9$$
    $$-x + 2y - z = 0$$

2. Use Cramer's rule to solve the two 3×3 linear systems in Exercises 11.

Just as for 2×2 systems, a zero value for D means that a unique solution does not exist for the 3 × 3 system. To illustrate, let's show that $D = 0$ for the following dependent system introduced earlier:

$$x + y + z = 6$$
$$2x + 6y + 3z = 9$$
$$3x + 7y + 4z = 15$$

$$D = \begin{vmatrix} 1 & 1 & 1 \\ 2 & 6 & 3 \\ 3 & 7 & 4 \end{vmatrix} \begin{matrix} 1 & 1 \\ 2 & 6 \\ 3 & 7 \end{matrix} = 47 - 47 = 0$$

$$-(18 + 21 + 8) + (24 + 9 + 14)$$

Exercises 17

1. If the second equation above were $4x + 8y + 5z = 21$, the system would contain three dependent equations. Show that $D = 0$ for that case.
2. By computing the value of D, show whether the following systems are dependent or independent.
    a. $2x + 3y + z = 2$
        $2x - 4y - 3z = 3$
        $-2x - 10y - 5z = -1$
    b. $-3x + 4y - z = 5$
        $x + 2y - 3z = 1$
        $x - 3y + 2z = 4$

Generalization of Cramer's Rule

Cramer's rule can be used to solve for the unknowns in any size of square linear system. In general the solution for the ith unknown, which we can call x_i, is

$$x_i = \frac{D_i}{D}$$

where D is the determinant of the left-side coefficients, and D_i is the determinant identical to D except that the column associated with the ith unknown is replaced by the column of right-side constants, or k's.

For example, consider the input-output 4×4 linear system equations on page 269. The base determinant of left-side coefficients is

$$D = \begin{vmatrix} .7 & -.1 & -.2 & -.5 \\ -.3 & .8 & -.4 & -.1 \\ 0 & -.1 & .9 & 0 \\ -.2 & -.3 & -.1 & .9 \end{vmatrix}$$

LINEAR SYSTEM SOLUTION METHODS

To solve for one of the unknowns, say, the second one or transportation output, we would form the determinant D_2 by replacing the second column of D by the k column as

$$D_2 = \begin{vmatrix} .7 & 1 & -.2 & -.5 \\ -.3 & 3 & -.4 & -.1 \\ 0 & 5 & .9 & 0 \\ -.2 & 4 & -.1 & .9 \end{vmatrix}$$

Then, once these two 4×4 determinants were evaluated, we would solve for the transportation output as

$$T = \frac{D_2}{D}$$

This is fine, but how does one find the value of a 4×4 determinant?

There are methods (although they are not of the handy-dandy visual type we have grown to love for small determinants) for evaluating 4×4, 5×5, 6×6, and so on, determinants. However, things quickly begin to get messy. A 4×4 determinant requires the computation of $4! = 24$ products of 4 constants. A 5×5 determinant requires the computation of $5! = 120$ products of 5 constants. A 6×6 determinant requires the computation of $6! = 720$ products of 6 constants. In general, an $n \times n$ determinant requires the computation of $n!$ products of n constants. Since $n!$ quickly gets very large (for example, $10! = 3,628,800$), even computers are not always up to the task. T. Marll McDonald, in his book *Mathematical Methods for Social and Management Scientists,* estimates that it would take a computer 30 million years to solve a 20×20 linear system using Cramer's rule. Now that was using the "slow" computers of the 1970's. But even today's super computers would need to chug away for several years!!!

And so we must be content to solve only 2×2 or 3×3 linear systems using Cramer's rule and determinants. As for those "bigger and badder" linear systems, we will sic matrix methods on them in the next chapter.

Now let's use Cramer's rule and determinants to solve the applications developed in Chapter 8.

9.6 CRAMER'S RULE SOLUTION OF APPLICATIONS

We will now use Cramer's rule to solve the problems formulated on page 255 and in Section 8.3.

Car Rental Problem

Gertz: $1C - .10M = 10$

Bavis: $1C - .05M = 20$

$$C = \frac{\begin{vmatrix} 10 & -.1 \\ 20 & -.05 \end{vmatrix}}{\begin{vmatrix} 1 & -.1 \\ 1 & -.05 \end{vmatrix}} = \frac{-.5 + 2}{-.05 + .1} = \$30$$

$$M = \frac{\begin{vmatrix} 1 & 10 \\ 1 & 20 \end{vmatrix}}{.05} = \frac{20 - 10}{.05} = 200 \text{ miles}$$

Thus, 200 miles, for which both companies charge $30, is the breakeven point. If you plan to drive more than that, rent from Bavis; otherwise, rent from Gertz.

Exercise 18

If both companies raise their mileage charge by the same constant amount (K cents per mile), use Cramer's rule to show that the decision of when to rent from each will not change.

Gas Mileage Problem

$3600a + 60b + 1c = 20$

$4900a + 70b + 1c = 18$

$900a + 30b + 1c = 10$

$$a = \frac{\begin{vmatrix} 20 & 60 & 1 \\ 18 & 70 & 1 \\ 10 & 30 & 1 \end{vmatrix} \begin{matrix} 20 & 60 \\ 18 & 70 \\ 10 & 30 \end{matrix}}{\begin{vmatrix} 3{,}600 & 60 & 1 \\ 4{,}900 & 70 & 1 \\ 900 & 30 & 1 \end{vmatrix} \begin{matrix} 3{,}600 & 60 \\ 4{,}900 & 70 \\ 900 & 30 \end{matrix}}$$

$$= \frac{(1{,}400 + 600 + 540) - (700 + 600 + 1{,}080)}{(252{,}000 + 54{,}000 + 147{,}000) - (63{,}000 + 108{,}000 + 294{,}000)}$$

$$= \frac{2{,}540 - 2{,}380}{453{,}000 - 465{,}000}$$

$$= \frac{160}{-12{,}000}$$

$$= -.0133$$

$$b = \frac{\begin{vmatrix} 3{,}600 & 20 & 1 \\ 4{,}900 & 18 & 1 \\ 900 & 10 & 1 \end{vmatrix} \begin{matrix} 3{,}600 & 20 \\ 4{,}900 & 18 \\ 900 & 10 \end{matrix}}{-12{,}000}$$

$$= \frac{(64{,}800 + 18{,}000 + 49{,}000) - (16{,}200 + 36{,}000 + 98{,}000)}{-12{,}000}$$

$$= \frac{131{,}800 - 150{,}200}{-12{,}000}$$

LINEAR SYSTEM SOLUTION METHODS

$$= \frac{-18{,}400}{-12{,}000}$$

$$= 1.533$$

At this point, the simplest way to solve for c is to substitute the values of a and b in the third equation.

$$c = 10 - 900(-.0133) - 30(1.533)$$

$$= -24$$

Recall from page 230 that a, b, and c were parameters in the following gas mileage function,

$$m = f(v) = av^2 + bv + c$$

where m is miles per gallon of gas and v is velocity in miles per hour. With the above solution, we have fitted the function as

$$m = f(v) = -.0133v^2 + 1.533v - 24$$

Exercise 19

If improvements in engine efficiency result in an increase in miles per gallon to 25, 20, and 15, respectively, use Cramer's rule to fit the revised mileage function.

Investments Problem

$$x + y = 100$$
$$.04x + .08y = 5$$

$$x = \frac{\begin{vmatrix} 100 & 1 \\ 5 & .08 \end{vmatrix}}{\begin{vmatrix} 1 & 1 \\ .04 & .08 \end{vmatrix}} = \frac{8 - 5}{.08 - .04} = 75 \text{ thousand}$$

$$y = \frac{\begin{vmatrix} 1 & 100 \\ .04 & 5 \end{vmatrix}}{.04} = \frac{5 - 4}{.04} = 25 \text{ thousand}$$

Thus, Bernard should make a $75,000 investment in Gangland Industries and a $25,000 investment in C.A.R.T.E.L. Then he should pray real hard!

Exercise 20

Use Cramer's rule to solve the linear system that you developed for Bernard's investments in Exercise 3 on page 261.

Gridiron Romance

$1a + .5b = 9$

$3a + .25b = 2$

$$a = \frac{\begin{vmatrix} 9 & .5 \\ 2 & .25 \end{vmatrix}}{\begin{vmatrix} 1 & .5 \\ 3 & .25 \end{vmatrix}} = \frac{2.25 - 1}{.25 - 1.5} = \frac{1.25}{-1.25} = -1$$

$$b = \frac{\begin{vmatrix} 1 & 9 \\ 3 & 2 \end{vmatrix}}{-1.25} = \frac{2 - 27}{-1.25} = \frac{-25}{-1.25} = 20$$

Recall that a and b were parameters in the following weight loss function,

$$y = f(x) = ax + \frac{b}{x+1}$$

where y is the rate of weight loss (pounds per week) and x is the time in weeks since the start of dieting.

With the above solution we have fitted the function as

$$y = f(x) = -1x + \frac{20}{x+1}$$

Good luck, Bea!

Exercise 21

Use Cramer's rule to solve the linear system that you set up to find Bea's parabolic weight-loss function in Exercise 4 on page 262.

Demand and Supply Equilibrium

$Q + 5,000P = 2,000,000$

$Q - 10,000P = 500,000$

$$Q = \frac{\begin{vmatrix} 2,000,000 & 5,000 \\ 500,000 & -10,000 \end{vmatrix}}{\begin{vmatrix} 1 & 5,000 \\ 1 & -10,000 \end{vmatrix}} = \frac{2 \cdot 10^6(-10^4) - 5 \cdot 10^5(5 \cdot 10^3)}{-10,000 - 5,000}$$

$$= \frac{-20 \cdot 10^9 - 2.5 \cdot 10^9}{-15 \cdot 10^3}$$

LINEAR SYSTEM SOLUTION METHODS

$$= \frac{-22.5 \cdot 10^9}{-15 \cdot 10^3}$$

$$= 1.5 \cdot 10^6$$

$$= 1,500,000$$

$$P = \frac{\begin{vmatrix} 1 & 2,000,000 \\ 1 & 500,000 \end{vmatrix}}{-15 \cdot 10^3} = \frac{5 \cdot 10^5 - 20 \cdot 10^5}{-15 \cdot 10^3}$$

$$= \frac{-15 \cdot 10^5}{-15 \cdot 10^3}$$

$$= 100$$

Thus, demand equals supply at a price of $100, at which point 1,500,000 watches are sold.

Exercise 22

Use Cramer's rule to solve the linear system that you set up for the revised demand-and-supply conditions in Exercise 5 on page 263.

Production Problem

$1x + 4y + 2z = 12$

$3x + 1y + 1z = 10$

$2x + 3y + 0z = 7$

$$x = \frac{\begin{vmatrix} 12 & 4 & 2 \\ 10 & 1 & 1 \\ 7 & 3 & 0 \end{vmatrix}}{\begin{vmatrix} 1 & 4 & 2 \\ 3 & 1 & 1 \\ 2 & 3 & 0 \end{vmatrix}} \begin{matrix} 12 & 4 \\ 10 & 1 \\ 7 & 3 \\ 1 & 4 \\ 3 & 1 \\ 2 & 3 \end{matrix} = \frac{(0 + 28 + 60) - (14 + 36 + 0)}{(0 + 8 + 18) - (4 + 3 + 0)}$$

$$= \frac{38}{19}$$

$$= 2$$

$$y = \frac{\begin{vmatrix} 1 & 12 & 2 \\ 3 & 10 & 1 \\ 2 & 7 & 0 \end{vmatrix} \begin{matrix} 1 & 12 \\ 3 & 10 \\ 2 & 7 \end{matrix}}{19} = \frac{(0 + 24 + 42) - (40 + 7 + 0)}{19}$$

$$= \frac{19}{19}$$

$$= 1$$

From equation 2,

$z = 10 - 3x - y$

$= 10 - 3(2) - 1$

$= 3$

Thus, two units of product A, one unit of product B, and three units of product C should be produced in order to use up all the inventory.

Exercise 23

Use Cramer's rule to solve the linear system that you developed for the expanded inventory conditions in Exercise 6 on page 264.

Least-Squares Regression

For the simple regression model, where math grade is related only to hours studied,

$8i + 16s = 112$

$16i + 40s = 256$

$$i = \frac{\begin{vmatrix} 112 & 16 \\ 256 & 40 \end{vmatrix}}{\begin{vmatrix} 8 & 16 \\ 16 & 40 \end{vmatrix}} = \frac{4480 - 4096}{320 - 256}$$

$$= \frac{384}{64}$$

$$= 6$$

$$s = \frac{\begin{vmatrix} 8 & 112 \\ 16 & 256 \end{vmatrix}}{64} = \frac{2048 - 1792}{64}$$

$$= \frac{256}{64}$$

$$= 4$$

So the best-fitting line for the data is

$y = 6 + 4x$

where y is math quiz grade, and x is hours of study.

For the multiple regression model, where math grade is related to hours studied and math aptitude,

LINEAR SYSTEM SOLUTION METHODS

$$8a + 16b + 16c = 112$$
$$16a + 40b + 37c = 256$$
$$16a + 37b + 40c = 252$$

$$a = \frac{\begin{vmatrix} 112 & 16 & 16 \\ 256 & 40 & 37 \\ 252 & 37 & 40 \end{vmatrix}}{\begin{vmatrix} 8 & 16 & 16 \\ 16 & 40 & 37 \\ 16 & 37 & 40 \end{vmatrix}} = \frac{\begin{vmatrix} 112 & 16 \\ 256 & 40 \\ 252 & 37 \end{vmatrix}}{\begin{vmatrix} 8 & 16 \\ 16 & 40 \\ 16 & 37 \end{vmatrix}} = \frac{1488}{312} = 4.77$$

$$b = \frac{\begin{vmatrix} 8 & 112 & 16 \\ 16 & 256 & 37 \\ 16 & 252 & 40 \end{vmatrix}}{312} = \frac{\begin{vmatrix} 8 & 112 \\ 16 & 256 \\ 16 & 252 \end{vmatrix}}{312} = \frac{928}{312} = 2.97$$

From equation 1, we can solve for c as follows:

$$16c = 112 - 8a - 16b$$
$$c = 7 - \tfrac{1}{2}a - b$$
$$= 7 - \tfrac{1}{2}(4.77) - 2.97$$
$$= 1.64$$

So the best-fitting linear function is

$$y = 4.77 + 2.97x + 1.64z$$

where

- x is hours studied
- z is math aptitude
- y is math quiz grade

Exercises 24

Use Cramer's rule to fit the simple and multiple least-squares regression equations from the normal equations derived in Exercises 7 and 8 on pages 265 and 267.

PROBLEMS

TECHNICAL TRIUMPHS

1. Consider the linear system:

$$x - y = 6$$
$$-2x + 2y = -12$$

a. Does this linear system have a unique solution? Why?
b. If the constant in the second equation was +12, would your answer to part a. change? Why?

2. Consider the linear system as amended in 1b. Solve the system using
   a. graphical methods
   b. substitution methods
   c. elimination methods
   d. Cramer's rule.

3. Consider the linear system:

   ε_1: $2x + 4y + 3z = 4$

   ε_2: $x - y + 5z = -5$

   a. Find $\varepsilon_1 - 2\varepsilon_2$. What happens?
   b. What constant times ε_2 if added to ε_1 would eliminate y?

4. If one had 4 variables to solve for, how many independent equations would be needed to find a unique solution?

5. Suppose one adds two equations and gets a third equation that is 5 times one of the original equations. What can be said about the original equations?

6. If the following determinant has a value of 10, determine the missing value.

$$\begin{vmatrix} 3 & 2 \\ 4 & x \end{vmatrix} = 10$$

7. Find the value of the base determinant (D).

   $x + 3y - z = 10$

   $3x - 3y + 2z = 2$

   $-x + 2y + 4z = 11$

8. a. How many determinants are needed to solve the linear system in problem 7?
   b. Use Cramer's rule to solve that system.

9. If the constants to the right of the equal sign changed in problem 7, how many new determinants would be needed to solve the system?

10. How many determinants would be required to solve a 7×7 linear system?

CONFIDENCE BUILDERS

Refer to the Confidence Builders problems in Chapter 8. Use Cramer's rule to solve each of those problems.

MIND STRETCHERS

Refer to the Mind Stretchers problems in Chapter 8. Use Cramer's rule to solve each of those problems.

CHAPTER 10

Matrix Algebra

10.1 INTRODUCTION

In chapters 8 and 9, we developed the concept of linear systems and learned methods to solve them. Hampered by the lack of good notation and the complexity of the algebra, we were restricted to small systems with only two or three equations. In this chapter we will overcome these problems by developing a new kind of algebra, called *matrix algebra,* which provides a compact, easy-to-manipulate representation of linear systems and is conducive to computer solution.

The previous notation of using x, y, z for the unknowns and a, b, c, and k for the constants was fine for small systems, but can you imagine trying to continue this pattern for large systems? We could go backward in the alphabet using w, v, u, and so on for the unknowns, while using d, e, f, and so on for the constants, but before long there would be a terrible collision in the middle of the alphabet. Then what would we do?

A better way to handle the notation, which will be just dandy regardless of the number of unknowns, is to refer to the ith unknown as x_i. For example, x_1, x_2, x_3, x_4 would be the first, second, third, and fourth unknowns; x_{29} would be the 29th unknown. All the left-side coefficients in the system can be described by a_{ij}, where i is the number of the equation and j is the number of the unknown it multiplies. For example, a_{42} would be the coefficient in the fourth equation that multiplies the second unknown. The right-side constants are described by b_i, where i is the equation number. Using this notation the general linear system with n unknowns and n equations would look like this:

$a_{11}x_1 + a_{12}x_2 + a_{13}x_3 + a_{14}x_4 + \ldots + a_{1n}x_n = b_1$

$a_{21}x_1 + a_{22}x_2 + a_{23}x_3 + a_{24}x_4 + \ldots + a_{2n}x_n = b_2$

$a_{31}x_1 + a_{32}x_2 + a_{33}x_3 + a_{34}x_4 + \ldots + a_{3n}x_n = b_3$

$a_{41}x_1 + a_{42}x_2 + a_{43}x_3 + a_{44}x_4 + \ldots + a_{4n}x_n = b_4$

$\vdots \qquad\qquad\qquad\qquad \vdots$

$a_{n1}x_1 + a_{n2}x_2 + a_{n3}x_3 + a_{n4}x_4 + \ldots + a_{nn}x_n = b_n$

Examples

$$x_1 + 2x_2 = 5$$
$$3x_1 + 4x_2 = 11$$

Unknowns: x_1, x_2

$a_{11} = 1 \quad a_{12} = 2 \quad b_1 = 5$
$a_{21} = 3 \quad a_{22} = 4 \quad b_2 = 11$

Unknowns: x_1, x_2, x_3

$$x_1 + x_2 - 2x_3 = 3$$
$$x_2 - x_3 = 0$$
$$2x_1 - 4x_2 + 3x_3 = -1$$

$a_{11} = 1 \quad a_{12} = 1 \quad a_{13} = -2 \quad b_1 = 3$
$a_{21} = 0 \quad a_{22} = 1 \quad a_{23} = -1 \quad b_2 = 0$
$a_{31} = 2 \quad a_{32} = -4 \quad a_{33} = 3 \quad b_3 = -1$

We could "extract" the coefficients (a_{ij}) in a linear system in their natural arrangement. For example, this extraction for the first system above would be what we call **A** below

$$\mathbf{A} = \begin{pmatrix} 1 & 2 \\ 3 & 4 \end{pmatrix}$$

Such an arrangement is called a *matrix*. We could even extract the unknowns to form the matrix **X** and the right side constants to form the matrix **B** as

$$\mathbf{X} = \begin{pmatrix} x_1 \\ x_2 \end{pmatrix} \quad \mathbf{B} = \begin{pmatrix} 5 \\ 11 \end{pmatrix}$$

Exercise 1

Define the **A**, **X**, and **B** matrices for the second example system.

Now you probably won't believe this, but these systems can be described in terms of the following matrix equation.

AX = B

This matrix equation is read, "The **A** matrix times the **X** matrix equals the **B** matrix."

Such an equation should remind you of our old friend, the single equation with one unknown,

$ax = b$

where a and b are constants.

Back in the glory days of solving such equations, we merely multiplied both sides by the inverse of the number a, or a^{-1}, to isolate and solve for the single unknown.

$$a^{-1}ax = a^{-1}b$$
$$x = a^{-1}b$$

MATRIX ALGEBRA

Would it be too much to ask that matrices also have inverses? If that were only true, we could do as follows—

$$AX = B$$
$$A^{-1}AX = A^{-1}B$$
$$X = A^{-1}B$$

where A^{-1} is the symbol for the inverse of A, and isolate and solve for all the unknowns at once. The thought of this truly excites the mind!

Learning Plan

Yes, Virginia, there is a matrix inverse. It does indeed allow for the isolation and solution of a whole bunch of unknowns. But before you can apply these wonders, we must settle down into a deliberate development of matrix algebra, including the following areas of inquiry.

1. What *is* a matrix?
2. What can you do (like add or multiply) with matrices!
3. How is it possible that little old $AX = B$ is the same as all those big equations?
4. How can you find one of those A inverse things?
5. What do you do with an A inverse if you are lucky enough to find one?

We will go to the files of Cantaford Motor Co. and Mate Tricks University to illustrate these ideas in the everyday world.

Finally, with all this pent-up knowledge, we will burst forth to apply matrix methods to solve some linear systems. We conclude by showing how computers solve large systems.

10.2 MATRIX BASICS

In this section we define a matrix, develop a scheme for naming it and its elements, explain order and transpose, introduce the idea of matrix equality, and identify several important types of matrices.

What is a Matrix?

A matrix is a rectangular array of numbers or letters representing numbers. We name matrices with capital letters. Here are some examples, which we refer to throughout the chapter.

$$A = \begin{pmatrix} 1 & 2 \\ 3 & 4 \end{pmatrix} \qquad B = \begin{pmatrix} 1 & -1 & 3 \\ 6 & 5 & 4 \end{pmatrix}$$

$$C = \begin{pmatrix} 1 & 2 & 3 \\ 0 & 0 & -5 \\ 9 & 10 & \frac{1}{2} \end{pmatrix} \qquad E = \begin{pmatrix} 3 & 0 & 0 \\ 0 & -1 & 0 \\ 0 & 0 & 4 \end{pmatrix}$$

$$\mathbf{X} = \begin{pmatrix} x_1 \\ x_2 \\ x_3 \\ x_4 \end{pmatrix} \qquad \mathbf{W} = \begin{pmatrix} 5 & 0 \\ 2 & 1 \\ -1 & 3 \end{pmatrix}$$

$$\mathbf{Q} = \begin{pmatrix} -3 & 2 & 6 \end{pmatrix}$$

The elements of a matrix are named by the small letter of the naming capital letter. For example, each element in the **A** matrix is referred to as a. Since each element is situated at the intersection of a row and column, we must describe each with a double subscript; thus, a_{ij} is the general element of the **A** matrix at the intersection of the ith row and jth column. For example, a_{12} refers to the element in the **A** matrix located at the intersection of the first row and second column ($a_{12} = 2$); b_{23} refers to the element in the **B** matrix at the intersection of the second row and third column ($b_{23} = 4$).

Examples · Exercises 2

Refer to the matrices defined earlier in this chapter.

$a_{11} = 1$	$a_{22} =$
$b_{12} = -1$	$b_{21} =$
$c_{31} = 9$	$c_{32} =$
$e_{13} = 0$	$e_{11} =$
$x_{21} = x_2$	$x_{41} =$
$w_{22} = 1$	$w_{32} =$
$q_{13} = 6$	$q_{12} =$

Order

Matrices are characterized by their number of rows and columns, called their *order*. This is reminiscent of the characterization of linear systems. In fact, a matrix formed from the left-side coefficients of a 2×2 linear system would have order of 2×2. The order of **A** is 2×2 (two rows by two columns); **B**'s order is 2×3 (two rows by three columns); **C**'s order is 3×3 (three rows by three columns); and **X**'s order is 4×1 (four rows by one column).

Exercise 3

Determine the orders of **E**, **W**, and **Q**.

Matrices can be practically employed as a means to organize and cross-classify data.

MATRIX ALGEBRA

Examples

Cantaford lists its inventory by type of car and color as shown in the following 3×4 matrix, which we name **F**.

		Color		
Type	Blue	Green	Gold	Red
Focus	6	3	2	1
Taurus	4	7	3	2
Explorer	2	0	5	0

$$\begin{pmatrix} 6 & 3 & 2 & 1 \\ 4 & 7 & 3 & 2 \\ 2 & 0 & 5 & 0 \end{pmatrix} = \mathbf{F}$$

Mate Tricks University classifies students by level and major as shown in this 2×3 matrix, which we call **T**.

		Major	
Level	Business	Liberal arts	Math-science
Undergraduate	300	1600	100
Graduate	200	400	100

$$\begin{pmatrix} 300 & 1600 & 100 \\ 200 & 400 & 100 \end{pmatrix} = \mathbf{T}$$

The average number of courses taken by Mate Tricks University students in the three departments per semester is given in the following 3×3 matrix, which we call **N**.

		Courses	
Major	Business	Liberal arts	Math-science
Business	2	2	1
Liberal arts	$\frac{1}{2}$	4	$\frac{1}{2}$
Math-science	$\frac{1}{2}$	$1\frac{1}{2}$	$3\frac{1}{2}$

$$\begin{pmatrix} 2 & 2 & 1 \\ \frac{1}{2} & 4 & \frac{1}{2} \\ \frac{1}{2} & 1\frac{1}{2} & 3\frac{1}{2} \end{pmatrix} = \mathbf{N}$$

For example, business students average two business courses, two liberal arts courses, and one math-science course per semester.

Transpose

Every matrix **M** has a transpose, which we symbolize as **M'**. **M'** is formed by interchanging the rows and columns of **M**. For example

If $\mathbf{M} = \begin{pmatrix} 1 & 2 \\ 3 & 4 \end{pmatrix}$ then $\mathbf{M'} = \begin{pmatrix} 1 & 3 \\ 2 & 4 \end{pmatrix}$

Notice how the first row of **M** is the first column of **M'** and the second row of **M** is the second column of **M'**.

The transpose of the **F** matrix defined earlier is found as

$$\mathbf{F} = \begin{pmatrix} 6 & 3 & 2 & 1 \\ 4 & 7 & 3 & 2 \\ 2 & 0 & 5 & 0 \end{pmatrix} \quad \text{so } \mathbf{F}' = \begin{pmatrix} 6 & 4 & 2 \\ 3 & 7 & 0 \\ 2 & 3 & 5 \\ 1 & 2 & 0 \end{pmatrix}$$

The order of the transpose matrix is found by reversing the numbers for the order of the original matrix. For example, $\mathbf{F}$ is a 3×4 matrix, while $\mathbf{F}'$ is a 4×3 matrix.

Exercises 4

1. Determine $\mathbf{T}'$ and $\mathbf{N}'$ from the $\mathbf{T}$ and $\mathbf{N}$ matrices defined earlier. Compare the orders of the matrices and their respective transposes.
2. What happens when one takes the transpose of a transpose?

Matrix Equality

Two matrices are equal only if all the corresponding elements are equal.
Thus,

$$\begin{pmatrix} 1 & 2 \\ 3 & 4 \end{pmatrix} = \begin{pmatrix} 1 & 2 \\ 3 & 4 \end{pmatrix}$$

but

$$\begin{pmatrix} 1 & 2 \\ 3 & 4 \end{pmatrix} \neq \begin{pmatrix} 4 & 3 \\ 2 & 1 \end{pmatrix}$$

Exercises 5

1. Are any of the following pairs of matrices equal?

   $$(3 \quad 4) \quad (3 \quad 4 \quad 0)$$

   $$\begin{pmatrix} -5 & -3 \\ 7 & 9 \end{pmatrix} \quad \begin{pmatrix} -5 & -3 \\ 7 & -9 \end{pmatrix}$$

   $$\begin{pmatrix} 5 & 8 \\ 4 & 7 \end{pmatrix} \quad \begin{pmatrix} 4 & 7 \\ 5 & 8 \end{pmatrix}$$

2. Construct a matrix that is equal to its transpose.

Special Types of Matrices

Here we introduce special types of matrices that have wide applicability. These are square, vector, identity, elementary, diagonal, unity, and null matrices.

Square

A matrix with an equal number of rows and columns is called *square*. Matrices **A**, **C**, **E**, and **N**, defined earlier, are square matrices.

Only square matrices have a main diagonal, which is formed by the elements in the (1, 1), (2, 2), (3, 3), ..., (n, n) positions.

$$\begin{pmatrix} m & & & \\ & a & & \\ & & i & \\ & & & n \end{pmatrix}$$

Vector

Matrices with one row (e.g., **Q**) are called *row vectors*. Matrices with one column (e.g., **X**) are called *column vectors*. Every matrix can be considered a composite of one or more vectors. For example, matrix **B** contains two row vectors (1 −1 3) and (6 5 4). Can you describe the three column vectors that make up **B**?

Identity

Square matrices with 1's on the main diagonal and 0's elsewhere are called *identity matrices* and are symbolized by **I**. Some examples are

$$\mathbf{I}_{2\times 2} = \begin{pmatrix} 1 & 0 \\ 0 & 1 \end{pmatrix} \qquad \mathbf{I}_{4\times 4} = \begin{pmatrix} 1 & 0 & 0 & 0 \\ 0 & 1 & 0 & 0 \\ 0 & 0 & 1 & 0 \\ 0 & 0 & 0 & 1 \end{pmatrix}$$

Every possible order of square matrix would have an identity matrix. Can you write the 3×3 identity matrix?

Elementary

Elementary matrices are square and are formed by changing one element of an identity matrix to a nonzero number. Some examples are

$$\begin{pmatrix} -2 & 0 \\ 0 & 1 \end{pmatrix} \qquad \begin{pmatrix} 1 & 1 & 0 \\ 0 & 1 & 0 \\ 0 & 0 & 1 \end{pmatrix} \qquad \begin{pmatrix} 1 & 0 & 0 & 0 \\ 0 & 1 & 0 & 0 \\ 4 & 0 & 1 & 0 \\ 0 & 0 & 0 & 1 \end{pmatrix}$$

Diagonal

Diagonal matrices are square and have nonzero values only on the main diagonal. Identity matrices are diagonal. Examples of others are

$$\begin{pmatrix} 2 & 0 \\ 0 & \frac{1}{2} \end{pmatrix} \quad \begin{pmatrix} 5 & 0 & 0 \\ 0 & \frac{1}{4} & 0 \\ 0 & 0 & -7 \end{pmatrix} \quad \begin{pmatrix} 1 & 0 & 0 & 0 \\ 0 & 2 & 0 & 0 \\ 0 & 0 & 3 & 0 \\ 0 & 0 & 0 & 4 \end{pmatrix}$$

Unity

Unity matrices contain elements that are all 1's. Some examples are

$$\begin{pmatrix} 1 \\ 1 \end{pmatrix} \quad (1 \ 1 \ 1) \quad \begin{pmatrix} 1 & 1 & 1 \\ 1 & 1 & 1 \end{pmatrix}$$

Null

A *null matrix* is a big nothing—zeros everywhere. Some examples are

$$\begin{pmatrix} 0 & 0 \\ 0 & 0 \end{pmatrix} \quad \begin{pmatrix} 0 & 0 \\ 0 & 0 \\ 0 & 0 \end{pmatrix} \quad (0 \ 0 \ 0 \ 0)$$

Exercises 6

1. Explain why

$$\begin{pmatrix} 1 & 0 & 0 \\ 0 & 1 & 0 \end{pmatrix}$$

   is not an identity matrix.
2. Determine the order and type of the following matrices.

   a. $\begin{pmatrix} 5 & 10 \\ 8 & -2 \end{pmatrix}$

   b. $(1 \ \ 2 \ \ 3 \ \ -4)$

   c. $\begin{pmatrix} 1 & 0 & 0 \\ 0 & 1 & 2 \\ 0 & 0 & 1 \end{pmatrix}$

   d. $\begin{pmatrix} 6 & 0 & 0 & 0 \\ 0 & 1 & 0 & 0 \\ 0 & 0 & 8 & 0 \\ 0 & 0 & 0 & 4 \end{pmatrix}$

   e. $\begin{pmatrix} 0 & 0 & 0 \\ 0 & 0 & 0 \end{pmatrix}$

   f. $\begin{pmatrix} 1 \\ 1 \end{pmatrix}$

MATRIX ALGEBRA

g. $\begin{pmatrix} 1 & 0 & 0 & 0 & 0 \\ 0 & 1 & 0 & 0 & 0 \\ 0 & 0 & 1 & 0 & 0 \\ 0 & 0 & 0 & 1 & 0 \\ 0 & 0 & 0 & 0 & 1 \end{pmatrix}$

h. $\begin{pmatrix} 1 & 0 & 0 \\ 0 & 5 & 0 \\ 0 & 0 & 1 \end{pmatrix}$

i. $\begin{pmatrix} 1 & -2 & 4 \\ 0 & 0 & 1 \end{pmatrix}$

j. $\begin{pmatrix} 0 & 0 \\ 0 & 1 \end{pmatrix}$

3. Make up matrices with the following characteristics.
   a. 2×2 diagonal
   b. 1×6 row vector
   c. 4×1 column vector
   d. 6×6 identity
   e. 3×2 unity
   f. 4×3 null
   g. 3×3 elementary
   h. 2×2 square

10.3 MATRIX OPERATIONS

We have now defined matrices, but what can we do with them? In this section we will answer this question by showing how to multiply a matrix by a constant and how to add, subtract, and multiply two matrices.

Multiplication of a Matrix by a Constant

Any matrix can be multiplied by a constant, k, which results in a new matrix, the elements of which are all k times the elements in the original matrix.

Examples

1. $2 \begin{pmatrix} 1 & 2 \\ 3 & 4 \end{pmatrix} = \begin{pmatrix} 2 & 4 \\ 6 & 8 \end{pmatrix}$

2. $3 \begin{pmatrix} x_1 \\ x_2 \\ x_3 \end{pmatrix} = \begin{pmatrix} 3x_1 \\ 3x_2 \\ 3x_3 \end{pmatrix}$

3. $-1\begin{pmatrix} 5 & -2 & 3 \\ 0 & 1 & -4 \end{pmatrix} = \begin{pmatrix} -5 & 2 & -3 \\ 0 & -1 & 4 \end{pmatrix}$

4. $4(-3 \quad 2) = (-12 \quad 8)$

Exercises 7

1. $4\begin{pmatrix} 2 & 0 \\ 0 & 2 \end{pmatrix} =$

2. $2(4 \quad 1 \quad 0) =$

3. $-2\begin{pmatrix} 1 & -1 \\ 2 & -2 \\ 3 & -3 \end{pmatrix} =$

If Cantaford wanted to determine the total number of tires on its various categories of cars, it could multiply the current inventory matrix **F** by 5 (spare tire included).

$$5\begin{pmatrix} 6 & 3 & 2 & 1 \\ 4 & 7 & 3 & 2 \\ 2 & 0 & 5 & 0 \end{pmatrix} = \begin{pmatrix} 30 & 15 & 10 & 5 \\ 20 & 35 & 15 & 10 \\ 10 & 0 & 25 & 0 \end{pmatrix}$$

Exercise 8

If each Mate Tricks University course is three credits, convert the **N** matrix to one that gives average number of credits taken.

Addition and Subtraction of Matrices

Two matrices,[1] **L** and **R**, can be added to give a sum matrix, **S**

L + R = S

only if they are of the same order. The sum matrix has the same order as the two matrices that are added.

If l_{ij} is the general element of the **L** matrix and r_{ij} is the general element of the **R** matrix, then s_{ij}, the general element of the **S** matrix, is

$s_{ij} = l_{ij} + r_{ij}$ for all ij possibilities

---

[1] We define these matrices as **L** (for left) and **R** (for right). Although the arrangement doesn't matter here, it is a good idea to start thinking about it since the arrangement does matter for matrix multiplication.

MATRIX ALGEBRA

Examples

1. $\begin{pmatrix} 1 & 2 \\ 3 & 4 \end{pmatrix} + \begin{pmatrix} 5 & 6 \\ 7 & 8 \end{pmatrix} = \begin{pmatrix} 1+5 & 2+6 \\ 3+7 & 4+8 \end{pmatrix} = \begin{pmatrix} 6 & 8 \\ 10 & 12 \end{pmatrix}$

2. $\begin{pmatrix} 3 & -2 \\ 0 & 4 \\ -5 & 1 \end{pmatrix} + \begin{pmatrix} \frac{1}{2} & 1 \\ 7 & 2 \\ 5 & 3 \end{pmatrix} = \begin{pmatrix} 3+\frac{1}{2} & -2+1 \\ 0+7 & 4+2 \\ -5+5 & 1+3 \end{pmatrix} = \begin{pmatrix} 3\frac{1}{2} & -1 \\ 7 & 6 \\ 0 & 4 \end{pmatrix}$

3. $(1 \quad 2 \quad 3) + (8 \quad 7 \quad 6) = (1+8 \quad 2+7 \quad 3+6) = (9 \quad 9 \quad 9)$

Exercises 9

1. $\begin{pmatrix} 1 & 2 \\ 3 & 4 \end{pmatrix} + \begin{pmatrix} 1 & 0 \\ 0 & 1 \end{pmatrix} =$

2. $\begin{pmatrix} 6 & 3 & -1 \\ 2 & -4 & 5 \end{pmatrix} + \begin{pmatrix} 0 & 2 & 4 \\ 7 & 1 & -1 \end{pmatrix} =$

3. $\begin{pmatrix} 5 \\ 5 \end{pmatrix} + \begin{pmatrix} 10 \\ 10 \end{pmatrix} =$

Matrix subtraction is similar to addition. It, too, can be accomplished only with matrices of equal order. Then **L** minus **R** gives a difference matrix **D**.

L − **R** = **D**

The order for **D** is the same as for the other two matrices. Its general element, d_{ij}, is equal to

$d_{ij} = l_{ij} - r_{ij}$ for all *ij* possibilities

Thus, matrix subtraction is a subtraction of corresponding elements.

Examples

1. $\begin{pmatrix} 1 & 2 \\ 3 & 4 \end{pmatrix} - \begin{pmatrix} 2 & 1 \\ 3 & 2 \end{pmatrix} = \begin{pmatrix} 1-2 & 2-1 \\ 3-3 & 4-2 \end{pmatrix} = \begin{pmatrix} -1 & 1 \\ 0 & 2 \end{pmatrix}$

2. $(7 \quad -3 \quad 5) - (-8 \quad 2 \quad 0) = (7-(-8) \quad -3-2 \quad 5-0)$
$= (15 \quad -5 \quad 5)$

Exercises 10

1. $(1 \quad 3 \quad -2 \quad 4) - (1 \quad 2 \quad 5 \quad -3) =$

2. $\begin{pmatrix} 1 & 0 & 0 \\ 0 & 1 & 0 \\ 0 & 0 & 1 \end{pmatrix} - \begin{pmatrix} .2 & .4 & .3 \\ .1 & .1 & .5 \\ .4 & .3 & .3 \end{pmatrix} =$

Cantaford expects a shipment of cars with breakdown by type and color as given in the **G** matrix.

$$\mathbf{G} = \begin{pmatrix} 0 & 0 & 3 & 3 \\ 1 & 0 & 2 & 1 \\ 2 & 3 & 1 & 4 \end{pmatrix}$$

When the shipment arrives, the total inventory matrix **H** can be found by adding **G** to **F** defined earlier.

$$\mathbf{H} = \mathbf{F} + \mathbf{G} = \begin{pmatrix} 6 & 3 & 2 & 1 \\ 4 & 7 & 3 & 2 \\ 2 & 0 & 5 & 0 \end{pmatrix} + \begin{pmatrix} 0 & 0 & 3 & 3 \\ 1 & 0 & 2 & 1 \\ 2 & 3 & 1 & 4 \end{pmatrix} = \begin{pmatrix} 6 & 3 & 5 & 4 \\ 5 & 7 & 5 & 3 \\ 4 & 3 & 6 & 4 \end{pmatrix}$$

If the sales breakdown during the next month is forecasted in the **J** matrix

$$\mathbf{J} = \begin{pmatrix} 3 & 3 & 2 & 2 \\ 1 & 2 & 4 & 0 \\ 2 & 0 & 5 & 2 \end{pmatrix}$$

then the remaining inventory breakdown **K** is forecasted to be

$$\mathbf{K} = \mathbf{H} - \mathbf{J} = \begin{pmatrix} 6 & 3 & 5 & 4 \\ 5 & 7 & 5 & 3 \\ 4 & 3 & 6 & 4 \end{pmatrix} - \begin{pmatrix} 3 & 3 & 2 & 2 \\ 1 & 2 & 4 & 0 \\ 2 & 0 & 5 & 2 \end{pmatrix} = \begin{pmatrix} 3 & 0 & 3 & 2 \\ 4 & 5 & 1 & 3 \\ 2 & 3 & 1 & 2 \end{pmatrix}$$

Exercise 11

Mate Tricks University forecasts their graduating (**Y**) and fall semester incoming student (**Z**) matrices, classified by level and major, to be

$$\mathbf{Y} = \begin{pmatrix} 50 & 500 & 25 \\ 100 & 200 & 50 \end{pmatrix} \quad \mathbf{Z} = \begin{pmatrix} 150 & 400 & 25 \\ 150 & 100 & 25 \end{pmatrix}$$

Find the fall semester forecasted student body breakdown matrix by the appropriate matrix addition and subtraction operations.

Combining Addition (Subtraction) and Constant Multiplication

The matrix algebra operations we have studied thus far can be combined to add or subtract matrices multiplied by constants. These combined operations are especially important in the process of finding an inverse to a matrix. There we will apply these combined operations to row vectors, as illustrated now.

1. $(2 \quad 4) - 2(1 \quad 3) = (2 \quad 4) - (2 \quad -6)$
$= (0 \quad 10)$

2. $(-3 \quad 0 \quad 3) + 3(1 \quad 1 \quad 2) = (-3 \quad 0 \quad 3) + (3 \quad 3 \quad 6)$
$= (0 \quad 3 \quad 9)$

3. $(0 \quad 5 \quad -2) - 5(0 \quad 1 \quad 6) = (0 \quad 5 \quad -2) - (0 \quad 5 \quad 30)$
$= (0 \quad 0 \quad -32)$

Exercises 12

1. $(2 \quad 3 \quad -2 \quad 4) - 2(1 \quad 1 \quad 7 \quad \tfrac{1}{2}) =$
2. $(0 \quad -5 \quad 1 \quad 9) + 5(0 \quad 1 \quad -2 \quad 3) =$

Matrix Multiplication

Since matrix addition and subtraction involved addition and subtraction of corresponding elements, it would make sense for matrix multiplication to involve multiplication of corresponding elements, such as

$$\begin{pmatrix} 1 & 2 \\ 3 & 4 \end{pmatrix} \cdot \begin{pmatrix} 0 & -1 \\ 2 & 4 \end{pmatrix} = \begin{pmatrix} 0 & -2 \\ 6 & 16 \end{pmatrix}$$ WARNING: Cover your eyes. Do not look at this. It is WRONG!!

If matrix multiplication were so simple, students would love studying it and get 100% on all their exams. But, the results would be useless. No linear systems could be solved. Many real world problems would go unanswered. So what to do? Well we must learn a complicated way to multiply matrices. Could this just be a diabolical plot by mathematicians to blow our minds? Or are we experiencing the old adage, "No pain, no gain"? Of course, the latter is true. But wait, there is good news. What will start out as looking complicated will soon, with a little effort, turn out to be easy. So, be patient!

Two matrices, **L** and **R** (we denote a left (**L**) and a right (**R**) matrix, since the arrangement is important), can be multiplied *only* if the number of columns of **L** equals the number of rows of **R**. The result is a product matrix, **P**, having the number of rows of **L** and the number of columns of **R**. (See Figure 10.1)

Figure 10.1

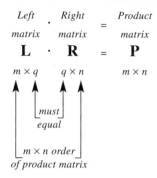

Examples

1. $(L_{2\times 4}) \cdot (R_{4\times 3}) = P_{2\times 3}$
2. $(L_{1\times 5}) \cdot (R_{5\times 1}) = P_{1\times 1}$
3. $(L_{6\times 3}) \cdot (R_{3\times 4}) = P_{6\times 4}$
4. $(L_{3\times 4}) \cdot (R_{6\times 3})$ Can't be multiplied since the number of columns of **L** doesn't equal the number of rows of **R**.

Exercises 13

Which of the following matrix multiplications can be performed, and what is the order of the product matrix?

1. $(L_{4\times 7}) \cdot (R_{7\times 2})$
2. $(L_{5\times 1}) \cdot (R_{1\times 5})$
3. $(L_{3\times 2}) \cdot (R_{3\times 2})$
4. $(L_{1\times 4}) \cdot (R_{4\times 1})$
5. $(L_{2\times 9}) \cdot (R_{2\times 8})$

Elements of Product Matrix

Once it is established that two matrices can be multiplied, the next question is, "What are the values of the $m \times n$ elements in the product matrix?" This is answered by the seemingly complicated formula,

$$p_{ij} = \sum_{k=1}^{q} l_{ik} r_{kj}$$ where q, the multiplication length, equals the number of columns of **L** and the number of rows of **R**

$$= l_{i1} r_{1j} + l_{i2} r_{2j} + l_{i3} r_{3j} + \cdots + l_{iq} r_{qj}$$

which gives the product matrix element located at the intersection of the ith row and jth column. Don't let this formula scare you. We need it to learn the basics; however, once you get the hang of it, you'll be multiplying matrices beautifully without it.

MATRIX ALGEBRA

Figure 10.2

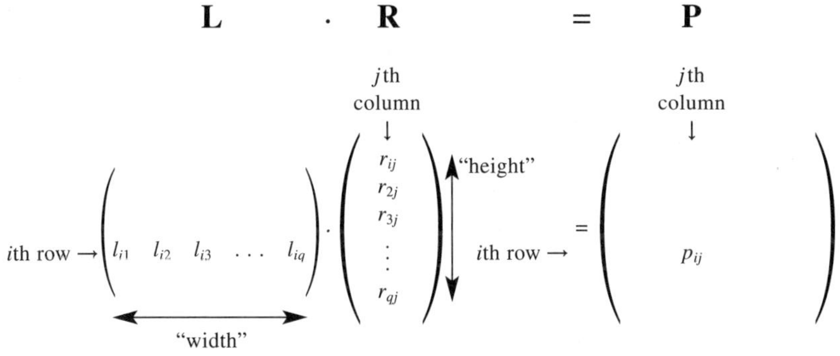

$$\mathbf{L} \cdot \mathbf{R} = \mathbf{P}$$

ith row $\to \begin{pmatrix} l_{i1} & l_{i2} & l_{i3} & \cdots & l_{iq} \end{pmatrix}$ · $\begin{pmatrix} r_{1j} \\ r_{2j} \\ r_{3j} \\ \vdots \\ r_{qj} \end{pmatrix}$ $\updownarrow$ "height" $= \begin{pmatrix} & & \\ & p_{ij} & \\ & & \end{pmatrix}$ ith row $\to$

$\xleftarrow{\text{"width"}}$

jth column $\downarrow$ (above R); jth column $\downarrow$ (above P)

Note: The "width" of **L** (number of columns) must equal the "height" of **R** (number of rows). But the other "dimensions" need not be equal. So our figure shows **L** to be "wider" than **R** and **R** to be "taller" than **L**.

The formula (see Figure 10.2) dictates that the first element in the ith row of **L**, or l_{i1}, is multiplied by the first element in the jth column of **R**, or r_{1j}, to give $l_{i1}r_{1j}$. Then the second element of the ith row of **L**, or l_{i2}, is multiplied by the second element of the jth column of **R**, or r_{2j}, to give $l_{i2}r_{2j}$. The third element of the ith row of **L**, or l_{i3}, is multiplied by the third element of the jth column of **R**, or r_{3j}, to give $l_{i3}r_{3j}$. The process of multiplying successive elements of the ith row of **L** by successive elements of the jth column of **R** is continued until the last element of the ith row of **L**, or l_{iq}, is multiplied by the last element of the jth column of **R**, or r_{qj}, to give $l_{iq}r_{qj}$. These q products are then summed to give the value of p_{ij}.

If you survived that maze of i's, j's, k's, and q's, you can probably see the sense of the matrix multiplication rule, which states that the number of columns of **L** must equal the number of rows of **R**. Since the successive q elements in a row of **L** (meaning that **L** has q columns) must be multiplied in turn by the successive elements in a column of **R**, it follows that each column of **R** must have q numbers (meaning that **R** has q rows).

Let's take a few concrete examples to clarify these matrix multiplication procedures.

Example 1

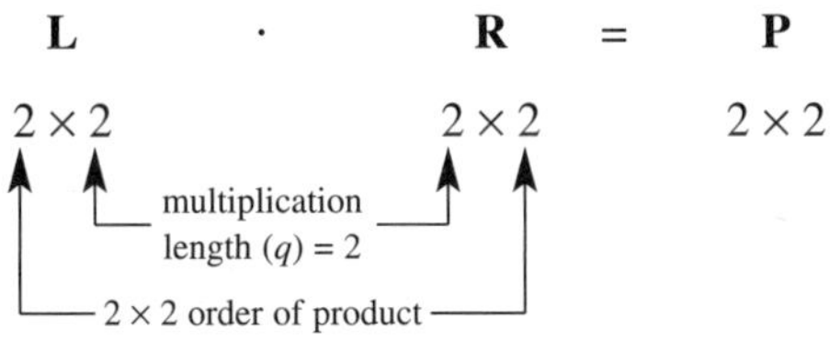

$$\mathbf{L} \cdot \mathbf{R} = \mathbf{P}$$
$$2 \times 2 \quad\quad 2 \times 2 \quad\quad 2 \times 2$$

multiplication length (q) = 2

2×2 order of product

$$\begin{pmatrix} 1 & 2 \\ 3 & 4 \end{pmatrix} \cdot \begin{pmatrix} 5 & 6 \\ 7 & 8 \end{pmatrix} = \begin{pmatrix} p_{11} & p_{12} \\ p_{21} & p_{22} \end{pmatrix}$$

Since q, the multiplication length, is 2, the formula for finding the elements in the product matrix is

$$p_{ij} = \sum_{k=1}^{2} l_{ik}r_{kj}$$

The value of the element at the intersection of the first row ($i = 1$) and the first column ($j = 1$) of the product matrix is found by substituting these values for i and j in the general formula.

$$\begin{pmatrix} i = 1 \\ j = 1 \end{pmatrix} \quad p_{11} = \sum_{k=1}^{2} l_{1k} r_{k1}$$

$$= \overbrace{l_{11} r_{11}}^{k=1} + \overbrace{l_{12} r_{21}}^{k=2}$$

$$= 1 \cdot 5 + 2 \cdot 7 = 19$$

$$\begin{pmatrix} 1 & 2 \\ 3 & 4 \end{pmatrix} \begin{pmatrix} 5 & 6 \\ 7 & 8 \end{pmatrix}$$

The value of the element at the intersection of the first row ($i = 1$) and the second column ($j = 2$) of the product matrix is found by substituting these values for i and j in the general formula.

$$\begin{pmatrix} i = 1 \\ j = 2 \end{pmatrix} \quad p_{12} = \sum_{k=1}^{2} l_{1k} r_{k2}$$

$$= \overbrace{l_{11} r_{12}}^{k=1} + \overbrace{l_{12} r_{22}}^{k=2}$$

$$= 1 \cdot 6 + 2 \cdot 8 = 22$$

$$\begin{pmatrix} 1 & 2 \\ 3 & 4 \end{pmatrix} \begin{pmatrix} 5 & 6 \\ 7 & 8 \end{pmatrix}$$

The value of the element at the intersection of the second row ($i = 2$) and the first column ($j = 1$) of the product matrix is found by substituting these values for i and j in the general equation.

$$\begin{pmatrix} i = 2 \\ j = 1 \end{pmatrix} \quad p_{21} = \sum_{k=1}^{2} l_{2k} r_{k1}$$

$$= \overbrace{l_{21} r_{11}}^{k=1} + \overbrace{l_{22} r_{21}}^{k=2}$$

$$= 3 \cdot 5 + 4 \cdot 7 = 43$$

$$\begin{pmatrix} 1 & 2 \\ 3 & 4 \end{pmatrix} \begin{pmatrix} 5 & 6 \\ 7 & 8 \end{pmatrix}$$

Exercise 14

Beginning with the general formula, show the steps necessary to find $p_{22} = 50$.

Thus, the 2×2 product matrix is

MATRIX ALGEBRA

$$\mathbf{P} = \begin{pmatrix} p_{11} & p_{12} \\ p_{21} & p_{22} \end{pmatrix} = \begin{pmatrix} 19 & 22 \\ 43 & 50 \end{pmatrix}$$

Let's summarize our methods and see the big picture in Figure 10.3.

Figure 10.3

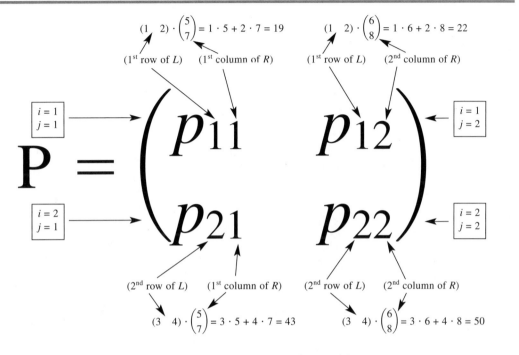

Keep alert for one more example, and then let's discuss it.

Example 2

$$\begin{array}{ccccc} \mathbf{L} & \cdot & \mathbf{R} & = & \mathbf{P} \\ 2 \times 3 & & 3 \times 2 & & 2 \times 2 \end{array}$$

multiplication length (q) = 3

2 × 2 order of product

$$\begin{pmatrix} 1 & 2 & 3 \\ 4 & 5 & 6 \end{pmatrix} \cdot \begin{pmatrix} 7 & 8 \\ 9 & 10 \\ 11 & 12 \end{pmatrix} = \begin{pmatrix} p_{11} & p_{12} \\ p_{21} & p_{22} \end{pmatrix}$$

Since q, the multiplication length, is 3, the formula for finding the elements in the product matrix is

$$p_{ij} = \sum_{k=1}^{3} l_{ik} r_{kj}$$

The value of the element at the intersection of the first row ($i = 1$) and the first column ($j = 1$) of the product matrix is found by substituting these values for i and j in the general formula.

$$\begin{pmatrix} i = 1 \\ j = 1 \end{pmatrix} \quad p_{11} = \sum_{k=1}^{3} l_{1k} r_{k1}$$

$$p_{11} = \overbrace{l_{11} r_{11}}^{k=1} + \overbrace{l_{12} r_{21}}^{k=2} + \overbrace{l_{13} r_{31}}^{k=3}$$

$$= 1 \cdot 7 + 2 \cdot 9 + 3 \cdot 11 = 58$$

$$\begin{pmatrix} 1 & 2 & 3 \\ 4 & 5 & 6 \end{pmatrix} \begin{pmatrix} 7 & 8 \\ 9 & 10 \\ 11 & 12 \end{pmatrix}$$

The value of the element at the intersection of the second row ($i = 2$) and the second column ($j = 2$) of the product matrix is found by substituting these values for i and j in the general formula.

$$\begin{pmatrix} i = 2 \\ j = 2 \end{pmatrix} \quad p_{22} = \sum_{k=1}^{3} l_{2k} r_{k2}$$

$$= \overbrace{l_{21} r_{12}}^{k=1} + \overbrace{l_{22} r_{22}}^{k=2} + \overbrace{l_{23} r_{32}}^{k=3}$$

$$= 4 \cdot 8 + 5 \cdot 10 + 6 \cdot 12 = 154$$

$$\begin{pmatrix} 1 & 2 & 3 \\ 4 & 5 & 6 \end{pmatrix} \begin{pmatrix} 7 & 8 \\ 9 & 10 \\ 11 & 12 \end{pmatrix}$$

Exercises 15

Verify that

$p_{12} = 1 \cdot 8 + 2 \cdot 10 + 3 \cdot 12 = 64$

$p_{21} = 4 \cdot 7 + 5 \cdot 9 + 6 \cdot 11 = 139$

So the completed matrix multiplication is

$$\begin{pmatrix} 1 & 2 & 3 \\ 4 & 5 & 6 \end{pmatrix} \cdot \begin{pmatrix} 7 & 8 \\ 9 & 10 \\ 11 & 12 \end{pmatrix} = \begin{pmatrix} 58 & 64 \\ 139 & 154 \end{pmatrix}$$

If your eagle eyes have been paying piercing attention to our complicated formula, and the examples of its use, you will now be rewarded as we jettison it forever.

Recall that each product matrix element results from using some row of the **L** matrix along with some column of the **R** matrix. The row (i) and column (j) location of the product matrix

MATRIX ALGEBRA

Figure 10.4

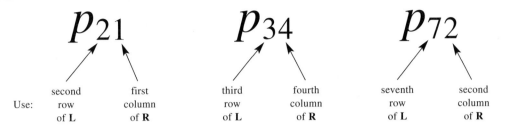

element tells us which row of **L** (the ith one) and which column of **R** (the jth one) to use. For example, Figure 10.4 illustrates how various p_{ij} would be calculated.

Let's now try our new simplified method on an example.

Example 3

$$\begin{pmatrix} 1 & 2 \\ 3 & 4 \end{pmatrix} \cdot \begin{pmatrix} 5 & 6 & 7 \\ 8 & 9 & 10 \end{pmatrix} = \mathbf{P}$$

$$\mathbf{L}_{2\times 2} \cdot \mathbf{R}_{2\times 3} = \mathbf{P}_{2\times 3}$$

same so can multiply

2 × 3 order of product matrix

Thus we must find the elements of a 2 × 3 product matrix. The simplified multiplication rule (for *ij* element, use the *i*th row of **L** along with the *j*th column of **R**) will quickly get us the product matrix, as developed in Figure 10.5.

Matrix Multiplication is Not Commutative

Matrix multiplication is not generally commutative. Thus, **LR** is generally not equal to **RL**. We can observe this by reversing the arrangement of the matrices in Example 1.

$$\begin{pmatrix} 5 & 6 \\ 7 & 8 \end{pmatrix} \begin{pmatrix} 1 & 2 \\ 3 & 4 \end{pmatrix} = \begin{pmatrix} p_{11} & p_{12} \\ p_{21} & p_{22} \end{pmatrix}$$

$$= \begin{pmatrix} 5 \cdot 1 + 6 \cdot 3 & 5 \cdot 2 + 6 \cdot 4 \\ 7 \cdot 1 + 8 \cdot 3 & 7 \cdot 2 + 8 \cdot 4 \end{pmatrix}$$

$$= \begin{pmatrix} 23 & 34 \\ 31 & 46 \end{pmatrix} \longleftarrow \text{This was } \begin{pmatrix} 19 & 22 \\ 43 & 50 \end{pmatrix} \text{ when we multiplied in reverse order}$$

Consider Example 3, which involved the following orders:

$\mathbf{L}_{2\times 2} \cdot \mathbf{R}_{2\times 3} = \mathbf{P}_{2\times 3}$

Figure 10.5

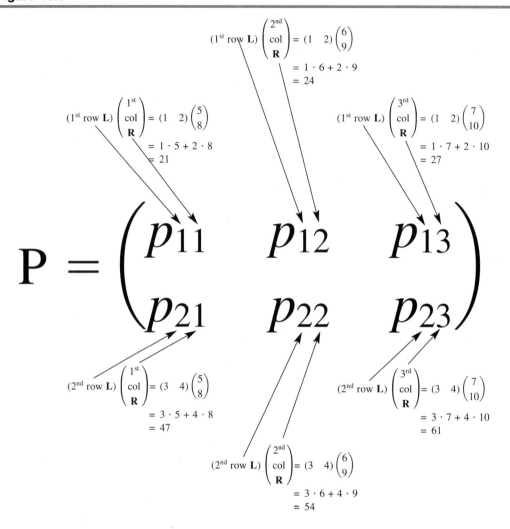

Now, if we reverse these matrices, we see below that multiplication is not possible.

$$R_{2\times 3} \cdot L_{2\times 2}$$

not equal, can't multiply

Reversing matrices can result in different products with different orders, as this example shows.

$$L_{2\times 3} \cdot R_{3\times 2} = P_{2\times 2}$$

$$\begin{pmatrix} 1 & 2 & 3 \\ 4 & 5 & 6 \end{pmatrix} \begin{pmatrix} 7 & 8 \\ 9 & 10 \\ 11 & 12 \end{pmatrix} = \begin{pmatrix} 7+18+33 & 7+18+33 \\ 28+45+66 & 32+50=72 \end{pmatrix} = \begin{pmatrix} 58 & 64 \\ 139 & 154 \end{pmatrix}$$

MATRIX ALGEBRA

$$\mathbf{R}_{3\times 2} \cdot \mathbf{L}_{2\times 3} = \mathbf{P}_{3\times 3}$$

$$\begin{pmatrix} 7 & 8 \\ 9 & 10 \\ 11 & 12 \end{pmatrix} \begin{pmatrix} 1 & 2 & 3 \\ 4 & 5 & 6 \end{pmatrix} = \begin{pmatrix} 7+32 & 14+40 & 21+48 \\ 9+40 & 18+50 & 27+60 \\ 11+48 & 22+60 & 33+72 \end{pmatrix} = \begin{pmatrix} 39 & 54 & 69 \\ 49 & 68 & 87 \\ 59 & 82 & 105 \end{pmatrix}$$

So, overall, reversing matrices and multiplying could result in three things:

1. a different product matrix of the same order
2. a different product matrix with different order
3. no product matrix possible.

Such is the wacky world of matrix multiplication.

Exercises 16

Consider the following matrices:

$$\mathbf{A} = \begin{pmatrix} 1 & 2 \\ 3 & 4 \end{pmatrix} \quad \mathbf{B} = \begin{pmatrix} 1 & 2 & 3 \\ 4 & 5 & 6 \end{pmatrix}$$

$$\mathbf{X} = \begin{pmatrix} x_1 \\ x_2 \\ x_3 \end{pmatrix} \quad \mathbf{T} = (1 \quad -1)$$

Form all possible arrangements of these matrices, taking two at a time (**AB**, **BA**, **AX**, **XA**, etc.). For each multiplication that is legitimate (*Hint:* there are five), compute the product matrix. For example, **TB** can be multiplied and results in the product

$$(1 \quad -1)\begin{pmatrix} 1 & 2 & 3 \\ 4 & 5 & 6 \end{pmatrix} = (1\cdot 1 + -1\cdot 4 \quad 1\cdot 2 + -1\cdot 5 \quad 1\cdot 3 + -1\cdot 6)$$

$$= (-3 \quad -3 \quad -3)$$

Application of Matrix Multiplication

Now, let's apply our matrix multiplication prowess in a real world situation.

Cantaford wishes to know the total value of its current inventory. If each Focus is worth 15 (thousand), each Taurus is worth 20 (thousand), and each Explorer is worth 30 (thousand), we can set up a 1×3 cost matrix **C** as

$$\mathbf{C} = (15 \quad 20 \quad 30)$$

and multiply it by the current inventory matrix **F**

$$\mathbf{C} \cdot \mathbf{F}$$

$$(15 \quad 20 \quad 30)\begin{pmatrix} 6 & 3 & 2 & 1 \\ 4 & 7 & 3 & 2 \\ 2 & 0 & 5 & 0 \end{pmatrix} = (p_{11} \quad p_{12} \quad p_{13} \quad p_{14})$$

$p_{11} = 15 \cdot 6 + 20 \cdot 4 + 30 \cdot 2 = 230$

$p_{12} = 15 \cdot 3 + 20 \cdot 7 + 30 \cdot 0 = 185$

$p_{13} = 15 \cdot 2 + 20 \cdot 3 + 30 \cdot 5 = 240$

$p_{14} = 15 \cdot 1 + 20 \cdot 2 + 30 \cdot 0 = 55$

These results are, respectively, the total value of the different colors of cars. To get the total value of all the cars, we add the four numbers. In matrix terms this is done by multiplying the above product matrix by a 4 × 1 unity matrix:

$$(230 \quad 185 \quad 240 \quad 55) \begin{pmatrix} 1 \\ 1 \\ 1 \\ 1 \end{pmatrix} = 230 \cdot 1 + 185 \cdot 1 + 240 \cdot 1 + 55 \cdot 1$$

$$= 710 \text{ (thousand)}$$

Exercise 17

The yearly costs of undergraduate and graduate tuition at Mate Tricks University are given in the following **C** matrix.

$\mathbf{C} = (10{,}000 \quad 15{,}000)$

Perform the proper matrix multiplication, using **C** and the **T** matrix (see page 311), to find total tuition revenue by major. Then multiply this result by the appropriate unity matrix to find the total tuition revenue for the entire university.

Multiplication by Identity Matrix

Let's observe what happens when we multiply various matrices by the identity matrix.

$$\mathbf{IA} = \begin{pmatrix} 1 & 0 \\ 0 & 1 \end{pmatrix} \begin{pmatrix} 1 & 2 \\ 3 & 4 \end{pmatrix} = \begin{pmatrix} 1 \cdot 1 + 0 \cdot 3 & 1 \cdot 2 + 0 \cdot 4 \\ 0 \cdot 1 + 1 \cdot 3 & 0 \cdot 2 + 1 \cdot 4 \end{pmatrix}$$

$$= \begin{pmatrix} 1 & 2 \\ 3 & 4 \end{pmatrix} = \mathbf{A}$$

$$\mathbf{AI} = \begin{pmatrix} 1 & 2 \\ 3 & 4 \end{pmatrix} \begin{pmatrix} 1 & 0 \\ 0 & 1 \end{pmatrix} = \begin{pmatrix} 1 \cdot 1 + 2 \cdot 0 & 1 \cdot 0 + 2 \cdot 1 \\ 3 \cdot 1 + 4 \cdot 0 & 3 \cdot 0 + 4 \cdot 1 \end{pmatrix}$$

$$= \begin{pmatrix} 1 & 2 \\ 3 & 4 \end{pmatrix} = \mathbf{A}$$

$$\mathbf{IX} = \begin{pmatrix} 1 & 0 \\ 0 & 1 \end{pmatrix} \begin{pmatrix} x_1 \\ x_2 \end{pmatrix} = \begin{pmatrix} 1x_1 + 0x_2 \\ 0x_1 + 1x_2 \end{pmatrix}$$

$$= \begin{pmatrix} x_1 \\ x_2 \end{pmatrix} = \mathbf{X}$$

MATRIX ALGEBRA

$$\mathbf{IB} = \begin{pmatrix} 1 & 0 \\ 0 & 1 \end{pmatrix}\begin{pmatrix} 1 & 2 & 3 \\ 4 & 5 & 6 \end{pmatrix} = \begin{pmatrix} 1\cdot 1 + 0\cdot 4 & 1\cdot 2 + 0\cdot 5 & 1\cdot 3 + 0\cdot 6 \\ 0\cdot 1 + 1\cdot 4 & 0\cdot 2 + 1\cdot 5 & 0\cdot 3 + 1\cdot 6 \end{pmatrix}$$

$$= \begin{pmatrix} 1 & 2 & 3 \\ 4 & 5 & 6 \end{pmatrix} = \mathbf{B}$$

$$\mathbf{BI} = \begin{pmatrix} 1 & 2 & 3 \\ 4 & 5 & 6 \end{pmatrix}\begin{pmatrix} 1 & 0 & 0 \\ 0 & 1 & 0 \\ 0 & 0 & 1 \end{pmatrix} = \begin{pmatrix} 1+0+0 & 0+2+0 & 0+0+3 \\ 4+0+0 & 0+5+0 & 0+0+6 \end{pmatrix}$$

$$= \begin{pmatrix} 1 & 2 & 3 \\ 4 & 5 & 6 \end{pmatrix} = \mathbf{B}$$

Exercises 18

Consider the auto inventory matrix **F** and the Mate Tricks University student body matrix **T**. Determine the appropriate identity matrices and perform the following multiplications.

1. **IF**
2. **FI**
3. **IT**
4. **TI**

Lo and behold, we always got back our original matrix as the product! This effect generalizes, so that for any matrix **M**

IM = MI = M

Identity multiplication in matrix algebra is analogous to multiplication by the number 1 in ordinary algebra.

$1a = a1 = a$

This effect will come in handy when we solve linear systems with matrix algebra.

Multiplication of Matrix by Its Inverse

On page 309 we introduced the idea of a matrix inverse and promised big things from them. Now we'll investigate what happens when a matrix **A** is multiplied by its inverse, denoted by $\mathbf{A}^{-1}$.

$\mathbf{AA}^{-1} = ?$

$\mathbf{A}^{-1}\mathbf{A} = ?$

It is a fact that if

$$\mathbf{A} = \begin{pmatrix} 1 & 2 \\ 3 & 4 \end{pmatrix}$$

then

$$\mathbf{A}^{-1} = \begin{pmatrix} -2 & 1 \\ \frac{3}{2} & -\frac{1}{2} \end{pmatrix}$$

Don't worry if you can't understand why $\mathbf{A}^{-1}$ is that. You shouldn't know why, since the methods to find an inverse come later in this chapter. At this point though it is very helpful to see the net result of multiplying a matrix by its inverse.

$$\mathbf{A}\mathbf{A}^{-1} = \begin{pmatrix} 1 & 2 \\ 3 & 4 \end{pmatrix}\begin{pmatrix} -2 & 1 \\ \frac{3}{2} & -\frac{1}{2} \end{pmatrix} = \begin{pmatrix} 1 \cdot -2 + 2 \cdot \frac{3}{2} & 1 \cdot 1 + 2 \cdot -\frac{1}{2} \\ 3 \cdot -2 + 4 \cdot \frac{3}{2} & 3 \cdot 1 + 4 \cdot -\frac{1}{2} \end{pmatrix}$$

$$= \begin{pmatrix} 1 & 0 \\ 0 & 1 \end{pmatrix} = \mathbf{I}$$

$$\mathbf{A}^{-1}\mathbf{A} = \begin{pmatrix} -2 & 1 \\ \frac{3}{2} & -\frac{1}{2} \end{pmatrix}\begin{pmatrix} 1 & 2 \\ 3 & 4 \end{pmatrix} = \begin{pmatrix} -2 \cdot 1 + 1 \cdot 3 & -2 \cdot 2 + 1 \cdot 4 \\ \frac{3}{2} \cdot 1 + -\frac{1}{2} \cdot 3 & \frac{3}{2} \cdot 2 + -\frac{1}{2} \cdot 4 \end{pmatrix}$$

$$= \begin{pmatrix} 1 & 0 \\ 0 & 1 \end{pmatrix} = \mathbf{I}$$

Both multiplications resulted in the identity matrix. This result generalizes to the rule that any square matrix $\mathbf{A}$ times its inverse $\mathbf{A}^{-1}$ (*Note:* not all square matrices have an inverse) equals $\mathbf{I}$.

$$\mathbf{A}\mathbf{A}^{-1} = \mathbf{I} \quad \text{and} \quad \mathbf{A}^{-1}\mathbf{A} = \mathbf{I}$$

Exercise 19

If

$$\mathbf{A} = \begin{pmatrix} 2 & 4 \\ -1 & 3 \end{pmatrix}$$

then

$$\mathbf{A}^{-1} = \begin{pmatrix} .3 & -.4 \\ .1 & .2 \end{pmatrix}$$

Multiply these matrices as $\mathbf{A}\mathbf{A}^{-1}$ and $\mathbf{A}^{-1}\mathbf{A}$. What is the product matrix in both cases?

10.4 AX = B REALLY IS ALL THOSE EQUATIONS

Earlier the claim was made that $\mathbf{AX} = \mathbf{B}$ in matrix notation duplicates the cumbersome algebraic representation of the general linear system presented on page 307. Now that we can multiply matrices, we are in a position to see why.

Let $\mathbf{A}$ be the $n \times n$ matrix of left-side, or a_{ij} coefficients; $\mathbf{X}$ be the $n \times 1$ matrix of unknowns; and $\mathbf{B}$ be the $n \times 1$ matrix of right-side constants.

MATRIX ALGEBRA

$$\mathbf{A}_{n \times n} = \begin{pmatrix} a_{11} & a_{12} & a_{13} & \cdots & a_{1n} \\ a_{21} & a_{22} & a_{23} & \cdots & a_{2n} \\ a_{31} & a_{32} & a_{33} & \cdots & a_{3n} \\ \vdots & & & & \vdots \\ a_{n1} & a_{n2} & a_{n3} & \cdots & a_{nn} \end{pmatrix} \quad \mathbf{X}_{n \times 1} = \begin{pmatrix} x_1 \\ x_2 \\ x_3 \\ \vdots \\ x_n \end{pmatrix} \quad \mathbf{B}_{n \times 1} = \begin{pmatrix} b_1 \\ b_2 \\ b_3 \\ \vdots \\ b_n \end{pmatrix}$$

Multiplying $\mathbf{A} \cdot \mathbf{X}$ gives a product matrix $\mathbf{P}$

$$\begin{pmatrix} a_{11} & a_{12} & a_{13} & \cdots & a_{1n} \\ a_{21} & a_{22} & a_{23} & \cdots & a_{2n} \\ a_{31} & a_{32} & a_{33} & \cdots & a_{3n} \\ \vdots & & & & \vdots \\ a_{n1} & a_{n2} & a_{n3} & \cdots & a_{nn} \end{pmatrix} \begin{pmatrix} x_1 \\ x_2 \\ x_3 \\ \vdots \\ x_n \end{pmatrix} = \begin{pmatrix} a_{11}x_1 + a_{12}x_2 + a_{13}x_3 + \cdots + a_{1n}x_n \\ a_{21}x_1 + a_{22}x_2 + a_{23}x_3 + \cdots + a_{2n}x_n \\ a_{31}x_1 + a_{32}x_2 + a_{33}x_3 + \cdots + a_{3n}x_n \\ \vdots \\ a_{n1}x_1 + a_{n2}x_2 + a_{n3}x_3 + \cdots + a_{nn}x_n \end{pmatrix}$$

The product matrix $\mathbf{P}$ is an $n \times 1$ matrix containing the left sides of the n equations of our linear system. Since

$$a_{11}x_1 + a_{12}x_2 + a_{13}x_3 + \cdots + a_{1n}x_n = b_1$$
$$a_{21}x_1 + a_{22}x_2 + a_{23}x_3 + \cdots + a_{2n}x_n = b_2$$

and so on, the elements of $\mathbf{P}$ are equal to the corresponding elements of the $\mathbf{B}$ matrix defined above. So

$\mathbf{P} = \mathbf{B}$

$\mathbf{AX} = \mathbf{P} = \mathbf{B}$

$\mathbf{AX} = \mathbf{B}$

represents our linear system.

10.5 FINDING THE INVERSE OF A MATRIX

If you are convinced that $\mathbf{AX} = \mathbf{B}$ really represents a linear system of equations, you can now turn your undivided attention toward isolating the $\mathbf{X}$ matrix of unknowns. This could be accomplished if $\mathbf{A}$'s inverse, which we symbolize as $\mathbf{A}^{-1}$ (analogous to a^{-1}, the inverse of a number) were found and used as follows.

		Comment
$\mathbf{AX} = \mathbf{B}$		Linear system in matrix notation
$\mathbf{A}^{-1}\mathbf{AX} = \mathbf{A}^{-1}\mathbf{B}$		Premultiply both sides by $\mathbf{A}^{-1}$
$\mathbf{IX} = \mathbf{A}^{-1}\mathbf{B}$	$\mathbf{A}^{-1}\mathbf{A} = \mathbf{I}$	
$\mathbf{X} = \mathbf{A}^{-1}\mathbf{B}$	$\mathbf{IX} = \mathbf{X}$	

You should understand all these steps. The only missing link in the above approach is, "How do we find $\mathbf{A}^{-1}$?" We attack this problem now.

We begin our search for A^{-1} theoretically by showing that matrix inversion really involves multiplying A by a series of elementary matrices. At the same time we show the similarity between this process and the algebraic elimination method of solving linear systems. Finally, with this good theoretical base, we learn how to obtain the effects of elementary matrix multiplication in much simpler fashion. This latter approach, using elementary row operations, is the one we will normally employ to find A^{-1}.

Some Theory of Matrix Inversion

Suppose that a square matrix A is premultiplied by E_1, an elementary matrix (defined on page 313)

$$E_1 A = A_1$$

and the product A_1 is premultiplied by another elementary matrix, E_2

$$E_2 A_1 = E_2(E_1 A) = A_2$$

and the product A_2 is premultiplied by another elementary matrix, E_3

$$E_3 A_2 = E_3(E_2 E_1 A) = A_3$$

and on and on. If, finally, after m of these multiplications we arrive at the identity matrix

$$(E_m E_{m-1} \cdots E_3 E_2 E_1) A = I$$

it then follows from the definition of a matrix inverse (the matrix that premultiplies another matrix to give a product of I) that the product of all the matrices to the left of A is the inverse of A, or

$$E_m E_{m-1} \cdots E_2 E_1 E_1 = A^{-1}$$

That will take a little digesting. So before we go on, let's take a look at the kind of results obtained when premultiplying by elementary matrices.

Elementary matrices are first cousins to identity matrices. They are identical to an I matrix, except that one element of I is replaced by some nonzero value (we hereafter call this the *tampered element*). Some examples of this breed of matrices are

$$\begin{pmatrix} 1 & 1 \\ 0 & 1 \end{pmatrix} \quad \begin{pmatrix} 1 & 0 \\ 3 & 1 \end{pmatrix} \quad \begin{pmatrix} 5 & 0 \\ 0 & 1 \end{pmatrix}$$

$$\begin{pmatrix} 1 & 0 \\ -2 & 1 \end{pmatrix} \quad \begin{pmatrix} 1 & 0 & 0 \\ 0 & 1 & 0 \\ 0 & 0 & 4 \end{pmatrix} \quad \begin{pmatrix} 1 & 0 & 0 \\ 0 & 1 & 2 \\ 0 & 0 & 1 \end{pmatrix}$$

Observe what happens when we premultiply matrices with them.

MATRIX ALGEBRA

	Changed row	Nature of change

1. $\begin{pmatrix} 1 & 1 \\ 0 & 1 \end{pmatrix} \begin{pmatrix} 1 & 2 \\ 3 & 4 \end{pmatrix}$ 1   1 times row 2 added to it

$= \begin{pmatrix} 1 + (1 \cdot 3) & 2 + (1 \cdot 4) \\ 3 & 4 \end{pmatrix}$

$= \begin{pmatrix} 4 & 6 \\ 3 & 4 \end{pmatrix}$

2. $\begin{pmatrix} 1 & 0 \\ 3 & 1 \end{pmatrix} \begin{pmatrix} 1 & 2 \\ 3 & 4 \end{pmatrix}$ 2   3 times row 1 added to it

$= \begin{pmatrix} 1 & 2 \\ (3 \cdot 1) + 3 & (3 \cdot 2) + 4 \end{pmatrix}$

$= \begin{pmatrix} 1 & 2 \\ 6 & 10 \end{pmatrix}$

3. $\begin{pmatrix} 5 & 0 \\ 0 & 1 \end{pmatrix} \begin{pmatrix} 1 & 2 \\ 3 & 4 \end{pmatrix}$ 1   multiplied by 5

$= \begin{pmatrix} 5 \cdot 1 & 5 \cdot 2 \\ 3 & 4 \end{pmatrix} = \begin{pmatrix} 5 & 10 \\ 3 & 4 \end{pmatrix}$

4. $\begin{pmatrix} 1 & 0 \\ -2 & 1 \end{pmatrix} \begin{pmatrix} 1 & 2 \\ 3 & 4 \end{pmatrix}$ 2   −2 times row 1 added to it (or 2 times row 1 subtracted from it)

$= \begin{pmatrix} 1 & 2 \\ (-2 \cdot 1) + 3 & (-2 \cdot 2) + 4 \end{pmatrix}$

$= \begin{pmatrix} 1 & 2 \\ 1 & 0 \end{pmatrix}$

5. $\begin{pmatrix} 1 & 0 & 0 \\ 0 & 1 & 0 \\ 0 & 0 & 4 \end{pmatrix} \begin{pmatrix} 1 & 2 & 3 \\ 4 & 5 & 6 \\ 7 & 8 & 9 \end{pmatrix}$ 3   multiplied by 4

$= \begin{pmatrix} 1 & 2 & 3 \\ 4 & 5 & 6 \\ 28 & 32 & 36 \end{pmatrix}$

6. $\begin{pmatrix} 1 & 0 & 0 \\ 0 & 1 & 2 \\ 0 & 0 & 1 \end{pmatrix} \begin{pmatrix} 1 & 2 & 3 \\ 4 & 5 & 6 \\ 7 & 8 & 9 \end{pmatrix}$ 2   2 times row 3 added to it

$= \begin{pmatrix} 1 & 2 & 3 \\ 4 + (2 \cdot 7) & 5 + (2 \cdot 8) & 6 + (2 \cdot 9) \\ 7 & 8 & 9 \end{pmatrix}$

$= \begin{pmatrix} 1 & 2 & 3 \\ 18 & 21 & 24 \\ 7 & 8 & 9 \end{pmatrix}$

Have you noticed elementary matrix premultiplication represented by

EA = P

results in a product matrix with one row different than the **A** matrix? Observe that the row in **E** containing the tampered element is always the same as the row in **P** that differs from **A**.

When the tampered element lies on the main diagonal of **E**, the effect is to multiply that row of **A** by its value (see examples 3 and 5 above).

Type I effect $\quad i$th row $\rightarrow \begin{pmatrix} 1 & & & & \\ & 1 & & & \\ & & \ddots & & \\ & & & k & \\ & & & & \ddots \\ & & & & & 1 \end{pmatrix}$ **A** = **P**   **P**'s ith row is k times **A**'s ith row

When the tampered element, having a value of k, is off the main diagonal, say in the ith row, jth column position, the effect is to add k times row j to row i (see examples 1, 2, 4, 6).

Type II effect $\quad i$th row $\rightarrow \begin{pmatrix} 1 & & & & & \\ & 1 & & & & \\ & & 1 & k & & \\ & & & \ddots & & \\ & & & & 1 & \\ & & & & & 1 \end{pmatrix}$ $\begin{array}{c} j\text{th column} \\ \downarrow \end{array}$ **A** = **P**   **P**'s ith row is **A**'s ith row plus k times **A**'s jth row

Exercises 20

Predict the outcome of the following matrix multiplications, and then check your prediction by carrying out the multiplications:

1. $\begin{pmatrix} 1 & 0 \\ 0 & -3 \end{pmatrix} \begin{pmatrix} 1 & 2 \\ 3 & 4 \end{pmatrix} =$

2. $\begin{pmatrix} 1 & 0 & 0 \\ -1 & 1 & 0 \\ 0 & 0 & 1 \end{pmatrix} \begin{pmatrix} 1 & 2 & 3 \\ 4 & 5 & 6 \\ 7 & 8 & 9 \end{pmatrix} =$

3. $\begin{pmatrix} 2 & 0 & 0 \\ 0 & 1 & 0 \\ 0 & 0 & 1 \end{pmatrix} \begin{pmatrix} 1 & 2 & 3 \\ 4 & 5 & 6 \\ 7 & 8 & 9 \end{pmatrix} =$

4. $\begin{pmatrix} 1 & 0 & 0 & 0 \\ 0 & 1 & 0 & 0 \\ 0 & 2 & 1 & 0 \\ 0 & 0 & 0 & 1 \end{pmatrix} \begin{pmatrix} 1 & 1 \\ 2 & 2 \\ 3 & 3 \\ 4 & 4 \end{pmatrix} =$

Do these effects ring a distant bell? Don't you remember that back in the happy days of solving linear systems by the elimination method, we multiplied equations by constants and added a constant times one equation to another equation? Well, elementary matrix premultiplication is just a sophisticated way of performing those same operations. If you are not convinced of that yet, don't worry, since this link will be illustrated later by an example.

MATRIX ALGEBRA

Let's now see how elementary matrix premultiplication[2] can be systematically employed to find an inverse. The process involves converting **A** to **I**. The following scheme generally proves to be the most efficient.

1. Convert the first column of **A** to the first column of **I**. First, make the diagonal element 1, then use that row to make the other elements zero in any order.
2. Convert the second column of **A** to the second column of **I**. Again, first make the diagonal element 1 and use that row to zero the other elements of the column.
3. Convert successive columns—3, 4, 5, ..., n—to the corresponding **I** matrix column. In all cases first make the diagonal element 1, then use that row to zero all the other elements of the column.

An example will illustrate this scheme. At the same time we can see the parallel between matrix inversion methods and the algebraic elimination method described in the previous chapter.

Elimination method

Consider the linear system

$1x_1 + 2x_2 = 5$
$3x_1 + 4x_2 = 11$

To eliminate x_1 from the second equation, 3 times equation 1 is subtracted from equation 2.

$3x_1 + 4x_2 = 11$
$-3x_1 - 6x_2 = -15$
$\overline{0x_1 - 2x_2 = -4}$

Thus, we have the revised linear system

$1x_1 + 2x_2 = 5$
$0x_1 - 2x_2 = -4$

Matrix inverse method

which in matrix notation is

$$\begin{matrix} \mathbf{A} & \mathbf{X} & = & \mathbf{B} \end{matrix}$$

$$\begin{pmatrix} 1 & 2 \\ 3 & 4 \end{pmatrix} \begin{pmatrix} x_1 \\ x_2 \end{pmatrix} = \begin{pmatrix} 5 \\ 11 \end{pmatrix}$$

Beginning with the first column of the **A** matrix, the diagonal element is already 1. So that column will be set once the 3 is converted to zero. The elementary matrix

$$\mathbf{E}_1 = \begin{pmatrix} 1 & 0 \\ -3 & 1 \end{pmatrix}$$

does this conversion as

$$\begin{matrix} \mathbf{E}_1 & & \mathbf{A} \end{matrix}$$
$$\begin{pmatrix} 1 & 0 \\ -3 & 1 \end{pmatrix} \begin{pmatrix} 1 & 2 \\ 3 & 4 \end{pmatrix}$$
$$= \mathbf{A}_1$$

---

[2] Aren't you curious as to what happens upon postmultiplication by elementary matrices?

	Changed column	Effect
$\begin{pmatrix} 1 & 2 \\ 3 & 4 \end{pmatrix}\begin{pmatrix} 10 & 0 \\ 0 & 1 \end{pmatrix} = \begin{pmatrix} 10 & 2 \\ 30 & .4 \end{pmatrix}$	1	column 1 multiplied by 10
$\begin{pmatrix} 1 & 2 \\ 3 & 4 \end{pmatrix}\begin{pmatrix} 1 & 1 \\ 0 & 1 \end{pmatrix} = \begin{pmatrix} 1 & (1 + 2 \cdot 1) \\ 3 & (3 + 4 \cdot 1) \end{pmatrix} = \begin{pmatrix} 1 & 3 \\ 3 & 7 \end{pmatrix}$	2	column 1 added to column 2

Lo and behold, postmultiplication changes columns.
Although it would be a simple matter to figure out the detailed effects of postmultiplication by elementary matrices, we need not concern ourselves with them. This is because finding an inverse can be done completely with premultiplication. In fact, mixing up the pre- and postmultiplications is almost as bad as a kosher cook mixing up milk and meat!

$$= \begin{pmatrix} 1 & 2 \\ -3+3 & -6+4 \end{pmatrix}$$

$$= \begin{pmatrix} 1 & 2 \\ 0 & -2 \end{pmatrix} = \mathbf{A}_1$$

Notice how the left-side linear system coefficients are identical to the elements in the $\mathbf{A}_1$ matrix.

Multiplying the second equation by $-\frac{1}{2}$ gives

$0x_1 + 1x_2 = 2$

and so the revised linear system is

$1x_1 + 2x_2 = 5$
$0x_1 + 1x_2 = 2$

Next we begin work on the second column. First we convert the diagonal element (–2) to 1. This job is made to order for the elementary matrix

$$\mathbf{E}_2 = \begin{pmatrix} 1 & 0 \\ 0 & -\frac{1}{2} \end{pmatrix}$$

which does this conversion as $\mathbf{E}_2(\mathbf{E}_1\mathbf{A}) = \mathbf{E}_2\mathbf{A}_1$

$$\begin{array}{ccc} \mathbf{E}_2 & \mathbf{A}_1 & = \mathbf{A}_2 \end{array}$$

$$\begin{pmatrix} 1 & 0 \\ 0 & -\frac{1}{2} \end{pmatrix} \begin{pmatrix} 1 & 2 \\ 0 & -2 \end{pmatrix} = \begin{pmatrix} 1 & 2 \\ 0 & 1 \end{pmatrix}$$

Again notice the equivalence between the left-side coefficients in the linear system and the elements in the $\mathbf{A}_2$ matrix.

We can eliminate x_2 from the first equation by adding –2 times the second equation to the first equation.

$1x_1 + 2x_2 = 5$
$\underline{0x_1 - 2x_2 = -4}$
$1x_1 + 0x_2 = 1$

The revised linear system is then

$1x_1 + 0x_2 = 1$
$0x_1 + 1x_2 = 2$

The 2 in the second column of $\mathbf{A}_2$ is converted to a zero by the elementary matrix

$$\mathbf{E}_3 = \begin{pmatrix} 1 & -2 \\ 0 & 1 \end{pmatrix}$$

which does the conversion as $\mathbf{E}_3(\mathbf{E}_2\mathbf{E}_1\mathbf{A}) = \mathbf{E}_3\mathbf{A}_2$

$$\begin{array}{ccc} \mathbf{E}_3 & \mathbf{A}_2 & = \mathbf{A}_3 \end{array}$$

$$\begin{pmatrix} 1 & -2 \\ 0 & 1 \end{pmatrix} \begin{pmatrix} 1 & 2 \\ 0 & 1 \end{pmatrix} = \begin{pmatrix} 1 & 0 \\ 0 & 1 \end{pmatrix}$$

And since $\mathbf{A}_3 = \mathbf{I}$, we have completed the procedure.

Once more notice the equivalence of left-side coefficients and elements of the revised $\mathbf{A}$ matrix. If you are getting the idea that the two methods are parallel, you are right.

Before we move on to find $\mathbf{A}^{-1}$, let's dispell some of the mystery behind the choice of elementary matrices. We can do this by recapping the various conversions required and observing the pattern of elementary matrices that accomplish the result.

MATRIX ALGEBRA

Matrix	Conversion required	Elementary matrix needed
$\begin{pmatrix} 1 & 2 \\ 3 & 4 \end{pmatrix}$	Convert 3 to 0	$\begin{pmatrix} 1 & 0 \\ -3 & 1 \end{pmatrix}$
$\begin{pmatrix} 1 & 2 \\ 0 & -2 \end{pmatrix}$	Convert -2 to 1	$\begin{pmatrix} 1 & 0 \\ 0 & -\frac{1}{2} \end{pmatrix}$
$\begin{pmatrix} 1 & 2 \\ 0 & 1 \end{pmatrix}$	Convert 2 to 0	$\begin{pmatrix} 1 & -2 \\ 0 & 1 \end{pmatrix}$

Can you see the pattern? The elementary matrix tampered element is in the same location as the element to be converted. When an off-diagonal element must be converted, the tampered element is the additive inverse of that number. For example, if 3 must be converted to zero, the tampered element is -3. If a main diagonal element must be converted, the tampered element is the multiplicative inverse of that number. For example, if -2 must be converted to 1, the tampered element is $1/-2 = -\frac{1}{2}$.

Exercises 21

What elementary matrices accomplish the following conversions?

Matrix	Conversion
$\begin{pmatrix} 4 & 7 \\ -1 & 5 \end{pmatrix}$	Convert 4 to 1
$\begin{pmatrix} 1 & -5 \\ 0 & 1 \end{pmatrix}$	Convert -5 to 0

Carry out the elementary matrix premultiplications to check your answers.

Now let's find the inverse. Because the product of the three elementary premultiplications of **A** was **I**

$$(E_3 E_2 E_1)A = I$$

and since $A^{-1}A = I$, it follows that

$$E_3 E_2 E_1 = A^{-1}$$

For further development it is better to throw an **I** matrix in this expression (don't worry since identity multiplication leaves it unchanged).

$$E_3 E_2 E_1 I = A^{-1}$$

Now we carry out the various multiplications to find A^{-1}.

$E_1 I = E_1$

$$E_2 E_1 = \begin{pmatrix} 1 & 0 \\ 0 & -\frac{1}{2} \end{pmatrix}\begin{pmatrix} 1 & 0 \\ -3 & 1 \end{pmatrix} = \begin{pmatrix} 1 & 0 \\ \frac{3}{2} & -\frac{1}{2} \end{pmatrix}$$

$$A^{-1} = E_3(E_2 E_1) = \begin{pmatrix} 1 & -2 \\ 0 & 1 \end{pmatrix}\begin{pmatrix} 1 & 0 \\ \frac{3}{2} & -\frac{1}{2} \end{pmatrix}$$

$$= \begin{pmatrix} -2 & 1 \\ \frac{3}{2} & -\frac{1}{2} \end{pmatrix}$$

To check whether this is indeed A^{-1}, let's perform the multiplication $A^{-1}A$

$$\begin{pmatrix} -2 & 1 \\ \frac{3}{2} & -\frac{1}{2} \end{pmatrix}\begin{pmatrix} 1 & 2 \\ 3 & 4 \end{pmatrix} = \begin{pmatrix} 1 & 0 \\ 0 & 1 \end{pmatrix}$$

The identity product verifies that we have indeed found the inverse.

Elementary Row Operation Method

There is no need to specify the various elementary matrices when finding an inverse. We need only to perform operations that duplicate the effects of premultiplying by elementary matrices. These operations (which we call *elementary row operations,* or ERO's) are of two types:

Type I   Multiply a row by a nonzero constant
Type II Add to a row a constant times another row

The theory of the previous section showed us that if we can convert A to I by elementary matrix premultiplications, these same operations convert I to A^{-1}.

If

$E_3 E_2 E_1 A = I$

then

$E_3 E_2 E_1 I = A^{-1}$

We performed these multiplications on A first, then on I. However, they could have been done simultaneously. Here we employ the simultaneous approach, using elementary row operations. We can illustrate this with the same matrix so that you can see the equivalence of methods.

$$A = \begin{pmatrix} 1 & 2 \\ 3 & 4 \end{pmatrix} \quad I = \begin{pmatrix} 1 & 0 \\ 0 & 1 \end{pmatrix}$$

Add -3 times row 1 to row 2.

$$\begin{pmatrix} 1 & 2 \\ 0 & -2 \end{pmatrix} \quad \begin{pmatrix} 1 & 0 \\ -3 & 1 \end{pmatrix}$$

Multiply[3] row 2 by $-\frac{1}{2}$.

$$\begin{pmatrix} 1 & 2 \\ 0 & 1 \end{pmatrix} \qquad \begin{pmatrix} 1 & 0 \\ \frac{3}{2} & -\frac{1}{2} \end{pmatrix}$$

Add -2 times row 2 to row 1.

$$\begin{pmatrix} 1 & 0 \\ 0 & 1 \end{pmatrix} = \mathbf{I} \qquad \begin{pmatrix} -2 & 1 \\ \frac{3}{2} & -\frac{1}{2} \end{pmatrix} = \mathbf{A}^{-1}$$

The elementary row operations that worked so beautifully to find the inverse of a 2×2 matrix will do equally as well if given the chance on larger matrices. Let's illustrate with a 3×3.

$$\begin{array}{cc} \mathbf{A} & \mathbf{I} \\ \begin{pmatrix} 1 & 1 & -2 \\ 0 & 1 & -1 \\ 2 & -4 & 1 \end{pmatrix} & \begin{pmatrix} 1 & 0 & 0 \\ 0 & 1 & 0 \\ 0 & 0 & 1 \end{pmatrix} \end{array}$$

The first column of $\mathbf{A}$ is in "identity" form except for the number 2 in row 3. This can be zeroed by forming a new row 3 by subtracting 2 times row 1 from the old row 3. Since things are getting more complicated, let's establish a precise notation for our ERO's. Letting $\mathbf{R}_i$ stand for row i and $\mathbf{R}_i^*$ stand for the revised row i, we can write this particular ERO as

$\mathbf{R}_3^* = \mathbf{R}_3 - 2\mathbf{R}_1$

Performing this ERO on both matrices gives

$\mathbf{R}_3^* = (2 \ -4 \ 1) - 2(1 \ 1 \ -2) \qquad \mathbf{R}_3^* = (0 \ 0 \ 1) - 2(1 \ 0 \ 0)$

$\phantom{\mathbf{R}_3^*} = (0 \ -6 \ 5) \qquad \phantom{\mathbf{R}_3^*} = (-2 \ 0 \ 1)$

Thus, the revised matrices are

$$\begin{pmatrix} 1 & 1 & -2 \\ 0 & 1 & -1 \\ 0 & -6 & 5 \end{pmatrix} \qquad \begin{pmatrix} 1 & 0 & 0 \\ 0 & 1 & 0 \\ -2 & 0 & 1 \end{pmatrix}$$

With the first column of the revised $\mathbf{A}$ matrix in identity form, we turn our attention to the second column. Beginning with the main diagonal element, we are in luck again, since the 1 we need is already there. Now we can use the second row with Type II ERO's to zero the other elements in the column.

---

[3] At this stage, we could have gotten a 1 on the diagonal by adding $\frac{3}{2}$ times row 1 to row 2

$$\begin{pmatrix} 1 & 2 \\ \frac{3}{2} & 1 \end{pmatrix}$$

but in the process, we lost a hard-earned zero in column 1. The moral of this story is, "Always use Type I ERO's to get 1's on the main diagonal."

$\mathbf{R}_1^* = \mathbf{R}_1 - \mathbf{R}_2$

$\mathbf{R}_1^* = (1 \quad 1 \quad -2) - (0 \quad 1 \quad -1)$ $\quad\quad \mathbf{R}_1^* = (1 \quad 0 \quad 0) - (0 \quad 1 \quad 0)$

$\quad\quad = (1 \quad 0 \quad -1)$ $\quad\quad\quad\quad\quad\quad\quad\quad\quad\quad = (1 \quad -1 \quad 0)$

$$\begin{pmatrix} 1 & 0 & -1 \\ 0 & 1 & -1 \\ 0 & -6 & 5 \end{pmatrix} \quad\quad\quad \begin{pmatrix} 1 & -1 & 0 \\ 0 & 1 & 0 \\ -2 & 0 & 1 \end{pmatrix}$$

$\mathbf{R}_3^* = \mathbf{R}_3 + 6\mathbf{R}_2$

$\mathbf{R}_3^* = (0 \quad -6 \quad 5) + 6(0 \quad 1 \quad -1)$ $\quad \mathbf{R}_3^* = (-2 \quad 0 \quad 1) + 6(0 \quad 1 \quad 0)$

$\quad\quad = (0 \quad 0 \quad -1)$ $\quad\quad\quad\quad\quad\quad\quad\quad\quad\quad = (-2 \quad 6 \quad 1)$

$$\begin{pmatrix} 1 & 0 & -1 \\ 0 & 1 & -1 \\ 0 & 0 & -1 \end{pmatrix} \quad\quad\quad \begin{pmatrix} 1 & -1 & 0 \\ 0 & 1 & 0 \\ -2 & 6 & 1 \end{pmatrix}$$

With the second column of the revised **A** matrix in identity form, we turn our attention to the third column. As always, we begin with the main diagonal element. For previous columns we didn't have to do anything, since the diagonal elements were 1 already. Here we will have to perform an ERO to get it that way. Only ERO's of Type I should be used on the main diagonal elements. Here we need the following one:

$\mathbf{R}_3^* = -1\mathbf{R}_3$

$\mathbf{R}_3^* = -1(0 \quad 0 \quad -1)$ $\quad\quad\quad\quad\quad \mathbf{R}_3^* = -1(-2 \quad 6 \quad 1)$

$\quad\quad = (0 \quad 0 \quad 1)$ $\quad\quad\quad\quad\quad\quad\quad\quad\quad\quad = (2 \quad -6 \quad -1)$

$$\begin{pmatrix} 1 & 0 & -1 \\ 0 & 1 & -1 \\ 0 & 0 & 1 \end{pmatrix} \quad\quad\quad \begin{pmatrix} 1 & -1 & 0 \\ 0 & 1 & 0 \\ 2 & -6 & -1 \end{pmatrix}$$

Now we can use row 3 and Type II ERO's to zero the other elements in the third column and arrive at the identity matrix.

$\mathbf{R}_1^* = \mathbf{R}_1 + \mathbf{R}_3$

$\mathbf{R}_1^* = (1 \quad 0 \quad -1) + (0 \quad 0 \quad 1)$ $\quad \mathbf{R}_1^* = (1 \quad -1 \quad 0) + (2 \quad -6 \quad -1)$

$\quad\quad = (1 \quad 0 \quad 0)$ $\quad\quad\quad\quad\quad\quad\quad\quad\quad\quad = (3 \quad -7 \quad -1)$

$$\begin{pmatrix} 1 & 0 & 0 \\ 0 & 1 & -1 \\ 0 & 0 & 1 \end{pmatrix} \quad\quad\quad \begin{pmatrix} 3 & -7 & -1 \\ 0 & 1 & 0 \\ 2 & -6 & -1 \end{pmatrix}$$

$\mathbf{R}_2^* = \mathbf{R}_2 + \mathbf{R}_3$

$\mathbf{R}_2^* = (0 \quad 1 \quad -1) + (0 \quad 0 \quad 1)$ $\quad \mathbf{R}_2^* = (0 \quad 1 \quad 0) + (2 \quad -6 \quad -1)$

$\quad\quad = (0 \quad 1 \quad 0)$ $\quad\quad\quad\quad\quad\quad\quad\quad\quad\quad = (2 \quad -5 \quad -1)$

$$\begin{pmatrix} 1 & 0 & 0 \\ 0 & 1 & 0 \\ 0 & 0 & 1 \end{pmatrix} = \mathbf{I} \quad\quad\quad \begin{pmatrix} 3 & -7 & -1 \\ 2 & -5 & -1 \\ 2 & -6 & -1 \end{pmatrix} = \mathbf{A}^{-1}$$

MATRIX ALGEBRA

Since **A** was converted to **I** by elementary row operations, the simultaneous conversion of **I** should be $\mathbf{A}^{-1}$. The check reveals we have succeeded.

$$\overset{\mathbf{A}^{-1}}{\begin{pmatrix} 3 & -7 & -1 \\ 2 & -5 & -1 \\ 2 & -6 & -1 \end{pmatrix}} \overset{\mathbf{A}}{\begin{pmatrix} 1 & 1 & -2 \\ 0 & 1 & -1 \\ 2 & -4 & 1 \end{pmatrix}} = \begin{pmatrix} 1 & 0 & 0 \\ 0 & 1 & 0 \\ 0 & 0 & 1 \end{pmatrix} = \mathbf{I}$$

To review the procedure, the transformation from **A** to **I** by elementary row operations was performed one column at a time, in this order; first column, second column, third column. For each column, the main diagonal element was first made equal to 1 by Type I ERO's. Then the row with the main diagonal element was used along with Type II ERO's to zero out the rest of the column.

This procedure is the most efficient because the "identitizing" work previously accomplished is never lost. For example, once the first column and the diagonal element in the second column were set, the revised **A** matrix was

$$\begin{pmatrix} 1 & 1 & -2 \\ 0 & 1 & -1 \\ 0 & -6 & 5 \end{pmatrix}$$

We then zeroed the nondiagonal elements of the second column by subtracting row 2 from row 1 and adding 6 times row 2 to row 3. However, we could have transformed the -6 to zero by adding 6 times row 1 to row 3. But look what would have happened.

$\mathbf{R}_3^* = \mathbf{R}_3 + 6\mathbf{R}_1$

$\phantom{\mathbf{R}_3^*} = (0 \quad -6 \quad 5) + 6(1 \quad 1 \quad -2)$

$\phantom{\mathbf{R}_3^*} = (6 \quad 0 \quad -7)$

The revised matrix would have been

$$\begin{pmatrix} 1 & 1 & -2 \\ 0 & 1 & -1 \\ 6 & 0 & -7 \end{pmatrix}$$

So we gained a zero in column 2, but lost a zero in column 1. You could say this was a net gain of zero, except that we didn't get any more zeros. This is really confusing! Anyway, you probably see that this is not the way to do it.

Application to Investments Problem

Consider the investments linear system developed in Chapter 8. (See pages 260–261.)

$x_1 + x_2 = 100$

$.04x_1 + .08x_2 = 5$

We now use ERO's to find $\mathbf{A}^{-1}$.

$$\mathbf{A} = \begin{pmatrix} 1 & 1 \\ .04 & .08 \end{pmatrix} \qquad \mathbf{I} = \begin{pmatrix} 1 & 0 \\ 0 & 1 \end{pmatrix}$$

$\mathbf{R}_2^* = \mathbf{R}_2 - .04\mathbf{R}_1$

$\mathbf{R}_2^* = (.04 \quad .08) - .04(1 \quad 1) \qquad \mathbf{R}_2^* = (0 \quad 1) - .04(1 \quad 0)$
$\phantom{\mathbf{R}_2^*} = (0 \quad .04) = (-.04 \quad 1)$

$$\begin{pmatrix} 1 & 1 \\ 0 & .04 \end{pmatrix} \qquad \begin{pmatrix} 1 & 0 \\ -.04 & 1 \end{pmatrix}$$

$\mathbf{R}_2^* = 25\mathbf{R}_2$

$\mathbf{R}_2^* = 25(0 \quad .04) = (0 \quad 1) \qquad \mathbf{R}_2^* = 25(-.04 \quad 1) = (-1 \quad 25)$

$$\begin{pmatrix} 1 & 1 \\ 0 & 1 \end{pmatrix} \qquad \begin{pmatrix} 1 & 0 \\ -1 & 25 \end{pmatrix}$$

$\mathbf{R}_1^* = \mathbf{R}_1 - \mathbf{R}_2$

$\mathbf{R}_1^* = (1 \quad 1) - (0 \quad 1) \qquad \mathbf{R}_1^* = (1 \quad 0) - (-1 \quad 25)$
$\phantom{\mathbf{R}_1^*} = (1 \quad 0) = (2 \quad -25)$

$$\begin{pmatrix} 1 & 0 \\ 0 & 1 \end{pmatrix} = \mathbf{I} \qquad \begin{pmatrix} 2 & -25 \\ -1 & 25 \end{pmatrix} = \mathbf{A}^{-1}$$

Exercises 22

Use ERO's to find the inverse of the following matrices:

1. $\begin{pmatrix} 1 & -4 \\ -1 & 3 \end{pmatrix}$

2. $\begin{pmatrix} 2 & 3 \\ 1 & -4 \end{pmatrix}$

3. $\begin{pmatrix} 1 & 1 & 4 \\ 2 & 3 & -1 \\ 0 & -2 & 5 \end{pmatrix}$

4. $\begin{pmatrix} 2 & 0 & 4 \\ 1 & -1 & 3 \\ 3 & 2 & -1 \end{pmatrix}$

5. The $\mathbf{A}$ matrix for the demand and supply problem formulated on page 262.

10.6 SOLVING LINEAR SYSTEMS BY MATRIX INVERSION

Recall that the linear system, expressed in matrix notation,

$\mathbf{AX} = \mathbf{B}$

MATRIX ALGEBRA

has the solution

$$X = A^{-1}B$$

Now that we know all about A^{-1}, let's use this knowledge to solve some linear systems.

Example 1

The linear system

$$x_1 + 2x_2 = 4$$
$$3x_1 + 4x_2 = 10$$

in matrix notation is

$$\begin{array}{ccc} A & X & = B \end{array}$$
$$\begin{pmatrix} 1 & 2 \\ 3 & 4 \end{pmatrix} \begin{pmatrix} x_1 \\ x_2 \end{pmatrix} = \begin{pmatrix} 4 \\ 10 \end{pmatrix}$$

We previously found the inverse of this A matrix to be

$$A^{-1} = \begin{pmatrix} -2 & 1 \\ \frac{3}{2} & -\frac{1}{2} \end{pmatrix}$$

So the solution is

$$X = A^{-1}B = \begin{pmatrix} -2 & 1 \\ \frac{3}{2} & -\frac{1}{2} \end{pmatrix} \begin{pmatrix} 4 \\ 10 \end{pmatrix} = \begin{pmatrix} 2 \\ 1 \end{pmatrix}$$

$$X = \begin{pmatrix} x_1 \\ x_2 \end{pmatrix} = \begin{pmatrix} 2 \\ 1 \end{pmatrix}$$

Two matrices are equal only if all their corresponding elements are equal. So $x_1 = 2$ and $x_2 = 1$ is the solution.

Example 2

The linear system

$$x_1 + 2x_2 = 1$$
$$3x_1 + 4x_2 = -3$$

has the same A matrix (and thus A^{-1} matrix) as the previous example. So the solution is simply

$$X = A^{-1}B = \begin{pmatrix} -2 & 1 \\ \frac{3}{2} & -\frac{1}{2} \end{pmatrix} \begin{pmatrix} 1 \\ -3 \end{pmatrix} = \begin{pmatrix} -5 \\ 3 \end{pmatrix} = \begin{pmatrix} x_1 \\ x_2 \end{pmatrix}$$

Example 3

The linear system

$$x_1 + x_2 - 2x_3 = 3$$
$$x_2 - x_3 = 1$$
$$2x_1 - 4x_2 + x_3 = -4$$

in matrix notation is

$$\overset{\mathbf{A}}{\begin{pmatrix} 1 & 1 & -2 \\ 0 & 1 & -1 \\ 2 & -4 & 1 \end{pmatrix}} \overset{\mathbf{X}}{\begin{pmatrix} x_1 \\ x_2 \\ x_3 \end{pmatrix}} = \overset{\mathbf{B}}{\begin{pmatrix} 3 \\ 1 \\ -4 \end{pmatrix}}$$

We previously found the inverse of this **A** matrix to be

$$\mathbf{A}^{-1} = \begin{pmatrix} 3 & -7 & -1 \\ 2 & -5 & -1 \\ 2 & -6 & -1 \end{pmatrix}$$

So the solution is

$$\mathbf{X} = \mathbf{A}^{-1}\mathbf{B} = \begin{pmatrix} 3 & -7 & -1 \\ 2 & -5 & -1 \\ 2 & -6 & -1 \end{pmatrix} \begin{pmatrix} 3 \\ 1 \\ -4 \end{pmatrix} = \begin{pmatrix} 6 \\ 5 \\ 4 \end{pmatrix} = \begin{pmatrix} x_1 \\ x_2 \\ x_3 \end{pmatrix}$$

Example 2 points out one advantage of the matrix algebra solution method. As long as the left-side coefficients remain unchanged, the same inverse can be used to solve all possible cases of right-side constants. Cramer's rule would not be so kind to us. It would require the re-computation of all numerator determinants each time.

We will illustrate this solution flexibility in the next section with some linear systems.

Matrix Solution Flexibility

In example 2 above we illustrated how the same matrix inverse can be used to solve many linear systems—as long as the left-side coefficients remain the same. This flexibility of matrix solution methods is often important in everyday applications. We can show this with the investments problem (See pages 341–342).

Earlier we found the inverse matrix to be

$$\mathbf{A}^{-1} = \begin{pmatrix} 2 & -25 \\ -1 & 25 \end{pmatrix}$$

If Bernard Broke wishes to investigate three possible investment strategies, namely,

MATRIX ALGEBRA

$$\mathbf{B}_1 = \begin{pmatrix} 100 \\ 7 \end{pmatrix} \quad \mathbf{B}_2 = \begin{pmatrix} 100 \\ 6 \end{pmatrix} \quad \mathbf{B}_3 = \begin{pmatrix} 150 \\ 9 \end{pmatrix}$$

(recall that the top number is the total investment and the bottom number is the annual dividend income—both in thousands) then we can quickly solve for the three $\mathbf{X}$ matrices as follows:

$$\mathbf{X}_1 = \mathbf{A}^{-1}\mathbf{B}_1 = \begin{pmatrix} 2 & -25 \\ -1 & 25 \end{pmatrix} \begin{pmatrix} 100 \\ 7 \end{pmatrix} = \begin{pmatrix} 25 \\ 75 \end{pmatrix}$$

$$\mathbf{X}_2 = \mathbf{A}^{-1}\mathbf{B}_2 = \begin{pmatrix} 2 & -25 \\ -1 & 25 \end{pmatrix} \begin{pmatrix} 100 \\ 6 \end{pmatrix} = \begin{pmatrix} 50 \\ 50 \end{pmatrix}$$

$$\mathbf{X}_3 = \mathbf{A}^{-1}\mathbf{B}_3 = \begin{pmatrix} 2 & -25 \\ -1 & 25 \end{pmatrix} \begin{pmatrix} 150 \\ 9 \end{pmatrix} = \begin{pmatrix} 75 \\ 75 \end{pmatrix}$$

Recall that the results are the amounts that should be invested in the stock of Gangland and C.A.R.T.E.L., respectively.

Exercises 23

The inverse matrix for the production problem, defined in Chapter 8 (See page 263), is

$$\mathbf{A}^{-1} = \frac{1}{209} \begin{pmatrix} -33 & 66 & 22 \\ 22 & -44 & 55 \\ 77 & 55 & -121 \end{pmatrix}$$

Solve the system for the two inventory vectors

$$\mathbf{B}_1 = \begin{pmatrix} 12 \\ 10 \\ 7 \end{pmatrix} \quad \mathbf{B}_2 = \begin{pmatrix} 17 \\ 14 \\ 12 \end{pmatrix}$$

10.7 APPLICATIONS OF MATRIX ALGEBRA

Matrix algebra can be applied to solve linear systems such as the practical problems developed in Chapter 8. Also, it is widely used because it provides a compact way of expressing and manipulating mathematical models and is conducive to computer methods. We wait until section 10.8 to tackle computer solutions. In this section we will illustrate the nature of these applications with the Leontief input-output, Markov market share, student-load, and least squares or best fitting equation models.

Leontief Input-Output Model Revisited

The Leontief input-output model, which has waited almost three chapters for solution, was initially described by the following four equations back in Chapter 8. (See pages 267–269.)

$M = .3M + .1T + .2F + .5C + 1$

$T = .3M + .2T + .4F + .1C + 3$

$F = 0M + .1T + .1F + 0C + 5$

$C = .2M + .3T + .1F + .1C + 4$

This can be compactly written in matrix notation as

$$\begin{matrix} \mathbf{X} & = & (\mathbf{A} & \cdot & \mathbf{X}) & + & \mathbf{B} \end{matrix}$$

$$\begin{pmatrix} M \\ T \\ F \\ C \end{pmatrix} = \left[\begin{pmatrix} .3 & .1 & .2 & .5 \\ .3 & .2 & .4 & .1 \\ 0 & .1 & .1 & 0 \\ .2 & .3 & .1 & .1 \end{pmatrix} \begin{pmatrix} M \\ T \\ F \\ C \end{pmatrix} \right] + \begin{pmatrix} 1 \\ 3 \\ 5 \\ 4 \end{pmatrix}$$

In Chapter 8 we went through some cumbersome manipulations to get the system's equations in implicit form. Let's see how simple the matrix manipulations are.

	Comments
$\mathbf{IX} = \mathbf{AX} + \mathbf{B}$	$\mathbf{IX} = \mathbf{X}$
$\mathbf{IX} - \mathbf{AX} = \mathbf{AX} - \mathbf{AX} + \mathbf{B}$	Subtract $\mathbf{AX}$ from both sides
$\mathbf{IX} - \mathbf{AX} = \mathbf{N} + \mathbf{B}$	$\mathbf{N}$ is the null matrix. Any matrix minus itself equals the null matrix
$(\mathbf{I} - \mathbf{A})\mathbf{X} = \mathbf{B}$	$\mathbf{X}$ is factored out on the right side since matrix multiplication isn't commutative. Like the number zero, adding the null matrix doesn't change a matrix

By premultiplying both sides of the matrix equation by $(\mathbf{I} - \mathbf{A})^{-1}$, we can solve for $\mathbf{X}$.

$$(\mathbf{I} - \mathbf{A})\mathbf{X} = \mathbf{B}$$

$$\underbrace{(\mathbf{I} - \mathbf{A})^{-1}(\mathbf{I} - \mathbf{A})}_{\mathbf{I}} \mathbf{X} = (\mathbf{I} - \mathbf{A})^{-1} \mathbf{B}$$

$$\mathbf{I} \cdot \mathbf{X} = (\mathbf{I} - \mathbf{A})^{-1} \mathbf{B}$$

$$\mathbf{X} = (\mathbf{I} - \mathbf{A})^{-1} \mathbf{B}$$

The solution requires us to compute $(\mathbf{I} - \mathbf{A})^{-1}$, the inverse of a 4×4 matrix. ERO's will certainly do that job. But with such large matrices, it would be a tedious job. Thankfully, computers have been programmed to carry out ERO's and find inverses very quickly. We will do a computer solution for this and other problems in Section 10.8.

Markov Chain Analysis of Market Share

Markov chains provide a method to trace the market share of a product through the use of matrix algebra. We will illustrate these methods with the titanic marketing struggle between Glamourpuss and Flooride toothpastes.

For model illustration purposes we assume only these two toothpastes in the market, with Glamourpuss having 60 percent of the sales (thus, Flooride has a 40 percent market share). Fur-

MATRIX ALGEBRA

thermore, suppose that consumer shifts from one toothpaste to another because of advertising and other influences occur along these lines: (1) 70 percent of the people who bought Glamourpuss will repeat on their next purchase (30 percent will switch to Flooride), and (2) 80 percent of the people who bought Flooride will repeat on their next purchase (20 percent will switch to Glamourpuss). Assuming that a purchase is made every month. Glamourpuss' sales next month will be made up of repeats plus switches.

Repeats	.7 fraction of 60% =	42% of market
Switches	.2 fraction of 40% =	8% of market
		50% of market

Flooride sales would obviously be 50 percent also, but let's see the intermediate steps.

Repeats	.8 fraction of 40% =	32% of market
Switches	.3 fraction of 60% =	18% of market
		50% of market

These calculations can be systematized by way of the following matrix operations. But first we need to introduce the "brand switch" matrix (S) as given below.

$$\begin{array}{cc} & G \quad F \end{array}$$
$$\begin{array}{c} G \\ F \end{array}\begin{pmatrix} .7 & .3 \\ .2 & .8 \end{pmatrix} = S$$

The numbers in the matrix are the fractions, as given above, of the row toothpaste buyers that will buy the column toothpaste on the next purchase. For example, the first row tells the story for current users of Glamourpuss—70 percent (.7 fraction) of them will repeat buy, and 30 percent (.3 fraction) of them will switch.

Premultiplying the S matrix by the initial market share vector (M_1) as

$$M_1 \cdot S = M_2$$

$$(.6 \quad .4) \begin{pmatrix} .7 & .3 \\ .2 & .8 \end{pmatrix} = (.42 + .08 \quad .18 + .32) = (.5 \quad .5)$$

duplicates the operations above and gives the second month market share (50 percent each).

The third month's market share is found by multiplying the revised market share and brand switch matrices

$$M_2 \cdot S = M_3$$

$$(.5 \quad .5) \begin{pmatrix} .7 & .3 \\ .2 & .8 \end{pmatrix} = (.35 + .10 \quad .15 + .40) = (.45 \quad .55)$$

The fourth month market share (M_4) is

$$M_3 \cdot S = M_4$$

$$(.45 \quad .55) \begin{pmatrix} .7 & .3 \\ .2 & .8 \end{pmatrix} = (.425 \quad .575)$$

Exercises 24

1. Find the fifth-month market shares.
2. Find the sixth-month market shares.

If we accept the notion of a matrix raised to a power

A^2 means $A \cdot A$

A^3 means $A \cdot A \cdot A$

we can systematize these operations further.

$M_3 = M_2 \cdot S$

but

$M_2 = M_1 \cdot S$

so

$M_3 = (M_1 \cdot S) \cdot S = M_1 \cdot S^2$
$M_4 = M_3 \cdot S = (M_1 \cdot S^2) \cdot S = M_1 \cdot S^3$

Exercise 25

Show $M_5 = M_1 \cdot S^4$

Steady State

At steady state the market share remains constant from one month to the next. In matrix notation, this is

$M_i \cdot S = M_{i+1} = M_i$

where M_i is market share matrix in month i.

If we define the steady-state market shares as x and y, respectively, we have

$$(x \quad y) \begin{pmatrix} .7 & .3 \\ .2 & .8 \end{pmatrix} = (x \quad y)$$

This yields the two equations

$.7x + .2y = x$ or $-.3x + .2y = 0$
$.3x + .8y = y$ $.3x - .2y = 0$

MATRIX ALGEBRA

There is only one independent equation here, but since we have an additional equation $x + y = 1$, we can solve

$x + y = 1$

$.3x - .2y = 0$

For this baby linear system, Cramer's rule probably yields the quickest solution.

$$x = \frac{\begin{vmatrix} 1 & 1 \\ 0 & -.2 \end{vmatrix}}{\begin{vmatrix} 1 & 1 \\ .3 & -.2 \end{vmatrix}} = \frac{-.2 - 0}{-.2 - .3} = \frac{-.2}{-.5} = .4$$

Thus, eventually Glamourpuss will have 40 percent of the market, and Flooride will have 60 percent of the market.

Exercises 26

1. If Glamourpuss began with 70 percent of the market, what would the eventual market shares be?
2. If Glamourpuss began with 60 percent of the market, with the following brand switch matrix

$$S = \begin{pmatrix} .8 & .2 \\ .4 & .6 \end{pmatrix}$$

   what would the eventual market shares be?

Student Load Model

Consider the Mate Tricks University current enrollment **C** and expected future enrollment **E** vectors. (*Note:* The numbers are business, liberal arts, and math-science majors, respectively.)

$\mathbf{C} = (300 \quad 1600 \quad 100)$

$\mathbf{E} = (400 \quad 1500 \quad 100)$

When he saw these figures, the dean demanded that the chancellor give him funds to increase his business faculty by one-third to reflect the one-third increase in business majors. Professor Matt Maddix of the math-science department immediately countered the dean's argument with—you guessed it—matrix algebra. Maddix explained that the student load taught by a department can be figured by multiplying the number of various majors times the average number of courses per major taken in that department. For example, the current crop of students takes business courses as follows.

Student major	Number of students	Courses per student	Total business load
Business	300	2	600
Liberal Arts	1600	$\frac{1}{2}$	800
Math-Science	100	$\frac{1}{2}$	50
			1450

This procedure can be followed for each teaching department for current and expected students. But the following matrix multiplications, using the **N** matrix (average number of courses classified by teaching department and student major as defined on page 311), give the whole story at once.

Current student load by teaching department = **CN**

Expected student load by teaching department = **EN**

The student load results are

$$\text{Current load} = (300 \quad 1600 \quad 100) \begin{pmatrix} 2 & 2 & 1 \\ \frac{1}{2} & 4 & \frac{1}{2} \\ \frac{1}{2} & 1\frac{1}{2} & 3\frac{1}{2} \end{pmatrix} = (1450 \quad 7150 \quad 1450)$$

$$\text{Expected load} = (400 \quad 1500 \quad 100) \begin{pmatrix} 2 & 2 & 1 \\ \frac{1}{2} & 4 & \frac{1}{2} \\ \frac{1}{2} & 1\frac{1}{2} & 3\frac{1}{2} \end{pmatrix} = (1600 \quad 6950 \quad 1500)$$

So, business faculty will be teaching 1600 students as opposed to 1450 now. This is an increase of a little over 10 percent, which is far from the one-third increase suggested by just looking at the number of majors.

Exercise 27

Suppose that the **E** vector equals the **C** vector, except for the 400 business students. What percentage increase in teaching load would be experienced by the math-science faculty?

Matrix Solution for Least Squares Regression

We saw on page 217 that the so-called normal equations needed to find the least squares line that best fits a set of data were:

$$ni + (\Sigma x)s = \Sigma y$$
$$(\Sigma x)i + (\Sigma x^2)s = \Sigma xy$$

Remember that we must await calculus before we can understand where these mysterious equations came from.

For realistic situations, we usually have lots of observations and so computers are needed.

MATRIX ALGEBRA

In addition, we now see that posing such problems in matrix algebra is also useful. Then the combination of using computers to work on matrix equations will be simply dynamite. Let's see.

Suppose one sets up $\mathbf{X}$, $\boldsymbol{\beta}$, and $\mathbf{Y}$ matrices defined as follows:

$$\mathbf{X} = \begin{pmatrix} 1 & x_1 \\ 1 & x_2 \\ 1 & x_3 \\ \vdots & \vdots \\ 1 & x_n \end{pmatrix}_{n \times 2} \quad \boldsymbol{\beta} = \begin{pmatrix} i \\ s \end{pmatrix}_{2 \times 1} \quad \mathbf{Y} = \begin{pmatrix} y_1 \\ y_2 \\ y_3 \\ \vdots \\ y_n \end{pmatrix}_{n \times 1}$$

Then, the normal equations can be duplicated by the matrix equation:

$$(\mathbf{X}'\mathbf{X})\boldsymbol{\beta} = \mathbf{X}'\mathbf{Y}$$

Let's verify that

$$\begin{pmatrix} 1 & 1 & 1 & \cdots & 1 \\ x_1 & x_2 & x_3 & \cdots & x_n \end{pmatrix}_{2 \times n} \cdot \begin{pmatrix} 1 & x_1 \\ 1 & x_2 \\ 1 & x_3 \\ \vdots & \vdots \\ 1 & x_n \end{pmatrix}_{n \times 2} \cdot \begin{pmatrix} i \\ s \end{pmatrix} = \begin{pmatrix} 1 & 1 & 1 & \cdots & 1 \\ x_1 & x_2 & x_3 & \cdots & x_n \end{pmatrix} \begin{pmatrix} y_1 \\ y_2 \\ y_3 \\ \vdots \\ y_n \end{pmatrix}$$

First multiplying $\mathbf{X}'\mathbf{X}$ gives

$$\begin{pmatrix} 1+1+1+\cdots+1 & 1x_1 + 1x_2 + 1x_3 + \cdots + 1x_n \\ 1x_1 + 1x_2 + 1x_3 + \cdots + 1x_n & x_1^2 + x_2^2 + x_3^2 + \cdots + x_n^2 \end{pmatrix} = \begin{pmatrix} n & \Sigma x \\ \Sigma x & \Sigma x^2 \end{pmatrix}$$

Then multiplying this product by $\boldsymbol{\beta}$ gives

$$\begin{pmatrix} n & \Sigma x \\ \Sigma x & \Sigma x^2 \end{pmatrix} \begin{pmatrix} i \\ s \end{pmatrix} = \begin{pmatrix} ni + (\Sigma x)s \\ (\Sigma x)i + (\Sigma x^2)s \end{pmatrix}$$

Notice how these are the left sides of the normal equations. Now working on the right side

$$\mathbf{X}'\mathbf{Y} = \begin{pmatrix} 1 & 1 & 1 & \cdots & 1 \\ x_1 & x_2 & x_3 & \cdots & x_n \end{pmatrix}_{2 \times n} \begin{pmatrix} y_1 \\ y_2 \\ y_3 \\ \vdots \\ y_n \end{pmatrix}_{n \times 1} = \mathbf{P}_{2 \times 1}$$

$$= \begin{pmatrix} 1y_1 + 1y_2 + 1y_3 + \cdots + 1y_n \\ x_1 y_1 + x_2 y_2 + x_3 y_3 + \cdots + x_n y_n \end{pmatrix} = \begin{pmatrix} \Sigma y \\ \Sigma xy \end{pmatrix}$$

Notice how this is the right side of the normal equations. Yipee!!

Exercise 28

Use the data for the 8 observations on page 265 to carry out the matrix operations

$(X'X)\beta = X'Y$.

Solving for the intercept (i) and slope (s) of the least squares line requires us to isolate the β matrix which contains those values in the matrix equation

$(X'X)\beta = X'Y$

This is achieved by taking the inverse of $X'X$, which is symbolized as $(X'X)^{-1}$, and premultiplying both sides by that, or

$\underbrace{(X'X)^{-1}(X'X)}_{I}\beta = (X'X)^{-1}X'Y$

So $\quad\quad \beta = (X'X)^{-1}X'Y$

Exercise 29

Using the data for the 8 observations on page 265, set up the X', X, and Y matrices. Substitute in the above equation. Do not compute the inverse or complete the multiplication.

We will use the computer to complete this job in section 10.8.

10.8 COMPUTER MATRIX OPERATIONS

Computers and the internet have opened up a new world for those wishing to carry out matrix operations. Microsoft Excel and various internet web sites allow for easy calculations with matrices. Excel just gives the final result of a matrix operation. One neat user-friendly web site: www.calc101.com also gives you intermediate steps. Next, we illustrate both softwares.

Matrix Calculations with www.calc101.com

As their home page states (See Figure 10.6), many answers and supporting steps are FREE! Steps for some operations require you to pay for a password. Being that you may be real cheap (like me!), we illustrate only the FREE stuff. In addition, notice that this site does calculus too.

Clicking on matrix algebra gives you the page in Figure 10.7, which allows for the input of an **A** and a **B** matrix. Let's illustrate with these example matrices.

$$A = \begin{pmatrix} 1 & 2 \\ 3 & 4 \end{pmatrix} \quad B = \begin{pmatrix} 5 & 6 \\ 7 & 8 \end{pmatrix}$$

MATRIX ALGEBRA

Figure 10.6

Automatic Calculus and Algebra

derivatives, graphs, integrals, matrix algebra

live step-by-step math
- derivatives
- integrals
- matrix algebra *new!*
- long multiplication
- long division

You've come to the right place for step-by-step solutions...

live curve sketching *newer!*

stored derivatives

- Anytime, from anywhere, get solutions fast, automatically.
- See all the steps in plain English, just like a math textbook.
- See it in standard mathematics notation.
- Check your homework, study for quizzes, and review for exams.

- product rule
- quotient rule
- chain rule

stored integrals

Free answers and steps to get unstuck fast...

- substitution
- integration by parts
- trig powers
- trig products
- multiple angles
- trig substitutions
- trig rationals
- partial fractions
- using reduction formulas
- special trig integrals
- deriving reduction formulas

- Do any first or second derivative step-by-step.
- Plot your function.
- Get the answer to practically any freshman indefinite integral.
- Do over 80 step-by-step matrix algebra operations.
- Get the answers for determinants and matrix inverses.
- Do long multiplication and division of polynomials.

Buy a <u>p</u>assword to boost your algebra and calculus grades...

- Get all the steps for indefinite integrals.
- Get all the steps for determinants and matrix inverses.
- Get a complete graphical analysis of your function.

FAQs | legal | email us | français | español | Deutsch

The drop-down menus allow for matrix multiplication and matrix inversion, which we do for our example matrices and show the results in Figures 10.8a and b. Voilá! Those results are exactly what we got by old-fashioned hand calculation.

Exercise 30

Reverse the matrices in the above example. Then use www.calc101.com to calculate the product matrix and the inverse of the matrix on the left.

Hand calculations for 2 by 2 matrices are not difficult. But, try finding the inverse of a 4 by 4 or bigger matrices by hand. If you were foolish enough to do that, you would soon be very thankful for the power of computers. We use the input output problem on page 346 to illustrate how this web site can painlessly give us results for large matrices. We input the 4 by 4 $(\mathbf{I} - \mathbf{A})$ matrix in the cells for the $\mathbf{A}$ matrix and the final demands as the 4 by 1 $\mathbf{B}$ matrix, as shown at the bottom of page 354.

Figure 10.7

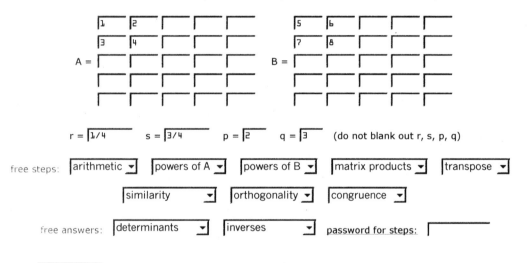

$$A = \begin{pmatrix} .7 & -.1 & -.2 & -.5 \\ -.3 & .8 & -.4 & -.1 \\ 0 & -.1 & .9 & 0 \\ -.2 & -.3 & -.1 & .9 \end{pmatrix} \quad B = \begin{pmatrix} 1 \\ 3 \\ 5 \\ 4 \end{pmatrix}$$

Then, from the drop down matrix multiplication menu, you would click on $\mathbf{A}^{-1}\,\mathbf{B}$, and voilá, almost instantaneously, you would get intermediate steps and the final solution below:

$$X = \begin{pmatrix} M \\ T \\ F \\ C \end{pmatrix} = \begin{pmatrix} 2400/157 \\ 2320/157 \\ 1130/157 \\ 2130/157 \end{pmatrix} = \begin{pmatrix} 15.29 \\ 14.78 \\ 7.20 \\ 13.57 \end{pmatrix}$$

MATRIX ALGEBRA

Figure 10.8a

$$A \cdot B = \begin{pmatrix} 1 & 2 \\ 3 & 4 \end{pmatrix} \cdot \begin{pmatrix} 5 & 6 \\ 7 & 8 \end{pmatrix}$$

Dot-multiply the rows of the first matrix with the columns of the second.

$$= \begin{pmatrix} (1\ 2) \cdot \begin{pmatrix} 5 \\ 7 \end{pmatrix} & (1\ 2) \cdot \begin{pmatrix} 6 \\ 8 \end{pmatrix} \\ (3\ 4) \cdot \begin{pmatrix} 5 \\ 7 \end{pmatrix} & (3\ 4) \cdot \begin{pmatrix} 6 \\ 8 \end{pmatrix} \end{pmatrix}$$

Dot-multiply each pair of vectors.

$$= \begin{pmatrix} 1 \times 5 + 2 \times 7 & 1 \times 6 + 2 \times 8 \\ 3 \times 5 + 4 \times 7 & 3 \times 6 + 4 \times 8 \end{pmatrix}$$

Simplify.

$$= \begin{pmatrix} 5 + 14 & 6 + 16 \\ 15 + 28 & 18 + 32 \end{pmatrix}$$

Simplify.

$$= \begin{pmatrix} 19 & 22 \\ 43 & 50 \end{pmatrix}$$

Figure 10.8b

The inverse is the matrix on the right.

$$A = \begin{pmatrix} 1 & 2 \\ 3 & 4 \end{pmatrix}$$

$$\begin{pmatrix} 1 & 2 \\ 3 & 4 \end{pmatrix}^{-1} = \begin{pmatrix} -2 & 1 \\ \frac{3}{2} & -\frac{1}{2} \end{pmatrix}$$

www.calc101.com is limited to matrices with order of 5 by 5 or smaller. For larger matrices, Microsoft Excel would be a good option, as we illustrate next.

Matrix Calculations with Microsoft Excel

To get a feel for Microsoft Excel, we first consider the same two example 2 by 2 matrices:

$$\mathbf{A} = \begin{pmatrix} 1 & 2 \\ 3 & 4 \end{pmatrix} \quad \mathbf{B} = \begin{pmatrix} 5 & 6 \\ 7 & 8 \end{pmatrix}$$

We will first illustrate the Excel procedure for multiplication and then we will find the inverse of **A**.

Multiplication of Matrices with Excel

The Excel procedure is as follows (See Figure 10.9a for the spreadsheet and results).

1. Input the two matrices in the spreadsheet. We can type in the name of the matrix right above each one. The location of the matrix is arbitrary. Let's put the **A** matrix in the second and third rows of the A and B columns and the **B** matrix in the second and third rows of the F and G columns. In Excel notation, the **A** matrix is located in the array space A2:B3 and the **B** matrix is located in the array space F2:G3. Notice that an array space is denoted by the cell in the upper left-hand corner, a colon, and the cell in the lower right-hand corner.
2. We need to allow space for the product matrix, which will be a 2 by 2 also. We can do that by highlighting the spreadsheet section or array space A8:B9. Right above that in cell A7 we can type in "A times B" so that we know what the answer means.
3. To actually calculate the product, we must go to the formula bar and type in: =MMULT (A2:B3,F2:G3).
4. To get the result printed in the array reserved for it, we must hold down the Shift and Control keys TOGETHER, while pressing the ENTER key. Notice that when the ENTER key is pressed, all three keys are pressed down.
5. Wahlah, the product matrix answer is ours and it appears in the reserved space of A8:B9.

Matrix Inverse on Excel

Let's compute $\mathbf{A}^{-1}$, or the inverse of the **A** matrix. The **A** matrix is already on the spreadsheet in the array section A2:B3. Then, the procedure is:

1. Set up an array space for the inverse. We can do that by highlighting the array space A15:B16. So that we can know what the result means, we type in "A inverse" right above that array in cell A14
2. Go to the formula bar and type in =MINVERSE(A2:B3)
3. Hold the Control and Shift keys down TOGETHER, while pressing the ENTER key.
4. Wahlah, **A** inverse appears in the reserved array space of A15:B16. Note that it is the same as we found earlier with the www.calc101.com software and we did by old-fashioned hand calculations.

Exercises 31

Use Excel to do the following:

1. Reverse the two matrices and find that product.
2. Find the inverse of the **B** matrix

Excel With Large Matrices

Recall that we need Excel to work with large matrices. So, let's take the following linear system with 6 equations and 6 unknowns:

$1x + 1y + 1z + 1w + .5v + 1u = 2$

$2x - 1y + 3z + 0w + 4v - 1u = 26$

$-1x - 1y - 1z + 2w + 5v + 0u = 12$

$5x - 1y + .5z - 3w - 1v + 1u = -3$

$4x + 3y + 2z + 1w - 1v + 0u = 12$

$6x + 1y - 1z - 2w - 3v - .4u = -1$

Solving this system, we use the procedure: (See Figure 10.9b)

1. Input the left-side coefficients as the **A** matrix in the 6 by 6 array section A2:F7.
2. Input the right side-coefficients as the **B** matrix in the 6 by 1 array section I2:I7.
3. Reserve a 6 by 6 array for **A** inverse in the array section A12:F17.
4. In the formula bar, type in: =MINVERSE(A2:F7).
5. Hold the Control and Shift keys down TOGETHER, while hitting ENTER.
6. Magically, A^{-1} is displayed in the section reserved for it.
7. Now to get the solution, we need to multiple $A^{-1}B$.
8. Reserve a 6 by 1 array in I12:I17 by highlighting those cells.
9. In the formula bar, type =MMULT(A12:F17,I12:I17)
10. Hold the Control and Shift keys down TOGETHER, while pressing ENTER.
11. Yippee, the solution is ours in the space provided for it.

Exercises 32

Use Excel to do the following:

1. Check the calculation of the inverse by finding: $A \cdot A^{-1}$. (Note: Your result should be **I**.)
2. Change the first element of the **B** matrix from 2 to 5. Solve the linear system with this change.

Exercises 33

1. Calculate by hand (and then you will appreciate computers more) the product of the two 6 by 6 matrices **A** and A^{-1}. Check to see if Excel did it correctly. P.S., It did. So if you get a different result, YOU made a mistake.

358   LINEAR SYSTEMS

Figure 10.9a

	A	B	C	D	E	F	G	H
1	A matrix					B matrix		
2	1	2				5	6	
3	3	4				7	8	
4								
5								
6								
7	A times B							
8	19	22						
9	43	50						
10								
11								
12								
13								
14	A inverse							
15	-2	1						
16	1.5	-0.5						
17								

Figure 10.9b

	A	B	C	D	E	F	G	H	I
1			A matrix						B matrix
2	1	1	1	1	0.5	1			2
3	2	-1	3	0	4	-1			26
4	-1	-1	-1	2	5	0			12
5	5	-1	0.5	-3	-1	1			-3
6	4	3	2	1	-1	0			12
7	6	1	-1	-2	-3	-0.4			-1
8									
9									
10									
11			A inverse						A inverse times B
12	1.113953	0.352941	-0.20943	-0.47855	-0.71839	0.706167			1
13	-3.98465	-1.29412	1.069054	1.815857	2.920716	-2.1867			3
14	0.708156	0.352941	-0.36885	-0.27565	-0.39955	0.19892			2
15	4.035237	1.176471	-0.95254	-1.9784	-2.66354	2.200909			-1
16	-2.0466	-0.58824	0.67917	1.003694	1.425973	-1.13669			4
17	0.150611	-0.29412	0.122194	0.414891	0.147769	-0.35095			-5

MATRIX ALGEBRA

2. There is no way you are going to check the inverse calculations for the 6 by 6 by hand using ERO's. So, input into Excel the 2 by 2 B matrix we used to illustrate the www.calc101.com methods earlier. Then use both ERO's and Excel to calculate its inverse. Compare the results.
3. Use Excel to solve for the least square line for the data on page 265.

PROBLEMS

TECHNICAL TRIUMPHS

1. What are the orders of the following matrices?

   a. $\begin{pmatrix} 9 & 6 & 2 & 8 & 5 \\ 1 & -1 & 1 & -1 & 6 \end{pmatrix}$

   b. $\begin{pmatrix} 3 \\ 2 \\ 0 \\ 5 \end{pmatrix}$

   c. $\begin{pmatrix} 1 & 1 & 1 \\ -2 & -2 & -2 \\ 3 & 3 & 3 \end{pmatrix}$

2. Construct the following matrices
   a. A 6 by 6 Identity matrix
   b. A 3 by 3 Null matrix
   c. A 4 by 4 Diagonal matrix with 7's on the diagonal
   d. A 1 by 6 row vector with a 3 in the (1, 1) position and 4's elsewhere
   e. A 4 by 2 Unity matrix

3. Consider the following 3 matrices.

   $\mathbf{A} = \begin{pmatrix} 1 & 2 & 3 \\ 4 & 5 & 6 \end{pmatrix} \quad \mathbf{B} = \begin{pmatrix} 8 \\ 6 \\ 4 \end{pmatrix} \quad \mathbf{C} = \begin{pmatrix} 9 & -8 & 7 \\ -6 & 5 & -4 \end{pmatrix}$

   a. Which two matrices can be added? Find that sum.
   b. Which matrices can be multiplied? Find those (there are more than one) products.
   c. Find the transpose of each matrix.
   d. Multiply each matrix by 2.
   e. Multiple the C matrix by the appropriate Identity matrix.

4. Determine the missing matrix.

   $\begin{pmatrix} 6 & 9 \\ -1 & 5 \end{pmatrix} + 2\begin{pmatrix} & \\ & \end{pmatrix} = \begin{pmatrix} 10 & 19 \\ -5 & 13 \end{pmatrix}$

5. Consider the matrices

   $\mathbf{A} = \begin{pmatrix} 1 & 2 & 3 \\ 4 & 5 & 6 \\ 7 & 8 & 9 \end{pmatrix} \quad \mathbf{D} = \begin{pmatrix} 3 & 0 & 0 \\ 0 & 3 & 0 \\ 0 & 0 & 3 \end{pmatrix}$

Show that $A \cdot D = D \cdot A$

6. Consider a matrix called **Q**. If $q_{11} = 4$, $q_{12} = 1$, $q_{21} = -5$, $q_{22} = 10$, $q_{31} = 0$, $q_{32} = 100$ what is **Q**?

7. If one has a 7 by 4 matrix called **M**.
   a. What order of a **N** matrix could be subtracted from **M**?
   b. What order(s) of a **N** matrix could be pre multiplied by **M**?
   c. What order(s) of an **N** matrix could be post multiplied by **M**?

8. Determine x and y.

$$\begin{pmatrix} 3 & 2 \\ 5 & 1 \end{pmatrix} \begin{pmatrix} x & -4 \\ 2 & y \end{pmatrix} = \begin{pmatrix} 7 & 6 \\ 7 & 6 \end{pmatrix}$$

9. If $A = \begin{pmatrix} 2 & 5 \\ -4 & 3 \end{pmatrix}$
   a. Find $A \cdot A'$ by hand.
   b. Use www.calc101 to check your answer.

10. Show that the matrix multiplication

    AX = B

    $$\begin{pmatrix} 1 & 1 \\ 3 & 2 \end{pmatrix} \begin{pmatrix} x_1 \\ x_2 \end{pmatrix} = \begin{pmatrix} 4 \\ 3 \end{pmatrix}$$

    is equivalent to the algebraic system

    $x_1 + x_2 = 4$
    $3x_1 + 2x_2 = 3$

11. Consider the two matrices.

    $$A = \begin{pmatrix} 3 & 2 \\ -4 & 1 \end{pmatrix} \quad B = \begin{pmatrix} 5 & 0 \\ 3 & -4 \end{pmatrix}$$

    Use both www.calc101.com and Microsoft Excel to find $A \cdot B$ and A^{-1}.

CONFIDENCE BUILDERS

1. A statistical study of a group of MBA students revealed the correlation matrix given below. The variables used were (M) for grade in their mathematics for management course, (S) for number of hours spent in studying mathematics, (A) for mathematics aptitude as measured by the graduate entrance examination, and (Y) for the number of years since taking the last mathematics course.

   $$C = \begin{matrix} & \begin{matrix} M & S & A & Y \end{matrix} \\ \begin{matrix} M \\ S \\ A \\ Y \end{matrix} & \begin{pmatrix} 1 & .4 & .5 & -.1 \\ .4 & 1 & -.2 & .6 \\ .5 & -.2 & 1 & -.3 \\ -.1 & .6 & -.3 & 1 \end{pmatrix} \end{matrix}$$

   In general, correlations range from –1 to +1, with positive correlations indicating that the variables move in the same direction, and negative correlations indicating that variables move in opposite directions. For example, more study is consistent with higher grades (+.4 correlation), but higher mathematics aptitude tends to mean lower number of years since the last math course (–.3 correlation).

MATRIX ALGEBRA

a. What is the order and type of the **C** matrix?
b. Describe the correlation embodied in c_{23}.
c. Why does $c_{ij} = c_{ji}$? Why are the diagonal elements all equal to 1?
d. If six variables instead of the four here were studied, describe the nature of the resulting correlation matrix.

2. If

$$\mathbf{M} = \begin{pmatrix} 2 & -1 & 3 & 2 \\ 1 & 4 & -5 & -3 \\ -3 & -2 & 6 & -1 \\ 1 & 2 & 2 & 1 \end{pmatrix}$$

then the inverse of M is

$$\mathbf{M}^{-1} = \frac{1}{150} \cdot \begin{pmatrix} 80 & 40 & 5 & -35 \\ -34 & -2 & -4 & 58 \\ 20 & 10 & 20 & 10 \\ -52 & -56 & -37 & 49 \end{pmatrix}$$

Find the solution to the linear system.

$$2x_1 - x_2 + 3x_3 + 2x_4 = 9$$
$$x_1 + 4x_2 - 5x_3 - 3x_4 = -1$$
$$-3x_1 - 2x_2 + 6x_3 - x_4 = 5$$
$$x_1 + 2x_2 + 2x_3 + x_4 = 3$$

3. Consider the following four linear systems:

$$x_1 + x_2 = 5 \qquad x_1 + x_2 = -2$$
$$2x_1 + 3x_2 = 11 \qquad 2x_1 + 3x_2 = -2$$

$$x_1 + x_2 = 1 \qquad x_1 + x_2 = 6$$
$$2x_1 + 3x_2 = 3 \qquad 2x_1 + 3x_2 = 16$$

a. If you only had to solve one of these linear systems, which solution method (Cramer's rule or matrix algebra using ERO's to find $\mathbf{A}^{-1}$, etc.) would you recommend? Why?
b. If you had to solve all four systems, which method (and why) would you recommend?

4. Consider the parallel lines case

$$-x_1 + 2x_2 = 3$$
$$-2x_1 + 4x_2 = 7$$

Recall that this linear system doesn't have a solution, as evidenced by the zero determinant of the left-side coefficients.

Now employ the ERO procedure to the **A** matrix and show that you get stuck (and thus **A** doesn't have an inverse) at the point when the revised **A** matrix is

$$\begin{pmatrix} 1 & -2 \\ 0 & 0 \end{pmatrix}$$

5. An airline reports the number of seats occupied on its flight by means of a 1 × 4 row vector. The first element represents the number of seats occupied by first-class passengers, the second element represents those occupied by coach passengers, the third element represents those occupied by students traveling at reduced rates, and the fourth element represents those occupied by nonpaying passengers.
   a. If **S** = (15 27 4 3) represents the vector of seats occupied on a flight as it leaves Los Angeles for New York via Chicago, **N** = (4 2 7 1) represents the vector of passengers getting on at Chicago, and **F** = (3 5 2 1) represents the vector of passengers getting off at Chicago, find the 1 × 4 vector that represents the seats occupied on this flight as it arrives in New York.
   b. If

$$\mathbf{A} = \begin{pmatrix} 200 \\ 180 \\ 100 \\ 0 \end{pmatrix} \quad \mathbf{B} = \begin{pmatrix} 100 \\ 90 \\ 50 \\ 0 \end{pmatrix} \quad \mathbf{C} = \begin{pmatrix} 250 \\ 220 \\ 125 \\ 0 \end{pmatrix}$$

   are matrices with elements of revenues per person for the four categories of passengers for the Los Angeles-to-Chicago trip (**A**), Chicago-to-New York trip (**B**), and Los Angeles-to-New York trip (**C**), set up the matrix expression for the total revenue for this plane trip.
   c. Calculate the total revenue for the flight.

6. A doll company makes "Pretty Penny" and "Gorgeous Gussie" dolls. Each "Pretty Penny" doll requires 1 unit of labor and 5 units of materials. Each "Gorgeous Gussie" doll requires 3 units of labor and 4 units of materials. The company has 106 units of labor and 200 units of materials available, all of which it wants to use up in the manufacture of the two types of dolls.

   If x_1 is the number of "Pretty Penny" dolls produced, and x_2 is the number of "Gorgeous Gussie" dolls produced,
   a. Set up this problem in matrix algebra terms.
   b. Using the matrix inversion procedure, find $\mathbf{A}^{-1}$.
   c. How many of each doll should be produced?

7. There you are, John Kleen, marketing manager of Tidy John Co. You are flushed with anger since your only competitor, Sparkle Pot, has just successfully introduced a new fluoride toilet cleaner. Your 80 percent market share may just go down the drain.

   Marketing research has just completed a study, which revealed Tidy John (*TJ*) and Sparkle Pot (*SP*) repeat and switch probabilities.

$$\mathbf{S} = \begin{matrix} & TJ & SP \\ TJ & \\ SP & \end{matrix} \begin{pmatrix} .7 & .3 \\ .1 & .9 \end{pmatrix}$$

   (Note the numbers in the matrix represent the fraction of row brand customers who will buy the column brand product next period. For example, 70 percent of Tidy John customers will repeat while 30 percent will switch to Sparkle Pot.)
   a. If we assume that consumers purchase one of the products every month, set up the matrix multiplication necessary to find the market shares at the end of one month.
   b. Compute the market shares at the end of one month.
   c. Set up the proper matrices and compute the market shares at the end of two months.
   d. What will the equilibrium market shares be?

8. Upon analyzing a four-industry input-output model, a student wrote the system of equations as

MATRIX ALGEBRA

$$-.90x_1 + .40x_2 + .05x_3 + .25x_4 = 0$$
$$.20x_1 - .85x_2 + .17x_3 + .22x_4 = -1000$$
$$.00x_1 + .01x_2 - .90x_3 + .05x_4 = -3000$$
$$.25x_1 + .10x_2 + .20x_3 - .85x_4 = -2000$$

The student then wrote this system in matrix terms as

$$\mathbf{AX} = -1 \cdot \mathbf{B}$$

where **X** is the matrix of industry outputs and **B** is the matrix of final demand defined as,

$$\mathbf{X} = \begin{pmatrix} x_1 \\ x_2 \\ x_3 \\ x_4 \end{pmatrix} \qquad \mathbf{B} = \begin{pmatrix} 0 \\ 1000 \\ 3000 \\ 2000 \end{pmatrix}$$

Using the computer, she found

$$\mathbf{A}^{-1} = -1 \cdot \begin{pmatrix} 1.5 & .8 & .4 & .7 \\ .5 & 1.5 & .4 & .5 \\ .03 & .04 & 1.1 & .1 \\ .5 & .4 & .4 & 1.5 \end{pmatrix}$$

and

$$\mathbf{X} = \begin{pmatrix} 3400 \\ 3700 \\ 3540 \\ 4600 \end{pmatrix}$$

She then wondered what the solution to the system would be if the final demand of the fourth industry was reduced from 2000 to 1000. Help the student solve her problem.

9. Slipshod Manufacturing Co. produces two products, A and B. The labor and materials required for one unit of each product is given by the following matrix.

	A	B
Labor	1	3
Materials	4	2

If 14 units of labor and 26 units of materials are available,
   a. Derive the equations required to find the amount of each product that should be produced in order to use up all the available labor and materials.
   b. Express the above system of equations in matrix form.
   c. Use www.calc101.com and Microsoft Excel to solve this problem.

10. Find the inverse of the matrix

$$\mathbf{A} = \begin{pmatrix} 1 & 1 & 1 \\ 1 & a & 0 \\ b & 0 & c \end{pmatrix}$$

where a, b, c are real numbers, but not equal to zero.

Use your result to solve for the breakdown of I. Hoarde's will (problem 3 on page 271).

11. Consider the following linear system:

$$x - 4y = -7$$
$$-x + 3y = 4$$

and the fact that:

$$\begin{pmatrix} 1 & -4 \\ -1 & 3 \end{pmatrix} \begin{pmatrix} -12 & -16 \\ -4 & -4 \end{pmatrix} + \begin{pmatrix} 2 & 2 \\ 2 & 2 \end{pmatrix} = \begin{pmatrix} 6 & 2 \\ 2 & 6 \end{pmatrix}$$

   a. Without using ERO's, determine $\mathbf{A}^{-1}$.
   b. Use matrix algebra (but not ERO's or computers) to solve the system.

12. A firm believes that its sales (S) will be the following function of time in years (x) from now.

$$S = f(x) = a + bx + c \cdot 2^x \quad (x = 0 \text{ now})$$

If sales 1, 2, and 5 years from now are expected to be 15, 20, and 57 respectively,
   a. Set up the 3 equations needed to solve for a, b, and c.
   b. Express this 3 by 3 linear system in matrix form as AX = B.
   c. Use the fact that

$$\mathbf{A}^{-1} = \frac{1}{22} \begin{pmatrix} 44 & -22 & 0 \\ -28 & 30 & -2 \\ 3 & -4 & 1 \end{pmatrix}$$

   to find the sales function.
   d. Use www.calc101 and Microsoft Excel to verify the inverse in part c.

MIND STRETCHERS

1. In problem 6 on page 272, you developed the equations to determine the outputs of the five industries in the hypothetical economy.
   a. Formulate this linear system in matrix notation.
   b. Find $\mathbf{A}^{-1}$ with the help of a computer.
   c. Solve the system.

2. In problem 4a on page 271, you formulated the linear system to determine the number of cars and trucks that crossed the Pollution River on a certain day.
   a. Formulate this linear system in matrix notation.
   b. Use ERO's to find $\mathbf{A}^{-1}$.
   c. Solve the system with matrix algebra.
   d. If, on another day, a total of 7000 vehicles paying a total of $3700 in tolls crossed the bridge, solve the revised system. Explain why you didn't have to find a new $\mathbf{A}^{-1}$ for this case. Compare the matrix solution to the one required if Cramer's rule and determinants were used.

3. In problem 4b on page 271, you incorporated the number of buses into the linear system.
   a. Repeat parts a, b, and c of problem 2 for that case.
   b. If, on another day, a total of 8000 vehicles paying a total of $4450 in tolls crossed the bridge, solve the revised system.

MATRIX ALGEBRA

4. Grade point average (**A**) in a given semester where you are taking n courses having total credit of T is computed as

$$\mathbf{A} = \frac{1}{T}\mathbf{CG}$$

where **C** is a $1 \times n$ matrix of credits per course, and **G** is an $n \times 1$ matrix of grades received (A grade = 4).

a. What is the order of the **A** matrix?
   Consider Stew Dent's case with

$$\mathbf{C} = (3 \ \ 3 \ \ 3 \ \ 4 \ \ 1) \qquad \mathbf{G} = \begin{pmatrix} 2 \\ 4 \\ 3 \\ 1 \\ 2 \end{pmatrix}$$

b. What is Stew's grade point average for the semester?
c. Consider your own case. Formulate your own **C** matrix and anticipated **G** matrix. What semester average grade do you predict for yourself?

5. Your overall grade point average (θ) is computed from the formula

$$\theta = \frac{1}{c_b + c_n}(c_b A_b + c_n A_n)$$

where c_b and c_n are credits amassed before and during this semester, respectively, and A_b and A_n are grade point averages before and during this semester, respectively.

a. Formulate this equation in matrix notation.
b. If Stew Dent had a 2.5 average based on 60 hours before this semester, what will his cumulative average be at the end of this semester? Use the information in problem 4 and the matrix equation formulated in part a.

6. Using elementary row operations, show that the inverse of the general 2×2 matrix

$$\mathbf{A} = \begin{pmatrix} a & b \\ c & d \end{pmatrix}$$

where a, b, c and d are constants, is

$$\mathbf{A}^{-1} = \begin{pmatrix} \dfrac{d}{D} & \dfrac{-b}{D} \\ \dfrac{-c}{D} & \dfrac{a}{D} \end{pmatrix}$$

where D is $ad - cb$, or the determinant of **A**.

7. Consider the matrices:

$$\mathbf{A} = \begin{pmatrix} 1 & 6 & -3 \\ 5 & -2 & 1 \end{pmatrix} \qquad \mathbf{B} = \begin{pmatrix} 1 & 4 & 1 \\ 3 & -3 & 5 \\ -2 & 6 & -2 \end{pmatrix}$$

Use www.calc101.com to find
a. **A′** (A transpose)
b. **A · B**
c. **B**$^{-1}$

8. Consider the following input-output table for a hypothetical 4 industry economy:

		M	T	F	C	Final Demand
Materials	(M)	.1	.4*	.05	.25	0
Transportation	(T)	.2	.15	.17	.22	1
Food	(F)	0	.01	.10	.05	3
Construction	(C)	.25	.10	.20	.15	2

*Note: Entries in table represent the amount of row industry input required per dollar of column industry output. For example, each dollar of transportation industry output requires $.4 or 40 cents of materials input.

a. Set up the 4 by 4 linear system of equations needed to solve for M, T, F, and C.
b. Define the matrices **A**, **X**, and **B** needed to express this system in matrix equation form as **AX** = **B**.
c. Incorporate the fifth industry, Leisure (L), into the economy. Make up reasonable input-output table entries and redefine matrices **A**, **X**, and **B** accordingly.
d. Use both www.calc101.com and Microsoft Excel to solve for the outputs of the 5 industries.

9. Show that for *any* 2 by 2 matrices

$$(AB)' = B'A'$$

10. Consider the linear system

$$x_1 + x_2 - x_3 = 4$$
$$2x_1 - 3x_3 = 3$$
$$3x_1 - 4x_2 + 5x_3 = 6$$

and the fact that

$$-\frac{1}{23} \cdot \begin{pmatrix} -12 & -1 & -3 \\ -19 & 8 & 1 \\ -8 & 7 & -2 \end{pmatrix} \cdot \begin{pmatrix} 1 & 1 & -1 \\ 2 & 0 & -3 \\ 3 & -4 & 5 \end{pmatrix} = \begin{pmatrix} 1 & 0 & 0 \\ 0 & 1 & 0 \\ 0 & 0 & 1 \end{pmatrix}$$

By manipulating the matrices and recognizing their nature (no ERO's or computers please), solve the linear system.

CHAPTER 11

Linear Programming

11.1 INTRODUCTION

In the last three chapters we have followed the continuing saga of Bernard Broke, who must decide how to invest $100 thousand between Gangland Industries, (x_1 thousand dollars) with a 4 percent dividend, and C.A.R.T.E.L., (x_2 thousand dollars) with an 8 percent dividend, in order to have an annual dividend income of $5 thousand. This problem resulted in the following linear system:

$$x_1 + x_2 = 100$$
$$.04x_1 + .08x_2 = 5$$

the solution of which, $x_1 = 75$ and $x_2 = 25$, occurred at the intersection of the two lines in Figure 11.1.

Figure 11.1

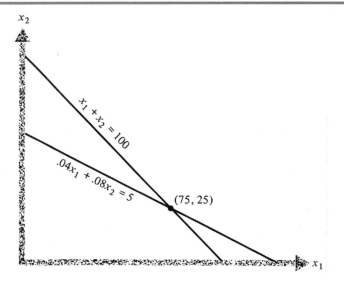

Now we will expand this into a more realistic problem by incorporating the possibility of higher income levels and then by incorporating risk. These adaptations will then bring us into the realm of linear programming.

The second equation in our system states that total dividends must equal exactly $5 thousand. But isn't this a little unrealistic? Why shouldn't Bernard want more dividends, other things being equal? For example, why shouldn't he want $6 thousand dividend income, in which case the linear system

$$x_1 + x_2 = 100$$
$$.04x_1 + .08x_2 = 6$$

with solution, $x_1 = 50$ and $x_2 = 50$, holds? But then, why shouldn't Bernard want even more dividends?

The various dividend income equations are all straight lines of the form

$$.04x_1 + .08x_2 = F$$

where F is the annual dividend income, with slope of $-.04/.08 = -\frac{1}{2}$, and intercept of $F/.08$. So, increasing F does change the intercept but not the slope of these straight lines. Consequently, the various dividend income lines are parallel, as shown in Figure 11.2.

Higher and higher income solutions are found by intersecting the capital line with successive parallel dividend income lines. The maximum income solution occurs at the upper left corner. Beyond that point negative solution values for x_1 occur. Since both x_1 and x_2 must be nonnegative,

$$x_1 \geq 0 \quad \text{and} \quad x_2 \geq 0$$

Figure 11.2

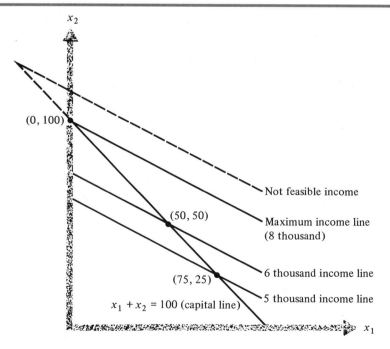

then only solution points in the first quadrant are possible (feasible).

You might say that we have discovered the obvious. If you want the highest dividend income, invest completely in the stock that pays the highest dividend. This is fine, but the idea of parallel straight lines leading to higher and higher income is a key concept in linear programming that just needed to be exposed.

Our second adaptation of the investment problem involves the incorporation of risk. Actual investment yields generally increase as the risks increase, so C.A.R.T.E.L.'s higher dividend rate probably reflects its riskier nature. Consequently, the blind channeling of all the money into C.A.R.T.E.L. in order to achieve the highest total dividend income may not be such a wise decision.

To allow for better investment decisions, we can develop a mathematical statement on risk and incorporate it into our system. Suppose the various business risks (bombings, employee rub-outs, police round-ups, etc.) combine to result in a 5 percent chance of bankruptcy for Gangland Industries and a 20 percent chance of bankruptcy for C.A.R.T.E.L. Thus, an "average" loss of $.05x_1$ on Gangland stock and $.20x_2$ on C.A.R.T.E.L. stock would occur in the long run. The total average loss can be limited to no more than, say, $10 thousand, as stated in the following inequality:

$.05x_1 + .20x_2 \leq 10$

The introduction of risk through the above equation may make investment of the entire $100 thousand unwise. For example, investment of all the money may exceed acceptable risk levels, whereas investment of some lesser amount may not. The mathematical statement that reflects the situation that total investment can be less than $100 thousand, but not more than that, is

$x_1 + x_2 \leq 100$

Now, finally we can gather and arrange all the pieces of this development as a complete problem in linear programming format.

	Linear programming format	
Maximize	$F = .04x_1 + .08x_2$	Objective function
Subject to		Constraints on
	$x_1 + x_2 \leq 100$	Capital
	$.05x_1 + .20x_2 \leq 10$	Risk
	$x_1 \geq 0$	Gangland investment
	$x_2 \geq 0$	C.A.R.T.E.L. investment

This formulation is typical of all linear programming problems. Each is characterized by an objective function. Here, the objective is to maximize total dividend income from the two investments. In other linear programming problems the objective could be to minimize a function—for example, to minimize costs.

In general, the maximization or minimization is subject to certain restrictions, called *constraints*, which are expressed as inequalities. Here, four inequalities incorporating the following constraining factors are present.

1. investment capital available
2. risk limitations
3. Gangland investment can't be negative
4. C.A.R.T.E.L. investment can't be negative

Inequalities of both the $\geq$ and $\leq$ type can generally serve as constraints. There is no limit to the number of such inequalities.

Finally, all equations contain variables raised to the first power. Beasts like x_1^2, log x_2, and e^{x_3} are no-no here.

The set of points that satisfy all four inequalities is called the *feasible solution space* or *feasible region*. Each point in this region (shaded in Figure 11.3) is a potential solution. Notice that the risk constraint has eliminated the possibility of an all-C.A.R.T.E.L. solution.

The solution to this linear programming problem is the point in the feasible region that results in the highest total dividend income (lies on the highest parallel income line passing through the feasible region). This point is $(66\frac{2}{3}, 33\frac{1}{3})$. We will see why later.

Learning Plan

With this rough idea of a linear programming problem in mind, we can now strike out to develop the topic systematically. We begin by developing the basics of inequalities. Then we formulate some representative linear programming problems found in the real world. These problems range from diet planning to production scheduling. Then we attack the basic theory needed to solve such problems. This means that we must delve into systems of inequalities, feasible re-

Figure 11.3

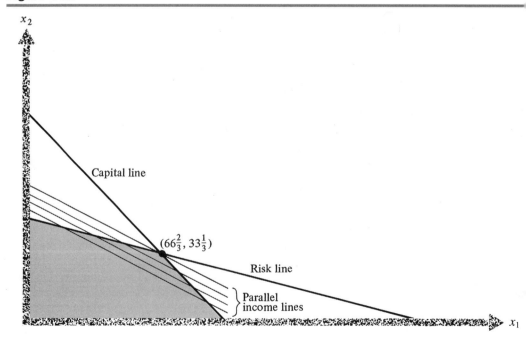

LINEAR PROGRAMMING

gion determination, and the corner point theorem, among others. Of course, this leaves us a little shaky. Thank goodness that graphical solution methods allow us to pull it all together. Then algebraic solution methods, systematized by the so-called *simplex method,* are unveiled. Meanwhile, matrix algebra and those lovable ERO's are doing their thing for the cause. Finally, we return, with the help of a computer, to solve the applications. We will use the Microsoft Excel computer software to do this job.

11.2 INEQUALITIES

Bernard Broke's investment problem exposed the central role that linear inequalities play in linear programming. Here we attempt to get a handle on them.

Sense and Strength

The real numbers are ordered. This means that for any two different numbers, a and b, only one of the following two inequalities can hold. Either

$a < b$ (read a is less than b)

or

$a > b$ (read a is greater than b)

For example,

$5 < 10$ $1 > -2$ $0 > -9$ $\frac{1}{4} < \frac{1}{2}$

"Less than" and "greater than" are called the *senses of the inequality.* The sense of an inequality can be reversed by merely reversing the numbers. For example,

$5 < 10$ but $10 > 5$

The above are examples of strong inequality. Weak inequality occurs when the equals case is included. For example,

$x \leq 10$ (read x is less than or equal to 10)

is a weak inequality, while

$x < 10$

is a strong inequality.

Manipulation Rules

In our travels we will have need to manipulate and simplify inequalities. Let's see the important rules of this process.

Addition and subtraction of the same number on both sides of the inequality do not change its sense. For example,

$5 < 10$

and

$5 + 7 < 10 + 7$ since $12 < 17$

$5 - 10 < 10 - 10$ since $-5 < 0$

Multiplication and division may or may not change the sense of an inequality. We'll investigate this with examples.

$5 < 10$

and

$3 \cdot 5 < 3 \cdot 10$ since $15 < 30$

but

$-3 \cdot 5 > -3 \cdot 10$ since $-15 > -30$

$3 < 9$

and

$\dfrac{3}{3} < \dfrac{9}{3}$ since $1 < 3$

but

$\dfrac{3}{-3} > \dfrac{9}{-3}$ since $-1 > -3$

So multiplication and division by positive numbers don't change the sense of an inequality, but multiplication and division by negative numbers do.

Exercises 1

Consider the inequality $4 > 1$. Perform the following operations to both sides of the inequality and observe what happens to its sense.

1. Add -1.
2. Multiply by 100.
3. Subtract 6.

LINEAR PROGRAMMING

4. Divide by $\frac{1}{2}$.
5. Multiply by -4.
6. Divide by -1.

Graphical Perspectives

A point divides the real number line into two parts, each described by an inequality (see Figure 11.4). All values of x_1 to the left of 5 are less than 5, so that this set of numbers can be described by the inequality, $x_1 < 5$. All values of x_1 to the right of 5 are greater than 5, so that this set of numbers can be described by the inequality, $x_1 > 5$.

A straight line divides a plane into two portions, called *half-planes*. We can illustrate this in Figure 11.5 with Bernard Broke's capital constraint.

Distinguishing the less than and greater than half-planes can always be done by testing for the sense at the origin. In this case, at the origin,

Figure 11.4

Figure 11.5

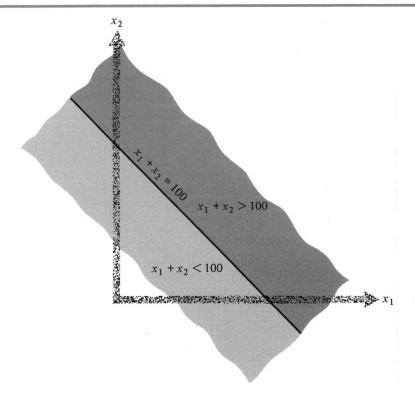

$x_1 + x_2 = 0 + 0 = 0$

which is less than 100. Thus for this example, the region that contains the origin is the less than half-plane.

Examples (see Figure 11.6)

Figure 11.6

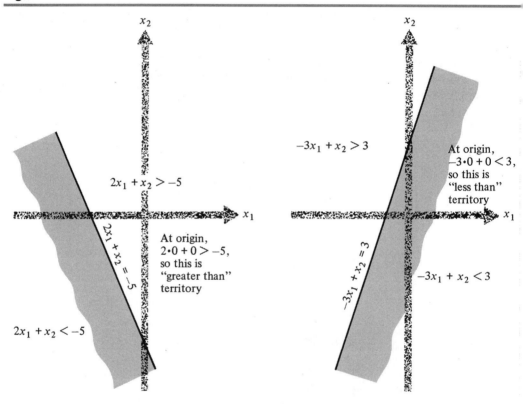

Exercises 2

Graph the following straight lines, and identify the half-planes by testing at the origin.

1. $3x_1 + 4x_2 = 5$
2. $x_1 - 3x_2 = -2$
3. $-2x_1 - x_2 = 4$
4. $-5x_1 + 2x_2 = -3$

Notice that the above inequalities allow many values of x_2 to be paired off with one value of x_1. Thus, inequalities represent relations, not functions, as defined in Chapter 5.

LINEAR PROGRAMMING

A plane divides a three-dimensional space into two "half-3D" regions, much like a room divider. In general a hyperplane divides an n-dimensional space into two "half-n dimensional" spaces.

Formulating Real World Inequalities

In this section we take examples of actual situations and model each with an inequality. Once you get the knack of it, you should be able to answer at least four of the exercises that follow. In this new inequality language, that is,

$x_1 \geq 4$ where x_1 is the number of correct answers.

Assuming no partial credit, x_1 can only be a natural number. Similar restrictions, such as that the number of children can only be a natural number, will hold in the following examples. Also $x_i \geq 0$ in all cases. However, for the time being, we won't dwell on these supplementary restrictions but rather will focus on the basic inequalities.

Examples

1. A couple want to get married and have at least six children.

    x = number of children

    x_1 = number of boys

    x_2 = number of girls

    $x \geq 6$ or $x_1 + x_2 \geq 6$ (see Figure 11.7)

Figure 11.7

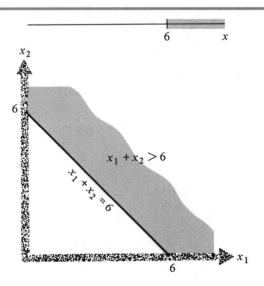

2. Joe Hardhat, who makes $10 an hour on straight time and $15 an hour on overtime wants to make at least $600 per week. His boss, Buck Pincher, wants Joe to make no more than $500 per week.

   x_1 = straight hours

   x_2 = overtime hours

   Joe's $10x_1 + 15x_2 \geq 600$

   Buck's $10x_1 + 15x_2 \leq 500$ (see Figure 11.8)

3. Joe "the Toe" Shmoe wants to score at least 100 points this season on touchdowns, field goals, and extra points.

   x_1 = touchdowns (6 points each)

   x_2 = field goals (3 points each)

   x_3 = extra points (1 point each)

   $6x_1 + 3x_2 + x_3 \geq 100$

4. Creaky Airlines needs to obtain at least $4000 revenue on a certain irregularly scheduled flight offering only seventh-class ($50 per seat) and ninth-class ($40 per seat) accommodations.

   x_1 = seventh-class passengers

   x_2 = ninth-class passengers

   $50x_1 + 40x_2 \geq 4000$ (see Figure 11.9)

Figure 11.8

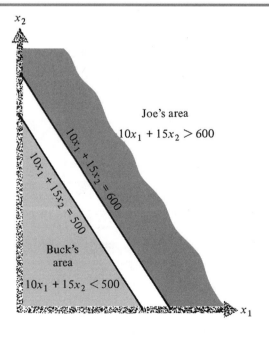

LINEAR PROGRAMMING

Figure 11.9

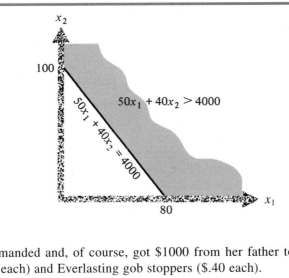

5. Veruka demanded and, of course, got $1000 from her father to buy Wonka chocolate bars ($.25 each) and Everlasting gob stoppers ($.40 each).

   x_1 = Wonka bars

   x_2 = Everlasting gob stoppers

   $.25x_1 + .40x_2 \leq 1000$ (see Figure 11.10)

Exercises 3

Define terms, describe, and graph (one- and two-variable cases only) the linear inequalities that result from the following situations.

1. A retired person must not earn more than $9,000 per year in order to preserve other desirable retirement benefits.

Figure 11.10

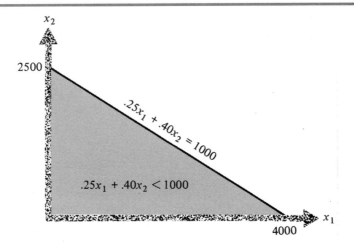

2. Madison A. Venue, a high-pressure salesman, makes $5 and $10 commission per unit, respectively, on the two products that he sells. He needs to earn at least $50,000 per year to support his swinger lifestyle.
3. "Lefty" Wright wants to limit the number of runners he allows on base (hits plus walks) to no more than 8.
4. Three different products cost $2, $7, and $12, respectively, to manufacture. Because of a liquidity problem, the company can't finance more than $100,000 of production.
5. Four foods contain (per ounce) 2, 10, 12, and 20 percent, respectively, of the minimum daily protein requirement set by the U.S. Department of Agriculture. A menu of these foods to provide at least the minimum daily requirement is desired.
6. Foreign and domestic crude oil cost $30 and $20 per barrel, respectively. The average cost of all oil used must be less than or equal to $24.

11.3 APPLICATIONS OF LINEAR PROGRAMMING

Applications of linear programming are found in many areas of human endeavor. Here, a representative sample, encompassing vacation planning, diet planning, production scheduling, and fuel blending, will be presented. In each case the various equations will be developed from the actual situation, then the complete problem will be stated in standard linear programming format. After the necessary theory is developed, each problem will be revisited for solution.

Vacation Planning

In planning a camping vacation, you should try to maximize enjoyment in the face of time, cost, and campsite limitations. Some typical conditions will reveal how a linear programming problem emerges.

Suppose that a total of 14 days is available for the camping vacation. If you travel at a rate of 400 miles per day and the camp is x_2 miles from home, it will take $x_2/400 = .0025x_2$ days to go one way and twice that amount or $.005x_2$ days to make the round trip. If x_1 is the number of days spent at the camp site, then the time constraint is

$$\begin{matrix}\text{Camping} \\ \text{days}\end{matrix} + \begin{matrix}\text{Travel} \\ \text{days}\end{matrix} \leq \begin{matrix}\text{Total days} \\ \text{available}\end{matrix}$$

$$x_1 + .005x_2 \leq 14$$

If your maximum vacation budget is $500, the daily camping fee is $10, and the travel cost is $.25 per mile ($.50 per one-way mile), the cost constraint is

$$\begin{matrix}\text{Camping} \\ \text{cost}\end{matrix} + \begin{matrix}\text{Travel} \\ \text{cost}\end{matrix} \leq \begin{matrix}\text{Total} \\ \text{budget}\end{matrix}$$

$$10x_1 + .5x_2 \leq 500$$

If there are no acceptable camping sites within 300 miles, the camping site constraint is

LINEAR PROGRAMMING

$$\frac{\text{One-way}}{\text{mileage}} \geq 300$$

$$x_2 \geq 300$$

Finally an enjoyment index is needed as an objective function. The development of this index, which would vary from person to person, requires some thinking on the virtues of camping days versus the added enjoyment of distant sites. The following enjoyment objective function describes the case where an extra camping day gives as much pleasure as a site 100 miles further away.

Enjoyment objective function $F = 100x_1 + 1x_2$

By pulling all the pieces together, we have the complete linear programming problem in standard form.

Maximize enjoyment $F = 100x_1 + 1x_2$

Subject to
$$x_1 + .005x_2 \leq 14 \quad \text{Time constraint}$$
$$10x_1 + .5x_2 \leq 500 \quad \text{Cost constraint}$$
$$x_2 \geq 300 \quad \text{Site constraint}$$
$$x_1 \text{ and } x_2 \geq 0 \quad \text{Nonnegativity}$$

Exercise 4

Formulate the revised linear programming problem if the rate of travel is 250 miles per day, the travel cost is 30¢ per mile (60¢ per one-way mile), and an extra day of camping gives as much enjoyment as a site 200 miles further away.

Production Scheduling

When products with unequal contributions to profit require varying time on different processes, each with limited capacity, you have a linear programming problem on your hands.

Noyes Products must decide on the number of bellwagers, x_1, ($5 profit per unit) and the number of honktinkers, x_2, ($4 profit per unit) to produce. Total profit from the two products, which they prefer to maximize, is

Bellwager profit + Honktinker profit = Total profit

$$5x_1 \quad + \quad 4x_2 \quad = \quad F$$

Product processing information is provided in the following table.

Process	Bellwagers	Honktinkers	Capacity
A (e.g., machining)	6 hours	2 hours	100 hours/weeks
B (e.g., assembling)	3	4	80
C (e.g., testing)	1	2	160

(The table is read as follows: Process A, which is limited to 100 hours per week of usage, is required 6 hours for each bellwager produced and 2 hours for each honktinker produced.)

An equation can be written to express the limitation on each process. For example, the process A constraint equation is

$$\begin{array}{c}\text{Bellwager}\\\text{time on A}\end{array} + \begin{array}{c}\text{Honktinker}\\\text{time on A}\end{array} \leq \begin{array}{c}\text{Process A}\\\text{capacity}\end{array}$$

$$6x_1 + 2x_2 \leq 100$$

Before peeking below, see if you can determine the process B and process C constraint equations.

The complete linear programming problem is

Maximize profit $F = 5x_1 + 4x_2$

Subject to
$$6x_1 + 2x_2 \leq 100 \quad \text{Process A}$$
$$3x_1 + 4x_2 \leq 80 \quad \text{Process B}$$
$$1x_1 + 2x_2 \leq 160 \quad \text{Process C}$$
$$x_1 \text{ and } x_2 \geq 0 \quad \text{Nonnegativity}$$

Exercise 5

Revise this linear programming problem to incorporate a fourth process, D, with 120 hours capacity, which is used 4 hours for each bellwager and 2 hours for each honktinker.

Diet Planning

If you want to meet nutritional goals at minimum cost by eating various foods, each supplying different amounts of nutrients at varying costs, you have a linear programming problem. For example, consider the big breakfast decision. Two cereals, Cornies, costing 8¢ per ounce, and Sweet Smackaroos, costing 6¢ per ounce, are available for eating separately or in combination.

The cereals contain the following percentages of U.S. Department of Agriculture recommended daily amounts of vitamin A and iron (per ounce):

	Vitamin A	Iron
Cornies	15%	25%
Sweet Smackaroos	30%	10%

If x_1 = ounces of Cornies and x_2 = ounces of Sweet Smackaroos, then to get at least 100 percent of the daily Vitamin A requirement, you can eat $6\frac{2}{3}$ ounces of Cornies, $3\frac{1}{3}$ ounces of Sweet Smackaroos, or various combinations of cereals (e.g., 3 ounces of Cornies and 2 ounces of Sweet Smackaroos) satisfying the inequality

$$\begin{array}{c}\text{Vitamin A from}\\\text{Cornies (\%)}\end{array} + \begin{array}{c}\text{Vitamin A from}\\\text{Sweet Smackaroos (\%)}\end{array} \geq \begin{array}{c}\text{Vitamin A}\\\text{requirement (\%)}\end{array}$$

$$15x_1 + 30x_2 \geq 100$$

LINEAR PROGRAMMING

See if you can formulate the iron constraint equation.

The total cost (in cents) of eating the two cereals, F, is

Total cost = Cost of Cornies + Cost of Sweet Smackaroos

$$F = 8x_1 + 6x_2$$

Meeting the nutritional requirements at minimum cost results in the complete linear programming problem.

Minimize cost $\quad F = 8x_1 + 6x_2$

Subject to
$$15x_1 + 30x_2 \geq 100 \quad \text{Vitamin A}$$
$$25x_1 + 10x_2 \geq 100 \quad \text{Iron}$$
$$x_1 \text{ and } x_2 \geq 0 \quad \text{Nonnegativity}$$

Exercise 6

A third food, costing 10¢ per ounce, supplies 5 percent of the daily vitamin A requirement and 20 percent of the daily iron requirement. Incorporate this food into a revised linear programming problem.

Fuel Blending

When fluid chemicals, each with restricted supply, are mixed to form products with different characteristics and contributions to profit, a linear programming problem arises. For example, three different octane blends with these particulars:

Blend	Minimum octane rating	Cost/gallons (¢)	Supplies (gallons)
A	99	85	2000
B	90	70	3000
C	80	65	1000

can be mixed to obtain two octane-rated brands of gasoline with the following particulars:

Brand	Price/gallons (¢)	Octane Rating
Supreme	80	96
Regular	75	91

Blending is assumed to result in a final product with the average octane level. For example, two gallons of 90-octane blend B mixed with three gallons of 80-octane blend C results in a gas with the average, or 84, octane level, as shown in the following calculation:

$$\frac{2 \cdot 90 + 3 \cdot 80}{2 + 3} = \frac{420}{5} = 84$$

Multiplying both sides of our example equation by the sum of the blend inputs, $(2 + 3)$, puts it in the preferred form for such constraint equations.

$2 \cdot 90 + 3 \cdot 80 = 84(2 + 3)$

Using this format, and defining x_{ij} as the amount of blend i in brand j (e.g., x_{32} is the amount of the third blend used to make the second brand), we can write general constraint equations insuring minimum octane ratings. For example, if only the second and third blends (B and C) are used to produce the hypothetical brand 3 gas with minimum 88-octane rating, the following inequality holds:

$x_{23} \cdot 90 + x_{33} \cdot 80 \geq 88(x_{23} + x_{33})$

or

$90x_{23} + 80x_{33} \geq 88(x_{23} + x_{33})$

By applying these ideas and notation to the actual problem, the constraint for the first (Supreme) gas is

Blend 1 octane in gas 1	+	Blend 2 octane in gas 1	+	Blend 3 octane in gas 1	$\geq$	Gas 1 octane requirement
$99x_{11}$	+	$90x_{21}$	+	$80x_{31}$	$\geq$	$96(x_{11} + x_{21} + x_{31})$

or

| $-3x_{11}$ | + | $6x_{21}$ | + | $16x_{31}$ | $\leq$ | 0 |

See if you can develop the constraint expression for the second (Regular) gas.

Now let's develop the profit function, which seems like a good thing to maximize.

On the revenue side the company gets 80¢ for each gallon of Supreme and 75¢ for each gallon of Regular. The total amount of Supreme is the sum of the gallons of blend 1 (x_{11}), blend 2 (x_{21}), and blend 3 (x_{31}) mixed to form it. The total amount of Regular is the sum of the gallons of blend 1 (x_{12}), blend 2 (x_{22}), and blend 3 (x_{32}) mixed to form it. Thus, total revenue from these two gases (R) is

Total revenue = Supreme revenue + Regular revenue

$R \quad = 80(x_{11} + x_{21} + x_{31}) + 75(x_{12} + x_{22} + x_{32})$

On the cost side we must determine the cost of each blend and sum these to get the total cost. Blend 1, which costs 85¢ per gallon and is used in the amount of $(x_{11} + x_{12})$ gallons, has a cost of $85(x_{11} + x_{12})$. Similar cost determinations for the other blends are made and summed to get the total cost (C) as follows.

Total cost = Blend 1 cost + Blend 2 cost + Blend 3 cost

$C \quad = 85(x_{11} + x_{12}) + 70(x_{21} + x_{22}) + 65(x_{31} + x_{32})$

Profit (F), which serves as the objective function, is found by taking total revenue (R) minus total cost (C). See if you can take these functions, simplify, and derive the profit function as

$F = -5x_{11} - 10x_{12} + 10x_{21} + 5x_{22} + 15x_{31} + 10x_{32}$

Blend supply is another factor in such problems. For example, if only 2000 gallons of the

LINEAR PROGRAMMING

first blend (A) are available, then the amount going into Supreme gas (x_{11}) plus the amount going into Regular gas (x_{12}) must be 2000 gallons or less. This constraint can be mathematically stated as

$$x_{11} + x_{12} \leq 2000$$

See if you can express the supply constraint equations if the second (B) and third (C) blends are respectively limited to 3000 and 1000 gallons. Finally, the complete linear programming problem is shown as follows.

Maximize profit

$$F = -5x_{11} - 10x_{12} + 10x_{21} + 5x_{22} + 15x_{31} + 10x_{32}$$

Subject to

$$-3x_{11} \quad\quad + 6x_{21} \quad\quad + 16x_{31} \quad\quad \leq \quad 0 \quad \text{Supreme}$$
$$8x_{12} + \quad\quad 1x_{22} \quad\quad + 11x_{32} \leq \quad 0 \quad \text{Regular}$$
$$x_{11} + x_{12} \quad\quad\quad\quad\quad\quad \leq 2000 \quad \text{Blend A}$$
$$x_{21} + x_{22} \quad\quad\quad\quad \leq 3000 \quad \text{Blend B}$$
$$x_{31} + x_{32} \leq 1000 \quad \text{Blend C}$$
$$x_{11} \text{ and } x_{12} \text{ and } x_{21} \text{ and } x_{22} \text{ and } x_{31} \text{ and } x_{32} \geq \quad 0 \quad \text{Nonnegativity}$$

Exercise 7

If the three blends cost 90¢, 75¢, and 60¢, respectively, and are limited in supply to 3000, 2000, and 4000 gallons, respectively, formulate the revised linear programming problem.

Having been exposed to some linear programming problems, you are no doubt eager to solve them. But constrain yourself. First we must develop the necessary theory.

11.4 BASIC THEORY AND GRAPHICAL INSIGHTS

In this section we will develop some basic theory needed to understand linear programming solution methods. This theory, spanning the topics of feasible solution space and the corner point theorem, will be developed with the aid of graphical methods, using a two-unknown case. On page 384, we define a reference problem and show it graphically in Figure 11.11. We will use that problem throughout the chapter to illustrate linear programming methods.

Feasible Solution Space

The constraint inequalities of a linear programming problem taken together define a region where solutions are possible. This region is called the *feasible solution space*.

Reference Problem

Maximize $\quad F = 4x_1 + 5x_2$

Subject to
$$\left.\begin{array}{r} x_1 + 3x_2 \leq 15 \\ 2x_1 + x_2 \leq 10 \\ x_1 \geq 0 \\ x_2 \geq 0 \end{array}\right\} \text{Constraints}$$

Figure 11.11

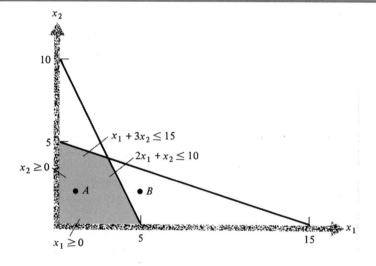

Each constraint defines a half plane, and taken together they define the feasible solution space (shaded region in Figure 11.11) for this problem.

Each point in the feasible solution space satisfies all four constraint inequalities. For example, point A, $(x_1, x_2) = (1, 2)$, satisfies each inequality as shown now:

$1 + 3 \cdot 2 = 7 < 15 \quad$ Constraint 1

$2 \cdot 1 + 2 = 4 < 10 \quad$ Constraint 2

$1 = 1 > 0 \quad$ Constraint 3

$ 2 = 2 > 0 \quad$ Constraint 4

Point B, $(5, 2)$, satisfies all constraints except the second one.

$2 \cdot 5 + 2 = 12 \nleq 10 \quad$ Note: $\nleq$ means not equal or less than

Thus, point B is not in the feasible solution space.

Exercises 8

1. Determine which of the following points are in the feasible solution space of the reference problem.

a. (3, 3)   b. (4, 3)
c. (2, 4)   d. (1, 5)
e. (5, 0)   f. (3, 4)

2. Graph the constraint inequalities and determine the feasible solution space for Bernard Broke's investment problem defined at the beginning of this chapter. Determine if the following points are in that space.
   a. (80, 20)   b. (40, 40)
   c. (100, 0)   d. (20, 70)
   e. (75, 25)   f. (0, 50)

Corner Point Theorem

The solution to a linear programming problem always occurs at a corner point, which is the intersection of constraint equations. We can illustrate this for our reference problem by superimposing the objective function, $F = 4x_1 + 5x_2$, on the feasible solution space for different values of F.

Different values of F yield a series of parallel lines

$$4x_1 + 5x_2 = F$$

$$\text{slope} = -\frac{4}{5} = -.8$$

$$\text{intercept} = \frac{F}{5}$$

since varying F changes the intercept, but not the slope, of these lines. By taking successively higher values of F (let's try 10, 20, 30, and 40), we generate parallel lines that move off in a northeasterly direction, as shown in Figure 11.12 on page 386.

F	Intercept	Slope
10	2	$-.8$
20	4	$-.8$
30	6	$-.8$
40	8	$-.8$

The lines for $F = 10$, $F = 20$, and $F = 30$ pass through the feasible solution space, but not the line for $F = 40$. You can see that the point (3, 4) "sticks out" the most in the direction of increasing F. Thus, the parallel line through that point (slope of $-.8$) will provide the highest feasible value of F. Using the point-slope formula, we determine that line as

$$x_2 - 4 = -.8(x_1 - 3)$$

$$x_2 = -.8x_1 + 6.4 \quad \text{(Explicit form)}$$

$$.8x_1 + x_2 = 6.4 \quad \text{(Implicit form)}$$

At the point (3, 4)

$$F = 4 \cdot 3 + 5 \cdot 4 = 32$$

Figure 11.12

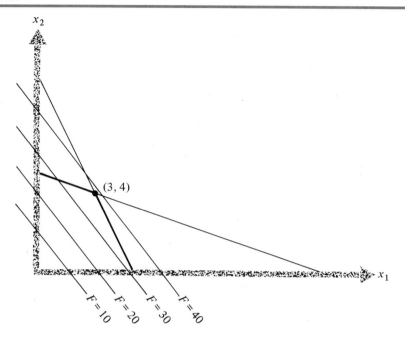

This is the highest value of F obtainable in the feasible solution space.

By viewing the geometry of the feasible solution space in Figure 11.13, you can see that another objective function with a different slope could result in maximization at a different corner point. For example, if the objective function slope was very large (in a negative sense), then a maximum would occur at corner point A. If the objective function has a low slope, then the maximum would occur at corner point B. If the objective function has the same slope as a constraint line edge of the feasible solution space, then a maximum occurs at all points (including the two corner points) along that edge line.

Exercises 9

Consider the following objective functions for our reference problem. By graphing the set of parallel lines that they represent, determine which corner provides the maximum F.

1. $F = x_1 + 4x_2$
2. $F = 3x_1 + x_2$
3. $F = 2x_1 + 3x_2$
4. $F = 4x_1 + 2x_2$

The results here have general applicability. The objective function for a linear programming problem is always maximized or minimized at a corner point. This means that we can always solve these problems by locating all corner points, determining the value of the objective function at each point, and identifying the corner point with the highest (or lowest) objective func-

LINEAR PROGRAMMING

Figure 11.13

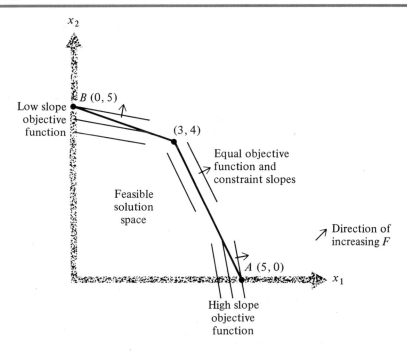

tion value. Even though this is a "brute force" solution method (later we will develop methods that efficiently locate the solution after considering only a few corner points), a little reflection on it will be beneficial.

A corner point represents the intersection of constraint equalities. For the two-variable case, every possible pair of constraint equations determines one corner point. This translates into ten corner points for a five-equation case. Let's apply the brute force method to such a case. See Figure 11.14 on page 388 and the table below.

Corner point	Coordinate	Intersecting lines	Status	Objective function value
1	(0, 0)	D, E	Feasible	0
2	(10, 0)	A, E	Feasible	20
3	(13, 0)	C, E	Not feasible	—
4	(25, 0)	B, E	Not feasible	—
5	(7, 3)	A, C	Feasible	17
6	(5, 5)	A, B	Not feasible	—
7	(1, 6)	B, C	Feasible	8
8	$(0, 6\frac{1}{4})$	B, D	Feasible	$6\frac{1}{4}$
9	$(0, 6\frac{1}{2})$	C, D	Not feasible	—
10	(0, 10)	A, D	Not feasible	—

The table shows that the objective function has its highest value at corner point 2. Thus, $(x_1, x_2) = (10, 0)$, with $F = 20$, is the solution.

Figure 11.14

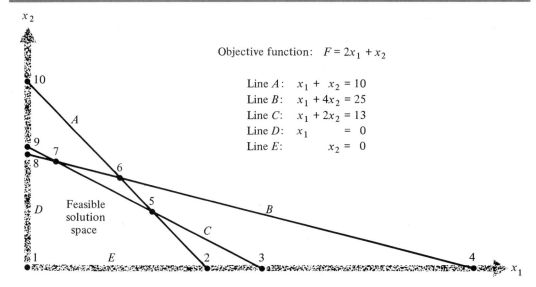

* This feasible solution space results from the $\leq$ sense of inequalities A, B, and C and the $\geq$ sense of inequalities D and E.

Exercise 10

Consider a sixth constraint equality ($\leq$ sense)

$$2x_1 + x_2 = 15$$

Intersect it with the other constraint equalities in the above case and determine the five additional corner points. Evaluate each new corner point for status and objective function value, if feasible. Reevaluate the five formerly feasible corner points. Did the new constraint change the solution?

11.5 ALGEBRAIC SOLUTION METHODS

We will now develop methods using ordinary algebra and matrix algebra to solve linear programming problems. In doing so we are laying the groundwork for systematizing these procedures into the so-called *simplex method*.

The algebraic methods presented here begin with the introduction of slack variables in the constraint equations. This allows for the determination of an initial corner point, referred to as the *basic solution*. This solution is usually far from optimum. However, by employing exchange formulas or elementary row operations (bet you thought you would never see those devils again!) along with certain rules, we can efficiently move from the basic solution through a sequence of other corner points to the solution.

LINEAR PROGRAMMING

We will illustrate these methods with our reference problem, restated and regraphed here with all its six corner points (see Figure 11.15).

Maximize $F = 4x_1 + 5x_2$

Subject to $x_1 + 3x_2 \leq 15$

$2x_1 + x_2 \leq 10$

$x_1 \geq 0$

$x_2 \geq 0$

Slack Variables

Slack variables, as the name implies, take up the slack between the two sides of an inequality. For example, the left side of the first constraint equation in our reference system can be less than the right side. But, by defining a slack variable, x_3, as the difference between the left side $(x_1 + 3x_2)$ and the right side (15), we can create the equality

$x_1 + 3x_2 + x_3 = 15$

For example, if $x_1 = 1$ and $x_2 = 2$, then $x_1 + 3x_2 = 7$, and so the slack taken on by x_3 must equal 8.

For cases where the left side of the equation can be more than the right side, we create an equality by subtracting off a slack variable. We can illustrate this with a couple's desire to have at least six children (See example 1 on page 375). By defining x_3 as the excess of children over six, and subtracting it, we get either

Figure 11.15

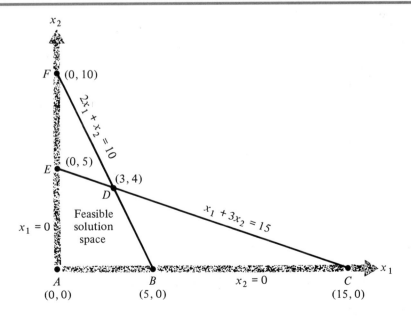

$x - x_3 = 6$ or $x_1 + x_2 - x_3 = 6$

depending on whether or not we consider the sex of the children.

To eliminate any slack in your mind regarding slack, let's put some slack in the other examples on pages 376 and 377.

Joe Hardhat	$10x_1 + 15x_2 - x_3$	$= 600$
Buck Pincher	$10x_1 + 15x_2 + x_3$	$= 500$
Joe "the Toe"	$6x_1 + 3x_2 + x_3 - x_4$	$= 100$
Creaky Airlines	$50x_1 + 40x_2 - x_3$	$= 4000$
Veruka	$.25x_1 + .40x_2 + x_3$	$= 1000$

Exercises 11

By introducing slack variables, turn the inequalities derived in the six exercises on pages 377 and 378 into equalities.

Returning now to our reference system, we can define a slack variable for each inequality (except the nonnegativity ones). Then we would have the following system of equalities:

$$x_1 + 3x_2 + x_3 = 15$$
$$2x_1 + x_2 + x_4 = 10$$

Here, x_3 and x_4 are the slack variables.

This system of constraint equalities can be written in matrix notation as

$$\begin{pmatrix}1\\2\end{pmatrix}x_1 + \begin{pmatrix}3\\1\end{pmatrix}x_2 + \begin{pmatrix}1\\0\end{pmatrix}x_3 + \begin{pmatrix}0\\1\end{pmatrix}x_4 = \begin{pmatrix}15\\10\end{pmatrix}$$

Using this notation, we can think of x_1, x_2, x_3, and x_4 as a set of constants, which are the respective coordinates at a corner point. Recall that a constant times a matrix results in every element of that matrix being multiplied by that constant. For example,

$$\begin{pmatrix}1\\2\end{pmatrix}x_1 = \begin{pmatrix}1x_1\\2x_1\end{pmatrix}$$

Thus, the system above is equivalent to

$$\begin{pmatrix}1x_1\\2x_1\end{pmatrix} + \begin{pmatrix}3x_2\\1x_2\end{pmatrix} + \begin{pmatrix}1x_3\\0x_3\end{pmatrix} + \begin{pmatrix}0x_4\\1x_4\end{pmatrix} = \begin{pmatrix}15\\10\end{pmatrix}$$

By adding the four matrices on the left side, we get

$$\begin{pmatrix}1x_1 + 3x_2 + 1x_3 + 0x_4\\2x_1 + 1x_2 + 0x_3 + 1x_4\end{pmatrix} = \begin{pmatrix}15\\10\end{pmatrix}$$

LINEAR PROGRAMMING

which is equivalent to the two constraint equalities.

The new notation developed above may seem like extra trouble and bother. However, it should prove very beneficial when we develop linear programming solution methods later.

Basic Solution

A corner point solution to a linear programming problem with only slack variables having nonzero values is called the *basic solution*. It can always be found by setting the respective slack variables equal to the right side constants. Here, it is $(x_1, x_2, x_3, x_4) = (0, 0, 15, 10)$. This is a point in four-dimensional space, which defies mortal graphing. However, in terms of the nonslack variables, this corresponds to the origin, or point A in Figure 11.15 on page 389.

Since the slack variables, x_3 and x_4, contribute nothing to the value of the objective function, the basic solution isn't going to maximize F. However, it gives us a necessary starting point from which we can move to better and better corner points, as will be shown next.

Moving on to Better Corners

We know from earlier discussion that the objective function is optimized (maximized or minimized depending on the case) at a corner point. The basic solution is one corner point, but not the best. However, starting there, we have procedures that can move us directly to another corner point, and from there we can move to another corner point, and so on. Each move, which results in a solution containing different variables, brings us closer to optimizing the objective function.

Basis

The nonzero variables at a corner point are called *basis* variables, and the set of these variables is called *the basis*. x_3 and x_4 comprise the basis for the basic solution. Introducing nonbasis variables, x_1 and x_2, allows us to increase the value of the objective function. However, to accomplish this x_3 and x_4 must be decreased in value. We can show this with our reference problem.

The following display shows that for every unit of x_1 introduced (x_2 held at zero), x_3 and x_4 must decrease by one and two units, respectively, in order to maintain the constraint equalities.

```
                   increase              initially
                   by 1      hold at 0   equal to 15
First                ↑           ↑            ↑
constraint         1x₁    +    3x₂    +     x₃       = 15
equation           ‿‿‿                     ‿‿‿
                   increase              must decrease
                   by 1                  by 1 to
                                         keep equality

                   increase              initially
                   by 1      hold at 0   equal to 10
Second               ↑           ↑            ↑
constraint         2x₁    +    x₂     +     x₄       = 10
equation           ‿‿‿                     ‿‿‿
                   increase              must decrease
                   by 2                  by 2 to
                                         keep equality
```

Exchange Equations

These relationships are embodied in the following exchange equation.

$1x_1 \Leftrightarrow 1x_3 + 2x_4$

This is a new type of equation. It literally says, "One unit of x_1 can be introduced in the solution in exchange for the removal of one unit of x_3 and two units of x_4." Be sure that you understand that this is *different* from $x_1 = 1x_3 + 2x_4 = 1 \cdot 15 + 2 \cdot 10 = 35$. The symbol $\Leftrightarrow$ identifies an equation to be of the exchange type.

For example, if three units of x_1 are introduced, then $1 \cdot 3 = 3$ units of x_3 and $2 \cdot 3 = 6$ units of x_4 must be removed from their basic solution values. These exchanges would result in the point $(3, 0, 15 - 3, 10 - 6) = (3, 0, 12, 4)$.

For every unit of x_2 introduced (x_1 held at zero) into the solution, the constraint equations require that x_3 and x_4 decrease by three and one, respectively.

First constraint equation:

$$\underset{\text{hold at 0}}{\underset{\uparrow}{1x_1}} + \underset{\text{increase by 1}}{\underset{\uparrow}{\underbrace{3x_2}_{\text{increase by 3}}}} + \underbrace{x_3}_{\substack{\text{must decrease} \\ \text{by 3 to} \\ \text{keep equality}}} = 15$$

Second constraint equation:

$$\underset{\text{hold at 0}}{\underset{\uparrow}{2x_1}} + \underset{\text{increase by 1}}{\underset{\uparrow}{\underbrace{x_2}_{\text{increase by 1}}}} + \underbrace{x_4}_{\substack{\text{must decrease} \\ \text{by 1 to} \\ \text{keep equality}}} = 10$$

This relationship is embodied in the following exchange equation:

$1x_2 \Leftrightarrow 3x_3 + 1x_4$

Literally, one unit of x_2 can be introduced in the solution in exchange for the removal of three units of x_3 and one unit of x_4.

You can see that the right side of each exchange equation contains the basis variables. The left side, so far, has contained only nonbasis variables. Let's see what happens when we form an exchange equation for a basis variable. A unit of x_3 can only be added to the solution if we subtract a unit of x_3 (from first constraint equation). A unit of x_4 can only be added to the solution if we subtract a unit of x_4 (from second constraint equation). These relationships, expressed in exchange equation form, are

$1x_3 \Leftrightarrow 1x_3 + 0x_4$

$1x_4 \Leftrightarrow 0x_3 + 1x_4$

LINEAR PROGRAMMING

If we compare the exchange equations with the matrix representation of the constraint equations at the basic solution, you will see something interesting.

$1x_1 \Leftrightarrow 1x_3 + 2x_4$

$1x_2 \Leftrightarrow 3x_3 + 1x_4$

$1x_3 \Leftrightarrow 1x_3 + 0x_4$

$1x_4 \Leftrightarrow 0x_3 + 1x_4$

$$\begin{pmatrix}1\\2\end{pmatrix}x_1 + \begin{pmatrix}3\\1\end{pmatrix}x_2 + \begin{pmatrix}1\\0\end{pmatrix}x_3 + \begin{pmatrix}0\\1\end{pmatrix}x_4 = \begin{pmatrix}15\\10\end{pmatrix}$$

Can you see that the exchange coefficients for each variable are the same as the column vector elements associated with that variable in the matrix representation?

Notice that the exchange coefficients for basis variables are a column of the identity matrix. Furthermore, since $x_1 = x_2 = 0$ in the basic solution, the matrix representation is equivalent to

$$\begin{pmatrix}1 & 0\\0 & 1\end{pmatrix}\begin{pmatrix}x_3\\x_4\end{pmatrix} = \begin{pmatrix}x_3\\x_4\end{pmatrix} = \begin{pmatrix}15\\10\end{pmatrix}$$

Entering Variable

Solution methods require that only one variable enter (while one leaves) the basis at a time. But which one should enter first? As a general rule, it makes sense to enter the variable that provides the greatest per unit change (increase for maximization or decrease for minimization) in the objective function. Here it is x_2.

Limitations on Entering Variable

Since increases in x_2 raise the value of the objective function, but decreases in x_3 and x_4 do not lower its value, it pays to add as much x_2 as possible. The second constraint allows the most x_2 ($x_2 = 10$ if x_4 is reduced to zero). But this would mean that $x_3 = -15$ in the first constraint equation. The "solution" would then be (0, 10, −15, 0). This is corner point F in Figure 11.15 on page 389. Unfortunately, it lies outside the feasible solution space. Oops, that happened because we violated the rule that all x's must be greater than or equal to zero. We desperately need a way to avoid this in the future.

Notice that the first constraint equation only allows $\frac{15}{3} = 5$ units of x_2 before x_3 turns negative. In effect, the smallest ratio

$$\frac{p_i}{a_{i2}}$$

where i is the constraint equation number, p symbolizes right side constants, and a symbolizes left side coefficients, determines how much x_2 can enter the solution before a former basis variable turns negative. We call these *limiting ratios,* and they are calculated as follows:

First constraint equation

$x_1 + 3x_2 + x_3 = 15$

$$\frac{p_1}{a_{12}} \Rightarrow \frac{15}{3} = 5$$

↑ Smallest

Second constraint equation

$2x_1 + 1x_2 + x_4 = 10$

$$\frac{p_2}{a_{22}} \Rightarrow \frac{10}{1} = 10$$

So while the second equation allows more x_2, the first equation effectively sets its limit at five units in order to prevent any variable from becoming negative.

Departing Variable

As more and more of the entering variable is added to the solution, a point is reached where a former basis variable first drops to zero. This variable is the one that departs the basis. For our reference problem, x_3 is reduced to zero first, so it is forced out of the basis.

Revision of Basis

For each basis change, we must develop a set of exchange equations in terms of the revised basis variables before we can determine the next corner point move. Here, to effect the basis change, we must determine the exchange equations that relate all variables to the revised basis variables, x_2 and x_4. We begin this process with the entering variable exchange equation, in terms of the old basis variables.

$x_2 \Leftrightarrow 3x_3 + 1x_4$

and solve for the departing basis variable, x_3.

$x_3 \Leftrightarrow \frac{1}{3}x_2 - \frac{1}{3}x_4$

Notice how the laws of algebra apply fully to exchange equations.
Next we take the old exchange equations and substitute x_3 from the above equation. This yields new exchange equations in terms of x_2 and x_4.

$x_1 \Leftrightarrow 1x_3 + 2x_4$

$\Leftrightarrow 1\left(\frac{1}{3}x_2 - \frac{1}{3}x_4\right) + 2x_4$

$\Leftrightarrow \frac{1}{3}x_2 + \frac{5}{3}x_4$

$x_2 \Leftrightarrow 3x_3 + 1x_4$

$\Leftrightarrow 3\left(\frac{1}{3}x_2 - \frac{1}{3}x_4\right) + 1x_4$

$\Leftrightarrow 1x_2 + 0x_4$

We previously found the x_3 exchange equation. See if you can show that

$x_4 \Leftrightarrow 0x_1 + 1x_4$

LINEAR PROGRAMMING

We can also use exchange equations to revise the solution. The basic solution was

solution $\Leftrightarrow 15x_3 + 10x_4$

This is read, "The solution contains 15 units of x_3 and 10 units of x_4." Substituting for x_3 gives the revised solution

solution $\Leftrightarrow 15(\frac{1}{3}x_2 - \frac{1}{3}x_4) + 10x_4$

$\Leftrightarrow 5x_2 + 5x_4$

Thus, the revised solution contains five units of x_2 and five units of x_4. This is the point (0, 5, 0, 5) when the zero nonbasis variables are included, which corresponds to point E in Figure 11.15 on page 389.

We can express the revised solution in matrix notation, with the ith variable premultiplied by the vector of exchange coefficients, or $\mathbf{P}_i$.

$$\mathbf{P}_1 x_1 + \mathbf{P}_2 x_2 + \mathbf{P}_3 x_3 + \mathbf{P}_4 x_4 = \mathbf{P}_0$$

$$\begin{pmatrix} \frac{1}{3} \\ \frac{5}{3} \end{pmatrix} x_1 + \begin{pmatrix} 1 \\ 0 \end{pmatrix} x_2 + \begin{pmatrix} \frac{1}{3} \\ -\frac{1}{3} \end{pmatrix} x_3 + \begin{pmatrix} 0 \\ 1 \end{pmatrix} x_4 = \begin{pmatrix} 5 \\ 5 \end{pmatrix}$$

Alternatively, we can accomplish the change of basis by employing elementary row operations (ERO's). This method requires us to make the exchange vector of the entering basis variable equal to that of the former basis variable, now departed. In our case we must make $\mathbf{P}_2$ equal to the former $\mathbf{P}_3 = \begin{pmatrix} 1 \\ 0 \end{pmatrix}$. This process begins with the latest set of constraint equations—basic solution ones in this case.

$1x_1 + 3x_2 + 1x_3 + 0x_4 = 15$

$2x_1 + 1x_2 + 0x_3 + 1x_4 = 10$

convert $\mathbf{P}_2$ to $\begin{pmatrix} 1 \\ 0 \end{pmatrix}$ by ERO's

Multiplying the first equation by $\frac{1}{3}$ gets the 1 into position.[1]

$\mathbf{R}_1{}^* = \frac{1}{3}\mathbf{R}_1$

$= \frac{1}{3}(1x_1 + 3x_2 + 1x_3 + 0x_4 = 15)$

$= \frac{1}{3}x_1 + 1x_2 + \frac{1}{3}x_3 + 0x_4 = 5$

We get the zero in position by the following ERO.

---

[1] In Chapter 9, we used ε_i to denote equation i. In Chapter 10, we thought of equations as rows in a matrix, and so we used $\mathbf{R}_i$ to denote row i. Here we are leading up to the use of ERO's on rows of a simplex tableau; thus we will continue to use the $\mathbf{R}_i$ notation.

$R_2^* = R_2 - R_1^*$

$$= (2x_1 + 1x_2 + 0x_3 + 1x_4 = 10)$$
$$\underline{-(\tfrac{1}{3}x_1 + 1x_2 + \tfrac{1}{3}x_3 + 0x_4 = 5)}$$
$$= \tfrac{5}{3}x_1 + 0x_2 - \tfrac{1}{3}x_3 + 1x_4 = 5$$

You should recognize the resulting system

$\tfrac{1}{3}x_1 + 1x_2 + \tfrac{1}{3}x_3 + 0x_4 = 5$

$\tfrac{5}{3}x_1 + 0x_2 - \tfrac{1}{3}x_3 + 1x_4 = 5$

or

$$\begin{pmatrix} \tfrac{1}{3} \\ \tfrac{5}{3} \end{pmatrix} x_1 + \begin{pmatrix} 1 \\ 0 \end{pmatrix} x_2 + \begin{pmatrix} \tfrac{1}{3} \\ -\tfrac{1}{3} \end{pmatrix} x_3 + \begin{pmatrix} 0 \\ 1 \end{pmatrix} x_4 = \begin{pmatrix} 5 \\ 5 \end{pmatrix}$$

as being identical to that which we found using exchange equations earlier.

From now on we will use ERO's rather than exchange equations to carry out basis revisions. However, since these two methods produce identical results, in doing so we will be generating exchange coefficients, too. Exchange coefficients and the ideas behind them will continue to be very important in our development of linear programming solution methods. In fact, we need these tools to decide whether to stop or move on to another corner point, as we will see later.

Objective Function Value

Our methodology has carried us from the basic solution (point A) to corner point E in Figure 11.15 on page 389. The value of the objective function (F), which is computed by multiplying the contribution vector $\mathbf{C}$ (coefficients in the objective function with zeros used for the slack variables) times the solution vector $\mathbf{S}$, was zero for the basic solution

$F = \mathbf{C} \cdot \mathbf{S}$

$$= (4 \quad 5 \quad 0 \quad 0) \begin{pmatrix} 0 \\ 0 \\ 15 \\ 10 \end{pmatrix} = 4 \cdot 0 + 5 \cdot 0 + 0 \cdot 15 + 0 \cdot 10 = 0$$

and increased to 25 at point E.

$$F = (4 \quad 5 \quad 0 \quad 0) \begin{pmatrix} 0 \\ 5 \\ 0 \\ 5 \end{pmatrix} = 4 \cdot 0 + 5 \cdot 5 + 0 \cdot 0 + 0 \cdot 5 = 25$$

LINEAR PROGRAMMING

Since the slack variables have zero contributions, we get the same results by a shortcut matrix multiplication that neglects them. For example,

$$F = (4 \quad 5)\begin{pmatrix}0\\5\end{pmatrix} = 4 \cdot 0 + 5 \cdot 5 = 25$$

Move on or Stop

We now ask the question of whether or not the objective function can be increased still further by moving on to another corner point. The answer rests with the idea of net gain from various exchanges. In other words, will the objective function increase as a result of introducing other variables into the basis? It doesn't make sense to introduce x_3 into the basis since that will undo the increase just accomplished. So we must investigate the introduction of x_1 into the basis.

The exchange coefficients for x_1, in terms of the current basis, x_2 and x_4, are given in the current $\mathbf{P}_1$ vector,

$$\mathbf{P}_1 = \begin{pmatrix}\frac{1}{3}\\\frac{5}{3}\end{pmatrix}$$

This means that for every one unit of x_1 introduced into the solution, $\frac{1}{3}$ unit of x_2 (loss of $5 \cdot \frac{1}{3}$) and $\frac{5}{3}$ units of x_4 (loss of $0 \cdot \frac{5}{3}$) must be removed. The objective function loss (per unit of x_1 introduced) as a result of reduction of former basis variables is summarized in the following matrix multiplication,

$$z_1 = \mathbf{C}_b\mathbf{P}_1 = (5 \quad 0)\begin{pmatrix}\frac{1}{3}\\\frac{5}{3}\end{pmatrix} = \frac{5}{3}$$

where $\mathbf{C}_b$ is the vector of basis variable contributions.

On the positive side, each unit of x_1 introduced into the solution adds $c_1 = 4$ to the objective function. Thus, the net change of the objective function for each unit of x_1 introduced is

$$c_1 - z_1 = 4 - \frac{5}{3} = \frac{7}{3}$$

So, introducing x_1 into the basis will increase the value of the objective function. But now this leaves us with the question of which variable should depart the basis.

Again, we add as much x_1 as possible before a former basis variable turns negative. Recall how we did this earlier with limiting ratios. The current constraint equations (see page 396) allow 15 and 3 units of x_1, respectively,

$$\frac{p_1}{a_{11}} = \frac{5}{1/3} = 15 \quad \frac{p_2}{a_{21}} = \frac{5}{5/3} = 3$$

before x_2 and x_4 turn negative. We are bound by the smallest of these ratios. (Notice that if we let x_1 be anything greater than 3, then x_4 would become negative—a real no-no in linear programming.) So, three units of x_1 will be introduced into the solution. At the same time we have found out that x_4 is the variable that will make a hasty exit from the basis.

We accomplish the change of basis by employing ERO's to make the exchange vector for x_1 equal to that of the departing variable, or

$$\frac{1}{3}x_1 + 1x_2 + \frac{1}{3}x_3 + 0x_4 = 5$$
$$\frac{5}{3}x_1 + 0x_2 - \frac{1}{3}x_3 + 1x_4 = 5$$

convert $\mathbf{P}_1$ to $\begin{pmatrix} 0 \\ 1 \end{pmatrix}$ by ERO's

The 1 is obtained by the following ERO on the second equation.

$\mathbf{R}_2^* = \frac{3}{5}\mathbf{R}_2$

$\quad = \frac{3}{5}(\frac{5}{3}x_1 + 0x_2 - \frac{1}{3}x_3 + 1x_4 = 5)$

$\quad = 1x_1 + 0x_2 - \frac{1}{5}x_3 + \frac{3}{5}x_4 = 3$

Then the zero is obtained by

$\mathbf{R}_1^* = \mathbf{R}_1 - \frac{1}{3}\mathbf{R}_2^*$

$\quad = (\frac{1}{3}x_1 + 1x_2 + \frac{1}{3}x_3 + 0x_4 = 5)$

$\quad -(\frac{1}{3}x_1 + 0x_2 - \frac{1}{15}x_3 + \frac{1}{5}x_4 = 1)$

$\quad = 0x_1 + 1x_2 + \frac{2}{5}x_3 - \frac{1}{5}x_4 = 4$

Thus the revised set of constraint equations is

$$0x_1 + 1x_2 + \frac{2}{5}x_3 - \frac{1}{5}x_4 = 4$$
$$1x_1 + 0x_2 - \frac{1}{5}x_3 + \frac{3}{5}x_4 = 3$$

or

$$\begin{pmatrix} 0 \\ 1 \end{pmatrix} x_1 + \begin{pmatrix} 1 \\ 0 \end{pmatrix} x_2 + \begin{pmatrix} \frac{2}{5} \\ -\frac{1}{5} \end{pmatrix} x_3 + \begin{pmatrix} -\frac{1}{5} \\ \frac{3}{5} \end{pmatrix} x_4 = \begin{pmatrix} 4 \\ 3 \end{pmatrix}$$

Since $x_3 = x_4 = 0$, the above is equivalent to

$$\begin{pmatrix} 1 \\ 0 \end{pmatrix} x_2 + \begin{pmatrix} 0 \\ 1 \end{pmatrix} x_1 = \begin{pmatrix} 1 & 0 \\ 0 & 1 \end{pmatrix}\begin{pmatrix} x_2 \\ x_1 \end{pmatrix} = \begin{pmatrix} x_2 \\ x_1 \end{pmatrix} = \begin{pmatrix} 4 \\ 3 \end{pmatrix}$$

Thus, the revised solution is $x_1 = 3$, $x_2 = 4$, or point D in Figure 11.15 on page 389. At this point the value of the objective function is

$$F = \mathbf{CS} = (4 \quad 5)\begin{pmatrix} 3 \\ 4 \end{pmatrix} = 32$$

Further revision of the solution to other corner points is unwise since either x_3 or x_4 (with zero contributions) would necessarily have to replace either x_1 or x_2 (with positive contributions).

So we have located the solution by a procedure that looks at corner points with successively higher objective function values. When it cannot find a corner point that enhances the objec-

LINEAR PROGRAMMING

tive function, the solution has been found. We will now systematize this procedure in the so-called *simplex method.*

11.6 SIMPLEX METHOD

The simplex method systematizes the steps in moving from corner point to corner point to eventual solution for linear programming problems. This method is organized around a tableau containing all the information necessary to evaluate a given basis. The tableau is revised for each new corner point using elementary row operations, and the resulting table provides the information needed to efficiently move on to the next corner point or to stop once the objective function has been optimized.

The simplex method employs five main steps, namely

1. Formation of initial tableau body.
2. Identification of variable to enter basis.
3. Identification of variable to leave basis.
4. "Pivoting": revision of tableau for new basis.
5. Recycle until optimum solution found.

These steps will be illustrated for a maximization case, our reference problem. At each step you should return to review the detailed algebraic operations and reasoning given earlier. Simplex procedures for minimization problems are nearly the same as for maximization problems. The slight differences will be highlighted in the next section when we solve the diet problem.

The reference problem is

Maximize $\quad F = 4x_1 + 5x_2$

Subject to $\quad 1x_1 + 3x_2 \leq 15$

$\quad\quad\quad\quad\quad 2x_1 + 1x_2 \leq 10$

$\quad\quad\quad\quad\quad$ all x's ≥ 0

Step 1. Formation of Initial Tableau Body

To form the initial tableau body, we first introduce the slack variables to get the following constraint equalities.

$1x_1 + 3x_2 + 1x_3 + 0x_4 = 15$

$2x_1 + 1x_2 + 0x_3 + 1x_4 = 10$

These equations give us the basic solution of $x_1 = 15$ and $x_2 = 10$.

This information is compactly arranged in the initial tableau on page 400. The left section of the tableau contains the initial basis along with its vector of contributions to the objective function (C_b). The top section lists each variable along with its contribution (c_j). The center section contains all the initial exchange coefficients (a_{ij}); these columns are headed by the variable symbols x_1, x_2, and so forth; however, the elements in the columns form the vectors: $\mathbf{P}_1$, $\mathbf{P}_2$, and so forth. The right section contains the solution ($\mathbf{P}_0$), which in this case is the basic solution. (In steps 2 and 3, the tableau will be amended to contain other information.)

Initial Tableau Body

Row	Basis	c_j C_b	4 x_1	5 x_2	0 x_3	0 x_4	P_0
A	x_3	0	1	3	1	0	15
B	x_4	0	2	1	0	1	10

Step 2. Identify Variable to Enter Basis

Recall that the nonbasis variable that provides the largest per unit change (increase for maximization problems) in the objective function is chosen to enter the basis.[2] This per unit change for the jth variable is computed as

$$c_j - z_j = c_j - (\mathbf{C}_b \mathbf{P}_j)$$

where

c_j is the contribution of jth variable
z_j is the loss from former basis variables because of introduction of the jth variable
$\mathbf{C}_b$ is the vector of basis contributions
$\mathbf{P}_j$ is the jth variable exchange vector

Strictly speaking, the $\mathbf{C}_b \mathbf{P}_j$ multiplication is done like any other good matrix multiplication. For example, z_1 is the single number which results from multiplying a 1×2 times a 2×1, or

$$z_1 = \mathbf{C}_b \mathbf{P}_1 = (0 \quad 0)\begin{pmatrix}1\\2\end{pmatrix} = 0 \cdot 1 + 0 \cdot 2 = 0$$

However, because of the simplex table arrangement, such multiplications are done column by column, as shown below.

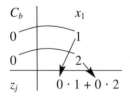

You should verify that $z_1 = z_2 = z_3 = z_4 = 0$ since both basis variables have zero contributions. Generally, we don't have to compute z for basis variables, since they are always zero (do you know why?). So for the initial tableau, each $c_j - z_j$ equals the corresponding contribution.

The tableau is expanded to include the z_j and $c_j - z_j$ information in the lower center section

---

[2] Curiously, the variable that causes the largest per unit increase in the objective function may not provide the largest total change in the objective function if the total amount of that variable allowed by step 3 is small.

LINEAR PROGRAMMING

as shown in step 3. Since $c_j - z_j$ tells us whether we can move to a better corner point, we will hereafter call it the *move index*. By looking at this index we can see here that a move is warranted (because positive $c_j - z_j$ exists), and that x_2, with the largest value, should enter the basis.

Step 3. Identify Variable to Leave Basis

Recall from the algebraic development that as much as possible of the entering variable is added to the solution. However, at some point of adding more, one of the former basis variables first drops to zero. It is this variable that leaves the basis. This departing variable is easy to find since it has the smallest of the ratios

$$\frac{p_{i0}}{a_{iJ}} \quad \text{where } J \text{ is the entering variable column}$$

These limiting ratios are incorporated into the simplex tableau at the top right. Notice that x_3, with the smallest limiting ratio, must say goodbye.

Amended Initial Tableau

Row	Basis	c_j C_b	4 x_1	5 x_2	0 x_3	0 x_4	P_0	Limiting ratios
A	x_3	0	1	③	1	0	15	$\frac{15}{3}$ ←Smallest
B	x_4	0	2	1	0	1	10	$\frac{10}{1}$
		z_j	0	0			$S = (0 \quad 0 \quad 15 \quad 10)$	
	Move index: $c_j - z_j$		4	5 ↑ Largest			$F = 0 \cdot 15 + 0 \cdot 10 = 0$	

At the bottom right of the tableau, we put the current solution (S) and the current value of the objective function (F).

At this point the initial simplex tableau is complete.

Step 4. Pivoting: Revise Tableau for New Basis

The tableau is revised by elementary row operations to make the exchange coefficients of the entering variable equal to the "identity" column of the departing variable. This process is called *pivoting*.

Before we go on, we need to establish some terminology. The row of the departing variable is called the *pivot row*. The column of the entering variable is called the *pivot column*. The element standing at the intersection of the pivot row and the pivot column is called the *pivot element*. We symbolize the value of the pivot element as ρ.

In general, the first step in pivoting is to revise the pivot row so as to have a 1 at the pivot element position. This is achieved by the following ERO.

$$\text{PR}^* = \frac{1}{\rho} \text{PR}$$

where **PR** is the pivot row, and **PR*** is the revised pivot row.

Then all the other coefficients in the pivot column must be made equal to zero. This is done for the ith row ($\mathbf{R}_i$) by using the revised pivot row in the following ERO.

$$\mathbf{R}_i^* = \mathbf{R}_i - a_{iJ}\mathbf{PR}^*$$

where $\mathbf{R}_i^*$ is the revised ith row, and a_{iJ} is the exchange coefficient in the ith row of the pivot (Jth) column.

For our reference problem, A is the pivot row, the x_2 column ($J = 2$) is the pivot column, and 3 (circled in the amended initial tableau) is the value of the pivot element. So, the specific ERO's needed to pivot are

$$\mathbf{PR}_A^* = \tfrac{1}{3}(1 \quad 3 \quad 1 \quad 0 \ | \ 15)$$
$$= (\tfrac{1}{3} \quad 1 \quad \tfrac{1}{3} \quad 0 \ | \ 5)$$

$$\mathbf{R}_B^* = \mathbf{R}_B - 1\mathbf{PR}_A^*$$
$$= (2 \quad 1 \quad 0 \quad 1 \ | \ 10) - 1(\tfrac{1}{3} \quad 1 \quad \tfrac{1}{3} \quad 0 \ | \ 5)$$
$$= (\tfrac{5}{3} \quad 0 \quad -\tfrac{1}{3} \quad 1 \ | \ 5)$$

The new basis and these new rows (step 4) are the first entries into the second tableau, given below. At this stage the tableau represents the solution at a corner point. By filling in the rest of the tableau, we lay the groundwork for proceeding to another corner point. Procedurally, the rest of the tableau is determined by repeating steps 2 and 3 in that order.

Second Tableau

Row	Basis	c_j / C_b	4 / x_1	5 / x_2	0 / x_3	0 / x_4	P_0	Limiting ratios
A	x_2	5	$\tfrac{1}{3}$	1	$\tfrac{1}{3}$	0	5	$\dfrac{5}{1/3}$
B	x_4	0	$\tfrac{3}{5}$	0	$-\tfrac{1}{3}$	1	5	$\dfrac{5}{5/3}$ ← Smallest
		z_j	$\tfrac{5}{3}$	5	$\tfrac{5}{3}$		$S = (0 \quad 5 \quad 0 \quad 5)$	
	Move index: $c_j - z_j$		$\tfrac{7}{3}$ ↑ Largest		$-\tfrac{5}{3}$		$F = 4 \cdot 0 + 5 \cdot 5 = 25$	

Step 4 first → (Basis column)
Step 3 last → (Limiting ratios)
Step 2 next → (Move index row)

Step 5. Recycle Until Optimum Solution Found

Successive tableaus are developed until the point is reached where no further improvement in the objective function is possible. Each tableau is derived from the former one by employing step 4, then step 2, then step 3 in that order.

The second tableau reveals that x_1 should enter the basis (highest positive move index) by forcing out x_4 (smallest limiting ratio). Thus, row B is the pivot row, the first column ($J = 1$) is the pivot column, and $\frac{5}{3}$ is the pivot element.

The ERO's that transform the pivot column to that of the departing variable, or $\binom{0}{1}$, are

$$\mathbf{PR}_B^* = \frac{1}{5/3} \mathbf{PR}_B$$

$$= \tfrac{3}{5}(\tfrac{5}{3} \quad 0 \quad -\tfrac{1}{3} \quad 1 \quad | \quad 5)$$

$$= (1 \quad 0 \quad \tfrac{1}{5} \quad \tfrac{3}{5} \quad | \quad 3)$$

$$\mathbf{R}_A^* = \mathbf{R}_A - \tfrac{1}{3}\mathbf{PR}_B^*$$

$$= (\tfrac{1}{3} \quad 1 \quad \tfrac{1}{3} \quad 0 \quad | \quad 5) - \tfrac{1}{3}(1 \quad 0 \quad -\tfrac{1}{5} \quad \tfrac{3}{5} \quad | \quad 3)$$

$$= (0 \quad 1 \quad \tfrac{2}{5} \quad -\tfrac{1}{5} \quad | \quad 4)$$

These revised rows, along with the new basis, initiate the third tableau. Then the step 2 calculations for nonbasis variables are repeated. Since there isn't any nonbasis variable with a positive move index, the search is over. We have found the optimum solution.

Third Tableau

Row	Basis	C_b	c_j	4 x_1	5 x_2	0 x_3	0 x_4	P_0	Limiting ratios
A	x_2	5		0	1	$\tfrac{2}{5}$	$-\tfrac{1}{5}$	4	
B	x_1	4		1	0	$-\tfrac{1}{5}$	$\tfrac{3}{5}$	3	
	z_j					$\tfrac{6}{5}$	$\tfrac{7}{5}$		$S = (3 \quad 4 \quad 0 \quad 0)$
Move index: $c_j - z_j$						$-\tfrac{6}{5}$	$-\tfrac{7}{5}$		$F = 4 \cdot 3 + 5 \cdot 4 = 32$

Repeat step 4 first

Step 3 unnecessary

Repeat step 2 next

No positive values, so optimum solution is

In reading a simplex tableau, you must be wary that the rows may not be in their natural order. In our reference case, the way the variables entered the basis resulted in the second variable being in the first row and the first variable being in the second row. So the optimum solution has $(x_1, x_2) = (3, 4)$, not the other way around. This unscrambled solution, including the zero solution values for the slack variables, is given in the lower right portion of the tableau.

Exercises 12

1. Employ the simplex method to solve our reference problem for the following different objective functions. In each case follow the movement of the solution from corner to corner on a graph. Explain how and why these movements differ from what we observed for our demonstration objective function of $F = 4x_1 + 5x_2$.

a. $F = 5x_1 + 4x_2$
b. $F = x_1 + 5x_2$
c. $F = 4x_1 + x_2$

Use the simplex method to solve the following linear programming problems:

2. Maximize $\quad F = 2x_1 + x_2$
   Subject to $\quad 3x_1 + x_2 \leq 15$
   $\quad\quad\quad\quad\quad x_1 + 2x_2 \leq 20$

3. Maximize $\quad F = x_1 + 3x_2$
   Subject to $\quad x_1 + x_2 \leq 10$
   $\quad\quad\quad\quad\quad 2x_1 + x_2 \leq 14$

11.7 SIMPLEX SOLUTION OF APPLICATIONS

We now apply the simplex method to solve Bernard Broke's investment problem and the everyday applications stated at the beginning of the chapter. We will proceed as follows. For the vacation problem we will solidify the simplex procedures by detailing the steps between tableaus. By then you should be well versed in tableau revisions. So, for the investments and production scheduling applications, we merely give the respective tableaus. Minimization problems require modifications to the maximization simplex procedures. These modifications will be explained while solving the diet problem. Finally, the many calculations needed to solve the fuel blending problem make it wise to call for computer help.

Vacation Planning

Maximize $\quad F = 100x_1 + 1x_2$

Subject to $\quad\quad\quad\quad\quad\quad\quad\quad\quad\quad$ Constraint for

$\quad\quad\quad\quad x_1 + .005x_2 \leq 14 \quad\quad$ Time

$\quad\quad\quad\quad 10x_1 + .5x_2 \leq 500 \quad\quad$ Cost

$\quad\quad\quad\quad\quad\quad\quad x_2 \geq 300 \quad\quad\quad\quad$ Site

$\quad\quad\quad\quad\quad$ all x's $\geq 0 \quad\quad\quad$ Nonnegativity

First, we introduce slack variables into all constraint equations except the nonnegativity ones.

$x_1 + .005x_2 + 1x_3 \quad\quad\quad\quad = 14$
$10x_1 + .5x_2 \quad\quad + 1x_4 \quad\quad = 500$
$\quad\quad - x_2 \quad\quad\quad\quad + 1x_5 = -300$

Notice that we make all slack variables have +1 coefficients. In the third equation this is accomplished by multiplying both sides of the original slack equation, $x_2 - x_5 = 300$, by -1.

LINEAR PROGRAMMING

This basic feasible solution information (step 1), along with the resulting move index (step 2) and limiting ratio (step 3) calculations, are organized to complete the initial tableau.

Initial Tableau

Row	Basis	C_b	c_j →	100	1	0	0	0	P_0	Limiting ratios
				x_1	x_2	x_3	x_4	x_5		
A	x_3	0		①	.005	1	0	0	14	$\frac{14}{1}$ ← Smallest
B	x_4	0		10	.5	0	1	0	500	$\frac{500}{10}$
C	x_5	0		0	−1	0	0	1	−300	

z_j: 0, 0
Move index: $c_j - z_j$: 100, 1
↑ Largest

$S = (0\ \ 0\ \ 14\ \ 500\ \ -300)$
$F = 100 \cdot 0 + 1 \cdot 0 = 0$

The simplex procedure calls for x_1 to replace x_3 in the basis. The ERO's needed for this revision are

$$PR_A^* = \frac{1}{1} PR_A$$

$$R_B^* = R_B - 10 PR_A^*$$

$$R_C^* = R_C - 0 PR_A^*$$

These ERO's leave rows A and C unchanged. The new row B is

$$R_B^* = (10\ \ .5\ \ 0\ \ 1\ \ 0\ \ |\ \ 500) - 10(1\ \ .005\ \ 1\ \ 0\ \ 0\ \ |\ \ 14)$$
$$= (0\ \ .45\ \ -10\ \ 1\ \ 0\ \ |\ \ 360)$$

The new basis information and supplementary move index and limiting ratio calculations are incorporated into the second tableau.

Second Tableau

Row	Basis	C_b	c_j →	100	1	0	0	0	P_0	Limiting ratios
				x_1	x_2	x_3	x_4	x_5		
A	x_1	100		1	.005	1	0	0	14	$\frac{14}{.005}$
B	x_4	0		0	.45	−10	1	0	360	$\frac{360}{.45}$
C	x_5	0		0	⊖1	0	0	1	−300	$\frac{-300}{-1}$ ← Smallest

z_j: .5, 100
Move index: $c_j - z_j$: .5, −100
↑ Largest

$S = (14\ \ 0\ \ 0\ \ 360\ \ -300)$
$F = 100 \cdot 14 + 1 \cdot 0 = 1400$

The second tableau cries out for x_2 to replace x_5 in the basis. This is done by dividing the pivot row (C) by -1 to get a 1 at the pivot point,

$$\mathbf{PR}_C^* = \frac{1}{-1} (0 \ \ -1 \ \ 0 \ \ 0 \ \ 1 \ \ | \ \ -300)$$

$$= (0 \ \ 1 \ \ 0 \ \ 0 \ \ -1 \ \ | \ \ 300)$$

and using the new pivot row to zero the rest of the pivot column as

$$\mathbf{R}_A^* = \mathbf{R}_A - .005\mathbf{PR}_C^*$$
$$= (1 \ \ .005 \ \ 1 \ \ 0 \ \ 0 \ \ | \ \ 14) - .005(0 \ \ 1 \ \ 0 \ \ 0 \ \ -1 \ \ | \ \ 300)$$
$$= (1 \ \ 0 \ \ 1 \ \ 0 \ \ .005 \ \ | \ \ 12.5)$$

$$\mathbf{R}_B^* = \mathbf{R}_B - .45\mathbf{PR}_C^*$$
$$= (0 \ \ .45 \ \ -10 \ \ 1 \ \ 0 \ \ | \ \ 360) - .45(0 \ \ 1 \ \ 0 \ \ 0 \ \ -1 \ \ | \ \ 300)$$
$$= (0 \ \ 0 \ \ -10 \ \ 1 \ \ .45 \ \ | \ \ 225)$$

These rows and the supplementary step 2 and step 3 calculations are incorporated into the third tableau.

Third Tableau

Row	Basis	c_j C_b	100 x_1	1 x_2	0 x_3	0 x_4	0 x_5	P_0	Limiting ratios
A	x_1	100	1	0	1	0	.005	12.5	$\frac{12.5}{.005}$
B	x_4	0	0	0	-10	1	㊺	225	$\frac{225}{.45}$ ← Smallest
C	x_2	1	0	1	0	0	-1	300	Negative
	z_j Move index: $c_j - z_j$				100 -100		$-.5$ .5 ↑ Largest	$S = (12.5 \ \ 300 \ \ 0 \ \ 225 \ \ 0)$ $F = 100 \cdot 12.5 + 1 \cdot 300$ $= 1550$	

The third tableau calls in x_5 from the bullpen and sends x_4 to the showers. The pivoting necessary to accomplish this is done by first dividing the pivot row (B) by the pivot element (.45) to get a 1 at the pivot point

$$\mathbf{PR}_B^* = \frac{1}{.45}(0 \ \ 0 \ \ -10 \ \ 1 \ \ .45 \ \ | \ \ 225)$$

$$= \left(0 \ \ 0 \ \ -\frac{10}{.45} \ \ \frac{1}{.45} \ \ 1 \ \ | \ \ 500\right)$$

and using the revised pivot row as follows to zero the rest of the pivot column.

R_A^* $= R_A - .005PR_B^*$

$= (1 \ 0 \ 1 \ 0 \ .005 \ | \ 12.5) - .005\left(0 \ 0 \ -\dfrac{10}{.45} \ \dfrac{1}{.45} \ 1 \ | \ 500\right)$

$= \left(1 \ 0 \ \dfrac{10}{9} \ -\dfrac{1}{90} \ 0 \ | \ 10\right)$

R_C^* $= R_C - (-1)PR_B^*$

$= (0 \ 1 \ 0 \ 0 \ -1 \ | \ 300) - (-1)\left(0 \ 0 \ -\dfrac{10}{.45} \ \dfrac{1}{.45} \ 1 \ | \ 500\right)$

$= \left(0 \ 1 \ -\dfrac{10}{.45} \ \dfrac{1}{.45} \ 0 \ | \ 800\right)$

These revised rows and the supplementary calculations are incorporated into the fourth tableau.

Fourth Tableau

		c_j	100	1	0	0	0		Limiting
Row	Basis	C_b	x_1	x_2	x_3	x_4	x_5	P_0	ratios
A	x_1	100	1	0	$\dfrac{10}{9}$	$-\dfrac{1}{90}$	0	10	
B	x_5	0	0	0	$-\dfrac{10}{.45}$	$\dfrac{1}{.45}$	1	500	
C	x_2	1	0	1	$-\dfrac{10}{.45}$	$\dfrac{1}{.45}$	0	800	
		z_j			$\dfrac{800}{9}$	$\dfrac{10}{9}$		$S = (10 \ 800 \ 0 \ 0 \ 500)$	
	Move index: $c_j - z_j$				$-\dfrac{800}{9}$	$-\dfrac{10}{9}$		$F = 100 \cdot 10 + 1 \cdot 800 = 1800$	

No positive values, so optimum solution is

It took four tableaus (corner points) to maximize the enjoyment function. The best vacation plan is to spend 10 days camping at a site 800 miles away. The other four days of the vacation are spent in traveling. Enjoy!

Exercise 13

Use the simplex method to solve the revised vacation planning problem that you formulated as an exercise on page 379.

Investments Problem

Maximize $\quad F = .04x_1 + .08x_2$

Subject to $\qquad\qquad\qquad\qquad$ Constraint for

$\qquad\qquad\qquad x_1 + \ \ x_2 \leq 100 \qquad$ Capital

$\qquad\qquad\qquad .05x_1 + .2x_2 \leq 10 \qquad$ Risk

$\qquad\qquad\qquad\quad$ all x's $\geq \ \ 0 \qquad$ Nonnegativity

Here we just present the sequence of tableaus. See if you can carry out all the intermediate steps.

Initial Tableau

Row	Basis	C_b	c_j .04 x_1	.08 x_2	0 x_3	0 x_4	P_0	Limiting ratios
A	x_3	0	1	1	1	0	100	$\frac{100}{1}$
B	x_4	0	.05	(.2)	0	1	10	$\frac{10}{2}$ ← Smallest
	z_j		0	0			$S = (0\ \ 0\ \ 15\ \ 10)$	
Move index: $c_j - z_j$			.04	.08 ↑ Largest			$F = .04 \cdot 0 + .08 \cdot 0 = 0$	

Second Tableau

Row	Basis	C_b	c_j .04 x_1	.08 x_2	0 x_3	0 x_4	P_0	Limiting ratios
A	x_3	0	(.75)	0	1	−5	50	$\frac{50}{.75}$ ← Smallest
B	x_2	.08	.25	1	0	5	50	$\frac{50}{.25}$
	z_j		.02			.4	$S = (0\ \ 50\ \ 50\ \ 0)$	
Move index: $c_j - z_j$			.02 ↑ Largest			−.4	$F = .04 \cdot 0 + .08 \cdot 50 = 4$	

Third Tableau

Row	Basis	C_b	c_j .04 x_1	.08 x_2	0 x_3	0 x_4	P_0	Limiting ratios
A	x_1	.04	1	0	$\frac{4}{3}$	$-\frac{20}{3}$	$\frac{200}{3}$	
B	x_2	.08	0	1	$-\frac{1}{3}$	$\frac{20}{3}$	$\frac{100}{3}$	
	z_j				$\frac{.08}{3}$	$\frac{.8}{3}$	$S = (\frac{200}{3}\ \ \frac{100}{3}\ \ 0\ \ 0)$	
Move index: $c_j - z_j$					$-\frac{.08}{3}$	$-\frac{.8}{3}$	$F = .04 \cdot \frac{200}{3} + .08 \cdot \frac{100}{3} = \frac{16}{3}$	

No positive values, so optimum solution is

The third tableau indicates that no further increase in Bernard's dividend income is possible. Thus, he should invest $66\frac{2}{3}$ thousand in Gangland Industries and $33\frac{1}{3}$ thousand in C.A.R.T.E.L. Then he should pray real hard!

Exercise 14

Before Bernard could invest, his broker called him up with the news that Gangland Industries, after losing some big "contracts," just cut their dividend to 1 percent. Using the simplex method help Bernard find the optimum solution under the new conditions.

Production Scheduling

Maximize $\quad F = 5x_1 + 4x_2$

Subject to $\qquad\qquad\qquad\qquad$ Constraint for

$\qquad\qquad 6x_1 + 2x_2 \le 100 \qquad$ Process A

$\qquad\qquad 3x_1 + 4x_2 \le 80 \qquad$ Process B

$\qquad\qquad 1x_1 + 2x_2 \le 160 \qquad$ Process C

$\qquad\qquad\quad$ all x's $\ge 0 \qquad\quad$ Nonnegativity

Here we just present the sequence of tableaus. See if you can carry out all the intermediate steps.

Initial Tableau

		c_j	5	4	0	0	0		Limiting
Row	Basis	C_b	x_1	x_2	x_3	x_4	x_5	P_0	ratios
A	x_3	0	⑥	2	1	0	0	100	$\frac{100}{6}$ ← Smallest
B	x_4	0	3	4	0	1	0	80	$\frac{80}{3}$
C	x_5	0	1	2	0	0	1	160	$\frac{160}{1}$
	z_j		0	0				$S = (0\ \ 0\ \ 100\ \ 80\ \ 160)$	
	Move index: $c_j - z_j$		5	4				$F = 5 \cdot 0 + 4 \cdot 0 = 0$	
			↑ Largest						

Second Tableau

		c_j	5	4	0	0	0		Limiting
Row	Basis	C_b	x_1	x_2	x_3	x_4	x_5	P_0	ratios
A	x_1	5	1	$\frac{1}{3}$	$\frac{1}{6}$	0	0	100/6	$\frac{100/6}{1/3}$
B	x_4	0	0	③	$-\frac{1}{2}$	1	0	30	$\frac{30}{3}$ ← Smallest
C	x_5	0	0	$\frac{5}{3}$	$-\frac{1}{6}$	0	1	430/3	$\frac{430/3}{5/3}$
	z_j		$\frac{5}{3}$	$\frac{5}{6}$				$S = (\frac{100}{6}\ \ 0\ \ 0\ \ 30\ \ \frac{430}{3})$	
	Move index: $c_j - z_j$		$\frac{7}{3}$	$-\frac{5}{6}$				$F = 5 \cdot \frac{100}{6} + 4 \cdot 0 = \frac{500}{6}$	
			↑ Largest						

Third Tableau

Row	Basis	c_j C_b	5 x_1	4 x_2	0 x_3	0 x_4	0 x_5	P_0	Limiting ratios
A	x_1	5	1	0	$\frac{2}{9}$	$-\frac{1}{9}$	0	$\frac{80}{6}$	
B	x_2	4	0	1	$-\frac{1}{6}$	$-\frac{1}{3}$	0	10	
C	x_5	0	0	0	$\frac{1}{9}$	$-\frac{5}{9}$	1	$\frac{380}{3}$	
	z_j				$\frac{4}{9}$	$\frac{7}{9}$		$S = (\frac{80}{6} \quad 10 \quad 0 \quad 0 \quad \frac{380}{3})$	
	Move index: $c_j - z_j$				$-\frac{4}{9}$	$-\frac{7}{9}$		$F = 5 \cdot \frac{80}{6} + 4 \cdot 10 = \frac{640}{6}$	

No positive values, so optimum solution is

The third tableau reveals that the objective function can't be increased any further. Thus, the final solution of the production scheduling problem is to produce $\frac{80}{6} = 13\frac{1}{3}$ bellwagers and 10 honktinkers. Since the market for partial bellwagers is nonexistent, we would have to settle for only 13 of them. So, in actuality, the maximum profit would be $5 \cdot 13 + 4 \cdot 10 = 105$.

Exercise 15

Use the simplex method to solve the revised production scheduling problem that you formulated as an exercise on page 380.

Diet Planning (A Minimization Case)

The diet planning problem involves a minimization of the objective function. Simplex procedures for doing this are slightly different from the maximization methods studied thus far. These differences and the use of the simplex method for solving minimization problems will be illustrated with the diet planning problem, restated as

Minimize cost $F = 8x_1 + 6x_2$

Subject to   Constraint for

$\qquad\qquad 15x_1 + 30x_2 \geq 100$ Vitamin A

$\qquad\qquad 25x_1 + 10x_2 \geq 100$ Iron

$\qquad\qquad$ all x's ≥ 0 Nonnegativity

The first difference is encountered when we introduce slack variables.

$15x_1 + 30x_2 - 1x_3 \qquad\quad = 100$

$25x_1 + 10x_2 \qquad - 1x_4 = 100$

LINEAR PROGRAMMING

Because the vitamin A and iron constraints are of the $\geq$ type, we must subtract the slack variables to create equalities. But then the basic feasible solution of $x_3 = -100$ and $x_2 = -100$ would violate the nonnegativity constraint. We can avoid this problem by creating artificial variables, x_5 and x_6, with positive coefficients, as defined by the equations

$$15x_1 + 30x_2 - 1x_3 \quad\quad + 1x_5 \quad\quad = 100$$
$$25x_1 + 10x_2 \quad\quad -1x_4 \quad\quad + 1x_6 = 100$$

Now we can form the following basic feasible solution with no negative elements: $(x_1\ x_2\ x_3\ x_4\ x_5\ x_6) = (0\ 0\ 0\ 0\ 100\ 100)$. This solution gives us a starting point from which the simplex method can take over and proceed to move to better corners.

The initial tableau is

Initial Tableau

Row	Basis	c_j C_b	8 x_1	6 x_2	0 x_3	0 x_4	M x_5	M x_6	P_0	Limiting ratios
A	x_5	M	15	30	-1	0	1	0	100	$\frac{100}{30}$ ← Smallest
B	x_6	M	25	10	0	-1	0	1	100	$\frac{100}{10}$
		z_j	40M	40M	-M	-M			$S = (0\ 0\ 0\ 0\ 100\ 100)$	
Move index: $c_j - z_j$			8 - 40M	6 - 40M	M	M			$F = 100M + 100M = 200M$	
				↑ Smallest (most negative)						

Notice that the cost contributions of the artificial variables are set at M, which represents a very large number. Since we wish to minimize costs, these settings will force the artificial variables out of the solution. In other words the artificial variables were invented to get us started, but we don't want them hanging around very long.

Observe that the z_j, move index, and limiting ratio calculations are the same as for maximization problems. However, the smallest or most negative move index value (as opposed to the largest positive value for maximization problems) determines the entering variable. The reasoning behind this difference is that $c_j - z_j$ is the net change in the objective function as a result of introducing the jth variable into the basis. Those variables with negative $c_j - z_j$ decrease the objective function. Since we want to minimize the objective function, it makes sense to choose the variable that decreases the objective function the most, or in other words, has the most negative move index.

The initial tableau calls for x_2 to enter the basis in place of x_5. Revising the tableau to incorporate the new basis is done exactly the same way as we did for maximization problems.

$$PR_A^* = \frac{1}{30} PR_A$$

$$= \frac{1}{30}(15\ \ 30\ \ -1\ \ 0\ \ 1\ \ 0\ |\ 100)$$

$$= \left(\frac{1}{2}\ \ 1\ \ -\frac{1}{30}\ \ 0\ \ \frac{1}{30}\ \ 0\ \middle|\ \frac{10}{3}\right)$$

$R_B^* = R_B - 10PR_A^*$

$= (25 \quad 10 \quad 0 \quad -1 \quad 0 \quad 1 \quad | \quad 100)$

$-10 \left(\dfrac{1}{2} \quad 1 \quad -\dfrac{1}{30} \quad 0 \quad \dfrac{1}{30} \quad 0 \quad \Big| \quad \dfrac{10}{3} \right)$

$= \left(20 \quad 0 \quad \dfrac{1}{3} \quad -1 \quad -\dfrac{1}{3} \quad 1 \quad \Big| \quad \dfrac{200}{3} \right)$

These new rows and the supplementary calculations result in the following second tableau.

Second Tableau

Row	Basis	c_j C_b	8 x_1	6 x_2	0 x_3	0 x_4	M x_5	M x_6	P_0	Limiting ratios
A	x_2	6	$\frac{1}{2}$	1	$-\frac{1}{30}$	0	$\frac{1}{30}$	0	$\frac{10}{3}$	$\frac{10/3}{1/2}$
B	x_6	M	⑳	0	$\frac{1}{3}$	-1	$-\frac{1}{3}$	1	$\frac{200}{3}$	$\frac{200/3}{20}$ ← Smallest
		z_j	$3 + 20M$		$-\frac{1}{5} + \frac{M}{3}$	$-M$	$\frac{1}{5} - \frac{M}{3}$		$S = (0 \quad \frac{10}{3} \quad 0 \quad 0 \quad 0 \quad \frac{200}{3})$	
Move index:		$c_j - z_j$	$5 - 20M$		$\frac{1}{5} - \frac{M}{3}$	M	$\frac{4M}{3} - \frac{1}{5}$		$F = 6 \cdot \frac{10}{3} + M\frac{200}{3}$	
			↑ Smallest (most negative)							

See if you can make the transition to the following third tableau.

Third Tableau

Row	Basis	c_j C_b	8 x_1	6 x_2	0 x_3	0 x_4	M x_5	M x_6	P_0	Limiting ratios
A	x_2	6	0	1	$-\frac{5}{120}$	$\frac{1}{40}$	$\frac{5}{120}$	$-\frac{1}{40}$	$\frac{5}{3}$	
B	x_1	8	1	0	$\frac{1}{60}$	$-\frac{1}{20}$	$-\frac{1}{60}$	$\frac{1}{20}$	$\frac{10}{3}$	
		z_j			$-\frac{7}{60}$	$-\frac{1}{4}$	$\frac{7}{60}$	$\frac{1}{4}$	$S = (\frac{5}{3} \quad \frac{10}{3} \quad 0 \quad 0 \quad 0 \quad 0)$	
Move index:		$c_j - z_j$			$\frac{7}{60}$	$\frac{1}{4}$	big	big	$F = 8 \cdot \frac{10}{3} + 6 \cdot \frac{5}{3} = \frac{110}{3}$	
					\multicolumn{4}{No negative values, so optimum solution is}		

The third tableau reveals that the objective function cannot be decreased further. Thus, the optimum solution is a bowl of cereal containing $\frac{10}{3}$ ounces of Cornies and $\frac{5}{3}$ ounces of Sweet Smackaroos, costing 110/3 or just under 37¢. Do you think that your stomach would agree on the optimality of this solution?

LINEAR PROGRAMMING

Exercise 16

Using the simplex procedure for minimization problems, solve the revised diet planning problem that you formulated as an exercise on page 381.

Fuel Blending

This problem, stated on page 381, brings us to a case where we can really appreciate computers. With six variables plus five slack variables, a hand-computed simplex procedure could be a little tedious. Fortunately, most computer systems have linear programming software that employ simplex procedures. Here, we will employ and describe the widely-used Microsoft Excel software. So be patient as we learn those computer methods in section 11.8. Then we will return to solve this fuel blending problem.

11.8 MICROSOFT EXCEL COMPUTER METHODS

In this section, we explain how to use Excel and its embedded software program, Solver, to solve linear programming problems. Solver was developed by the company, Frontline Systems, Inc. Microsoft then incorporated it into Excel.

Reference Problem

Before we attack the large fuel blending problem above that motivated us to seek computer help, let's start with our fairly simple reference problem, defined as:

Maximize $\quad F = 4x_1 + 5x_2$

Subject to

$\quad$ Constraints: $\quad 1x_1 + 3x_2 \leq 15$

$\quad\quad\quad\quad\quad\quad\quad\quad\, 2x_1 + 1x_2 \leq 10$

$\quad$ Non-negativity $\quad 1x_1 \quad\quad \geq 0$

$\quad\quad\quad\quad\quad\quad\quad\quad\quad\quad 1x_2 \geq 0$

Earlier, we solved this problem both algebraically and with the Simplex Method. Truthfully, those methods, although important for our learning process, were tedious. So it will be a pleasure to see a computer quickly whip off the same answer.

To solve this problem with the Solver software on Excel, we start by inputting the relevant information into the Excel spreadsheet, as shown in Figure 11.16a. Now we spell out the step by step development of this spreadsheet.

In spreadsheet columns C and D, we input the various coefficients associated with the x_1 and x_2 variables, respectively, as follows:

Figure 11.16a

	A	B	C	D	E	F	G	H
1								
2								
3	Descriptions:		x1	x2			Objective	
4							Function Value	
5	Objective Function		4	5			0	
6								
7	Units of x		0	0				
8								
9					Left Side	Sense	Right Side	
10	Constraints:		1	3	0	<=	15	
11			2	1	0	<=	10	
12			1		0	>=	0	
13				1	0	>=	0	
14								
15								

Figure 11.16b

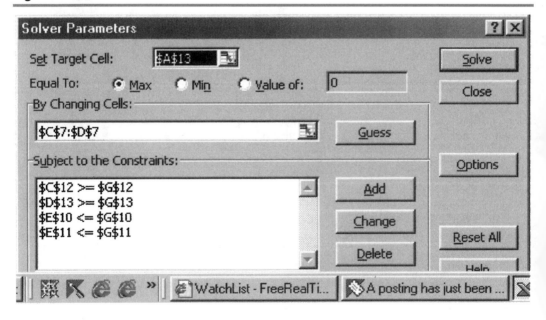

LINEAR PROGRAMMING

- Objective function coefficients: 4 for x_1 coefficient in cell C5 and 5 for x_2 coefficient in cell D5
- Units of variable: We begin with the initial solution, which is 0 units of x_1 in cell C7 and 0 units of x_2 in cell D7
- Constraint equations coefficients: The x_1 coefficients are 1 in cell C10 and 2 in cell C11 and the x_2 coefficients are 3 in cell D10 and 1 in cell D11.
- Non-negativity coefficients: We input 1 for x_1 in cell C12 and 1 for x_2 in cell D13.

At this point, the numbers duplicate the left sides of all the equations.

We then define the value of the objective function in cell G5 by typing in the Excel equation: = 4 * C7 + 5 * D7

Next we input the formulas for the overall values of the left sides of the equations in column E as follows:

Cell	Excel Formula
E10	= 1 * C7 + 3 * D7
E11	= 2 * C7 + 1 * D7
E12	= 1 * C7
E13	= 1 * D7

Finally, we show the sense of the inequalities in column F and the right side constants in column G.

At this point, we are ready for Solver to do its thing. So we click on Tools and click on Solver about two thirds of the way down the menu. This brings up the Solver dialog box, whose final appearance is shown in Figure 11.16b. Now we illustrate how to get to that point.

The Set Target Cell represents the value of the objective function. So we go to that box and either type in $G$5 or click on the cell G5.

Since this is a maximize problem, we click on Max

Next, we have to input the changing cells box. These represent the number of units of the variables x_1 and x_2. We can either type in $C$7 : $D$7 or highlight cells C7 and D7 and press ENTER.

Next, we must input the constraint equations. So we click in that box and click on Add. The first constraint equation has its left-side formula in cell E10 and its right-side constant in cell G10. Thus we can click on those cells to input them into the dialogue box. We also have to click on the proper sense, in this case <=. So we have completed the first constraint equation. To go on to the second one, click Add. The other constraint equations are similarly done.

When all the constraint equations are done, we merely click on "Solve". And we get the optimum solution of $x_1 = 3$ and $x_2 = 4$ with the maximum value of 32 as shown in Figure 11.17a. We can then request an Answer report, which is shown in Figure 11.17b.

Exercises 17

1. Change the objective function to

    $F = 6x_1 + 3x_2$

    Use Excel's Solver to find the maximum for this case.
2. Use Excel's Solver to *minimize* the Total Cost for the Diet Planning problem on pages 380–381.

Figure 11.17a

	A	B	C	D	E	F	G	H
1								
2								
3	Descriptions:		x1	x2			Objective	
4							Function Value	
5	Objective Function		4	5			32	
6								
7	Units of x		3	4				
8								
9					Left Side	Sense	Right Side	
10	Constraints:		1	3	15	<=	15	
11			2	1	10	<=	10	
12			1		3	>=	0	
13				1	4	>=	0	
14								
15								

Figure 11.17b

Microsoft Excel 8.0a Answer Report
Worksheet: [LPReference.xls]Sheet1
Report Created: 12/9/03 3:49:30 PM

Target Cell (Max)

Cell	Name	Original Value	Final Value
$G$5	Objective Function Function Value	0	32

Adjustable Cells

Cell	Name		Original Value	Final Value
$C$7	Units of x	x1	0	3
$D$7	Units of x	x2	0	4

Constraints

Cell	Name		Cell Value	Formula	Status	Slack
$D$13	x2		1	$D$13>=$G$13	Not Binding	1
$C$12	x1		1	$C$12>=$G$12	Not Binding	1
$E$10	Constraints:	Left Side	15	$E$10<=$G$10	Binding	0
$E$11	Left Side		10	$E$11<=$G$11	Binding	0

LINEAR PROGRAMMING

Fuel Blending Problem

Refer back to page 381, where this problem is defined. We set up the spreadsheet in the same manner as for the reference problem. The only difference is that we have more values due to the larger system.

In Figure 11.18a, we show the spreadsheet after Solver has found the maximum. You would begin the process with a spreadsheet that is the same as this final one, except that you would put zeros in the units of x cells. We also show the answer report in Figure 11.18b, which shows the solution as

$x_1 = 0$

$x_2 = 375$

$x_3 = 0$

$x_4 = 3000$

$x_5 = 0$

$x_6 = 0$

Note: The answer report does not employ double subscripts, as we did in defining the fuel problem. The variable correspondences are: $x_1 = x_{11}$, $x_2 = x_{12}$, $x_3 = x_{21}$, $x_4 = x_{22}$, $x_5 = x_{31}$, and $x_6 = x_{32}$.

The objective function has a maximum value of 11,250.

Exercise 18

Redo this problem if Blend A was limited to 1500.

PROBLEMS

TECHNICAL TRIUMPHS

1. What is the sense of the following inequalities?
   a. $10 < 20$
   b. $-5 > -6$
   c. $a < b + c$
   d. $0 > -4$

2. Graph the inequality:

   $x + y \leq 10$

3. Add this inequality to the one in problem 2:

   $2x - y \geq 5$

   What region is now feasible for the set of two inequalities?

4. What triangle is represented by:

Figure 11.18a

	A	B	C	D	E	F	G	H	I	J	K	L
1												
2												
3	Descriptions		x11	x12	x21	x22	x31	x32				Value of the
4												Obj Function
5	Objective Function		-5	-10	10	5	15	10				11250
6												
7	Units of x		0	375	0	3000	0	0				
8												
9										Left Side	Sense	Right Side
10	Subject to (Constraints)		-3	0	6	0	16	0		0	<=	0
11			0	-8	0	1	0	11		0	<=	0
12			1	1	0	0	0	0		375	<=	2000
13			0	0	1	1	0	0		3000	<=	3000
14			0	0	0	0	1	1		0	<=	1000
15												
16	Nonnegativity		1							0	>=	0
17				1						375	>=	0
18					1					0	>=	0
19						1				3000	>=	0
20							1			0	>=	0
21								1		0	>=	0
22												0
23												

Figure 11.18b

Microsoft Excel 8.0a Answer Report
Worksheet: [LPReference.xls]Sheet2
Report Created: 12/10/03 1:24:22 PM

Target Cell (Max)

Cell	Name	Original Value	Final Value
$L$5	Objective Function Obj Function	0	11250

Adjustable Cells

Cell	Name	Original Value	Final Value
$C$7	Units of x x1	0	0
$D$7	Units of x x2	0	375
$E$7	Units of x x3	0	0
$F$7	Units of x x4	0	3000
$G$7	Units of x x5	0	0
$H$7	Units of x x6	0	0

LINEAR PROGRAMMING

$x \geq 0$

$y \geq 0$

$x + y \leq 5$

5. Which of these two inequalities includes the origin?

    $-x + 2y \leq 3$

    $4x - 3y \geq -1$

6. If one wishes to maximize

    $F = 5x + 7y$

    a. How much does F increase for each unit increase in x?
    b. How much does F increase for each unit increase in y?

7. Consider the following maximization problem

    $F = 6x_1 + 4x_2$

    $x_1 + x_2 \leq 20$

    $3x_1 + 2x_2 \leq 15$

    a. Introduce slack variables.
    b. Which variable would enter the solution first? What slack variable would it replace? Explain why.
    c. Maximize F using the algebraic approach.
    d. Maximize F using the simplex method.
    e. Use Microsoft Excel to solve this problem.

CONFIDENCE BUILDERS

1. Bernard Broke received a late night telephone call with the following curt message, "If you know what's good for you, you'll invest at least 80 grand in Gangland Industries." Add this constraint to the problem defined at the beginning of the chapter.
    a. Graph all the constraint equations and identify the feasible solution space.
    b. Find all the corner points, identify which are feasible, and determine the income at each feasible corner.
    c. Draw successive parallel lines of the objective function form $F = .04x_1 + .08x_2$ for $F = 2, 3, 4, 5,$ and 6. Notice how the optimum corner identified in part b lies on a higher parallel line than any other corner in the feasible solution space.
2. A company has just introduced a new product and wishes to spread the word by television, radio, and newspaper advertising. Each dollar of television, radio, and newspaper advertising is expected to reach and inform 20, 15, and 10 people, respectively, of the wonders of the new product. Since they want a coupon-clipping campaign, newspaper advertising must be at least one-third of the total of television and radio advertising. Also, to reach a diversified audience (for example, truck drivers and traveling salespeople who listen extensively to radio while they travel), the company requires that radio advertising be no less than one-half of television advertising. If the advertising budget is $20 thousand
    a. Formulate the linear programming problem.
    b. Use Microsoft Excel to solve for the advertising budget allocation that reaches the most people under the given constraints.
3. The Carter Nut Company has 1000 pounds of peanuts and 1500 pounds of other nuts to be sold in two different mixtures. Mixture A, which contains a 50–50 mix of peanuts and other nuts, sells for $2.00 per pound. Mixture B, which contains a 25–75 mix of peanuts and other nuts, sells for $2.50 per

pound. Assuming that nuts not in one of the mixtures can't be sold, use a graphical approach to answer the following questions.
   a. How much of each mixture should be produced in order to maximize revenue?
   b. How would a drop in price to $1.50 per pound for mixture A affect the solution?
   c. If Carter could not sell more than 800 pounds of mixture B, determine the new solutions (assume the mixture A price of $2.00 per pound).
4. Reconsider part a of the problem above, but remove the assumption that nuts not in one of the two mixtures can't be sold. Use the simplex procedure to solve for the following separate price cases:
   a. $1.50 and $2.00 per pound, for peanuts and other nuts respectively.
   b. $1.50 and $3.00 per pound, for peanuts and other nuts respectively.
   c. Use Microsoft Excel to solve parts a. and b.

WALL BANGERS

1. With two days left (40 available hours for course work), Kay Ramm has a term paper in management and finals in accounting and mathematics to salvage. Accounting is the only course she is in danger of failing. She must get at least a 60 on that exam in order to pass accounting and stay in school. Kay can turn in the first draft of her term paper in management and get a 50; however, she can improve her grade by 3 points per hour, up to a maximum grade of 80, by improving the presentation of the paper. She is way behind in accounting and would only get a grade of 20 if she went into the final without studying. For each additional hour of accounting study, she feels she can gain 2 points on the accounting final. Kay could get a 40 on the math final with no further study; however, for each additional hour of study (up to a maximum of 15), she could improve her grade by 4 points. The three courses carry the same number of credits. The term paper is one-third the management course grade, while the accounting and mathematics finals are, respectively, one-quarter and one-half of the course grades.
   a. Formulate the linear programming problem.
   b. Solve, using Microsoft Excel, for the time allocation among the three courses that would be best for Kay Ramm's last-minute grade salvage effort.

2. Congressman Doolittle must allocate his time in the last ten days of the campaign against Manny Boondoggle. On any day, Doolittle can go on TV and reach viewers in the entire district (the whole day must be spent preparing for the tube) or go out among the people and campaign in one of the two major areas (A or B) in his congressional district. To please powerful party interests, which he will need for various favors, he can't spend more than one day more in A than in B. Doolittle is fearful of getting weary on the campaign trail and doing something stupid (he lost the election four years ago when the cameras caught him sleeping during the governor's speech).

    The cost, votes expected to be gained, and weariness index expected from one day at each campaigning possibility are

	Cost (thousands)	Votes (hundreds)	Weariness index
TV	10	10	2
A	5	7	3
B	3	2	1

    If Doolittle has a total of $50 thousand available for campaign costs, and wishes to have a total weariness index of no more than 15, how should he allocate his time during the last ten days of the campaign?

3. Red Corpuscles wants to have at least 100 percent of the minimum daily requirements for vitamins A, B, C, and D. To meet this goal, he wishes to choose the minimum cost diet from four foods, W, X, Y, and Z, having the following percentages (per ounce) of the minimum daily requirements and costs (per ounce). Use Microsoft Excel for this problem.

LINEAR PROGRAMMING

	A	B	C	D	Cost(¢)
W	10	5	15	20	10
X	30	10	10	10	25
Y	15	20	40	10	20
Z	5	15	5	5	15

a. If Red restricts his diet to foods W and X, formulate and solve the linear programming problem for the minimum cost diet.

b. If Red restricts his diet to foods W, X, and Y, formulate and solve the linear programming problem for the minimum cost diet.

4. The Willy Wonka Company makes two candy products, Wonka Bars and Everlasting Gob-Stoppers, which yield profits of $30 and $40 respectively for each production batch of 1000. Making each candy product requires three operations: blending, cooking, and packaging. The time (in minutes) for each operation required to produce 1000 of each product and the available operation capacities (minutes per day) are given in the following table.

	Wonka Bars	Gob-Stoppers	Operation capacity
Blending	20	30	600
Cooking	50	80	900
Packaging	30	10	400

If all candy output can be sold, how much of each product should be made in order to maximize profits?

5. If in problem 4 above

C is a 1×2 matrix of product contributions to profit

X is a 2×1 matrix of unknown production batch levels

A is a 3×2 matrix of operation requirements for product batches

B is a 3×1 matrix of operation capacities

show that the linear programming problem can be written in matrix terms as

Maximize $\mathbf{C} \cdot \mathbf{X}$

Subject to $\mathbf{A} \cdot \mathbf{X} \leq \mathbf{B}$

MODULE FIVE

Calculus

Calculus sorely needs a public relations effort. Many students are convinced it is very difficult if not impossible to grasp. Not so! Calculus is straightforward and rests solely on one mathematical concept. No greater step is necessary to learn calculus than was required to understand functions. With diligence you will undoubtedly master calculus, too.

Calculus is divided into two branches: *differential* and *integral.* Differential calculus (the subject of Chapters 12 through 15) allows us to transform a total function into a rate function. Integral calculus (Chapters 16 through 18) allows us to take this latter function and transform it back to the original function.

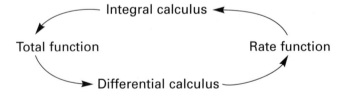

This circular path may seem like a lot of effort just to get back to the starting point. However, in the evolution of this cycle, literally thousands of applications arise, ranging from the investment multiplier in the economy, to maximizing profits, to finding the lifetime sales of a product. And it is such applications that make the round trip worthwhile.

CHAPTER 12

Introduction to Differential Calculus

12.1 INTRODUCTION

Differential calculus answers the question, "How fast does $f(x)$ change relative to changes in x?". Or in other words, "What is the slope of a function?" Such questions may seem too esoteric for a practical person, such as yourself. Maybe you are even thinking that you could live another 50 years or so untouched by calculus and such concerns. But alas, you are wrong! Calculus touches us when we read a speedometer, when we try to comprehend economic concepts such as marginal revenue and elasticity of demand, when we learn probability and statistics, when we graph functions, and in many other situations. We badly needed calculus earlier in the text. Recall from Chapters 6 and 7 how we developed the total revenue and total cost functions for S. Lumlord's apartment complex. Back then in B.C. time (Before Calculus), we were only able to determine the break-even point(s) by setting the two functions equal, or solving for the rent that results in zero total profits. Graphically, we had:

Figure 12.1

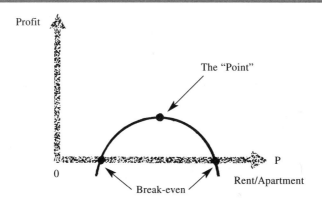

As any greedy capitalist would know, we missed "the point." It is interesting to know what price results in break even. However, it is crucial to determine the price that yields the maximum profit. Differential calculus is the tool that finds these vital points. Not wanting to "miss the point," let's embark on a study of differential calculus for "fun and profit."

The central idea of differential calculus is a *rate of change* (often referred to as a *rate*). The following episode from the life of a typical student is presented to illustrate this concept and also to compare it to the concept of a total, with which you already have some familiarity.

Rates Versus Totals

Upon hearing that he was required to take calculus, Al Cohol went to a bar to drown his sorrow in 10 drinks. Which should be the appropriate comment?

1. Who carried him home?
2. What a great time he must have had!
3. You sinner, mend your evil ways.

Any of these comments could be appropriate since the mere fact of 10 drinks—in total—isn't indicative of Al's alcoholic state. Rather, we need to relate total drinks to total time spent in drinking (this ratio is called a rate) to get an idea of his condition. If he had 10 drinks during the first hour, he would have a very high feeling and a very low body slope, as he lay stretched out. However, if he spread out the 10 drinks evenly over 5 hours (2 drinks per hour), he probably would be a little high, but quite sociable. Thus, the same total can result in quite different effects, depending on its relationship to another total, as shown in Figure 12.2.

It is the rate of consumption (drinks per hour) that really determines an inebriated state. A person with 4 drinks can be floored if they are all downed within 15 minutes (rate of 16 per hour), while the same person with 12 drinks over 12 hours (1 per hour) could be fine.

Figure 12.2

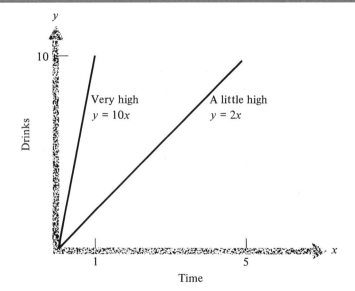

INTRODUCTION TO DIFFERENTIAL CALCULUS

Just as the number of drinks can be related to time as a rate (drinks per hour), it can be related to the number of potato chips eaten (10 drinks while having 50 chips would be a rate of $\frac{1}{5}$ drink per chip), the number of beautiful people who were impressed (10 drinks while impressing no one is an infinite rate), or the number of fights started (10 drinks while starting 2 fights, or a rate of 5 drinks per fight).

Rates of Change of Practical Significance

The list of possible rates that could be devised is endless. Here are some that have practical significance.

Description; name	*Typical units*
1. Velocity	Miles per hour
2. Marginal revenue; marginal cost	Dollars per unit of price; output
3. Birth rate	Births per 1000 families
4. Death rate	Deaths per 1000 people
5. Caloric intake rate	Calories per day
6. Pollution rate	Pounds per cubic mile
7. Disease rate	Infections per 1000 people
8. Snowfall; rainfall rates	Inches per hour
9. Multiplier	Dollars of GNP per dollar of investment
10. Sales rate	Units per month
11. Traffic flow rate	Cars per hour
12. Marginal consumption	Consumption expenditure per dollar of income
13. Interest rate	Interest per dollar invested
14. Earned run average	Runs per nine innings

Note that a rate is always expressed as something per unit of something else.

Exercises 1

1. It rained 1.11 inches in 10 minutes in New York City on July 5, 1973. *The New York Post* reported that a National Weather Service spokesperson called it "an incredible amount of rain." Do you think the word *amount* is appropriate here? Can you think of a different word that would more accurately describe the situation?
2. Dr. Calvin Kulus has a schedule of rates on his office wall. What would be typical units for such rates?
3. Using the concepts of rates and totals, discuss how a student with a poor mathematics background can learn at a high rate, yet end up with only a C grade.

Learning Plan

With a rough idea of a rate of change in mind, we can proceed to develop the topic of differential calculus in the following manner. First, we will take a little automobile trip and make the connection between rates of change and the slopes of straight lines on a graph. These constant slopes, which represent average rates of change, provide us with the basis for tackling the tougher problem of finding instantaneous rates of change, which graphically represent slopes of curving functions. The latter problem requires us to fire a rocket to find its velocity at various split seconds as it accelerates into space. This leads us to develop a limit process for finding instantaneous rates of change, which we learn to call *derivatives*. Limits and continuity concepts, which are crucial to finding derivatives, are then developed. To relieve the theoretical tension, we then introduce some limit applications. Next a programmed learning section attempts to get you intimately immersed in the derivative process.

Calculus is based on one (yes, just one!) basic concept, which we summarize at the end of the chapter. The goal is to permanently engrave that concept in your brain. If you master it, the rest of calculus is "Easy Street."

Most of this chapter is pure mathematics. But it is necessary to build a solid mathematical foundation in order to appreciate the applications that appear in later chapters.

12.2 RATES OF CHANGE AND GRAPHICAL SLOPE

Rates of change and slopes of lines on a graph are vitally tied together. Here, the two important types of rates—average and instantaneous—and their graphical counterparts will be illustrated with Cal Kulus' car trip to visit his girlfriend Sally Slope. For this case the rates happen to be velocities, so these terms are used interchangeably.

Cal leaves his garage, goes on local streets leading to the Quickway, which then leads to the Thruway. He exits the Thruway at Sally's town and travels on local streets until reaching her house. The total elapsed time (x) and distance (y) at various points in the trip were as follows:

Trip stage	Elapsed time (minutes)	x (hours)	Total distance y (miles)
Home streets	0	0	0
Quickway	10	$\frac{1}{6}$	3
Thruway	20	$\frac{2}{6}$	10
Away streets	50	$\frac{5}{6}$	40
	60	1	41

These points, connected by straight lines, each with its slope (s), are presented in Figure 12.3.

On the first portion of his trip, Cal traveled 3 miles in $\frac{1}{6}$ of an hour, which, expressed as a rate, is 3 miles per $\frac{1}{6}$ hour. We normally express such rates, which we call *velocities*, on an hourly basis. On that basis, had Cal continued his initial rate for a full hour (six periods of $\frac{1}{6}$ hour), he would have gone a total of $6 \cdot 3 = 18$ miles. In graphical terms, extending the first straight line out to one hour would result in an ordinate value of 18. So his velocity over the first portion of the trip was 18 miles per hour.

Notice that the slope of the first line segment is also 18.

$$s = \frac{\Delta y}{\Delta x} = \frac{3}{1/6} = 18$$

INTRODUCTION TO DIFFERENTIAL CALCULUS

Figure 12.3

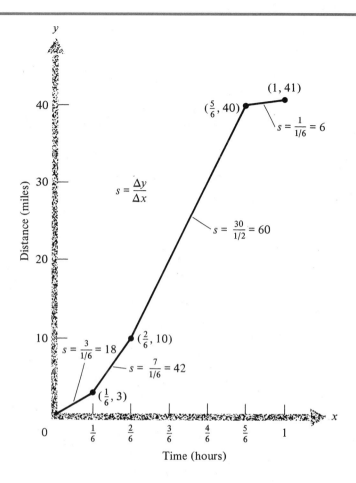

Could this be just a coincidence?

On the Quickway portion of the trip Cal went 7 miles in $\frac{1}{6}$ of an hour. Had he continued that rate for a full hour (six periods of $\frac{1}{6}$ hour), he would have gone $6 \cdot 7 = 42$ miles. So his velocity on the Quickway was 42 miles per hour.

In computing the slope of the Quickway line, notice how the steps are equivalent to finding the rate or velocity above.

$$s = \frac{\Delta y}{\Delta x} = \frac{7}{1/6} = 6 \cdot 7 = 42$$

Indeed, rates of change (velocities here) are equal to slopes on a graph.

Exercise 2

Verify that the Thruway and Away-streets velocities were 60 and 6 miles per hour, respectively.

Average Rates of Change

Drawing straight lines with their constant slopes on the graph implies constant velocity for each stage of the trip. For example the Thruway line with a constant slope of 60 means that for each 1-minute period, 1 mile (rate of 1/(1/60) = 60 miles per hour) was traveled. For this to be true, the speedometer had to be glued to 60, which is unlikely.

More likely each stage of the trip was characterized by different velocities fluctuating around some average level. In this light the constant slope of a line can be thought to reflect the average rate of change (velocity) of that stage of the trip. To get a better understanding of this, consider the following detailed breakdown of the first 10 minutes (home streets) of the trip.

Elapsed time		Total distance	Average velocity
(minutes)	x (hours)	y (miles)	$\bar{v} = \Delta y/\Delta x$ (mph)
0	0	0	
			$\dfrac{1/4}{1/60} = 15$
1	$\tfrac{1}{60}$	$\tfrac{1}{4}$	
			$\dfrac{1/4}{3/60} = 5$
4	$\tfrac{4}{60}$	$\tfrac{1}{2}$	
			$\dfrac{3/2}{3/60} = 30$
7	$\tfrac{7}{60}$	2	
			$\dfrac{0}{1/60} = 0$
8	$\tfrac{8}{60}$	2	
			$\dfrac{1}{2/60} = 30$
10	$\tfrac{10}{60}$	3	

Figure 12.4

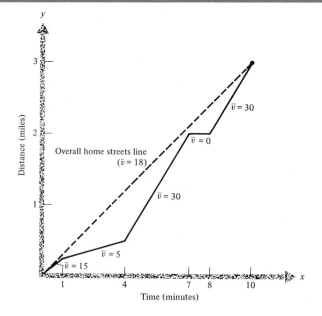

INTRODUCTION TO DIFFERENTIAL CALCULUS

The dotted line in Figure 12.4 is a blowup of the home streets's line of the previous graph. Its slope of 18 represents the average velocity of that stage of the trip. For smaller segments of that 10-minute period, we now see that average velocities ranging from zero (red light?) to 30 miles per hour were experienced. Note: we denote average velocity by the symbol $\bar{v}$.

By drawing straight lines for the short intervals of time, we only convey the average velocity for each. So it is a good bet that Joe went over 30 miles per hour during the two periods in which he averaged 30 miles per hour.

Exercise 3

If Cal's total distance traveled after the first, third, fifth, eight, and tenth minutes of the Quickway stage were 3.5, 5, 7.2, 8.2, and 10 miles, respectively, graph the points and find the various average velocities.

Given more detailed information about the trip, we could form even tinier line segments on a graph. For example, consider the following 10-second breakdown of the first minute of the trip, which is portrayed in the table below and graphed in Figure 12.5.

Elapsed time		Total distance	Average velocity
(seconds)	x (hours)	y (miles)	$\bar{v} = \Delta y/\Delta x$ (mph)
0	0	.00	
			$\dfrac{.01}{1/360} = 3.6$
10	$\frac{1}{360}$	.01	
			$\dfrac{.02}{1/360} = 7.2$
20	$\frac{2}{360}$	.03	
			$\dfrac{.03}{1/360} = 10.8$
30	$\frac{3}{360}$	.06	
			$\dfrac{.05}{1/360} = 18.0$
40	$\frac{4}{360}$	.11	
			$\dfrac{.06}{1/360} = 21.6$
50	$\frac{5}{360}$	.17	
			$\dfrac{.08}{1/360} = 28.8$
60	$\frac{6}{360}$	.25	

Again, the best we can do is to connect the points with straight lines, which are indicative of constant velocities. Yet, we know this first minute to be a period of acceleration, or continuously increasing velocity.

If we want the velocities to be more accurate, we must imagine getting detailed distance information second by second, or by even finer time segments. However, at some point, our large

Figure 12.5

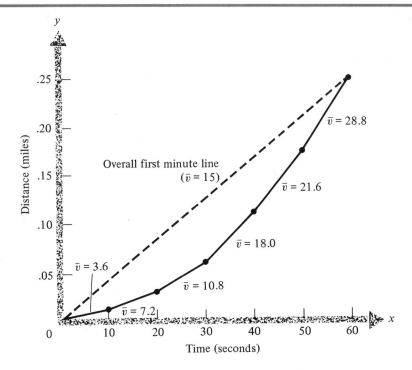

number of tiny straight lines would be indistinguishable from a curved line. But then what is the slope of a curved line?

Instantaneous Rates of Change

Since the slope is continuously changing along a curved line, we can speak only about its value at a given point, or instant. Consequently, it is referred to as an *instantaneous rate of change*. Finding instantaneous rates will be the subject of much of this chapter. For now, we merely want to conceptualize it.

Consider the set of points on the curved line, which represents the first minute of Cal's trip. Think of magnifying them with a super microscope to such a degree that we can see spaces between the points, as shown in Figure 12.6.

Focus on point A. The straight line that goes through that point, but just misses the two adjacent points, is called the *tangent line* (note that each point along the curve has its own unique tangent line).

At point A the tangent line and the curved line are changing at the same rate, so we define the tangent line slope to be the instantaneous rate of change (slope) of the curved line. Now this leaves us with the question of how to find the slope of curved functions.

Figure 12.6

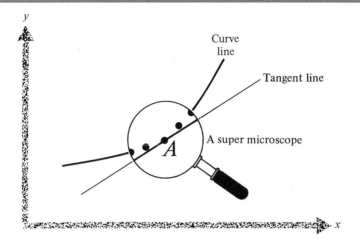

12.3 SLOPE OF CURVED FUNCTIONS

To gain insight into the above question, let's go to the rocket pad and watch Calculus I being shot into orbit. The rocket's distance (*y*) above ground in miles is the following parabolic function of time (*x*) in minutes since blast off.

$y = f(x) = x^2$

Rockets don't accelerate, decelerate, and then stop in traffic. Their velocity starts at zero and increases continuously until earth orbit is reached. Thus, velocity is different at each split second. Let's focus on the time of one minute after blast off and see if we can determine the velocity of a rocket. This translates into the problem of finding the slope of the tangent line to the distance function at the point (1, 1). (See Figure 12.7.)

The tangent line slope represents the instantaneous rate of change of the function. We will find it by employing a process of determining average rates of change of the function over shorter and shorter periods of time.

As we did for Cal, we find average velocity by connecting two points on the distance function with a straight line and determining the slope of that line. Hereafter, we call such lines *secant lines*. In each case we form the secant line by connecting the fixed point of $x = 1$, $y = 1$ with a second point further along the function. For example, if we select the second point at $x = 2$ where $y = 2^2 = 4$, the secant line has a slope of 3, which represents the average velocity of the rocket during that one-minute period.

Obviously, the instantaneous velocity at $x = 1$ was less than 3 miles per minute, as evidenced by the lower slope of the tangent line. By shortening the time interval we can get a better approximation of the instantaneous velocity. For example, if we take our second point at $x = 1.5$, where $y = 1.5^2 = 2.25$, and $\Delta y = 1.25$ and $\Delta x = .5$, the secant line slope is $\Delta y/\Delta x = 2.5$. So, the average velocity over this half minute of flight is 2.5 miles per minute. It seems as though we are heading in the right direction.

The process of taking a second point closer and closer to the fixed point is carried further and summarized in the table on page 434 and Figure 12.8.

Figure 12.7

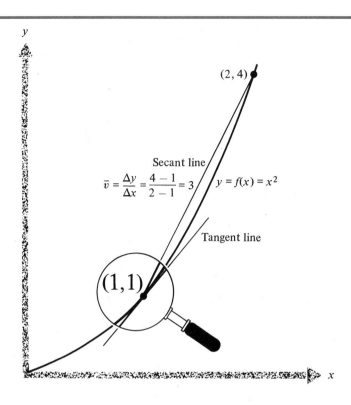

Line	Fixed point	Moving second point	Δy	Δx	Secant slope $\bar{v} = \Delta y/\Delta x$
A	(1, 1)	(2, 4)	3	1	3
B	(1, 1)	(1.5, 2.25)	1.25	.5	2.5
C	(1, 1)	(1.1, 1.21)	.21	.1	2.1
D	(1, 1)	(1.01, 1.0201)	.0201	.01	2.01
			↓	↓	↓
			Both Δy and Δx approaching zero		Seems to be approaching 2; or is it 2.001 or 1.99?

Exercise 4

Find the slope of the secant line (call it E) to the second point at $x = 1.001$.

Both Δy and Δx are approaching zero. In fact we can make them as close to zero as we wish by the choice of the second point. However, the ratio, $\Delta y/\Delta x$, which is always the slope of a secant line and represents the average velocity here, is not approaching zero. Graphically, we see this ratio is approaching the tangent line slope; yet, its exact value still eludes us.

INTRODUCTION TO DIFFERENTIAL CALCULUS

Figure 12.8

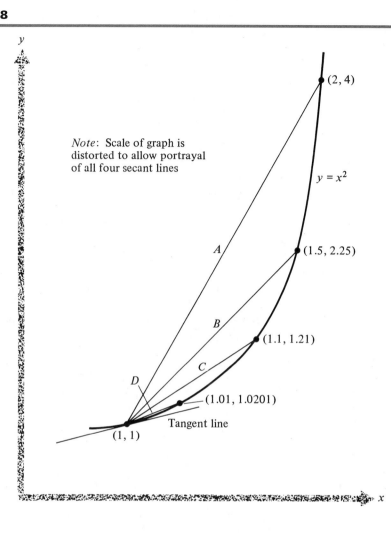

Exercise 5

Consider the rocket two minutes after liftoff. Fill in the following table to find average velocities over the time interval from $x = 2$ to $x = 2 + \Delta x$.

Fixed point	Moving second point	Δy	Δx	$\bar{v} = \Delta y/\Delta x$
(2, 4)	(3, 9)	5	1	5
(2, 4)	(2.5,)			
(2, 4)	(2.1,)			
(2, 4)	(2.01,)			
		Where are these headed?		Where is this headed?

Before proceeding, let's review our effort to date and see where we are heading. We went through considerable effort to determine the slopes of successive secant lines leading out of the point (1, 1). After all that work, we still only had an approximation to the tangent line slope. And our approximation was only good for that single point on the function. Then, you repeated that time-consuming process in an exercise to obtain an approximation for the tangent slope at the single point (2, 4). Now that we have this process down pat, we could repeat it over and over and over again for all the other points on the function. But that would be a long job since there are an infinite number of points on the function (let's amend that to an "infinitely long" job). I can just hear you bright people out there saying, "There must be a better way!" Of course, you are right again. We could have even started with this better (generalized) approach. But without the sweat and toil of our tedious approach, you never would have appreciated this better way or the subtleties of calculus. Like bad tasting medicine, everything we have done so far was good for your mathematical health.

We first apply the general approach to find the exact tangent line slope at the point (1, 1). Afterward, we use the general approach to find the exact tangent line slope at any point along the rocket function.

See Figure 12.9. Consider secant lines that emanate from the point (1, 1) to a second point that could be any other point on the function. If the x-axis position of the second point is Δx units greater than 1, then its x-coordinate is $1 + \Delta x$. Since ordinates for this function are always $y = x^2$, then the y-coordinate of the second point is $(1 + \Delta x)^2$. Connecting the two points as shown in Figure 12.9 gives a secant line with slope as follows:

$$\frac{\Delta y}{\Delta x} = \frac{(1 + \Delta x)^2 - 1}{\Delta x}$$

$$= \frac{1 + 2\Delta x + (\Delta x)^2 - 1}{\Delta x}$$

$$= \frac{\Delta x(2 + \Delta x)}{\Delta x}$$

$$= 2 + \Delta x$$

If we had this formula earlier, we could have saved a lot of time in computing secant line slopes. But don't fret, because you can't learn the process without going through it the hard way. For example, this formula tells us directly that line B with $\Delta x = .5$ has a slope of $2 + \Delta x = 2 + .5 = 2.5$.

Exercises 6

1. Use the above formula to quickly find the slopes for secant lines A, C, and D.
2. Consider the fixed point at $x = 2$. Show (using the same methods as for $x = 1$ above) that the general expression for the slope of secant lines emanating from this fixed point is

    $$\frac{\Delta y}{\Delta x} = 4 + \Delta x$$

3. Using the formula derived in exercise 2, determine the average velocity of the rocket during the five minutes of flight from $x = 2$ to $x = 7$.

INTRODUCTION TO DIFFERENTIAL CALCULUS

Figure 12.9

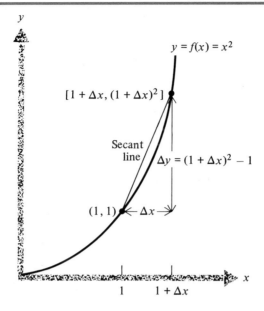

Next, we will introduce more generality so that we can find the tangent line slope at any point along the distance function.

Consider any time, x, and any subsequent time, Δx units later. These two times define the two coordinates (x, x^2) and $(x + \Delta x, (x + \Delta x)^2)$ on the distance function, as shown in Figure 12.10.

Figure 12.10

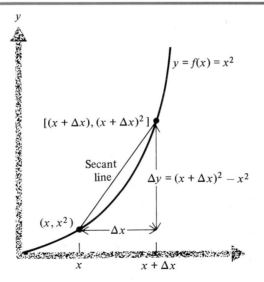

The secant line connecting these two points has slope

$$\frac{\Delta y}{\Delta x} = \frac{(x + \Delta x)^2 - x^2}{\Delta x}$$

$$= \frac{x^2 + 2x(\Delta x) + (\Delta x)^2 - x^2}{\Delta x}$$

$$= \frac{\Delta x(2x + \Delta x)}{\Delta x}$$

$$= 2x + \Delta x$$

This is the general formula for finding the slope of a secant line between any two points. For example, it would tell us that line A, studied earlier, between the points (1, 1) and (2, 4), with $x = 1$ and $\Delta x = 1$, has a slope of $2x + \Delta x = 2 \cdot 1 + 1 = 3$, which confirms our earlier finding. Since we know these slopes to be average velocities, we now can find average velocity over any period of time. For example, the average velocity of the rocket for the 5 minutes following $x = 3$ (thus $\Delta x = 5$) is

$$\bar{v} = \frac{\Delta y}{\Delta x} = 2x + \Delta x$$

$$= 2 \cdot 3 + 5 = 11 \text{ miles per minute}$$

Exercises 7

1. Determine the slope of the secant line between the points (2, 4) and (4, 16).
2. What is the average velocity over the first 7 minutes of the flight?
3. Explain why the average velocity from blastoff until some point in the trip equals the time since blastoff.
4. Beginning with the time $x = 4$, how many minutes must pass before the average velocity of the period equals 14 miles per minute?

Now, we are finally ready to reach our goal of finding tangent slope for any value of x. To do this, we merely observe what happens to the secant slope formula

Secant slope $= 2x + \Delta x$

as Δx approaches 0. x, the coordinate of the fixed point, does not change; so, of course $2x$ does not change. The only part of the secant slope formula that changes is Δx, which approaches zero. So,

Tangent slope $= 2x$

Despite the lack of fanfare, this was a momentous occasion. We just derived our first derivative function. Whoopie!

The function $2x$ is called *the derivative of the function* x^2. For the rocket problem, the derivative quickly gives us the instantaneous velocity at any time by merely multiplying the time by 2. For example, at $x = 1$, the instantaneous velocity is $2 \cdot 1 = 2$, which confirms our earlier, much more tedious finding. Other velocities are determined just as quickly. So, at 3, 7, and 8.5

INTRODUCTION TO DIFFERENTIAL CALCULUS

minutes into the flight, the instantaneous velocities are $2 \cdot 3 = 6$, $2 \cdot 7 = 14$, and $2 \cdot 8.5 = 17$ miles per minute, respectively.

Exercises 8

1. Determine the rocket's velocity at exactly $\frac{1}{2}$, 5, and 10 minutes into the flight.
2. At what time after blastoff will the instantaneous velocity be 12 miles per minute?
3. Graph the functions x^2 and $2x$. Can you see that the slope of x^2 is given by the ordinate value of $2x$?

In our effort to get across the basic unadulterated idea of a derivative, we have glossed over some important technicalities involving limits and continuity. This must be corrected. So without further ado, let's limit this introduction and continue.

12.4 LIMITS

In this section we focus on three topics. First, we show why the somewhat involved limit process of letting Δx approach zero must not be subverted to a quick and simple setting $\Delta x = 0$. Second, we develop the notation for limits and list the four laws of limits that will be needed in our next chapter. Last, we introduce two important limit applications. The first application leads us to discover a very important number in mathematics, e. The second application shows us how limits are used in everyday banking.

Quick Fix: Set $\Delta x = 0$?

Secant line slopes were always functions Δx. We would then observe what happens to these equations as Δx got closer and closer to 0. I bet you wondered about this "closer and closer" business. That sounds inefficient. Why not introduce some modern business efficiency by directly and immediately setting $\Delta x = 0$? The answer to that question will be obvious once we consider the rocket curve slope at $(1, 1)$. Substituting $\Delta x = 0$ in the initial secant line slope formula leaves us with the following undefined result

$$\frac{\Delta y}{\Delta x} = \frac{(1 + \Delta x)^2 - 1}{\Delta x} = \frac{(1 + 0)^2 - 1}{0} = \frac{0}{0} \text{ (undefined)}$$

This results from the fact that when $\Delta x = 0$, the fixed and second points are one and the same. Since an infinite number of straight lines, each with a different slope, can go through one point, a unique slope cannot be defined.

Limit Notation

We now see that the process of determining tangent line slope involves letting Δx get closer and closer (but never equal, heaven forbid) to zero. We symbolize this as

$$\lim_{\Delta x \to 0} \left[\frac{\Delta y}{\Delta x} \right]$$

where $\Delta y/\Delta x$ is called the *difference quotient*. We read this as "taking the limit of the difference quotient as Δx approaches zero."

In the particular cases we have been considering, the limit expressions would be:

$$\lim_{\Delta x \to 0} \left[\frac{\Delta y}{\Delta x} \right] = \lim_{\Delta x \to 0} [2 + \Delta x]$$

Operationally, the limit process requires us to look at the expression within the brackets and see what number it is approaching as Δx approaches zero (although never actually reaching, since we already know setting $\Delta x = 0$ is a no-no!).

This process is analogous to walking halfway to a wall, then walking halfway again, over and over again. You'll never reach the wall, but you can get as close as you wish (for example, it wouldn't be difficult to determine how many halfway trips would be required to get within 10^{-100} inches from the wall).

Limit Laws

Although, as above, we can often determine a limit by observing what happens to the bracketed term as Δx approaches zero, there are more complicated limit expressions that require the use of the laws for limits. Certainly as we will see later in this chapter, these four laws, stated here without proof, will come in mighty handy.

Type	Law
1. Constant	$\lim_{\Delta x \to 0} [c \cdot f(\Delta x)] = c \lim_{\Delta x \to 0} [f(\Delta x)]$
2. Sum	$\lim_{\Delta x \to 0} [f(\Delta x) + g(\Delta x)] = \lim_{\Delta x \to 0} [f(\Delta x)] + \lim_{\Delta x \to 0} [g(\Delta x)]$
3. Product	$\lim_{\Delta x \to 0} [f(\Delta x) \cdot g(\Delta x)] = \lim_{\Delta x \to 0} [f(\Delta x)] \cdot \lim_{\Delta x \to 0} [g(\Delta x)]$
4. Quotient	$\lim_{\Delta x \to 0} \left[\frac{f(\Delta x)}{g(\Delta x)} \right] = \frac{\lim_{\Delta x \to 0} [f(\Delta x)]}{\lim_{\Delta x \to 0} [g(\Delta x)]}$

$$\text{(if } g(\Delta x) \neq 0 \quad \text{and} \quad \lim_{\Delta x \to 0} [g(\Delta x)] \neq 0\text{)}$$

Let's apply these laws to the Calculus 1 difference quotient.

$$\lim_{\Delta x \to 0} \left[\frac{\Delta y}{\Delta x} \right] = \lim_{\Delta x \to 0} [2x + (\Delta x)]$$

Sum ↓ Law

$$= \lim_{\Delta x \to 0} [2x] + \lim_{\Delta x \to 0} [\Delta x]$$

Constant ↓ Law

$$= 2 \lim_{\Delta x \to 0} [x] + \lim_{\Delta x \to 0} [\Delta x]$$

$$= 2 \cdot x \quad\quad + 0$$

Note: x is not affected by $\Delta x \to 0$

$$= 2x$$

INTRODUCTION TO DIFFERENTIAL CALCULUS

It is true that we were able to find this result earlier without all this "Limit Law" business. However, we need to get familiar with the laws when they are really needed later for complicated cases.

Limit Applications

Now, let's go off on a tangent and look at two applications of the limit process. In the meanwhile, we will discover one of the most important numbers in mathematics, *e,* and we will see how limits play a role in everyday banking.

The Important Constant, Called "e"

The constant, called "*e*", which is an irrational number (2.718 . . .) is one of, if not "the" most important number in all of mathematics. It is possible you never even heard of it. Well, now is the time to correct for that "black hole" in your education.

e is defined from the following limit:

$$e = \lim_{\Delta x \to 0} [1 + \Delta x]^{\frac{1}{\Delta x}}$$

For one last time, let's subvert the limit process by setting $\Delta x = 0$.

$$\lim_{\Delta x \to 0} [(1 + 0)^{1/0}] = \lim_{\Delta x \to 0} [1^\infty] = 1$$

This is NOT the limit, as we will see next by properly carrying out the limit process. So, that was the limit to our subverting the limit process. Never again!!

We approach the limit process correctly now by taking smaller and smaller values of Δx, as shown in the following table.

Δx	$\dfrac{1}{\Delta x}$	$(1 + \Delta x)$	$(1 + \Delta x)^{\frac{1}{\Delta x}}$
1	1	2	$2^1 = 2$
$\frac{1}{2}$	2	1.5	$(1.5)^2 = 2.25$
$\frac{1}{4}$	4	1.25	$(1.25)^4 = 2.4414$
.1	10	1.10	$(1.10)^{10} = 2.5937$
.01	100	1.01	$(1.01)^{100} = 2.7048$
.001	1000	1.001	$(1.001)^{1000} = 2.7169$
↓	↓	↓	↓
0	∞	1	?

It does appear that $(1 + \Delta x)^{1/\Delta x}$ is approaching some limit as Δx approaches zero. In fact, this limit is the irrational number *e*, which approximated to three decimal places is 2.718. (You met *e* earlier in the exponential and normal probability distribution applications in Section 6.7. Later, you will meet *e* in other applications and as the base of the natural logarithms.)

Exercises 9

1. Find the value of $(1 + \Delta x)^{1/\Delta x}$ for $\Delta x = .0001$ and $.00001$.
2. If you begin 1 foot from a wall and move halfway to the wall on each step, how many steps will it take you to reach a point $(\frac{1}{2})^{100}$ feet from the wall? (*Hint:* geometric progression)

The following will be a welcome "application-oasis" for you practical people. Since this chapter is 95% basic concepts and theory of calculus, I am sure you welcome an application.

Banking (Continuous Compounding) Application

A "cousin" to the above limit for e forms the basis for compound interest calculations whenever banks give continuous compounding, which means that your account is credited with interest an infinite number of times per year. On the surface, this sounds too good to be true. Could this be the way to parlay some small change into multi-millions? Let's see by considering the case of Bank Ltd., which pays an annual 6% interest rate, compounded continuously.

Compound interest in general is based on the formula (Review Section 4.2 if necessary.)

$$\left(1 + \frac{i}{c}\right)^c$$

which gives the year-end balance of $1 invested at an annual rate of i (denoted as a decimal equivalent of the annual percentage rate) when interest is credited c times per year. For the case of Bank Ltd., $i = .06$, so its specific compounding formula is

$$\left(1 + \frac{.06}{c}\right)^c$$

Compounding annually ($c = 1$), semi-annually ($c = 2$), and quarterly ($c = 4$) give

Annually: $\left(1 + \frac{.06}{1}\right)^1 = 1.06$

Semi-annually: $\left(1 + \frac{.06}{2}\right)^2 = 1.0609$

Quarterly: $\left(1 + \frac{.04}{4}\right)^4 = 1.06136$

Exercise 10

Show that monthly compound results in 1.06168

The following table incorporates the above values and values for greater numbers of compounds.

INTRODUCTION TO DIFFERENTIAL CALCULUS

Period	Compounds/Year	Ending Balance
Annually	1	$\left(1 + \frac{.06}{1}\right)^1 = 1.06$
Semi-annually	2	$\left(1 + \frac{.06}{2}\right)^2 = 1.0609$
Quarterly	4	$\left(1 + \frac{.06}{4}\right)^4 = 1.06136$
Monthly	12	$\left(1 + \frac{.06}{12}\right)^{12} = 1.06168$
Daily	365	$\left(1 + \frac{.06}{365}\right)^{365} = 1.06183$
Hourly	$365 \cdot 24 = 8{,}760$	$\left(1 + \frac{.06}{8760}\right)^{8760} = 1.061836$

As you can see, more and more, even substantially more, compounds per year do not give you much more interest. In fact, it does appear that the ending balances appear to be approaching a limit.

Now, let's make an interesting calculation

$$e^{.06} = 2.718281828^{.06} = 1.0618365$$

Could it be? Yes indeed, this is the limit that the ending balances is approaching. Mathematically, we state this fact as:

$$\lim_{c \to \infty} \left(1 + \frac{.06}{c}\right)^c = e^{.06}$$

Or in general for any annual interest rate, i

$$\lim_{c \to \infty} \left(1 + \frac{i}{c}\right)^c = e^i$$

Keep in mind that we didn't prove this. But by looking at the table of ending balances, you probably can accept it as true. Now let's use this fact in applications.

Example

If Bank Ltd. paid an annual 7% interest rate compounded continuously, the year-ending balance of an initial $1 investment would be:

$$\lim_{c \to \infty} \left(1 + \frac{.07}{c}\right)^c = e^{.07} = 2.718281828^{.07} = 1.072508$$

Thus, the effective interest rate would be 7.2508%

Exercises 11

1. Two banks are competing for business. Bank A offers 8.25% compounded quarterly. Bank B offers 8% compounded continuously. Which bank actually gives you more interest over a year? What are the effective interest rates for each bank?
2. Bank C offers 9% compounded just once per year. Bank D wants to just beat out Bank C. If Bank D offers continuous compounding, what should their annual rate be (assume that interest rates are offered in increments of .25 (thus, 8, 8.25, 8.5, etc.)). Take note that a trial and error solution is needed here.

12.5 CONTINUITY

The limit process to find derivatives requires that the function be continuous. Our rocket function, $y = f(x) = x^2$, was continuous; so we had no problem with it. Now we want to see what it means to have a continuous function, and why such a condition is necessary (although not sufficient) for finding derivatives.

In a nutshell, continuous functions are those that can be graphed without lifting one's pencil from the paper. Examples of continuous and discontinuous functions are shown in Figure 12.11

Figure 12.11

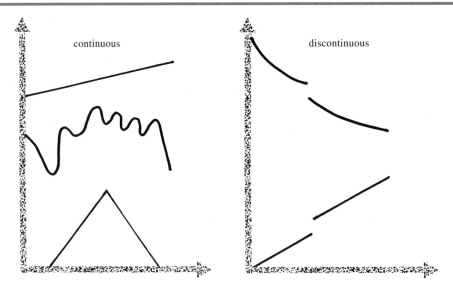

INTRODUCTION TO DIFFERENTIAL CALCULUS

Continuous functions can be quite orderly as is the linear case, have abrupt turns as is the "pyramid" case, or very squiggly as is the third case. Discontinuous functions can actually be continuous over wide portions of the domain. However, they would have at least one place where a "gap" exists.

Finding derivatives, which are based on limits, require that the limits be unique. As can be seen from the discontinuous "postage cost" step function (See page 160), the limit at $x = 1$ is not unique. The same can be said for the hyperbola, $y = \frac{1}{x}$, at $x = 0$.

The "pyramid" function is continuous. However, it does not have a unique slope at its peak. So it too doesn't have a derivative at that point.

In general, a derivative at a point, can only be found if the function is continuous (can be drawn without lifting one's pencil from the page). But, as the pyramid function shows, continuity is a necessary, but not sufficient, condition for a derivative to exist.

Exercises 12

1. Graph $|x|$ and explain why it doesn't have a derivative for $x = 0$.
2. At what x-coordinate would the function, $f(x) = \dfrac{x^2 + 4}{x - 2}$, be discontinuous?

12.6 PROGRAMMED LEARNING OF CALCULUS BASICS

A thorough understanding of the limit process to obtain instantaneous rates of change is essential for success in calculus. So, before we proceed to finalize the definition of a derivative and apply it to various functions, you should take one more crack at achieving this understanding by immersing yourself in the details of the process. The following programmed section, featuring the function $y = f(x) = 1/x$, is designed for that end. We will limit our function to the domain, $x > 0$, so as to avoid the point of discontinuity at $x = 0$.

Questions	Answers
1. Function of study: $y = f(x) = 1/x$ What type of function is this?	(1) hyperbola
2. Functional notation We need to review functional notation in order to determine ordinates corresponding to various x coordinates. $f(1) = 1/1 = 1$ $f(3) = $ _____ $f(1 + \Delta x) = 1/(1 + \Delta x)$ $f(3 + \Delta x) = $ _____	(2) $f(3) = 1/3$ $f(3 + \Delta x) = 1/(3 + \Delta x)$

3. Relation of secant and tangent lines

   Using the point at $x = 1$ as our fixed point, draw the secant and tangent lines on the following graph.

   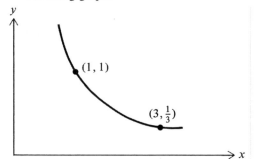

(3)

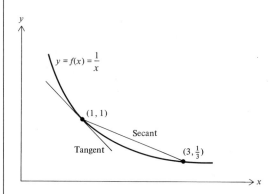

4. Slope of secant line

   The slope of the secant line in panel 3 is
   $$\frac{\Delta y}{\Delta x} =$$

(4) $\dfrac{\Delta y}{\Delta x} = \dfrac{\frac{1}{3} - 1}{3 - 1} = -\dfrac{1}{3}$

5. Series of secant lines

   Form secant lines to the second points with $x = 2$, 1.5, and 1.25 respectively. Call these lines B, C, and D, respectively. Call the line in panel 3 A.

(5)

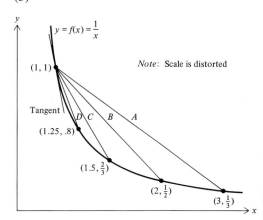

Note: Scale is distorted

6. Table of secant slopes

   Fill in the missing entries.

(6) B: 1 $-\frac{1}{2}$ $-\frac{1}{2}$
    D: .25 $-.2$ $-.8$

Line	Fixed point	Second point	Δx	Δy	$\Delta y/\Delta x$
A	(1, 1)	$(3, \frac{1}{3})$	2	$-\frac{2}{3}$	$-\frac{1}{3}$
B	(1, 1)	$(2, \frac{1}{2})$	___	___	___
C	(1, 1)	$(1.5, \frac{2}{3})$	$\frac{1}{2}$	$-\frac{1}{3}$	$-\frac{2}{3}$
D	(1, 1)	(1.25, .8)	___	___	___

INTRODUCTION TO DIFFERENTIAL CALCULUS

7. Limits of terms
 	
As the second point moves closer and closer to the fixed point in panel 6,
  a. Δx approaches
  b. Δy approaches
  c. Does $\Delta y/\Delta x$ seem to be approaching the same value as Δx and Δy?
	(7)
	  a. zero
	  b. zero
	  c. No

8. Generalized secant slope at $x = 1$
 	
Previously we have found secant line slope at $x = 1$ in a rather inefficient manner by treating each case individually. Here we generalize the process. With the second point Δx units on the x-axis to the right of the fixed point, we can write the expression
 	
$$\frac{\Delta y}{\Delta x} = \frac{f(1 + \Delta x) - f(1)}{\Delta x}$$
 	
  a. What do we call this expression?
  b. What does it represent
	(8)
	  a. Difference quotient
	  b. Slope of secant line

9. Algebraic simplification of difference quotient at $x = 1$ Substituting the functional values in the difference quotient in panel 8 gives $$\frac{\Delta y}{\Delta x} = \frac{\left(\frac{1}{1 + \Delta x}\right) - 1}{\Delta x}$$ Perform the algebraic steps to simplify the difference quotient and arrive at the expression $$\frac{\Delta y}{\Delta x} = \frac{-1}{1 + \Delta x}$$	(9) $$\frac{1 - 1(1 + \Delta x)}{(1 + \Delta x)}$$ $$\overline{\Delta x}$$ $$\frac{\frac{-\Delta x}{1 + \Delta x}}{\Delta x}$$ Then cancel the Δx's.
10. Quick determination of secant slopes at $x = 1$ We can use the panel 9 formula to quickly find the secant slope from $(1, 1)$ to any other point. For example, it verifies the	(10) a. $\dfrac{-1}{1 + 1} = -\dfrac{1}{2}$

earlier slope of $-\frac{1}{3}$ to the point $(3, \frac{1}{3})$. $\frac{\Delta y}{\Delta x} = \frac{-1}{1+2} = \frac{-1}{3} = -\frac{1}{3}$ Use the formula to find the secant slope to the points a. $(2, \frac{1}{2})$ b. point with $x = 1.1$	b. $\frac{-1}{1 + .1} = -\frac{1}{1.1}$
11. Notation for limit of difference quotient at $x = 1$ Express in standard notation the process of letting Δx get closer and closer to zero for the difference quotient at $x = 1$.	(11) $\displaystyle\lim_{\Delta x \to 0} \left[\frac{\Delta y}{\Delta x} \right] = \lim_{\Delta x \to 0} \left[\frac{-1}{1 + \Delta x} \right]$
12. Taking limit at $x = 1$ The value that the bracketed expression (answer to panel 11) approaches as Δx approaches zero is called *the limit of the difference quotient at $x = 1$*. What is the value of this limit?	(12) $\dfrac{\displaystyle\lim_{\Delta x \to 0} [-1]}{\displaystyle\lim_{\Delta x \to 0} [1 + \Delta x]} = \dfrac{-1}{1} = -1$

13. Meaning of panel 12 limit
The -1 result in panel 12 represents at $x = 1$ the
a. Slope of the _____ line
         (secant) or (tangent)
b. _____ rate of change
  (instantaneous) or (average)
In the remaining panels we generalize to find the instantaneous rate of change (tangent line slope) at any point on the function $y = f(x) = 1/x$.
	(13)
	a. Tangent
	b. Instantaneous

14. Specifying general points on $y = f(x) = 1/x$. Consider any fixed point on the function $y = f(x) = 1/x$. Its coordinate is a. $(x, \_\_)$. Consider a second (moveable) point on the function with an x coordinate that is Δx units greater than the fixed point. The coordinate of the second point is b. $(\_\_, \_\_)$. c. Explain why the y coordinate of the second point is not $(1/x) + \Delta x$.	(14) a. $(x, 1/x)$ b. $\left(x + \Delta x, \dfrac{1}{x + \Delta x} \right)$ c. Functional notation, $f(x) = 1/x$, requires substituting the value of x ($x + \Delta x$ here) in the denominator.

15. Graph of general conditions

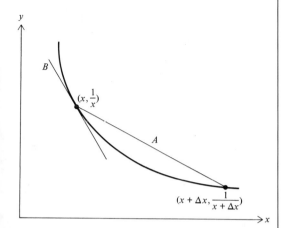

a. The tangent line is _____.
b. The secant line is _____.
c. The slope of A is the _____ rate of change.
d. The slope of B is the _____ rate of change.

(15)
a. B
b. A
c. Average
d. Instantaneous

16. General difference quotient

From panels 14 and 15, fill in the missing information to find the general difference quotient (slope of a secant line).

$$\frac{\Delta y}{\Delta x} = \frac{\begin{pmatrix} y \text{ coordinate of} \\ \text{second point} \end{pmatrix} - \begin{pmatrix} y \text{ coordinate} \\ \text{of fixed point} \end{pmatrix}}{\Delta x}$$

$$= \frac{(\quad) - (\quad)}{\Delta x}$$

(16)

$$\frac{\Delta y}{\Delta x} = \frac{\left(\frac{1}{x+\Delta x}\right) - \left(\frac{1}{x}\right)}{\Delta x}$$

17. Checking for particular cases

The general formula developed in panel 16 gives the secant line slope (average rate of change) for any two points on the function $y = f(x) = 1/x$. Check this for lines A and C in panel 6.

(17)

Line A
($x=1$)
($\Delta x=2$)

$$\frac{\Delta y}{\Delta x} = \frac{\left(\frac{1}{1+2}\right) - \left(\frac{1}{1}\right)}{2}$$

$$= -\frac{1}{3}$$

Line C $(x=1)$ $(\Delta x = \frac{1}{2})$	$\dfrac{\Delta y}{\Delta x} = \dfrac{\left(\dfrac{1}{1+\frac{1}{2}}\right) - \left(\dfrac{1}{1}\right)}{\frac{1}{2}}$ $= -\dfrac{2}{3}$

18. Limit notation

    Express in standard notation the process of letting Δx get closer and closer to zero for the general difference quotient found in panel 16.

(18)

$$\lim_{\Delta x \to 0} \left[\dfrac{\Delta y}{\Delta x}\right] = \lim_{\Delta x \to 0} \left[\dfrac{\left(\dfrac{1}{x+\Delta x}\right) - \left(\dfrac{1}{x}\right)}{\Delta x}\right]$$

19. Subverting limit process

    Subverting the limit process by letting $\Delta x = 0$ at the onset is usually doomed to failure. See why by substituting $\Delta x = 0$ into the expression found in panel 18.

(19)

$$\lim_{\Delta x \to 0} \left[\dfrac{\left(\dfrac{1}{1+0}\right) - \left(\dfrac{1}{1}\right)}{0}\right]$$

$$\lim_{\Delta x \to 0} \left[\dfrac{0}{0}\right] \quad \text{(Which is undefined)}$$

20. Algebraic simplification of difference quotient

    Generally, before limits are taken, we employ algebra to simplify the difference quotient. The result is a difference quotient much more amenable to the limit process.
    Start with the difference quotient in answer panel 18, and see if you can perform the algebraic simplifications to arrive at

    $$\dfrac{\Delta y}{\Delta x} = \dfrac{-1}{x(x+\Delta x)}$$

(20)

$$\dfrac{\Delta y}{\Delta x} = \dfrac{\left(\dfrac{1}{x+\Delta x}\right) - \left(\dfrac{1}{x}\right)}{\Delta x}$$

$$= \dfrac{\dfrac{x - 1(x+\Delta x)}{x(x+\Delta x)}}{\Delta x}$$

$$= \dfrac{\dfrac{-\Delta x}{x(x+\Delta x)}}{\Delta x}$$

$$= \dfrac{-1}{x(x+\Delta x)}$$

21. Meaning of simplified difference quotient

    Consider the difference quotient in the simplified form of panel 20.
    Is it equivalent (albeit in different form) to the difference quotient found in panel 16? ___a.___

(21)
a. Yes
b. Secant
c. Average
d. With $\Delta x = 1$ and $\Delta x = 2$

It represents the slope of the __b.__ line. It gives the __c.__ rate of change. d. Show how it can be used to find the slope of Line A in panel 6.	$\dfrac{\Delta y}{\Delta x} = \dfrac{-1}{1(1+2)}$ $= \dfrac{-1}{3}$
22. Limit of difference quotient Finally, we are in a position to take the limit (thank goodness, because you were probably at the limit of your patience). As we take the limit, or observe what happens to the difference quoatient as Δx gets closer and closer to zero, a. Does -1 change? b. Does x change? c. So, then, what is the limit, or $\lim\limits_{\Delta x \to 0}\left[\dfrac{\Delta y}{\Delta x}\right] = \lim\limits_{\Delta x \to 0}\left[\dfrac{-1}{x(x+\Delta x)}\right]$	(22) a. No b. No c. $\dfrac{\lim\limits_{\Delta x \to 0}[-1]}{(\lim\limits_{\Delta x \to 0}[x])(\lim\limits_{\Delta x \to 0}[x+\Delta x])}$ $= \dfrac{-1}{x \cdot x}$ $= \dfrac{-1}{x^2}$
23. Meaning of limit The limit just found a. represents an _____ rate of change. b. is the slope of the _____ line. c. holds for (*all*) or (*positive*) values of x along the function $y = f(x) = 1/x$.	(23) a. Instantaneous b. Tangent c. All (except $x = 0$, where the function is not defined)
24. Finding instantaneous rates on function $y = f(x) = 1/x$ By substituting a particular value of x in the formula, $-1/x^2$, we can find the instantaneous rate of change of the function, $1/x$, at that point. Try it for the following cases. a. $x = 1$ b. $x = 3$ c. $x = 2.5$ d. $x = -4$ e. $x = 10$	(24) a. $-1/1^2 = -1$ b. $-1/3^2 = -1/9$ c. $-1/2.5^2 = -1/6.25$ d. $-1/(-4)^2 = -1/16$ e. $-1/10^2 = -1/100$

12.7 THE DERIVATIVE

Now we can generalize the process of finding instantaneous rate of change, or slope of a tangent line. Consider any function, symbolized by $y = f(x)$, any fixed point on the function, $[x, f(x)]$, and any second point Δx units to the right of the fixed point, or $[(x + \Delta x, f(x + \Delta x))]$. These two points determine a secant line, as shown in Figure 12.12, with slope given by the difference quotient

$$\frac{\Delta y}{\Delta x} = \frac{f(x + \Delta x) - f(x)}{\Delta x}$$

By letting Δx get closer and closer to zero, the graphical effect is to have the second point approach the fixed point. In the process the slopes of successive secant lines approach the slope of the tangent line at the fixed point.

This limit process is symbolized as

$$\lim_{\Delta x \to 0} \left[\frac{f(x + \Delta x) - f(x)}{\Delta x} \right]$$

This symbolism is bulky, so the following three "simple" symbols have been devised over the years to have the same meaning.

$$\left. \begin{array}{c} \dfrac{dy}{dx} \\ D[f(x)] \\ f'(x) \end{array} \right\} \text{ all mean } \lim_{\Delta x \to 0} \left[\frac{f(x + \Delta x) - f(x)}{\Delta x} \right]$$

Figure 12.12

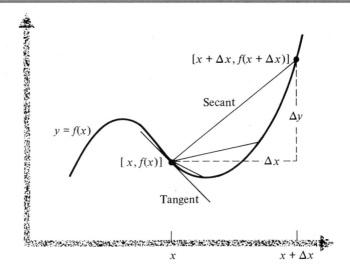

INTRODUCTION TO DIFFERENTIAL CALCULUS

Each symbolism form is best suited for certain situations, and so all will be used in this text.[1]

The result of taking the above limit is called a *derivative*. By applying this limit process to any function (no matter how complicated), we can determine (with a little luck, a little algebraic expertise, and a little prowess in taking limits) the derivative of that function. This process is called *differentiation*. Let's try out our new knowledge on a specific function.

Example

$y = f(x) = x^2 + x$

$f(x + \Delta x) = (x + \Delta x)^2 + (x + \Delta x)$

$\Delta y = f(x + \Delta x) - f(x)$

$\quad = [(x + \Delta x)^2 + (x + \Delta x)] - (x^2 + x)$

$\quad = x^2 + 2x(\Delta x) + (\Delta x)^2 + x + \Delta x - x^2 - x$

$\quad = 2x(\Delta x) + (\Delta x)^2 + \Delta x$

$\dfrac{\Delta y}{\Delta x} = \dfrac{2x(\Delta x) + (\Delta x)^2 + \Delta x}{\Delta x}$

$\quad = \dfrac{\Delta x(2x + \Delta x + 1)}{\Delta x}$

$\quad = 2x + \Delta x + 1$

Until now it has been all algebra. Now, we finally take the limit of this difference quotient to determine the derivative.

$\dfrac{dy}{dx} = \underset{\Delta x \to 0}{\text{limit}} \left[\dfrac{\Delta y}{\Delta x}\right] = \underset{\Delta x \to 0}{\text{limit}} \left[2x + \Delta x + 1\right]$

$\qquad = \underset{\Delta x \to 0}{\text{limit}} [2x] + \underset{\Delta x \to 0}{\text{limit}} [\Delta x] + \underset{\Delta x \to 0}{\text{limit}} [1]$

$\qquad = 2\underset{\Delta x \to 0}{\text{limit}} [x] + \underset{\Delta x \to 0}{\text{limit}} [\Delta x] + \underset{\Delta x \to 0}{\text{limit}} [1]$

Since x and the constant, 1, are not affected by $\Delta x \to 0$, then the derivative is:

$\qquad = 2x + 1$

---

[1] The symbols x and y are used to develop the theory of this chapter. However, when we get to the applications, we often use mnemonic symbols, such as R for revenue and P for price. The transformation to encompass such other variables is straightforward. For example, if $R = f(P)$

$\left.\begin{array}{l} \dfrac{dR}{dP} \\ D[f(P)] \\ f'(P) \end{array}\right\}$ all mean $\underset{\Delta P \to 0}{\text{limit}} \left[\dfrac{f(P + \Delta P) - f(P)}{\Delta P}\right]$

The process to obtain the derivative for this function could be used to differentiate any function, even a monster such as

$$y = f(x) = \frac{x^{77.5}\sqrt{x^2 - 10^x}}{\log_7(xe^x + x^3 - 4x^{-2})}$$

But you better believe that it would take a while.

Exercises 13

Find the derivatives of

1. $f(x) = 5x$
2. $f(x) = 1 + 3x^2$

In summary, derivatives are one of the greatest achievements of man. Evidence of that is that derivative applications abound in every field of endeavor. Now that we have mastered the basic definition, we can find the derivative of any specific function, one at a time. Of course, it now behooves us to generalize and systematize our knowledge. This is the task of Chapter 13, where we determine derivative rules for general classes of functions and for different configurations of functions.

PROBLEMS

TECHNICAL TRIUMPHS

1. Consider the following functions

    $y = f(x) = 4x$

    $y = f(x) = 2x^2$

    $y = f(x) = x^2 - 5x + 6$

    a. Find $f(x + \Delta x)$ for each.
    b. Determine the difference quotient for each.

2. Consider the following functions. For each, determine where (if at all) the function is discontinuous.

    a. $y = f(x) = \dfrac{x}{x - 5}$

    b. $y = f(x) = \dfrac{x^2}{(x - 1)(x + 2)}$

    c. $y = f(x) = \dfrac{2}{2^x}$

    d. $y = f(x) = 7x + 10 \quad \text{if } \{x \leq 4\}$
       $ = 38 \quad\quad\quad \text{if } \{x > 4\}$

3. Use the limit laws to break down these limits into more than one "simpler" limit. For example,

    $\displaystyle\lim_{\Delta x \to 0} \lfloor x + (\Delta x) \rfloor = \lim_{\Delta x \to 0} \lfloor x \rfloor + \lim_{\Delta x \to 0} \lfloor \Delta x \rfloor$

a. $\lim\limits_{\Delta x \to 0} \left[x^2(\Delta x) \right]$

b. $\lim\limits_{\Delta x \to 0} \left[\dfrac{\Delta x}{x^3} \right]$

c. $\lim\limits_{\Delta x \to 0} \left[2x^4(\Delta x)^5 \right]$

d. $\lim\limits_{\Delta x \to 0} \left[\dfrac{2^{\Delta x}}{3^{\Delta x}} \right]$

4. Consider the function

$$y = f(\Delta x) = 10^{100}(\Delta x)$$

a. For $\Delta x = 1, .1, .01, .001$, find y. From this alone, can you speculate on the limit of y as Δx approaches 0?

b. Apply the laws of limits to determine

$\lim\limits_{\Delta x \to 0} \left[10^{100}(\Delta x) \right]$

5. The sum (S) of the following geometric progression

$$S = 1 + \tfrac{1}{2} + \tfrac{1}{4} + \tfrac{1}{8} + \tfrac{1}{16} + \cdots$$

has a limit of 2 as the number of terms approaches infinity.
a. Make a table showing the value of S for 3, 4, 5, and 6 terms.
b. Use the geometric sum formula ($n = \infty$) on page 69 to verify that limit of 2.

6. Consider the function, $y = f(x) = 10 \ \{0 \leq x \leq 20\}$
a. Graph the function.
b. Determine Δy
c. Show that the limit of the difference quotient is zero.
d. Explain why you should have been able to predict the result in part c, given the graph.

7. A secant line between the x values of 2 and 4 has the equation

$$y = 6 - 3x$$

a. What was Δy?
b. What was Δx?
c. What is the slope of the secant line?

8. If the derivative is negative when $x = 10$, what can you say about the function at that point?

9. Suppose

$$\Delta y = x^2(\Delta x)^2 + x(\Delta x)$$

a. Find the difference quotient
b. Take the limit to find the derivative.

10. What is the effective interest rate for a bank that compounds continuously and pays a 7% annual interest rate?

11. New York City taxi cab meters go up 30 cents when a fifth of a mile is recorded. Discuss whether taxi cab fare is a continuous or discontinuous function of distance traveled.

CONFIDENCE BUILDERS

1. The following function forecasts total world population (P) in billions as a function of years since 1988 (for example, $x = 2$ in 1990).

   $P = f(x) = 5(1.015)^x$

   For the period 1990 to 2000,
   a. Find ΔP
   b. Find the difference quotient.
   c. What is the slope of the secant line? What interpretation can be given to that slope?

2. Consider the function

   $y = f(x) = \begin{cases} 5x & \text{for all } x, \text{ except } x = 2 \\ 15 & \text{for } x = 2 \end{cases}$

   a. What is the limit of this function as x approaches 2?
   b. Is this function continuous? Why?

3. Consider the more powerful rocket, Calculus II, whose distance above ground (y) in miles is the following function of time (x) in minutes since blast-off.

   $y = f(x) = 2x^2$

   a. Graph the distance function.
   b. For the fixed point (1, 2), find the slopes of secant lines to the second points with x-coordinates of 2, 1.5, 1.1, and 1.01. What value does these slopes seem to be approaching?
   c. Find the general equation for the difference quotient from the point (1, 2) to points with an x-coordinate of $1 + \Delta x$.
   d. Take the limit as $\Delta x \to 0$ of the difference quotient found in part c.
   e. Find the difference quotient for any two points on the function.
   f. Take the limit of your difference quotient in part e. to find the instantaneous velocity of Calculus II.
   g. What will Calculus II's instantaneous velocity be at 3 minutes after blast-off?
   h. At what time will Calculus II's velocity be 50 miles per minute?

4. Consider linear functions, with the general formula

   $y = f(x) = sx + i$

   a. Show that $\dfrac{\Delta y}{\Delta x} = \dfrac{f(x + \Delta x) - f(x)}{\Delta x} = s$

   b. Show that $\lim\limits_{\Delta x \to 0} \left[\dfrac{\Delta y}{\Delta x} \right] = s$

   c. Explain, with the help of a graphical analysis, why you got these results.

5. A company can produce 1 million units of product with its existing plant and workforce. Beyond that production level, additional plant buildings, machinery, and workers are needed. Roughly sketch the shape of the total cost curve facing this company. Is the curve continuous? Explain.

6. If the effective interest rate was 5.654% for a bank that compounds continuously, what was their annual interest rate.

7. The Chicago Board of Trade trades various agricultural commodities. They have limits on the price movements allowed in a given day. For example, during the mad cow scare of late 2003, the prices of beef closed lower at their limit for 4 days in a row.

INTRODUCTION TO DIFFERENTIAL CALCULUS

   a. Determine the current price of beef futures on that exchange.
   b. Discuss how such price limits either do or do not compare to the mathematical limits we learned.
8. Consider the following three practical uses of limits.
   a. Is there a limit to the velocity of an object? What is that limit?
   b. Is there a limit to the number of fouls that a college basketball player can have in one game? What usually happens when a player approaches that limit early in the game?
   c. Explain the relationship between limits and asymptotes.
9. Refer to the Calculus I blast-off table (page 434), where secant lines A, B, C, and D were defined for the fixed point at (1, 1).
   a. Determine the equation for each of these 4 secant lines.
   b. Determine the equation for the tangent line.
   c. We learned that the slopes of the secant lines approach the slope of the tangent line. Do the equations found in parts a and b confirm that?
   d. What happened to the intercept terms of the secant lines as Δx approached 0?
10. Do you have a quarter to spare buddy? If you do, get it out and start flipping. As you do this, keep record of the side that lands up and cumulative fraction of times that heads lands up. For example, here are the results of my first 10 flips.

Flip Number	Side Up	Cumulative Fraction of Heads
1	Tail	$0/1 = 0$
2	Head	$1/2 = .500$
3	Tail	$1/3 = .333$
4	Tail	$1/4 = .250$
5	Tail	$1/5 = .200$
6	Head	$2/6 = .333$
7	Tail	$2/7 = .286$
8	Head	$3/8 = .375$
9	Tail	$3/9 = .333$
10	Tail	$3/10 = .300$

   a. Plot the cumulative fraction of heads as a function of flip number.
   b. Continue by making flips 11 through 20. Amend the results to your graph in part a.
   c. Continue by making flips 21 through 50 and graph the cumulative fraction of heads.
   d. Of course, we know that the limit of the cumulative fraction is .5. Is your graph approaching that? If not, go out and flip 50 more times!!
11. Consider S. Lumlord's revenue function:

$$R = 1000P - 2P^2$$

Derive the difference quotient.

MIND STRETCHERS

1. Derivatives can be found by having the moving point on the left of the fixed point. Find the derivative in this way for the Calculus I function

$$y = f(x) = x^2$$

   a. At the point $x = 3$.
   b. at any point.

2. In the chapter we saw how derivatives could be found "from the right" or "from the left." Another equivalent way to find a derivative is to "straddle" the fixed point. Refer back to the Calculus I problem and

   a. Find $\dfrac{\Delta y}{\Delta x} = \dfrac{f(1 + \Delta x) - f(1 - \Delta x)}{2(\Delta x)}$

   Show that the limit of this difference quotient as Δx approaches 0 is 2.
   b. Graphically show what is involved in this "straddle" approach.
   c. Apply this straddle approach to any point to show that $\dfrac{dy}{dx} = 2x$.

3. The multiplication limit law when applied to $x^2 = x \cdot x$ gives

   $$\lim_{x \to k} [x \cdot x] = [\lim_{x \to k} x][\lim_{x \to k} x] = [\lim_{x \to k} x]^2$$

   Show how this law can be generalized to give

   $$\lim_{x \to k} [x^r] = [\lim_{x \to k} x]^r$$

4. The average cost per unit (y) of a product is given by the function of x, number of units produced

   $$y = f(x) = \dfrac{10,00 + 5x}{x}$$

   a. Show that the difference quotient is

   $$\dfrac{\Delta y}{\Delta x} = \dfrac{10,000}{x(x + \Delta x)}$$

   b. Use the limit laws to show that the derivative is

   $$\dfrac{dy}{dx} = \dfrac{-10,000}{x^2}$$

   c. Find the value of the derivative at $x = 10$. Interpret the meaning of this value.

5. The function

   $$y = f(x) = 16x^2$$

   represents the distance (y) in feet that a body falls in a vacuum in Earth's gravity as a function of the time (x) in seconds of fall.
   a. Graph this function over the domain $\{0 \leq x \leq 4\}$
   b. Consider the point at one second into the fall. Construct secant lines with $\Delta x = 1, .5, .01,$ and $.001$. Determine the slope of each of these secant lines.
   c. Find the general equation for the difference quotient at $x = 1$. What is the average velocity from $x = 1$ to $x = 3$?
   d. Determine the instantaneous velocity at $x = 1$.
   e. Derive the general difference quotient for any value of x. Use it to verify your average velocity result found in part c.
   f. Find the instantaneous velocity function.
   g. Calculate the instantaneous velocity for $x = 1$, $x = 2.5$, and $x = 4$.

INTRODUCTION TO DIFFERENTIAL CALCULUS

6. The Babylonians used the following limit to approximate $\sqrt{2}$.

$$\lim_{n \to \infty} [a_n] \quad \text{where} \quad a_n = \left[\frac{a_{n-1} + \frac{2}{a_{n-1}}}{2} \right]$$

Beginning with $a_1 = 1$, they would determine a_2 from the above recursive formula as

$$a_2 = \frac{a_1 + \frac{2}{a_1}}{2} = \frac{1 + \frac{2}{1}}{2} = 1.5$$

In turn, a_2 would be used in the recursive formula to determine a_3, etc.
a. Find a_3, a_4, and a_5.
b. Comment on whether this limit seems to be a good approximation for $\sqrt{2}$.

7. The present value (PV) of the constant $5 million per year returns from an investment of money discounted at a 20% discount rate is given by the formula

$$PV = \frac{5\left[1 - \left(\frac{1}{1.2}\right)^n\right]}{.2}$$

where n is the economic life of the asset.
a. Determine the limit of PV as n approaches infinity.
b. In general, if the annual returns (R) are constant, the investment has a present value given by the formula

$$PV = \frac{R\left[1 - \left(\frac{1}{1+i}\right)^n\right]}{i}$$

where i is the discount rate. Show that

$$\lim_{n \to \infty} [PV] = \frac{R}{i}$$

8. The present value (PV) of the dividends ($$D$ every year at year end) of a common stock is given by the formula

$$PV = \frac{D\left[1 - \left(\frac{1}{1+i}\right)^n\right]}{i}$$

where n is the number of years that the constant dividends are expected to continue, and i is the discount rate.

Explain why

$$\lim_{n \to \infty} [PV] = \frac{D}{i}$$

9. Total sales (S) of a company since its inception is given by the function of time (x) in years since its inception

   $S = f(x) = 10(2^x) - 10$

   a. Graph this function for $\{0 \leq x \leq 5\}$
   b. Determine the secant line slope between the fixed point (0, 0) and the general second point (x, S). What does this slope represent in an economic sense?
   c. Determine the difference quotient between any two points represented by (x, S) and $(x + \Delta x, f(x + \Delta x))$. For $x = 0$, show that you get the same result as in part b.

10. Consider the even more powerful rocket, Calculus III. Its distance (y) above ground is the following function of time (x) in minutes

    $y = f(x) = x^3$

    a. For the fixed point (1, 1), determine the slope of secant lines to the points with x-coordinates of 2, 1.5, 1.01.
    b. Determine secant slope in general at (1, 1). Use your formula to verify the secant slope found for the line to $x = 2$ above.
    c. Find the general equation for the difference quotient between any two points on the function.
    d. Take the limit of your difference quotient as Δx approaches 0.
    e. What is the velocity of the rocket at time $x = 2$?

CHAPTER 13

Derivative Rules

13.1 INTRODUCTION

Cinderivative: My evil stepmather makes me begin with the derivative definition, carry out the algebraic simplification, and take the limit for every function. I wish I didn't have to do all this hard work each time.

Hairy Godmather: Cinderivative, your wish is granted. I will show you how to bypass the drudgery by developing derivative rules that handle general classes of functions.

In the previous chapter, our purpose was to learn what a derivative really represents. The most effective way to achieve that goal was to use specific and simple functions, such as $f(x) = x^2$ and $f(x) = 1/x$. Of course, to make further progress in our study of calculus, we must generalize our efforts. Then, we even made a start in this generalization mode. Recall that we first investigated the instantaneous slope (velocity) of the rocket function, $f(x) = x^2$, at the single point of $x = 1$. Discovering that the velocity was 2 at that point was nice; but it didn't tell us the velocity at any of the other points. We soon corrected that by generalizing to determine that the slope at any value of x was $2x$. This general result confirmed the slope of $2 \cdot 1 = 2$ at $x = 1$; but more importantly, it enabled us to find the slope at any of the other infinite number of values of x along the function.

Learning Plan

In this chapter, we go much further in this generalization effort. For example, the rocket function, $y = f(x) = x^2$, is just a specific case of a general class of functions called power functions. They are symbolized as

$y = f(x) = x^n$ where n is any real number

For our rocket example, $n = 2$. But, there are an infinite number of other values of n, both positive and negative, that could be considered. Believe it or not, we can and will determine a rule for finding derivatives for the entire class of functions. Once this is accomplished, we can use the rule to find derivatives for an infinite number of power functions, including such cases as x^3, $x^{5.4}$, x^{-4}, and $x^{-147.9}$.

Besides finding derivatives for entire classes of functions, there is need to know how to handle various configurations of more than one function. For example, how does one take the derivative of the sum of two functions?

Finally, with an arsenal of derivative rules for classes and configurations of functions, it will be possible to "build up" the derivative of complicated functions, even this "Monster"

$$y = f(x) = \frac{x^{77.5}\sqrt{x^2 - 10^x}}{\log_7(xe^x + x^3 - 4x^{-2})}$$

from the derivatives of their component functions.

In the above spirit, we first develop derivative rules for the univariate function cases listed in Table 1. This is a purely theoretical development. You will be glad you endured it, once you see how we can then find derivatives of complex functions from their simpler component parts. We complete our theoretical development by learning how to find derivatives (called partial derivatives) of multivariate functions. So, Seymour Bottoms (page 132), we haven't forgotten your needs.

We close the chapter on a bright applications note by applying our newfound derivative prowess to help S. Lumlord (page 154) decide on the rent to charge his apartment dwellers. Finally, we introduce a neat web site (www.calc101.com) that can painlessly find derivatives.

We will use the 4-step procedure on page 463 to derive derivative rules. It is so nice to have such a well-defined procedure to find derivatives. It is good to know that this procedure would work on any function or configurations of more than one function. It would even work on a newly discovered function in the far reaches of the mathematical world.

Table 1

General Case	Description
$y = f(x) = k$	Constant function
$y = f(x) = x^n$	Power function
$y = f(x) = \log_b g(x)$	Logarithmic function
$y = f(x) = b^{g(x)}$	Exponential function
$y = f(x) = k \cdot g(x)$	Constant times a function
$y = f(x) = g(x) + h(x)$	Sum of functions
$y = f(x) = g(x) \cdot h(x)$	Product of functions
$y = f(x) = \dfrac{g(x)}{h(x)}$	Quotient of functions
$y = f(x) = g[h(x)]$	Chain rule: function of a function

The symbols k and n stand for any real numbers, and b stands for any positive number.

DERIVATIVE RULES

Step 1: Derivative Definition:

$$\frac{dy}{dx} = \lim_{\Delta x \to 0}\left[\frac{\Delta y}{\Delta x}\right] = \lim_{\Delta x \to 0}\left[\frac{f(x + \Delta x) - f(x)}{\Delta x}\right]$$

Step 2: Functional Values:
Here, we determine

$$f(x)$$
$$f(x + \Delta x)$$

and substitute them into the difference quotient.

Step 3: Simplification of Difference Quotient:
Algebra is KING in this step.

Step 4: Take Limit:
The laws of limits and some common sense will prove successful here. At the end of this step, the derivative rule will be revealed.

After deriving each rule, we illustrate its workings by some examples and test your understanding by some exercises.

13.2 CONSTANT RULE

Derivative definition:

$$\frac{dy}{dx} = \lim_{\Delta x \to 0}\left[\frac{\Delta y}{\Delta x}\right] = \lim_{\Delta x \to 0}\left[\frac{f(x + \Delta x) - f(x)}{\Delta x}\right]$$

Functional values:

$f(x) = k$ where k (a constant) can be any real number.

$f(x + \Delta x) = k$

Simplification of difference quotient:

$$\frac{\Delta y}{\Delta x} = \frac{k - k}{\Delta x} = \frac{0}{\Delta x}$$

Take limit:

$$\frac{dy}{dx} = \lim_{\Delta x \to 0}\left[\frac{0}{\Delta x}\right] = 0$$

Notice that zero divided by any positive number—even a very, very, very small one like Δx (remember that Δx can't equal zero in taking the limit)—is zero.

The result confirms that the slope of a horizontal line is zero. Ho-hum, we knew that already.

Examples

$$\frac{dy}{dx}$$

1. $y = f(x) = 7$ 0
2. $y = f(x) = -12$ 0
3. $y = f(x) = \frac{1}{2}$ 0

This is monotonous! Now try these real hardies!

Exercises 1

$$\frac{dy}{dx}$$

1. $y = f(x) = -2.9$ _____
2. $y = f(x) = 123$ _____
3. $y = f(x) = 0$ _____

13.3 POWER RULE

Derivative definition:

$$\frac{dy}{dx} = \lim_{\Delta x \to 0}\left[\frac{\Delta y}{\Delta x}\right] = \lim_{\Delta x \to 0}\left[\frac{f(x + \Delta x) - f(x)}{\Delta x}\right]$$

Functional values:

$f(x) = x^n$ where n is a positive integer.

$f(x + \Delta x) = (x + \Delta x)^n$

Simplification of difference quotient:

$$\frac{\Delta y}{\Delta x} = \frac{(x + \Delta x)^n - x^n}{\Delta x}$$

We expand $(x + \Delta x)^n$ by the binomial expansion formula (see page 38).

$$(x + \Delta x)^n = x^n + nx^{n-1}(\Delta x)^1 + \frac{n!}{2!(n-2)!}x^{n-2}(\Delta x)^2 + \frac{n!}{3!(n-3)!}x^{n-3}(\Delta x)^3 + \cdots + (\Delta x)^n$$

DERIVATIVE RULES

We have left out a whole host of terms, but as you'll see later, we don't need their exact expression.

$$\frac{\Delta y}{\Delta x} = \frac{\left(x^n + nx^{n-1}(\Delta x)^1 + \frac{n!}{2!(n-2)!}x^{n-2}(\Delta x)^2 + \frac{n!}{3!(n-3)!}x^{n-3}(\Delta x)^3 + \cdots + (\Delta x)^n\right) - x^n}{\Delta x}$$

Cancelling the x^n terms and factoring out Δx in the numerator gives

$$\frac{\Delta y}{\Delta x} = \frac{\Delta x \left(nx^{n-1} + \frac{n!}{2!(n-2)!}x^{n-2}(\Delta x)^1 + \frac{n!}{3!(n-3)!}x^{n-3}(\Delta x)^2 + \cdots + (\Delta x)^{n-1}\right)}{\Delta x}$$

Cancelling the Δx's results in the "simplified" difference quotient.

$$\frac{\Delta y}{\Delta x} = nx^{n-1} + \frac{n!}{2!(n-2)!}x^{n-2}(\Delta x)^1 + \frac{n!}{3!(n-3)!}x^{n-3}(\Delta x)^2 + \cdots + (\Delta x)^{n-1}$$

Take limit:

$$\frac{dy}{dx} = \lim_{\Delta x \to 0} \left[\frac{\Delta y}{\Delta x}\right]$$

$$= \lim_{\Delta x \to 0}\left[nx^{n-1} + \frac{n!}{2!(n-2)!}x^{n-2}(\Delta x)^1 + \frac{n!}{3!(n-3)!}x^{n-3}(\Delta x)^2 + \cdots + (\Delta x)^{n-1}\right]$$

The limit laws for a sum of functions states that we can take the limit of each function separately. Except for the first function, x^{n-1}, each function has Δx raised to a positive power (the power increases by one each term as you go down the line) and can be characterized as

$$\frac{n!}{p!(n-p)!}x^{n-p}(\Delta x)^q$$

where $(n-p) + q = n - 1$.

The values of n, p, q, and x do not change as a result of changes in Δx; thus, they act together as one big constant (K) in the limit process. So, we can characterize these terms as $K(\Delta x)^q$ where

$$K = \frac{n!}{p!(n-p)!}x^{n-p}$$

Regardless of how big K is, the limit of $K(\Delta x)^q$ is equal to zero. Skeptical? OK, perhaps going through the limit process table for $K = 10^{1000}$ (is that big enough?) and $q = 1$ will convince you.

Δx	$10^{1000}(\Delta x)^1$	Comment
1	10^{1000}	Mighty big!
.1	10^{999}	I told you it won't approach zero.
10^{-1000}	1	Gulp!
10^{-2000}	10^{-1000}	So what!
↓	↓	
0	0	

For terms where the exponent of Δx is greater than one, this process leading to a limit of zero is hastened, since a small number raised to a high power gets extremely small quickly.

Only the first limit doesn't go to zero as Δx approaches zero; it remains unchanged. Thus, the derivative is:

$$\frac{dy}{dx} = nx^{n-1} + 0 + 0 + \cdots + 0$$

$$= nx^{n-1}$$

This is known as the *power rule*. It generalizes for any real number value of n. Thus, if

$$y = f(x) = x^n$$

where n is any real number, then

$$\frac{dy}{dx} = nx^{n-1}$$

Or, alternatively stated,

$$D[x^n] = nx^{n-1}$$

Examples

	$f(x)$	$\frac{dy}{dx}$
1.	x^5	$5x^4$
2.	x^{-9}	$-9x^{-10}$
3.	$x^{2.6}$	$2.6x^{1.6}$
4.	$x^{1/2}$	$\frac{1}{2}x^{-(1/2)}$
5.	x^0	$0x^{-1} = 0$

Exercises 2

	$f(x)$	$\frac{dy}{dx}$
1.	x^3	_____
2.	$x^{-(1/2)}$	_____
3.	x^{1000}	_____
4.	$x^{-1.2}$	_____
5.	$x^{9/4}$	_____

DERIVATIVE RULES

13.4 LOGARITHMIC RULE

Derivative definition:

$$\frac{dy}{dx} = \lim_{\Delta x \to 0}\left[\frac{\Delta y}{\Delta x}\right] = \lim_{\Delta x \to 0}\left[\frac{f(x + \Delta x) - f(x)}{\Delta x}\right]$$

Functional values:

$f(x) = \log_b x$ where b is a positive number.

$f(x + \Delta x) = \log_b(x + \Delta x)$

Simplification of difference quotient:

$$\frac{\Delta y}{\Delta x} = \frac{\log_b(x + \Delta x) - \log_b x}{\Delta x}$$

$$= \frac{\log_b \frac{(x + \Delta x)}{x}}{\Delta x}$$

$$= \frac{1}{\Delta x}\log_b\left(1 + \frac{\Delta x}{x}\right)$$

This step was brought to you by the second law of logarithms. Now watch what wonders multiplying by 1 or x/x can perform.

$$\frac{\Delta y}{\Delta x} = \left(\frac{x}{x} \cdot \frac{1}{\Delta x}\right)\log_b\left(1 + \frac{\Delta x}{x}\right)$$

$$= \left(\frac{1}{x}\right)\frac{x}{\Delta x}\log_b\left(1 + \frac{\Delta x}{x}\right)$$

$$= \frac{1}{x}\log_b\left(1 + \frac{\Delta x}{x}\right)^{x/\Delta x}$$

Which logarithmic law brought you the previous step?

Let $h = \Delta x/x$, so the difference quotient becomes

$$\frac{\Delta y}{\Delta x} = \frac{1}{x}\log_b(1 + h)^{1/h}$$

The algebraic simplifications and what may appear to be mysterious manipulations of the difference quotient have brought us as far as they could. Now it's time to take the limit. Take limit:

$$\frac{dy}{dx} = \lim_{\Delta x \to 0}\left[\frac{1}{x}\log_b(1 + h)^{1/h}\right]$$

Recall from page 441 that $(1 + h)^{1/h}$ approaches the number e as h approaches zero. And from its definition above, h does approach zero whenever Δx approaches zero. Meanwhile, x and b do not change as Δx approaches zero. So the limit is

$$\frac{dy}{dx} = \frac{1}{x} \log_b e$$

Since $\log_b e = \dfrac{1}{\log_e b}$ (see page 41), an alternative version of the rule just derived is

$$D[\log_b x] = \frac{1}{x \log_e b} \quad \text{or} \quad \frac{1}{x \ln b}$$

We won't prove it now (see chain rule section for proof), but the general rule to find the derivative of the logarithm of any function $g(x)$ is

$$D[f(x) = \log_b g(x)] = \frac{D[g(x)]}{g(x) \cdot \ln b}$$

Notice how these rules simplify for natural logarithms (base $b = e$), since $\ln e = 1$.

$$D[\log_e x = \ln x] = \frac{1}{x \ln e} = \frac{1}{x}$$

$$D[\log_e g(x) = \ln g(x)] = \frac{D[g(x)]}{g(x) \cdot \ln e} = \frac{D[g(x)]}{g(x)}$$

Examples

	$y = f(x)$	$\dfrac{dy}{dx}$
1.	$\log_{10} x$	$\dfrac{1}{x \ln 10}$
2.	$\log_{1/2} x$	$\dfrac{1}{x \ln 1/2}$
3.	$\log_5 x^2$	$\dfrac{D[x^2]}{x^2 \ln 5} = \dfrac{2x}{x^2 \ln 5} = \dfrac{2}{x \ln 5}$
4.	$\ln x^3$	$\dfrac{D[x^3]}{x^3} = \dfrac{3x^2}{x^3} = \dfrac{3}{x}$

Exercises 3

Find the derivatives of the following functions.

1. $y = \log_3 x$
2. $y = \ln x^2$

DERIVATIVE RULES

3. $y = \log_{1/4} x$
4. $y = \log_{10} x^4$

13.5 EXPONENTIAL RULE

We could develop the rule for differentiating exponential functions by the same old route that we have used several times already. However, this time we'll try another way that allows us to learn the inverse rule of derivatives.

The inverse rule states

$$\frac{dx}{dy} = \frac{1}{dy/dx}$$

It is merely saying, for example, that if velocity, dy/dx, is 60 miles per hour, then dx/dy (which would be inverse velocity) is $\frac{1}{60}$ hour per mile.

Consider the function $y = \frac{1}{2}x$, which has a derivative of $\frac{1}{2}$. Solving for x gives $x = 2y$, which has a derivative of 2 (see Figure 13.1) We could have found this result by the inverse rule once dy/dx was known.

$$\frac{dx}{dy} = \frac{1}{dy/dx} = \frac{1}{\frac{1}{2}} = 2$$

Now, back to the exponential rule. If y is a logarithmic function of x so that $y = \log_b x$, we just derived the fact that

$$\frac{dy}{dx} = \frac{1}{x \ln b}$$

But, solving the logarithmic function explicitly for x

$b^y = b^{\log_b x} = x$ (See page 44 for why $b^{\log_b x} = x$)

reveals that x is an exponential function of y.

Figure 13.1

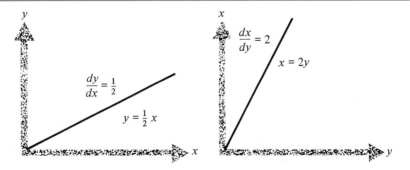

We can find the derivative of the exponential function, dx/dy, by using the inverse rule as follows.

$$\frac{dx}{dy} = \frac{1}{\frac{dy}{dx}} = \frac{1}{\frac{1}{x \ln b}} = x \ln b = b^y \ln b$$

Switching variables to conform to our usual notation, this means that if

$$y = b^x$$

where b is a positive real number, then

$$\frac{dy}{dx} = b^x \ln b$$

We won't prove it now (see chain rule section for proof), but the general rule for differentiating exponential functions is

$$D[f(x) = b^{g(x)}] = b^{g(x)}(\ln b)D[g(x)]$$

Just as they did for logarithmic functions, the exponential derivative rules simplify when $b = e$.

$$D[e^x] = e^x \ln e = e^x$$
$$D[e^{g(x)}] = e^{g(x)}(\ln e)D[g(x)] = e^{g(x)}D[g(x)]$$

Examples

1. $D[10^x] = 10^x \ln 10$
2. $D[(\frac{1}{2})^x] = (\frac{1}{2})^x \ln \frac{1}{2}$
3. $D[e^{x^2}] = e^{x^2}(\ln e)2x = 2xe^{x^2}$
4. $D[8^{x^{-1}}] = 8^{x^{-1}}(\ln 8)(-1x^{-2})$

Exercises 4

Differentiate the following functions.

1. $y = 5^x$
2. $y = e^{x-1}$
3. $y = (\frac{1}{4})^x$
4. $y = 3^{x^2}$

13.6 CONSTANT TIMES A FUNCTION RULE

At this point in our derivative rule development, we can handle $y = x^2$ but not $y = 3x^2$, $y = \log_6 x$ but not $y = 2 \log_6 x$, $y = 4^x$ but not $y = 7 \cdot 4^x$. These gaps in our knowledge are reflective of the general problem of finding the effect that multiplying a function by a constant has on the overall derivative. In other words, if we know the derivative of $g(x)$, can we then find the derivative of $f(x) = k \cdot g(x)$, where k is any constant? And where do you suppose we begin to tackle this problem? You guessed it!

$$\frac{dy}{dx} = \lim_{\Delta x \to 0} \left[\frac{f(x + \Delta x) - f(x)}{\Delta x} \right]$$

Since (see Figure 13.2)

$$f(x) = k \cdot g(x)$$
$$f(x + \Delta x) = k \cdot g(x + \Delta x)$$

Figure 13.2

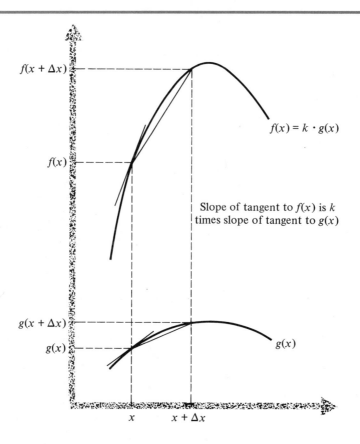

So

$$\frac{dy}{dx} = \lim_{\Delta x \to 0} \left[\frac{k \cdot g(x + \Delta x) - k \cdot g(x)}{\Delta x} \right]$$

$$= \lim_{\Delta x \to 0} \left[\frac{k}{1} \left(\frac{g(x + \Delta x) - g(x)}{\Delta x} \right) \right] = \lim_{\Delta x \to 0} [k] \cdot \lim_{\Delta x \to 0} \left[\frac{g(x + \Delta x) - g(x)}{\Delta x} \right]$$

As $\Delta x \to 0$, k remains pat, but $\left[\dfrac{g(x + \Delta x) - g(x)}{\Delta x} \right]$ approaches the slope along $g(x)$. In the limit it is $D[g(x)]$. Thus,

$$D[k \cdot g(x)] = k \cdot D[g(x)]$$

This rule in effect says that we can bring out the constant and multiply it by the derivative of the function.

Examples

1. $D[3x^2] = 3D[x^2] = 3(2x) = 6x$
2. $D[-5x^{-3}] = -5D[x^{-3}] = -5(-3x^{-4}) = 15x^{-4}$
3. $D[4x] = 4D[x^1] = 4(1x^0) = 4$
4. $D[2x^{.5}] = 2D[x^{.5}] = 2(.5x^{-.5}) = x^{-.5}$
5. $D[\frac{3}{7}x^{9.1}] = \frac{3}{7}D[x^{9.1}] = \frac{3}{7}(9.1)x^{8.1}$
6. $D[4 \log_5 x] = 4D[\log_5 x] = \dfrac{4}{x \ln 5}$
7. $D[10 \ln x] = 10D[\ln x] = \dfrac{10}{x}$
8. $D[6(3^x)] = 6D[3^x] = 6(3^x \ln 3)$
9. $D[-2e^x] = -2D[e^x] = -2e^x$
10. $D[\frac{1}{2}(3^{x^2})] = \frac{1}{2}D[3^{x^2}] = \frac{1}{2} \cdot 3^{x^2}(\ln 3)2x = x \cdot 3^{x^2} \cdot \ln 3$

Exercises 5

1. $D[9x^3] =$
2. $D[20 \log_4 x] =$
3. $D[(-5)7^x] =$
4. $D[1.83(x^{-1})] =$
5. $D[-7 \log_{4.5} x] =$
6. $D[3 \ln x] =$
7. $D[-4e^x] =$
8. $D[5x^{3.5}] =$
9. $D[\frac{1}{2}x^{-(1/2)}] =$
10. $D[9 \log_2 x^3] =$

DERIVATIVE RULES

13.7 SUM OF FUNCTIONS RULE

At this point we can differentiate many functions, but not these:

1. $x^2 + 7$
2. $8x - 2^x$
3. $\ln x + e^x$

These are specific examples of a sum of two functions, which can be generally expressed as

$f(x) = g(x) + h(x)$

So, let's develop the rule that gives $D[f(x)]$ in terms of the derivatives of $g(x)$ and $h(x)$. (See Figure 13.3.)

$$D[f(x)] = \lim_{\Delta x \to 0} \left[\frac{f(x + \Delta x) - f(x)}{\Delta x} \right]$$

$f(x + \Delta x) = g(x + \Delta x) + h(x + \Delta x)$

So

Figure 13.3

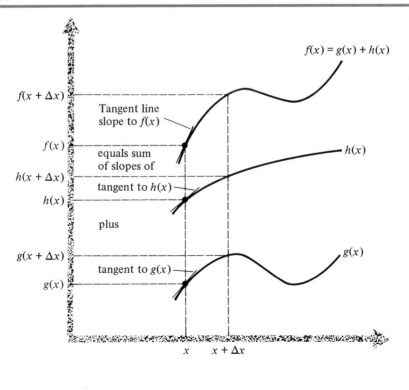

$$D[f(x)] = \lim_{\Delta x \to 0} \left[\frac{[g(x + \Delta x) + h(x + \Delta x)] - [g(x) + h(x)]}{\Delta x} \right]$$

$$= \lim_{\Delta x \to 0} \left[\underbrace{\frac{g(x + \Delta x) - g(x)}{\Delta x}}_{\text{This limit is } D[g(x)]} + \underbrace{\frac{h(x + \Delta x) - h(x)}{\Delta x}}_{\text{This limit is } D[h(x)]} \right]$$

So

$$D[f(x)] = D[g(x)] + D[h(x)]$$

In words, to find the derivative of the sum of functions, add the derivatives of each function taken separately.

Examples

1. $D[x^2 + 7] = D[x^2] + D[7] = 2x + 0 = 2x$
2. $D[x - x^{1/2}] = D[x] - D[x^{1/2}] = 1 - (\frac{1}{2})x^{-(1/2)}$
3. $D[\ln x + e^x] = D[\ln x] + D[e^x] = \frac{1}{x} + e^x$
4. $D[3^x + x^{-4}] = D[3^x] + D[x^{-4}] = 3^x \ln 3 - 4x^{-5}$
5. $D[\log_5 x - x^3] = D[\log_5 x] - D[x^3] = \frac{1}{x \ln 5} - 3x^2$

Exercises 6

1. $D[3 - x^3] =$
2. $D[\log_2 x + x^{-2}] =$
3. $D[e^x - 2^x] =$
4. $D[\log_5 x - \log_4 x] =$
5. $D[x^{-2} + x^6] =$

The summation rule generalizes to as many functions as necessary. So, for example, if we have the sum of three functions

$$f(x) = g(x) + h(x) + i(x)$$

then

$$D[f(x)] = D[g(x)] + D[h(x)] + D[i(x)]$$

Examples

1. $D[x^2 + e^x + \log_5 x] = D[x^2] + D[e^x] + D[\log_5 x]$

$$= 2x + e^x + \frac{1}{x \ln 5}$$

DERIVATIVE RULES

2. $D[x^4 + x^3 + 10] = D[x^4] + D[x^3] + D[10]$
   $= 4x^3 + 3x^2 + 0$
3. $D[x^2 + x + 5 - x^{-1}] = D[x^2] + D[x] + D[5] - D[x^{-1}]$
   $= 2x + 1 + 0 + x^{-2}$

Exercises 7

1. $D[2^x - x^{-(1/2)} + x^3] =$
2. $D[\ln x + e^x + x + x^2] =$
3. $D[50 + x^4 - x^{1/3}] =$

13.8 PRODUCT RULE

We are really making good progress in our derivative rule development. But we still can't handle something like $f(x) = xe^x$, which is an example of a product of two functions, x and e^x in this case. In general, we wish to know how to differentiate

$$f(x) = g(x) \cdot h(x)$$

The same process that was used to derive the previous derivative rules could be used to derive the product rule. However, since in this case simplification of the difference quotient involves some fancy algebraic moves, we will just state the final result.

$$D[g(x) \cdot h(x)] = g(x) \cdot D[h(x)] + h(x) \cdot D[g(x)]$$

Examples

1. $D[\overset{g(x)}{x} \cdot \overset{h(x)}{e^x}] = (x)D[e^x] + (e^x)D[x]$
   $= (x)(e^x) + (e^x)(1)$
   $= e^x(x + 1)$
2. $D[x^2 \cdot \log_3 x] = (x^2)D[\log_3 x] + (\log_3 x)D[x^2]$
   $= x^2 \left(\dfrac{1}{x \ln 3} \right) + (\log_3 x)(2x)$
   $= x \left(\dfrac{1}{\ln 3} + 2 \log_3 x \right)$
3. $D[x^{-1} \cdot 2^x] = (x^{-1})D[2^x] + (2^x)D[x^{-1}]$
   $= (x^{-1})(2^x \ln 2) + (2^x)(-1x^{-2})$
   $= 2^x[(x^{-1} \ln 2) - x^{-2}]$
4. $D[5x^2] = (5)D[x^2] + (x^2)D[5]$
   $= (5)(2x) + (x^2)(0) = 10x$

Note: The product rule works for this case; however, the constant times a function rule is more efficient.

Exercises 8

1. $D[e^x \cdot \ln x] =$
2. $D[x^3 \cdot 10^x] =$
3. $D[x^{-4} \cdot \log_3 x] =$
4. $D[x^5 \cdot x^{-3}] =$
5. Use the exponent laws to put exercise 4 into a form so that the power rule can be used. Do both the product and power rules give the same answer?

13.9 QUOTIENT RULE

Differentiation of a function divided by another function, or

$$f(x) = \frac{g(x)}{h(x)}$$

requires the quotient rule, which we won't prove but rather just state as

$$D\left[\frac{g(x)}{h(x)}\right] = \frac{h(x) \cdot D[g(x)] - g(x) \cdot D[h(x)]}{[h(x)]^2}$$

Examples

1. $D\left[\dfrac{x}{\ln x}\right] = \dfrac{(\ln x)D[x] - (x)D[\ln x]}{(\ln x)^2}$

   where x is $g(x)$ and $\ln x$ is $h(x)$.

   $= \dfrac{(\ln x)(1) - (x)\left(\dfrac{1}{x}\right)}{(\ln x)^2} = \dfrac{\ln x - 1}{(\ln x)^2}$

2. $D\left[\dfrac{e^x}{x^2}\right] = \dfrac{(x^2)D[e^x] - (e^x)D[x^2]}{(x^2)^2}$

   $= \dfrac{(x^2)(e^x) - (e^x)(2x)}{x^4}$

   $= \dfrac{xe^x(x - 2)}{x^4} = \dfrac{e^x(x - 2)}{x^3}$

3. $D\left[\dfrac{1}{x}\right] = \dfrac{(x)D[1] - (1)D[x]}{x^2}$

   $= \dfrac{(x)(0) - (1)(1)}{x^2} = \dfrac{-1}{x^2}$

DERIVATIVE RULES

Note: The quotient rule works, but it would be easier to express $f(x)$ as x^{-1} and use the power rule.

4. Sometimes a quotient can be made into a product so that the simpler product rule can be used. For example,

$$D\left[\frac{e^x}{x}\right] = D[x^{-1}e^x] = (x^{-1})D[e^x] + (e^x)D[x^{-1}]$$
$$= (x^{-1})(e^x) + (e^x)(-1x^{-2})$$
$$= x^{-1}e^x(1 - x^{-1})$$

See if you get the same result by using the quotient rule.

Exercises 9

1. $D\left[\dfrac{x^2}{e^x}\right] =$

2. $D\left[\dfrac{5}{x^3}\right] =$

3. $D\left[\dfrac{x^3}{\log_5 x}\right] =$

4. $D\left[\dfrac{x^{1/2}}{x^{-2}}\right] =$

13.10 CHAIN RULE

Our development of derivative rules has brought us very far, yet there are functions such as

$y = 3^{-x^2}$

$y = \log_{10}(4x^3)$

$y = e^{2x}$

$y = (x^3 + 7)^{2.5}$

that our rules can't handle. Such functions can be characterized as functions of functions. This can be seen by writing the above functions as

$y = g(u) = 3^u \quad$ where $\quad u = h(x) = -x^2$

$y = g(u) = \log_{10} u \quad$ where $\quad u = h(x) = 4x^3$

$y = g(u) = e^u$ where $u = h(x) = 2x$

$y = g(u) = u^{2.5}$ where $u = h(x) = x^3 + 7$

We have created an intermediate variable, u, such that y is a function of u, which in turn is a function of x. We shall now develop the chain rule for differentiating such beasts.

Let's consider $y = e^{2x}$ to gain insights into the chain rule. Letting $u = 2x$, we have $y = e^u$ (see Figure 13.4).

Notice that these two functions can be differentiated by already known rules as

$$\frac{du}{dx} = 2 \qquad \frac{dy}{du} = e^u$$

Figure 13.4

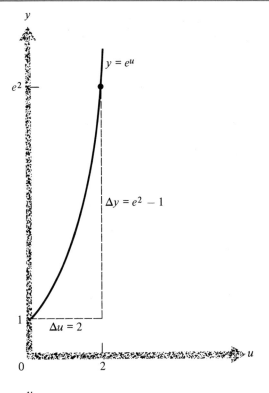

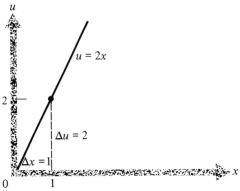

DERIVATIVE RULES

But the real question, "What is dy/dx?" requires the chain rule for resolution.

We can obtain some insight here by viewing Figure 13.4. If we take a fixed point at $x = 0$ and let $\Delta x = 1$, then $\Delta u = 2$ and $\Delta y = e^2 - e^0 = e^2 - 1$. So

$$\frac{\Delta y}{\Delta x} = \frac{e^2 - 1}{1} \qquad \frac{\Delta y}{\Delta u} = \frac{e^2 - 1}{2}$$

Notice that for the same Δy, Δx is $\frac{1}{2} \Delta u$. Thus,

$$\frac{\Delta y}{\Delta x} = \frac{\Delta y}{\frac{1}{2}\Delta u} = 2 \frac{\Delta y}{\Delta u}$$

So the rate of change of y with respect to x is twice that with respect to u. We get this same result in another way that is relevant to later development:

$$\frac{\Delta y}{\Delta x} = \frac{\Delta y}{\Delta x} \cdot 1$$
$$= \frac{\Delta y}{\Delta x} \cdot \frac{\Delta u}{\Delta u}$$
$$= \frac{\Delta u}{\Delta x} \cdot \frac{\Delta y}{\Delta u} = 2 \cdot \frac{\Delta y}{\Delta u}$$

Now let's generalize to derive the chain rule. The definition of the derivative of y with respect to x is

$$\frac{dy}{dx} = \lim_{\Delta x \to 0} \left[\frac{\Delta y}{\Delta x} \right]$$

We can introduce the intermediate variable change, Δu, in the difference quotient without affecting its magnitude by multiplying by $\Delta u/\Delta u = 1$.

$$\frac{dy}{dx} = \lim_{\Delta x \to 0} \left[\left(\frac{\Delta y}{\Delta x}\right)\left(\frac{\Delta u}{\Delta u}\right) \right] = \lim_{\Delta x \to 0} \left[\left(\frac{\Delta y}{\Delta u}\right)\left(\frac{\Delta u}{\Delta x}\right) \right] = \underbrace{\lim_{\Delta x \to 0} \left[\frac{\Delta y}{\Delta u} \right]}_{\frac{dy}{du}} \cdot \underbrace{\lim_{\Delta x \to 0} \left[\frac{\Delta u}{\Delta x} \right]}_{\frac{du}{dx}}$$

By taking the two limits, we get the chain rule

$$\frac{dy}{dx} = \left(\frac{dy}{du}\right)\left(\frac{du}{dx}\right)$$

Using this rule on our illustrative function, we get

$$y = e^u \quad \text{where} \quad u = 2x$$
$$\frac{dy}{du} = e^u \qquad \frac{du}{dx} = 2$$

so

$$\frac{dy}{dx} = \left(\frac{dy}{du}\right)\left(\frac{du}{dx}\right) = e^u \cdot 2 = 2e^{2x}$$

Examples

1. $y = 3^{-x^2}$ so $y = 3^u$ where $u = -x^2$

$$\frac{dy}{dx} = \left(\frac{dy}{du}\right)\left(\frac{du}{dx}\right) = (3^u \ln 3)(-2x) = (3^{-x^2} \ln 3)(-2x)$$

2. $y = \log_{10}(4x^3)$ so $y = \log_{10} u$ where $u = 4x^3$

$$\frac{dy}{dx} = \left(\frac{dy}{du}\right)\left(\frac{du}{dx}\right) = \left(\frac{1}{u \ln 10}\right)(12x^2) = \frac{12x^2}{4x^3 \ln 10} = \frac{3}{x \ln 10}$$

3. $y = (x^3 + 7)^{2.5}$ so $y = u^{2.5}$ where $u = x^3 + 7$

$$\frac{dy}{dx} = \left(\frac{dy}{du}\right)\left(\frac{du}{dx}\right) = (2.5u^{1.5})(3x^2) = 7.5(x^3 + 7)^{1.5}(x^2)$$

Exercises 10

Use the chain rule to differentiate

1. $y = \ln x^{1/2}$
2. $y = 5^{x^3}$
3. $y = (x^2 + x)^{1/2}$
4. $y = e^{5x}$
5. $y = (2^x + x)^2$

Earlier, we stated generalized exponential and logarithmic differentiation rules without proof. Now with the chain rule, we are in a position to show why they are so.
Exponential:

$$y = b^{g(x)}$$

Let $y = b^u$ where $u = g(x)$

$$\frac{dy}{dx} = \left(\frac{dy}{du}\right)\left(\frac{du}{dx}\right)$$

since $\dfrac{dy}{du} = b^u \ln b = b^{g(x)} \ln b$

$\dfrac{du}{dx}$ means $D[g(x)]$

so

$$D[b^{g(x)}] = (b^{g(x)} \ln b) D[g(x)]$$

Logarithmic:

$$y = \log_b g(x)$$

Let $y = \log_b u$ where $u = g(x)$

$$\frac{dy}{dx} = \left(\frac{dy}{du}\right)\left(\frac{du}{dx}\right)$$

since $\dfrac{dy}{du} = \dfrac{1}{u \ln b} = \dfrac{1}{g(x) \ln b}$ and $\dfrac{du}{dx}$ means $D[g(x)]$

so

$$D[\log_b g(x)] = \frac{1}{g(x) \ln b} D[g(x)]$$

13.11 SUMMARY OF DERIVATIVE RULES

At this point we have completed our development of derivative rules. Table 2 (where $g(x)$ and $h(x)$ represent any functions of x, k and n represent any real numbers, and b represents a positive real number) summarizes our effort.

Table 2 Derivative Rules

Name	Rule
Constant	$D[k] = 0$
Power	$D[x^n] = nx^{n-1}$
Exponential	$D[b^{g(x)}] = b^{g(x)} \ln b \, D[g(x)]$
Logarithmic	$D[\log_b g(x)] = \dfrac{1}{g(x) \ln b} D[g(x)]$
Multiplying constant	$D[k \cdot g(x)] = k \cdot D[g(x)]$
Summation	$D[g(x) + h(x)] = D[g(x)] + D[h(x)]$
Product	$D[g(x) \cdot h(x)] = g(x) \cdot D[h(x)] + h(x) \cdot D[g(x)]$
Quotient	$D\left[\dfrac{g(x)}{h(x)}\right] = \dfrac{h(x) \cdot D[g(x)] - g(x) \cdot D[h(x)]}{[h(x)]^2}$
Chain rule	If $y = g(u)$ and $u = h(x)$, then $\dfrac{dy}{dx} = \left(\dfrac{dy}{du}\right)\left(\dfrac{du}{dx}\right)$

With this complete set of rules, you are equipped to handle any function with management applications, no matter how hairy. Often, when taming a complicated function, we need to employ a sequence of these rules. To illustrate, consider the function

$$y = f(x) = \frac{5 + x + 3x^2}{e^x}$$

For an opener we employ the quotient rule

$$D[f(x)] = \frac{(e^x)D[5 + x + 3x^2] - (5 + x + 3x^2)D[e^x]}{(e^x)^2}$$

Then we need the summation rule

$$D[5 + x + 3x^2] = D[5] + D[x] + D[3x^2]$$

This leads us to the constant, power, and multiplying constant rules

$$D[5] = 0$$
$$D[x^1] = 1x^0 = 1$$
$$D[3x^2] = 3D[x^2] = 3(2x) = 6x$$

not to mention the exponential rule

$$D[e^x] = e^x$$

Finally, we bring it all together.

$$D[f(x)] = \frac{(e^x)(0 + 1 + 6x) - (5 + x + 3x^2)(e^x)}{e^{2x}}$$

$$= \frac{e^x(1 + 6x - 5 - x - 3x^2)}{e^{2x}} = \frac{-4 + 5x - 3x^2}{e^x}$$

Exercises 11

Now see if you can put it all together by finding the derivatives of the following complicated functions. In doing so refer to the table of derivative rules (never memorize them, although with extensive use they will come naturally) and state the sequence used.

1. $y = f(x) = \dfrac{x + x^2}{x^3}$

2. $y = f(x) = e^x(1 + x)^{1/2}$

3. $y = f(x) = \dfrac{x \log_2 x}{1 + x}$

4. $y = f(x) = 5x^2(1 + 10^x)^x$

DERIVATIVE RULES

5. $y = f(x) = \dfrac{3x^5}{e^{2x}x^{3.5}}$

13.12 PARTIAL DERIVATIVES

Until now all our differentiation has been done on functions of one variable. But can we differentiate functions of more than one variable? If so, what does the result mean? To get a feel for this problem, let's return to the Sinema Theater revenue function developed on page 132.

$$y = f(x, z) = 6x + 3z$$

where

- x is male attendees
- z is female attendees
- y is total revenue

It is evident from the formula that every additional male attendee (holding the number of females constant) increases revenue by \$6, while every additional female (holding the number of males constant) increases revenue by \$3. These rates of change (each with the other variable held constant) are called *partial derivatives* and are specified by the notation

$$\frac{\partial y}{\partial x} = 6 \qquad \frac{\partial y}{\partial z} = 3$$

The different derivative symbol is used to alert you to the fact that all other variables are held constant.

In general, partial derivatives are instantaneous rates of change with respect to one variable, holding all other variables constant.

Geometrically, the revenue function is a plane in three-dimensional space as shown in Figure 13.5. The partial derivatives tell us that for every one unit in a direction parallel to the x-axis, the plane increases by three units; for every one unit in the direction of the z-axis, the plane increases by two units. Partial derivatives do not explicitly tell us the slope in directions not parallel to the axes.

All the derivative rules for functions of one variable apply for partial derivatives, as long as we remember to treat all the other variables as if they were constant. Let's illustrate with the function for the present value (V) of a long-lived project that we derived on pages 105–106.

$$V = f(R, i) = \frac{R}{i}$$

where R is annual return and i is the discount rate.

Treating i as a constant, the partial derivative with respect to R is

$$\frac{\partial V}{\partial R} = D\left[\frac{1}{i} \cdot R\right] = \frac{1}{i} D[R] = \frac{1}{i}(1) = \frac{1}{i}$$

Figure 13.5

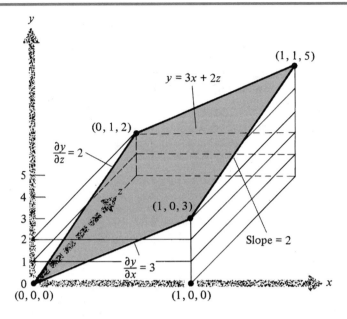

Treating R as a constant, the partial derivative with respect to i is

$$\frac{\partial V}{\partial i} = D[Ri^{-1}] = RD[i^{-1}] = R(-1i^{-2}) = -\frac{R}{i^2}$$

Examples

1. $y = f(x, z) = x^2 z$

   $$\frac{\partial y}{\partial x} = zD[x^2] = z(2x) = 2xz$$

   $$\frac{\partial y}{\partial z} = x^2 D[z] = x^2(1) = x^2$$

2. $y = f(x, z) = 5^x \ln z$

   $$\frac{\partial y}{\partial x} = (\ln z)D[5^x] = (\ln z)(5^x \ln 5)$$

   $$\frac{\partial y}{\partial z} = 5^x D[\ln z] = 5^x \left(\frac{1}{z}\right) = \frac{5^x}{z}$$

Exercises 12

Find $\dfrac{\partial y}{\partial x}$ and $\dfrac{\partial y}{\partial z}$ for each of the following functions.

DERIVATIVE RULES

1. $y = 5x - 4z$
2. $y = x^{-4}z^3$
3. $y = \dfrac{e^x}{\log_4 z}$
4. $y = (2x + 3z)^{1/2}$
5. $y = 7 + xz - x^2 z^3$

13.13 COMPUTER DIFFERENTIATION

We first met the neat software, www.calc101.com, on page 352, where we used it to perform matrix operations. As the home page indicates, it does derivatives too. By clicking on that, you get the web page shown at the top of Figure 13.6. Notice that the page gives notation to correctly input the function to be differentiated.

First let's try this software out on a simple function, $y = f(x) = x^2$, whose derivative we know to be $2x$. In the box headed by, "Take the derivative of", we input x^2 (Note that ^ means raise to the power of"). Once we click on "Do it", out comes the correct derivative, as shown at the bottom of Figure 13.6.

Next, let's get more adventuresome by inputting the more complex function, $y = f(x) = x \cdot e^x$. This function was differentiated in Example 1 on page 475. We input this function as: x*e^x (See Figure 13.7). The result, $e^x(x + 1)$, is exactly what we got the old-fashioned way.

Now, let's do one more example involving logarithms. The software is set up to deal with natural or base e logarithms. When either ln x or log x are inputted or outputted, the natural logarithm is implied. To use some other logarithmic base, one must use the notation, log [b, x], where b is the base. The software will automatically convert that logarithm to a natural or base e logarithm by the conversion relationship (see page 46).

$$\log_b x = \frac{\ln x}{\ln b} = \left(\frac{1}{\ln b}\right) \ln x$$

and then work with the natural logarithm expression. Let's see how by inputting the function, $y = \log_{10} x$. (See Figure 13.8) The result, $\dfrac{1}{x \ln 10}$, is exactly what we got by old-fashioned methods.

Now, I know what you are thinking. Why did we go through all that hard work to derive derivative rules when this neat software can differentiate without all that sweat? The answer is four-fold.

1. Your professor won't let you bring a computer into the exam room.
2. The development of rules gave you a very good basic understanding of what derivatives are all about.
3. No pain, no gain!
4. Stop complaining!

One last point: This software can serve as a nice check on your old-fashioned determination of derivatives.

Figure 13.6

Step-by-Step Differentiation

Plot your function!

home | FAQs | info@calc101.com

polynomial multiplication | polynomial division | graphs *new!* | matrix algebra *new!* | integrals

Use the right input format to get good results.	Wrong	Right
Use square brackets [] for functions.	sin(x), sqrt x, lnx, sqrt[ln x]	sin[x], sqrt[x], ln[x], sqrt[ln [x]]
Use parentheses () for grouping.	[x+1]^3, e^[x+1], x^2/[1+x^3]	(x+1)^3, e^(x+1), x^2/(1+x^3)
Use spaces or * for multiplication.	xe^-x, xsin[pix], bx^2	x e^-x, x sin[pi x], b x^2

Take the derivative of
```
x^2
```

with respect to
```
x
```

For the second derivative as well as the first, replace 1 by 2:
```
1
```

▶ DO IT

$$\frac{d}{dx}(x^2)$$

The derivative of x^n is $n\, x^{n-1}$.

$= 2x$

This material is subject to copyright.
Any unauthorized use, copying, or mirroring is prohibited.
©2000–2003 calc101.com

POWERED BY
web*MATHEMATICA*

Figure 13.7

Step-by-Step Differentiation

Plot your function!

home | FAQs | info@calc101.com

polynomial multiplication | polynomial division | graphs *new!* | matrix algebra *new!* | integrals

Use the right input format to get good results.	Wrong	Right
Use square brackets [] for functions.	sin(x), sqrt x, lnx, sqrt[ln x]	sin[x], sqrt[x], ln[x], sqrt[ln [x]]
Use parentheses () for grouping.	[x+1]^3, e^[x+1], x^2/[1+x^3]	(x+1)^3, e^(x+1), x^2/(1+x^3)
Use spaces or * for multiplication.	xe^-x, xsin[pix], bx^2	x e^-x, x sin[pi x], b x^2

Take the derivative of

x*e^x

with respect to

x

For the second derivative as well as the first, replace 1 by 2:

1

DO IT ▶

$\frac{d}{dx}(e^x x)$

Use the product rule
$\frac{d(uv)}{dx} = \frac{du}{dx}v + u\frac{dv}{dx}$,
where $u = e^x$ and $v = x$.

$= x \frac{d}{dx}(e^x) + e^x \frac{d}{dx}(x)$

The derivative of e^x is e^x.

$= e^x x + e^x \frac{d}{dx}(x)$

The derivative of x^n is $n x^{n-1}$.

$= e^x x + e^x$

Simplify, assuming all variables are positive.

$= e^x (x + 1)$

Figure 13.8

Step-by-Step Differentiation

Plot your function!

home | FAQs | info@calc101.com

polynomial multiplication | polynomial division | graphs *new!* | matrix algebra *new!* | integrals

Use the **right input format** to get good results.	Wrong	Right
Use square brackets [] for functions.	sin(x), sqrt x, lnx, sqrt[ln x]	sin[x], sqrt[x], ln[x], sqrt[ln [x]]
Use parentheses () for grouping.	[x+1]^3, e^[x+1], x^2/[1+x^3]	(x+1)^3, e^(x+1), x^2/(1+x^3)
Use spaces or * for multiplication.	xe^-x, xsin[pix], bx^2	x e^-x, x sin[pi x], b x^2

Take the derivative of
`log[10,x]`

with respect to
`x`

For the second derivative as well as the first, replace 1 by 2:
`1`

▮ DO IT ▶

$$\frac{d}{dx}\left(\frac{\log(x)}{\log(10)}\right)$$

The derivative of a constant times a function is the constant times the derivative of the function.

$$= \frac{1}{\log(10)} \frac{d}{dx}(\log(x))$$

The derivative of $\log(x)$ is $\frac{1}{x}$.

$$= \frac{1}{x \log(10)}$$

Exercises 13

Use www.calc101.com to differentiate the following functions.

1. $y = f(x) = x^3$
2. $y = f(x) = 5^x$
3. $y = f(x) = \ln x^3$
4. $y = f(x) = x^4 e^{x^3}$

13.14 WHOA! LET'S STOP AND SEE WHAT WE HAVE ACCOMPLISHED

Don't let the Δy's, Δx's, derivatives, rules, limits, and so on cloud what we have accomplished in this chapter. Believe it or not, we have laid a solid foundation for the many applications of succeeding chapters.

We began the previous chapter by recognizing that totals, such as total distance traveled, total sales of a product, total pollution in a lake, and total cost of a production run, are relevant to various applications. Often we can relate one total variable to another in functional form. But then this brings up the question of the effect of one variable on the other, or how do changes in one variable affect the value of the other variable. Such questions can only be answered by knowing the rate of change or derivative of the function. So this started us on a lengthy two-chapter process to determine derivatives for various important functions. This "pure math" had to be done (take my word for it) before you could really understand the applications in the next two chapters. We can illustrate this contention, while providing a glimpse of the nature of the applications, by returning to the S. Lumlord problem.

Recall that the total number of apartments rented (Q) as a function of the total rent per apartment (P) is

$$Q = f(P) = -2P + 1000$$

and so the total revenue (R) function is

$$R = PQ = P(-2P + 1000) = -2P^2 + 1000P$$

Now, this raises two types of questions for S. Lumlord to ponder, namely (1) given that he currently charges rent of $225, would he increase or decrease his total revenue by raising the rent? and (2) how can he achieve the maximum possible total rent revenue? The first question is typical of the applications in Chapter 14, and the second question is typical of the applications in Chapter 15. Both questions require finding the derivative of the revenue function first. And since it is a fairly complex function requiring three derivative rules for resolution, this is where all the sweat expended in Chapters 12 and 13 begin to reap applied dividends.

$$R = f(P) = -2P^2 + 1000P$$

$$\frac{dR}{dP} = f'(P) = D[-2P^2 + 1000P]$$

(Notice how we used all three symbols for the derivative on one line. Whee!)

$$\frac{dR}{dP} = f'(P) = D[-2P^2] + D[1000P] \quad \text{Summation rule}$$

$$= -2D[P^2] + 1000D[P] \quad \text{Constant multiplying rule}$$

$$= -2(2P^1) + 1000(1P^0) \quad \text{Power rule}$$

$$= -4P + 1000$$

This is as far as we can go here. In the next two chapters we will return to this problem and answer the two questions.

PROBLEMS

TECHNICAL TRIUMPHS

1. The Calculus II rocket's distance above ground (y) in miles is the following function of time since blast-off (x) in minutes.

   $$y = f(x) = x^3$$

   a. What rule is needed to find the derivative?
   b. Use that rule to find the derivative.
   c. How fast is the rocket traveling when $x = 2$?

2. Consider the function:

   $$y = f(x) = e^x + \ln x$$

   a. List the series of rules needed to find the derivative
   b. Use those rules to show that the derivative is $e^x + \frac{1}{x}$.

3. If a learning curve is described by the function:

   $$y = f(x) = 20 + 40\log_{10} x$$

   where y is the grade on a mathematics exam and x is the number of hours one studies.
   a. Find the derivative. Specify the rules that you used in this process.
   b. Graph the derivative for the domain $\{1 \leq x \leq 100\}$.

4. Consider the multivariate function:

   $$y = f(x,z) = x^4 e^z$$

   a. Find $\dfrac{\partial y}{\partial x}$
   b. Find $\dfrac{\partial y}{\partial z}$

DERIVATIVE RULES

5. x^{-1}, a power function, can also be expressed in quotient function form as $\frac{1}{x}$. Use both the power and quotient rules to find the derivative. Do both rules give the same result?

6. Use the derivative rules to show that s is the slope of the general linear function, $y = f(x) = sx + i$, where s and i are constants.

7. The demand (Q) for apartments is a function of rent level (P) for S. Lumlord

$$Q = f(P) = 1000 - 2P$$

Use the rule that you derived in problem #6 to find the derivative of the demand function.

8. If the derivative turns out to be the same as the original function, what was that function?

9. What is the derivative of $y = f(x) = \log_{10} 3$? Why?

10. Use the chain rule to find the derivative of

$$y = f(x) = (x^2 + 10)^5$$

CONFIDENCE BUILDERS

1. Consider the function

$$y = f(x) = x^2 e^x$$

   a. Differentiate this function, starting with the product rule.
   b. Put the function in the form of a quotient and differentiate it, starting with the quotient rule.
   c. Verify that both starting points in differentiation lead to the same final results.

2. Show that both the exponential rule and the chain rule as starting points lead to the same derivative for

$$y = f(x) = 3^{x^2 - x}$$

3. Show that both the logarithmic rule and the chain rule as starting points lead to the same derivative for

$$y = f(x) = \log_{10}(4 + x^5)$$

4. Use the inverse rule to find $\frac{dy}{dx}$, if $x = y^2$.

5. The following function is the result of taking the derivative of some unknown function, $f(x)$.

$$D[f(x)] = 5x^4$$

   a. What was $f(x)$?
   b. Could there be other functions that have that same derivative? Explain.

6. Consider the function

$$y = f(x) = e^x \log_{10} x$$

   a. What derivative rule should be used first to get started?
   b. Complete the process of finding $\frac{dy}{dx}$. For each step, give the derivative rule used.

7. Suppose that the segmented linear velocity functions for Roger Rannister's Mathachusetts Marathon are given as

$$v = f(x) = 10 - 2x \quad for \quad \{x \le 2\}$$
$$v = g(x) = 6 + 5(x - 2) \quad for \quad \{2 < x \le 3\}$$

   a. Find the derivative for each segment.
   b. Explain why these derivatives could be referred to as "second" derivatives.

8. On page 227, the liquidity trap is explained by the model

$$R = a + \frac{b}{c + i}$$

   where R is total cash reserves in the economy, i is the interest rate, and a, b, and c are constants.

   a. Use the derivative rules to find $\frac{dR}{di}$.
   b. What happens to the value of $\frac{dR}{di}$ as i gets very large?

9. A 1973 plan to limit the California budget (spurred by the budget-cutting philosophy of the then-governor Ronald Reagan) restricted the state budget (B) in billions of dollars over time (t) in years since 1973 according to the function

$$B = f(t) = (8.75 - .1t).08^t$$

   a. Find $\frac{dB}{dt}$.
   b. Graph the derivative over the domain $\{0 \le t \le 100\}$.

10. Consider the function

$$y = e^x \ln y$$

   a. Find the partial derivatives
   b. If $x = 0$ and $y = 1$, find the values of those partial derivatives.
   c. For large values of y, which partial would be bigger. Why?

MIND STRETCHERS

1. On page 194??, we saw that total world population (P) is approximately the following function of time (t) in years since 1990.

$$P = f(t) = 5(1.015)^t$$

   a. Find $\frac{dP}{dt}$.
   b. The percentage growth of population (ρ) is $\dfrac{100\left(\dfrac{dP}{dt}\right)}{P}$.
   c. Determine $\dfrac{d\rho}{dt}$

DERIVATIVE RULES

2. Consider the function

   $$y = f(x) = (x - 2)^4$$

   a. Use the chain rule to find its derivative.
   b. Expand the function using the binomial expansion. Starting with this form of the function, find the derivative.
   c. Show that both derivatives found above are equivalent.

3. The Mean Value Theorem of Calculus (we will need it later in Chapter 16 to develop integral calculus) states that for any continuous function, $f(x)$, over the interval from $x = a$ to $x = b$

   $$\frac{f(b) - f(a)}{b - a} = f'(c) \quad \text{where } c \text{ is some value of } x \text{ within the interval.}$$

   In plain English, it says that the slope of the secant line connecting the end points of an interval is equaled by the instantaneous rate or derivative of the function somewhere within the interval.

   a. Consider the Calculus Rocket I distance function

      $$y = f(x) = x^2$$

      over the interval, $x = 2$ to $x = 4$. At what point is the Mean Value Theorem satisfied?
   b. Show that c is always at the midpoint of any interval for this function.
   c. For the function $g(x) = x^3$ that describes Calculus II's blast-off, determine the value of c for the interval $x = 1$ to $x = 5$.

4. Using the Mean Value Theorem described in problem 3, relate the value of c to the values of a and b for the function

   $$y = f(x) = \ln x$$

5. The following function is a result of using the derivative rules on some unknown function, $f(x)$.

   $$D[f(x)] = 6^{x^2} 2 \ln x$$

   a. What was $f(x)$?
   b. Could there be other functions that have this same derivative? Explain.

6. Using the four-step process for finding derivatives, show that

   $$D[f(x) + g(x) + h(x)] = D[f(x)] + D[g(x)] + D[h(x)]$$

7. In Chapter 4 we derived the following formula for the value (V) of a common stock in terms of its current dividend (D), dividend growth rate (g), and the discount rate (i).

   $$V = \frac{D}{i - g}$$

   a. Find the partial derivatives $\partial V/\partial D$, $\partial V/\partial i$, and $\partial V/\partial g$.
   b. If $D = \$2$, $i = .10$, and $g = .05$, what is the rate of change of value with respect to dividend growth rate? By approximately how much does the stock value increase for a 1 percentage point increase in dividend growth rate?

8. The hyperbolic revenue function on page 239 was

   $$R = f(P) = \frac{14P}{4 + P}$$

a. Find $\dfrac{dR}{dP}$.

b. What does the value of this derivative approach as P goes to infinity?

9. Find the derivative of

$$y = f(x) = a^{\log_b x^c}$$ where a, b, and c are constants

10. If S. Lumlord's total revenue (R) and total cost (C) functions of rent (P) are:

$$R = 1000P - 2P^2$$
$$C = 64{,}000 - 40P$$

Show that the profit function has a zero slope where $\dfrac{dR}{dP} = \dfrac{dC}{dP}$.

11. The perfect gas law states that for one mole of gas

$$PV = .082T$$

where P is the pressure, V is the volume, and T is the temperature. Determine the following partial derivatives.

a. $\dfrac{\partial P}{\partial V}$

b. $\dfrac{\partial P}{\partial T}$

c. $\dfrac{\partial T}{\partial P}$

d. $\dfrac{\partial T}{\partial V}$

12. Find the derivative of the "monster" function on page 462.

CHAPTER 14

Applications of the Derivative

14.1 INTRODUCTION

Here you stand with a brain full of differential calculus goodies (if this isn't so, you are in trouble!), ready to be unleashed for the good of mankind! But, you may be a little skeptical and wondering what it is all good for. So this chapter (and the next one too) is designed to provide you with these much-needed applications.

For a starter let's return to the continuing saga of S. Lumlord. Recall that in the last episode we left Scrooge standing there with the derivative of the revenue function, called *marginal revenue*,

$$\frac{dR}{dP} = f'(P) = -4P + 1000$$

as he pleaded for answers to two pressing questions, which we restate here:

1. Given that he currently charges rent of $225 per apartment, would be increase or decrease total revenue for his 600 apartment complex by raising rent?
2. How can he achieve the maximum possible total rent revenue for the apartment complex?

The answer to Scrooge's first question rests on the sign of the derivative. If the revenue function is increasing at the point of interest (positive marginal revenue), it pays to gouge those tenants some more. However, if the revenue function is decreasing there (negative marginal revenue), he gouges himself by trying to gouge them. Let's see which case is true at $P = \$225$.

$$f'(225) = -4(225) + 1000 = +100$$

This result means that at the $225 rent level, revenue is increasing at the instantaneous rate of $100 per dollar increase in rent. Even Scrooge Lumlord understands the meaning of that; in fact, the rent increase notices are in the mail already.

Exercise 1

Compute the exact total revenue at rents of $225 and $226. Explain why the higher rent level doesn't result in $100 more revenue.

Had the rent been $285, marginal revenue would have been negative.

$$f'(285) = -4(285) + 1000 = -140$$

Further rent increases would be foolish. In fact, rent decreases are in order if Scrooge wants to achieve higher total revenue.

Figure 14.1 summarizes the information that the derivative has provided so far.

Exercise 2

Evaluate marginal revenue for rent levels of $240 and $275. What would you advise Scrooge to do (raise or lower rent) for each case?

Scrooge's second question involved finding the rent that results in maximum revenue. This could be done by trying various rents until we found one with a zero marginal revenue. However, we will wait until the next episode (chapter) for a systematic search of such questions. Tune in then for the exciting climax to Scrooge's story.

Learning Plan

In this chapter we illustrate the widely varied applications of the derivative. These applications fall into four categories.

1. *Direction of total function.* S. Lumlord gave us an example of this type of application. Here, the derivative is used to find the direction of the total function. Depending on

Figure 14.1

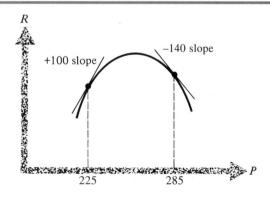

APPLICATIONS OF THE DERIVATIVE

whether the derivative is positive or negative and whether we wish to achieve higher or lower totals, we can then make a wise decision.

2. *Specification of concept or model.* The second type of application occurs when a derivative is part of the definition of some concept (for example, elasticity of demand) or included in the specification of some mathematical model (for example, the Harrod-Domar economic growth model).

3. *Special meaning.* The third type of application occurs when the derivative takes on a special meaning. Recall that the derivative of distance is velocity, an important concept in itself. Acceleration and probability are two other examples whereby derivatives have taken on special and important meaning.

4. *Graphing aid.* Since derivatives give the direction of a curve, they can be put to good use as a graphing aid.

To illustrate these varied uses of derivatives, we present in this chapter applications relating to microeconomics, general economic models, sports, public administration, velocity and acceleration, probability theory, science and engineering, and graphing.

You probably won't relate well to or even need some of these applications. For example, business students wouldn't enhance their careers by mastering the science and engineering applications; gals would be content to be ignorant of the sports applications (except those with "sports nuts" boyfriends/husbands). Nevertheless, seeing such a wide expanse of derivative applications will surely awe you and thus strengthen your appreciation of their applications in your own area of interest.

14.2 MICROECONOMICS

Microeconomics is the study of economic activity at the small-unit level (individual, household, firm, educational institution, etc.). Many of the applications already discussed in this text are a part of this subject matter. Here, we will extend our knowledge of this area by presenting derivative applications relating to elasticity of demand and individual consumption functions.

Elasticity of Demand

By using the marginal revenue concept, S. Lumlord was able to tell whether raising the rent was a wise move. An alternative approach, using the concept of elasticity of demand (ε), defined in terms of a derivative as

$$\varepsilon = -\frac{P}{Q}\left(\frac{dQ}{dP}\right)$$

where P is price per unit, and Q is quantity sold, can answer S. Lumlord's question, while also providing greater insight into such pricing decisions.

First of all it is important to show that elasticity of demand is not the brainchild of a mathematician wanting to have a formula containing a derivative. Rather, as we show now, the formula complete with derivative just happens to result as a byproduct from a common-sense approach toward pricing. Let's see.

Unfortunately, when the price of a product is increased, the sales go down. This in itself should not deter price increases. After all, who (excluding the consumers) would mind a 2 per-

Figure 14.2

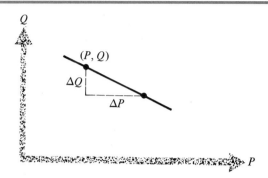

cent sales drop after a 20 percent price increase? So you can see that it is the relationship of sales and price percentage changes between two points on the demand curve as captured by the so-called *arc elasticity* formula[1]

$$-\frac{\% \text{ change in } Q}{\% \text{ change in } P} = -\frac{100\left(\frac{\Delta Q}{Q}\right)}{100\left(\frac{\Delta P}{P}\right)}$$

that counts (see Figure 14.2). For example, if the price increases by 10 percent and sales only decrease 5 percent [arc elasticity of $-(-5)/10 = +\frac{1}{2}$], that is good. If a 10 percent price increase results in a 20 percent sales drop [arc elasticity of $-(-20)/10 = +2$], that is bad.

By manipulating our common sense arc elasticity formula, we get

$$\text{Arc elasticity} = -\frac{100\left(\frac{\Delta Q}{Q}\right)}{100\left(\frac{\Delta P}{P}\right)} = -\frac{\Delta Q}{Q} \cdot \frac{P}{\Delta P} = -\frac{P}{Q}\left(\frac{\Delta Q}{\Delta P}\right)$$

To consider price increases in general at any point, we let $\Delta P \to 0$. The result is the so-called point elasticity of demand formula given earlier.

$$\varepsilon = \text{Point elasticity} = \lim_{\Delta P \to 0}\left[-\frac{P}{Q}\left(\frac{\Delta Q}{\Delta P}\right)\right]$$

$$= -\frac{P}{Q}\left(\frac{dQ}{dP}\right)$$

Point elasticity (ε) must be a positive number since dQ/dP, the slope of the demand curve, is negative. When $\varepsilon < 1$, demand is said to be inelastic; when $\varepsilon > 1$, demand is said to be elas-

---

[1] Since the percentage change in quantity sold is negative when the percentage change in price is positive (and vice versa), a negative sign is included in the definition so as to always get positive elasticities.

APPLICATIONS OF THE DERIVATIVE

tic. For a given product both elastic and inelastic portions of the demand curve may exist, as we illustrate now for the case of a linear demand curve.

Elasticity for Linear Demand

The general negative-sloping linear demand curve can be expressed as

$$Q = f(P) = -sP + i$$

where $-s$ is the slope ($s > 0$), and i is the intercept. We vary from normal convention by explicitly giving s a negative sign because it makes subsequent results more understandable.

Finding the derivative of the demand function and substituting it into the elasticity formula gives

$$\frac{dQ}{dP} = D[-sP + i] = -s$$

$$\varepsilon = -\frac{P}{Q}\left(\frac{dQ}{dP}\right) = -\frac{P}{-sP + i}(-s) = \frac{sP}{i - sP}$$

The dividing point between elastic and inelastic portions of the demand curve is $\varepsilon = 1$. Solving for P at this point gives

$$1 = \frac{sP}{i - sP} \quad \text{so} \quad i - sP = sP$$

$$i = 2sP$$

$$P = \frac{i}{2s}$$

It is instructive to show (while providing another application of the derivative) that for $P > \frac{i}{2s}$, $\varepsilon > 1$ (elastic), and for $P < \frac{i}{2s}$, $\varepsilon < 1$ (inelastic). We do this by using the quotient rule to find the rate of change of ε with respect to P.

$$\frac{d\varepsilon}{dP} = D\left[\frac{sP}{i - sP}\right] = \frac{(i - sP)D[sP] - (sP)D[i - sP]}{(i - sP)^2}$$

$$= \frac{(i - sP)(s) - (sP)(-s)}{(i - sP)^2}$$

$$= \frac{is}{(i - sP)^2}$$

Since the numerator is positive ($i > 0$, $s > 0$) and the denominator is positive (the square of any number is positive), $d\varepsilon/dP$ is always positive. This means that ε increases whenever P increases. Since we know that $\varepsilon = 1$ when $P = i/2s$, we now know that $\varepsilon > 1$ when $P > i/2s$, or conversely, that $\varepsilon < 1$ when $P < i/2s$. See Figure 14.3.

Figure 14.3

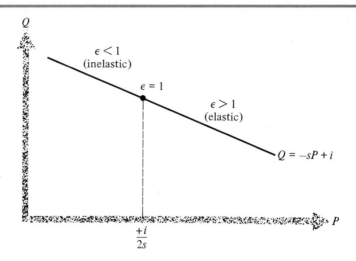

Relationship of Elasticity and Marginal Revenue

Elasticity and marginal revenue are closely linked, as we illustrate for the linear demand case.

$Q = f(P) = -sP + i$

$R = PQ = P(-sP + i) = -sP^2 + iP$

Marginal revenue $\left(\dfrac{dR}{dP}\right)$	Elasticity case
$\dfrac{dR}{dP} = -2sP + i$	
$\dfrac{dR}{dP} = 0$ if $2sP = i$ or $P = \dfrac{i}{2s}$	$\varepsilon = 1$
$\dfrac{dR}{dP} > 0$ if $i > 2sP$ or $P < \dfrac{i}{2s}$	$\varepsilon < 1$ (Inelastic)
$\dfrac{dR}{dP} < 0$ if $i < 2sP$ or $P > \dfrac{i}{2s}$	$\varepsilon > 1$ (Elastic)

So, revenue increases with price increases in the inelastic section and decreases with price increases in the elastic section. When $\varepsilon = 1$, revenue momentarily remains unchanged.

Example

The linear demand curve facing S. Lumlord is

$Q = -2P + 1000$

The elasticity of demand for this case is

APPLICATIONS OF THE DERIVATIVE

$$\varepsilon = \frac{sP}{i - sP} = \frac{sP}{-sP + i} = \frac{2P}{-2P + 1000}$$

and

$$\varepsilon = 1 \quad \text{and} \quad \frac{dR}{dP} = 0 \quad \text{when} \quad P = \frac{i}{2s} = \frac{1000}{2 \cdot 2} = \$250$$

For the linear demand case, as exemplified here, elasticity is a hyperbolic function and revenue is a parabolic function (check them out with the quadratic identification rules) of price, which graph as shown in Figure 14.4.

Figure 14.4

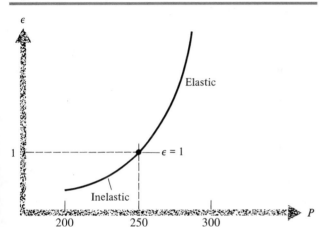

P	$\varepsilon = \dfrac{2P}{-2P + 1000}$
200	$\dfrac{400}{600} = .67$
225	$\dfrac{450}{550} = .82$
250	$\dfrac{500}{500} = 1$
275	$\dfrac{550}{450} = 1.22$
300	$\dfrac{600}{400} = 1.5$

P	$R = PQ = P(-2P + 1000)$
200	$200 \cdot 600 = 120{,}000$
225	$225 \cdot 550 = 123{,}750$
250	$250 \cdot 500 = 125{,}000$
275	$275 \cdot 450 = 123{,}750$
300	$300 \cdot 400 = 120{,}000$

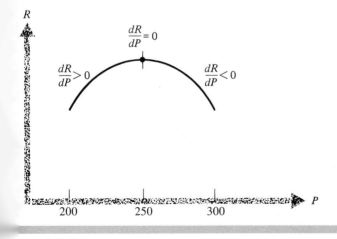

Exercise 3

Consider the linear following demand curve facing the Maxi Taxi Co as derived on page 244.

$$Q = f(P) = -.4P + 90$$

    a. Find the elasticity function.
    b. Find the marginal revenue function.
    c. At what price is $\varepsilon = 1$?
    d. At what price is marginal revenue equal to zero?
    e. Explain why the answers to parts c. and d. are the same.
    f. Graph the demand curve (delineating the different elasticity sections) and the total revenue curve (delineating the different marginal revenue sections). What do you observe?

The elasticity and marginal revenue relationships derived above for the linear demand case hold for any realistic demand curve. We illustrate the proof, in part, of this by showing that for any demand function, $\varepsilon = 1$ and marginal revenue is zero at the same price.

$$Q = f(P) \qquad \text{Note: } \frac{dQ}{dP} = D[f(P)]$$

$$R = PQ = P \cdot f(P)$$

Elasticity   *Marginal revenue*

$$\varepsilon = -\frac{P}{Q}\left(\frac{dQ}{dP}\right) \qquad \frac{dR}{dP} = D[P \cdot f(P)]$$

$$= P \cdot D[f(P)] + f(P) \cdot D[P]$$

$$1 = -\frac{P}{f(P)} \cdot D[f(P)] \qquad 0 = P \cdot D[f(P)] + f(P) \cdot 1$$

$$f(P) = -P \cdot D[f(P)] \qquad f(P) = -P \cdot D[f(P)]$$

$$P = -\frac{f(P)}{D[f(P)]} \xleftarrow{\text{same}}_{\text{price}} \rightarrow P = -\frac{f(P)}{D[f(P)]}$$

Exercise 4

Show that $\varepsilon = 1$ and marginal revenue equals zero at the same price for parabolic demand curves of the form

$$Q = f(P) = a - bP^2$$

where a and b are constants.

APPLICATIONS OF THE DERIVATIVE

Unitary Elasticity

Now consider the hyperbolic demand curve

$$Q = \frac{k}{P}$$

where k is a constant. This is the so-called unitary demand curve described in many economics texts. It is famous for its unusual elasticity and marginal revenue functions, which we now derive.

$$\varepsilon = \frac{-P}{Q}\left(\frac{dQ}{dP}\right)$$

Since

$$\frac{dQ}{dP} = D[kP^{-1}] = -kP^{-2} = \frac{-k}{P^2}$$

then

$$\varepsilon = \frac{-P}{kP^{-1}}\left(\frac{-k}{P^2}\right) = 1$$

$$R = PQ = P\left(\frac{k}{P}\right) = k$$

so

$$\frac{dR}{dP} = D[k] = 0$$

Thus, for the unitary demand curve, elasticity of demand is 1 and marginal revenue is zero for all prices. This results from the fact that total revenue is a constant.

Exercise 5

Find the elasticity and marginal revenue for the demand function

$$Q = f(P) = \frac{5000}{P}$$

From a total revenue standpoint, does it matter what price is charged? If costs and, subsequently, profits are considered, do you think that the price decision can make a difference?

Constant Elasticity

Hyperbolic-like demand functions of the form.

$$Q = f(P) = \frac{k}{P^n} = kP^{-n}$$

where k and n are positive real numbers, have constant elasticities as we show now.

$$\varepsilon = -\frac{P}{Q}\left(\frac{dQ}{dP}\right) = -\frac{P}{kP^{-n}}(-nkP^{-n-1}) = n$$

Unitary elasticity is a special case of these functions with $n = 1$. The marginal revenue of such functions is

$$\frac{dR}{dP} = D[P(kP^{-n})] = PD[kP^{-n}] + kP^{-n}D[P]$$

$$= P(-nkP^{-n-1}) + kP^{-n}(1)$$

$$= kP^{-n}(-n + 1)$$

Notice that for $n \neq 1$, marginal revenue is never equal to zero, and elasticity is never equal to 1.

For the case of $n > 1$, elasticity is greater than 1 (elastic), and marginal revenue is negative. Thus, any price increase results in a revenue decrease. At first this may seem impossible. But don't scoff at it yet. This may well be the case for mass transit systems. Raises in bus, rail, or subway fares generally cause such a large decrease in riders that revenue drops. Yet, isn't it true that such revenue declines produce demands to increase the fares?

For the case of $n < 1$, elasticity is less than 1 (inelastic) and marginal revenue is always positive. Thus, any increase in price increases total revenue. This could be the case for the so-called prestige items. The higher they are priced, the more people like them, since it is felt that the products are better.

Exercises 6

1. Consider the following demand curves:
   a. $Q_1 = \dfrac{1000}{P^{1/2}}$
   b. $Q_2 = \dfrac{1000}{P^2}$

   Find the elasticity and marginal revenue functions for each. What are the pricing implications of each demand curve?

2. Consider the hyperbolic demand function:

   $$Q = f(P) = \frac{14}{4 + P}$$

   Find the elasticity and marginal revenue functions. Should the price be raised from its current level of $10?

Individual Consumption and Saving Functions

Disposable income (Y) is defined as consumption (C) plus saving (S).

$$Y = C + S$$

APPLICATIONS OF THE DERIVATIVE

By observing individual (or family) spending patterns, we would observe that consumption and saving are related to disposable income (except for your Uncle Harry who came to visit five years ago and forgot to leave or go to work). We will see how understanding derivatives allows for the formulation of a model to describe consumption and saving.

If we hypothesize that consumption is linearly related to disposable income

$$C = f(Y) = sY + i$$

then the derivative (called *marginal propensity to consume*) is

$$\frac{dC}{dY} = D[sY + i] = s$$

Marginal propensity to consume, which in this case is constant, is the fraction of additional disposable dollars earned that are spent in consumption. The intercept term, i, represents the "minimum subsistence level" of consumption, or that level required just to survive even if no income is made.

With savings as the following function of disposable income,

$$S = Y - C = Y - (sY + i) = (1 - s)Y + i$$

the derivative of the savings function (called *marginal propensity to save*)

$$\frac{dS}{dY} = D[(1 - s)Y + i] = 1 - s$$

is a constant, meaning that the same fraction of each additional disposable dollar earned is saved.

Marginal propensity to consume and marginal propensity to save are related by the following equation of derivatives.

$$\frac{dS}{dY} = 1 - \frac{dC}{dY}$$

Exercise 7

For the linear consumption function

$$C = 3000 + .8Y$$

find

    a. the "minimum subsistence level."
    b. marginal propensity to consume.
    c. marginal propensity to save.
    d. the amount consumed and saved from a $1000 Christmas bonus.

The constant marginal propensity to consume and marginal propensity to save of the linear consumption function do not agree with reality. After all, don't we satiate more and more of our consumption wants as we go up the income ladder? So it makes sense that we consume a smaller fraction of each successive dollar we earn. Mathematically, this means that dC/dY decreases as Y increases.

A logarithmic consumption function of the form

$$C = a + c \log_b Y$$

where a and c are constants, satisfies this requirement since its derivative

$$\frac{dC}{dY} = \frac{c}{Y \ln b}$$

does decrease as Y increases.

Exercise 8

Verify for the above logarithmic consumption function that the savings function and marginal propensity to save are

$$S = f(Y) = Y - a - c \log_b Y$$

and

$$\frac{dS}{dY} = 1 - \frac{c}{Y \ln b}$$

What is the relationship between marginal propensity to consume and marginal propensity to save?

The logarithmic function's declining marginal propensity to consume, which makes sense for middle- and upper-income folks, may not adequately portray the situation for low-income folks. So perhaps a linear function up to some income level, followed by a logarithmic function, best explains the overall consumption function of individuals or families. Of course, each would have different parameters. We'll show this case for one family.

Example

Suppose that the following consumption function holds for the Juan Tamore family.

$C = 6 + .8Y$ for $\{Y < 25\}$

$C = -24 + 25 \log_5 Y$ for $\{Y \geq 25\}$

Marginal propensity to consume for the two regions is

APPLICATIONS OF THE DERIVATIVE

Figure 14.5

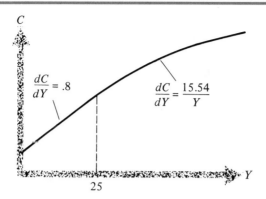

$$\frac{dC}{dY} = .8 \quad \text{for} \quad \{Y < 25\}$$

$$\frac{dC}{dY} = \frac{25}{Y \ln 5}$$

$$= \frac{15.54}{Y} \quad \text{for} \quad \{Y \geq 25\} \quad \text{(See Figure 14.5.)}$$

For example,

at $Y = 10 \quad \dfrac{dC}{dY} = .8$

at $Y = 30 \quad \dfrac{dC}{dY} = \dfrac{15.54}{30} = .52$

at $Y = 50 \quad \dfrac{dC}{dY} = \dfrac{15.54}{50} = .31$

Exercise 9

If the Juan Tamore family increases its standard of living as measured by the revised consumption function

$C = 6 + .9Y \quad \text{for} \quad \{Y < 25\}$

$C = -31.5 + 30 \log_5 Y \quad \text{for} \quad \{Y \geq 25\}$

find the marginal propensity to consume at disposable income levels of $10, $30, and $50 thousand.

14.3 GENERAL ECONOMIC MODELS

> Giant mathematical models of the United States economy—which attempt to predict the future on the basis of hundreds of equations drawn from the past—have developed recently into a $10 million-a-year business.... Top corporate economists inside the nation's large corporations now pay the model builders from $5,000 to more than $25,000 a year to tell them what seems to lie ahead....
>
> "Economic Models Now Big Business," *The New York Times,* April 8, 1975

This quote was from an article published about 30 years ago. To get a sense of today's activity, you can multiply each number cited by 10. Economic modeling by large corporations is big business indeed.

We can't give you (nor do I suspect you want) the hundreds-of-equations models here. However, the simple economic models of the economy presented here will give you a flavor for the real thing, while illustrating how calculus helps in the development and interpretation of such models.

Simple Model of Economy

Neglecting government expenditures, the sum of goods and services produced in the economy, called the *gross national product* (GNP) or *national income* (Y), is the sum of consumption (C) and investment (I) expenditures.

$$Y = C + I$$

Just as for individuals, as we discussed in the previous section, consumption in an economy is related to total income in the economy. Studies have shown that a linear consumption function, with constant marginal propensity to consume

$$C = f(Y) = sY + i$$

$$\frac{dC}{dY} = s$$

holds reasonably well for the overall economy.

Substituting the linear consumption function into the national income equation gives national income as a function of investment.

$$Y = C + I = (sY + i) + I$$
$$Y - sY = i + I$$
$$Y = \frac{i}{1-s} + \frac{I}{1-s}$$

Multiplier

Changes in investment have a multiple effect on changes in national income. We can see this by taking the derivative of national income with respect to investment.

APPLICATIONS OF THE DERIVATIVE

$$\frac{dY}{dI} = D\left[\frac{i}{1-s}\right] + D\left[\frac{I}{1-s}\right]$$

$$= 0 + \frac{1}{1-s} D[I]$$

$$= \frac{1}{1-s} (1)$$

$$= \frac{1}{1 - \frac{dC}{dY}}$$

Since the marginal propensity to consume is less than 1 for the economy, the derivative is greater than 1. So changes in investment have a multiple effect on changes in national income. Thus, this derivative is called *the multiplier*. For example, if the marginal propensity to consume is .8, then the multiplier is

$$\frac{dY}{dI} = \frac{1}{1 - .8} = 5$$

So, a $1 billion increase in investment results in a $5 billion increase in national income.

Exercises 10

1. For the above conditions, what would be the effect of a $1 billion decrease in investment?
2. If the marginal propensity to consume is .9, what is the effect of a $1 billion increase in investment? of a $1 billion decrease in investment?

Harrod-Domar Growth Model

The Harrod-Domar model is an example of models that investigate the nature of economic growth. This model hypothesizes that savings (S) are a constant fraction (a) of national income (Y), and investment (I) is a constant (b) times the rate of change of national income over time (t). These relationships are specified by the following equations:

$$\frac{S}{Y} = a$$

$$I = b\left(\frac{dY}{dt}\right)$$

Notice how a derivative is used in specifying this model.
To solve this model we must find the equation that gives the pattern of national income (Y) over time. This requires the methods of differential equations, which are beyond the scope of this book. So we just state the result as

$$Y = Y_0 e^{(a/b)t}$$ where Y_0 is the initial national income

Isn't it amazing how that confounded number e appears again!

By taking the derivative and evaluating it at $t = 0$, we can find the economic growth rate

$$\frac{dY}{dt} = f'(t) = Y_0 D[e^{(a/b)t}] = Y_0 e^{(a/b)t}\left(\frac{a}{b}\right)$$

$$f'(0) = Y_0 \left(\frac{a}{b}\right)$$

The initial absolute growth rate in national income is $Y_0(a/b)$, while (a/b) gives the percentage growth rate. For example, if $Y_0 = \$1{,}000$ billion, $a = .07$ and $b = 2.33$, then the economy initially experiences a $1000\left(\frac{.07}{2.33}\right) = 1000(.03) = \$30$ billion absolute growth rate in national income, which is a 3 percent growth rate.

Exercises 11

What does the Harrod-Domar model predict will happen to economic growth rates if:

a. The number a increases.
b. The number b increases.
c. Both a and b double.

14.4 SPORTS

Hello sports fans. This is Howard Cobuy. You are cognizant of the veracity that no one can comprehend me without a dictionary (you perspicacious fellow, Howard). In addition, I would be remiss in not informing you of the concomitant need for calculus. Let me elucidate.

Sports Statistics as Rates of Change

The sports world is filled with statistics that are really rates of change. Examples are batting averages (hits per time at bat), passing effectiveness (yards gained per pass), foul shot accuracy (good foul shots per attempt), bowling averages (pins per game), astronomical salaries (exhorbitant pay per year), and many others.

Most of the sports statistics you read about in the newspaper are average rates of change, as they represent the complete effort since the beginning of the season. Sports announcers, with their knowledge of recent performances, can augment these with something akin to instantaneous rates of change. For example, Howard may say, "Although he hit on only 40 percent of his foul shots this season, Moneybags Green is hitting a sensational 90 percent from the foul line in the playoffs thus far." Here, 90 percent is akin to an instantaneous rate of change, while 40 percent is an average rate of change. We can illustrate this in more detail with the case of Lefty Screwball, a baseball pitcher who always has a bad spring but comes on strong in the warm summer months.

APPLICATIONS OF THE DERIVATIVE

Baseball experts use the earned run average (ERA) to measure the ability of a pitcher. The ERA, as commonly used, is an average rate of change defined as the number of earned runs (those not a result of errors) allowed (R) per game (G) or number of innings pitched divided by 9.

$$\text{ERA} = \frac{R}{G}$$

For example, if Lefty gave up 18 earned runs in the first 27 innings of the season, he would have an ERA = $18/(27/9) = 18/3 = 6$.

In Lefty's case the following logarithmic function describes his pattern of slow start but improving performance throughout the season.

$$R = 50 \log_9 G \quad \{G \geq 1\}$$

Lefty's conventional ERA is

$$\text{ERA} = \frac{R}{G} = \frac{50 \log_9 G}{G}$$

This measure, because it is an average that includes the early setbacks, does not adequately evaluate Lefty's ability in the stretch run of the pennant race. The instantaneous ERA, which is the derivative of the runs function, does provide this proper evaluation.

$$\frac{dR}{dG} = D[50 \log_9 G] = 50 D[\log_9 G]$$

$$= 50 \left(\frac{1}{G \ln 9} \right)$$

$$= \frac{22.76}{G}$$

The two types of earned run averages are related as follows:

$$\text{ERA} = \frac{50 \log_9 G}{G} \left(\frac{\ln 9}{\ln 9} \right)$$

$$= \frac{dR}{dG} (\ln 9)(\log_9 G) \quad \text{since } \frac{dR}{dG} = \frac{50}{G \ln 9}$$

$$= \frac{dR}{dG} \left(\frac{\log_9 G}{\log_9 e} \right) \quad \text{since } \log_b B = \frac{1}{\log_B b} \quad \text{(see page 46)}$$

As long as G is greater than e (2.718 . . .), the fraction $(\log_9 G)/(\log_9 e)$ is greater than 1, and thus the average (ERA) is greater than the instantaneous (dR/dG) earned run rate, as illustrated in Figure 14.6. *Note:* Difficulty in defining R for $G < 1$ (actually, Lefty pitched a shutout in the first game since $R = 0$ when $G = 1$) allowed ERA to be smaller for a while. This would not have occurred had we defined $R = 50 \log_9 (G + 1)$, but that would have complicated the mathematics.

Figure 14.6

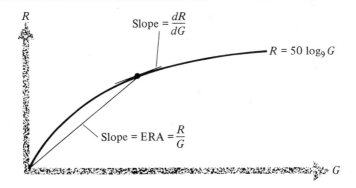

Exercises 12

1. Find the average and instantaneous earned run averages after 3, 9, and 27 games for Lefty.
2. Joe Detow's punting improves throughout the season, as reflected by his total punt yardage (Y) as a function of the number of kicks (K)

   $$Y = f(K) = 40K + .1K^2$$

   a. Determine Joe's average and instantaneous punt yardage rate.
   b. Show that the latter rate is always greater than the average rate.
   c. Find both rates after 10, 30, and 50 kicks.

Batting Championship Dynamics

On the last day of the baseball season when two players, Don Money and Royce Rolls, are slugging it out for the batting championship, Howard Cobuy needs to be able to determine quickly the change in batting average as a result of a hit or an out. Here we will provide this understanding—which, of course, illustrates another derivative application—to Howard.

Batting average (A) is defined as the number of hits (H) divided by the number of official times (walks and certain other plays don't count) at bat (B) or

$$\text{Batting average} = A = \frac{H}{B}$$

Now, if a player subsequently goes into a streak with x hits in a row, the batting average will be

$$A = \frac{H + x}{B + x}$$

To find the effect of x, we take the derivative.

APPLICATIONS OF THE DERIVATIVE

$$\frac{dA}{dx} = D\left[\frac{H+x}{B+x}\right]$$

$$= \frac{(B+x)D[H+x] - (H+x)D[B+x]}{(B+x)^2}$$

$$= \frac{(B+x)(1) - (H+x)(1)}{(B+x)^2}$$

$$\frac{dA}{dx} = \frac{B-H}{(B+x)^2}$$

Defining θ as the number of outs, where $\theta = B - H$, and θ/B as the "outing" average, and recognizing that B is much much greater than x (so $B + x \cong B$), we can approximate the derivative as

$$\frac{dA}{dx} \cong \frac{\theta}{B^2} = \left(\frac{\theta}{B}\right)\left(\frac{1}{B}\right)$$

$$= \left(\begin{array}{c}\text{outing}\\ \text{average}\end{array}\right)\left(\frac{1}{B}\right)$$

Thus, the instantaneous rate of change of a batting average (itself an average rate of change) is approximately the "outing" average times the multiplicative inverse of the number of bats. For example, a 300 hitter[2] (700 "outer") who has batted 400 times has an approximate rate of

$$\frac{dA}{dx} = (700)\left(\frac{1}{400}\right) = 1.75$$

Thus, each hit raises his batting average by about 1.75 points.

Exercises 13

1. If Rolls has a 350 average going into his last game (500 bats), find the approximate rate of change of his batting average with respect to x, number of successive hits.
2. If Money finished the season hitting 357, can Rolls win the batting championship if he bats four times in the final game? five times?

Now, let's determine the effect of outs on the batting average. If no hits are recorded in z times at bat, the batting average as a function of z is

$$A = \frac{H}{B+z}$$

The derivative of this function, which gives the instantaneous rate of change of a batting average with respect to the number of consecutive outs, is

---

[2] Here we express batting average in points, or 1000 times the average defined above. For example, a player that hits 30 percent of the time has a batting average of .300, which corresponds to 300 points.

$$\frac{dA}{dz} = D\left[\frac{H}{B+z}\right] = \frac{(B+z)D[H] - HD[B+z]}{(B+z)^2}$$

$$= \frac{(B+z)(0) - H(1)}{(B+z)^2}$$

$$= -\frac{H}{(B+z)^2}$$

Since B is much much greater than z (so $B + z \cong B$), we can approximate the derivative as

$$\frac{dA}{dz} \cong -\left(\frac{H}{B}\right)\left(\frac{1}{B}\right)$$

So the approximate drop in batting average per out is the previous average (before the bad streak) times the multiplicative inverse of the number of bats. For example, the 300 hitter, who has batted 400 times, sees his average drop by about $\frac{3}{4}$ point

$$\frac{dA}{dz} = -(300)\left(\frac{1}{400}\right) = -\frac{3}{4}$$

for each out.

Exercises 14

1. If Rolls has a 350 average going into the last game (500 bats), find the approximate rate of change of his batting average with respect to the number of consecutive outs.
2. If Money finished the season hitting 347, can Rolls lose the batting championship if he bats four times in the final game? five times?

14.5 PUBLIC ADMINISTRATION

Derivatives have many applications in the area of public administration, two of which involving state budgets and tax equalization are illustrated here.

State Budgets

In 1973 Governor Reagan's administration in California proposed a plan to limit the state budget, while still allowing the total budget to increase due to growth in California's total personal income. Assuming a $100 billion personal income for California in 1973, the formula, which relates the state budget (B) and the number of years since 1973, (t), is

$$B = (8.75 - .1t)(1 + g)^t$$

APPLICATIONS OF THE DERIVATIVE

where g is the growth rate of personal income in California (decimal). The October 16, 1973, *Wall Street Journal*, which reported this plan, quoted sources that said the state budget would increase threefold over the next 15 years. Let's investigate the direction of change of B with respect to t by taking the derivative and seeing how long it remains positive for different personal income growth rates.

$$\frac{dB}{dt} = (8.75 - .1t)D[(1+g)^t] + (1+g)^t D[8.75 - .1t]$$

$$= (8.75 - .1t)(1+g)^t \ln(1+g) + (1+g)^t(-.1)$$

$$= (1+g)^t \ln(1+g)\left[8.75 - .1t - \frac{.1}{\ln(1+g)}\right]$$

An approximation (that we won't prove) is

$$\ln(1+g) \cong g$$

if g is small (the case here). So our derivative simplifies to

$$\frac{dB}{dt} \cong (1+g)^t g \left[8.75 - .1t - \frac{.1}{g}\right]$$

Now we'll see how long the budget will increase, or in technical terms, how long the derivative will remain positive. The first two terms, $(1+g)^t$ and g are always positive for any positive value of g. The bracketed factor can be positive or negative, depending on the values of t and g. Initially it is positive, but over time it declines to zero. So by setting it equal to zero, we can find how long the budget will increase.

$$8.75 - .1t - \frac{.1}{g} = 0$$

$$t \cong \frac{8.75 - .1/g}{.1} = 87.5 - \frac{1}{g}$$

For example, if an 8 percent personal income growth rate is forecasted ($g = .08$), then the budget would continue to increase for approximately

$$t = 87.5 - \frac{1}{.08} = 87.5 - 12.5 = 75 \text{ years}$$

Exercises 15

1. How long would the budget increase if California had only a 5 percent annual growth of personal income?
2. What growth rate would result in a decline in the budget after only 37.5 years?
3. If a more stringent limit were placed on the budget, as reflected in the revised budget function,

$$B = (8.75 - .2t)(1 + g)^t$$

how long would the budget increase under the assumption of an 8 percent growth in personal income? 5 percent growth?

Tax Equalization Rates

Many counties apportion taxes to meet their budgets according to the total value of taxable property in the various districts within their jurisdiction. It is a long involved statistical process to estimate these totals (like estimating the price tag for a whole city); however, at the risk of oversimplification, the following procedure is used.

A random sample of properties from important classes of property (e.g., residential, commercial) is selected. Each property has an assessed value (figure on which property taxes are levied) and a market value, which is estimated by a team of appraisers. Assessed values as a percentage of market values vary widely over a district (this is a big gripe of taxpayers who feel that their assessments are too high). So, for each sampled class of property, an average ratio of assessed to market value is determined. Then the various sampled class results are averaged to give an overall ratio, called *the equalization rate*. Finally, the market value for the entire district is estimated by projecting the equalization rate to all property.

The author is involved in estimating equalization rates and total market values for all the taxing districts in one New York State county. Recently, one district disputed the appraisal of one large property (treated as a class by itself) in the sample. It claimed that an error in appraisal changed its equalization rate so as to result in more taxes for that district. The relevant tax equalization and total market value formulas to investigate this claim are

$$R = \frac{A_s}{M_s + m}$$

$$M = \frac{A}{R} = \frac{A}{A_s/(M_s + m)} = \frac{A}{A_s}(M_s + m)$$

where

A_s is the total assessed value of all the sampled classes
M_s is the estimated market value of all the undisputed sampled classes
m is the disputed appraisal
A is the total assessed value of all property
R is the equalization rate
M is the estimated market value for the entire district

The rate of change of the equalization rate with respect to changes in the disputed appraisal is

$$\frac{dR}{dm} = D\left[\frac{A_s}{M_s + m}\right] = \frac{(M_s + m)D[A_s] - A_s D[M_s + m]}{(M_s + m)^2}$$

$$= \frac{(M_s + m)(0) - A_s(1)}{(M_s + m)^2} = \frac{-A_s}{(M_s + m)^2}$$

APPLICATIONS OF THE DERIVATIVE

The actual values (rounded to ease calculation) were

$A_s = 200$ million

$M_s = 390$ million

$m = 10$ million

so

$$R = \frac{200}{400} = \frac{1}{2}$$

and

$$\frac{dR}{dm} = \frac{-200}{400^2} = \frac{200}{400}\left(-\frac{1}{400}\right) = \frac{1}{2}\left(-\frac{1}{400}\right)$$

The actual equalization rate was $\frac{1}{2}$ (property is assessed at half the market value). Now we see that the instantaneous rate of change is $-\frac{1}{400}$ times the equalization rate. So a decrease of $1 million in the disputed appraisal would mean only about a $\frac{1}{800}$ increase in equalization rate—from .500 to .50125.

The effect on total district market value can be determined by taking the derivative of M with respect to m

$$\frac{dM}{dm} = D\left[\frac{A}{A_s}(M_s + m)\right] = D\left[\left(\frac{A}{A_s}\right)M_s\right] + D\left[\left(\frac{A}{A_s}\right)m\right]$$

$$= 0 + \left(\frac{A}{A_s}\right)(1)$$

$$= \frac{A}{A_s}$$

With 80 percent of the property sampled, $A = 250$, so the rate of change of total value estimate with respect to change in appraisal of the disputed property is

$$\frac{dM}{dm} = \frac{250}{200} = 1.25$$

So, each additional million of appraised value of the disputed property adds $1.25 million to the estimate of district value. Conversely, and more relevant since the district thought that the disputed property had lower value, each drop of a million subtracts $1.25 million from the total value estimate.

Exercises 16

If the disputed property had a greater share of the district's assessed value, appraisal error would have had a greater effect. Determine dR/dm, dM/dm, and the changes in equalization rate and total market value for the following sets of conditions (keep $m = 10$).

a. $A_s = 100$ $M_s = 190$ $A = 125$
b. $A_s = 50$ $M_s = 90$ $A = 62.5$

14.6 VELOCITY AND ACCELERATION

Velocity, as we know, is the derivative of distance with respect to time. Velocity itself can change over time. The rate at which it changes, called *acceleration,* is found by taking the derivative of velocity with respect to time. Acceleration is thus the derivative of a derivative. We say that velocity is the first derivative and acceleration is the second derivative. This is our first exposure to second derivatives. They will play a prominent role in our study of optimization in the next chapter. Here, we will begin to love and understand them through applications relating to falling and propelled bodies.

Falling Bodies

Are you having so much trouble with derivatives that you are thinking about jumping from the nearest tall building? Well, stop, because you'll only come under the influence of other derivatives—velocity and acceleration—that describe the movement of falling bodies.

Scientists have long known that a body falling in earth's gravity will fall a distance (y) in feet given by the following function of time (x) in seconds.

$$y = f(x) = 16x^2$$

This formula, which was derived using arithmetic progressions on page 65 assumes frictionless fall, such as in a vacuum.

Velocity (v) and acceleration (g) are found by respectively taking the first and second derivatives of the distance function.

$$v = \frac{dy}{dx} = D[16x^2] = 32x$$

$$g = \frac{dv}{dx} = D[32x] = 32$$

Earth's gravitational constant is called g. It tells us that the velocity of a falling body increases 32 feet per second every second. Thus, the units of acceleration must be feet per second per second, or feet per second^2. Didn't you always wonder, as I did in high school physics, why acceleration had those crazy units? Well, thanks to differential calculus, you'll be able to sleep nights again.

When we derived the distance function using arithmetic progressions, the common difference was 32. It is interesting to see that g, or the second derivative, equals the common difference.

By graphing the distance, velocity, and acceleration functions, as we have done in Figure 14.7, we can see their interrelationships.

Now we'll return to that question of whether to jump or not. If you choose the observation tower of the Empire State Building (approximately 1024 feet) as your launching site, we can solve for x in the distance function to find the flying time.

APPLICATIONS OF THE DERIVATIVE

Figure 14.7

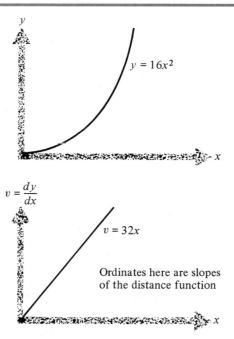

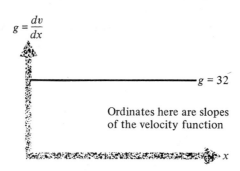

$1024 = 16x^2$ so $x = \sqrt{\dfrac{1024}{16}} = 8$ seconds

Thus, the impact velocity is

$v = 32 \cdot 8 = 256 \text{ ft/sec} \cong 175 \text{ mph}$

Whew! Derivatives aren't that bad after all!

Exercises 17

What would the impact velocities be if you jumped from the

1. Eiffel Tower (984 feet)?
2. Sear's building (1454 feet)?

Other celestial bodies have different gravitational forces. For example, on the moon the free-fall distance (y) in feet as a function of time (x) in seconds is approximately

$$y = f(x) = 3x^2$$

so

$$v = \frac{dy}{dx} = D[3x^2] = 6x$$

$$g_{moon} = \frac{dv}{dx} = D[6x] = 6$$

Exercises 18

1. Graph the moon's distance, velocity, and acceleration functions.
2. Find the distance fallen in moon free fall *during* the first, second, third, fourth, and so on seconds. Show that these numbers form an arithmetic progression with a common difference of 6, which equals g_{moon}.
3. How long would it take to jump off a 1000-foot cliff on the moon? What would the impact velocity be?
4. On Mars the free-fall distance function is approximately $y = f(x) = 6.08x^2$.
   a. Find the Martian velocity and acceleration functions.
   b. How long would it take to jump off a 1000-foot cliff on Mars? What would the impact velocity be?

Propelled Bodies

Calculus I Rocket

Calculus I propelled us into the world of derivatives in Chapter 12. Now let's see if we can propel it out of this world.

The escape velocity from earth is approximately 32,000 miles per hour, or 533 miles per minute, which, let's assume, must be attained by the time the rocket is 900 miles up. Recall that Calculus I had the distance function

$$y = f(x) = x^2$$

so we can find how long it takes to go up 900 miles.

$$900 = x^2 \quad \text{or} \quad x = 30 \text{ minutes}$$

The velocity at this point can be found by differentiating the distance function and evaluating the result at $x = 30$.

$$v = f'(x) = D[x^2] = 2x$$

$f'(30) = 2 \cdot 30 = 60$ miles per minute

APPLICATIONS OF THE DERIVATIVE

Not good enough! Watch out, it's coming back down!

If the Calculus I rockets imparted it with the following distance function

$$y = f(x) = x^4$$

so that the velocity function was

$$v = f'(x) = D[x^4] = 4x^3$$

then the time to reach 900 miles would be

$$900 = x^4 \quad \text{so} \quad x = \sqrt[4]{900}$$
$$\text{or} \quad x = 5.48 \text{ minutes}$$

At that point, the rocket will be whizzing upward at a velocity of

$$v = f'(5.48) = 4(5.48)^3 = 658.27 \text{ miles per minute}$$

Goodbye!

Exercise 19

Show that if Calculus I had the distance function $y = f(x) = x^3$, it would not be able to escape the earth's gravitational pull.

Hot Rods

Hot-Rod Rodney peels out from the red light and accelerates according to the distance function

$$y = f(x) = 6x^2$$

so

$$v = \frac{dy}{dx} = D[6x^2] = 12x$$

Rodney's acceleration would then be

$$a = \frac{dv}{dx} = D[12x] = 12$$

At this pace Rodney will reach 60 miles per hour (88 feet per second) in

$$88 = 12x \quad \text{so} \quad x = 7\tfrac{1}{3} \text{ seconds}$$

Exercise 20

Find Hot-Rod Rodney's velocity, acceleration, and time necessary to reach 60 miles per hour if

$y = f(x) = 7x^2$

14.7 PROBABILITY

Entire books are written on probability theory. Our only purpose here is to illustrate probabilities as rates, and to show how derivatives play a role in probability theory.

Probabilities are long-run rates of occurrence of some phenomena—in other words, the fraction of trials that the phenomenon occurs. For example, the probability of flipping heads with a fair coin is 1/2, since an infinite number of flips would result in a head appearing $\frac{1}{2}$ the time. The function giving the probability (rate of occurrence) of all possible results to a given experiment is called *a probability distribution*. Discrete and continuous probability distributions exist, as we now illustrate.

Discrete Probability

Some probability distributions are discrete, meaning that only a finite number of different results are possible. For example, the experiment of flipping a coin twice has four possible and equally likely outcomes—HH, HT, TH, TT—where H symbolizes head and T symbolizes tail. If we categorize results in terms of the number of heads, then the possible outcomes are 0, 1, and 2, and the probability distribution is

Number of heads	Probability
0	1/4
1	2/4
2	1/4

So, in the long run, zero and 2 heads each occur at the rate of once per four trials; 1 head occurs at the rate of twice per four trials. Overall, the sum of the rates is one, which is always the case when all the outcomes of a probability experiment are considered.

Exercise 21

For the three-flip experiment, the probability distribution is

Number of heads	Probability
0	1/8
1	3/8
2	3/8
3	1/8

APPLICATIONS OF THE DERIVATIVE

What is

    a. The rate of occurrence of two heads?
    b. The rate of occurrence of two or more heads?
    c. The sum of all the rates?

Continuous Probability

Some probability distributions are continuous, meaning that an infinite number of different outcomes is possible. For example, consider the height of an adult male. Theoretically, this variable can take on an infinite number of real number values. So, a man can be 70.829567 inches tall. Only the inaccurate measuring devices we use prevent us from knowing the "exact" height.

The normal probability distribution is a good continuous model to portray the rate of occurrence of physical and mental characteristics in large human populations. The general normal probability formula is

$$\frac{dy}{dx} = \left(\frac{1}{2.5\sigma}\right) e^{-.5[(x-\mu)/\sigma]^2}$$

where μ is the population average (mean), and σ is the population standard deviation.

For the specific case of the heights of adult males in the United States, with a 70-inch average height and a 4-inch standard deviation of heights, the formula reduces to

$$\frac{dy}{dx} = .1 e^{-.5[(x-70)/4]^2}$$

Here, the rate is measured in fraction of the population per inch of height. So, for example, at $x = 70$ inches.

$$\frac{dy}{dx} = .1 e^{-.5(0)} = .1$$

This means that about one-tenth of the population has heights between 70 and 71 inches. Of course, the rate continually changes, so our statement should be taken only as an approximation.

Exercises 22

1. What is the rate of occurrence of heights at
   a. $x = 74$ inches.
   b. $x = 66$ inches.
   c. $x = 62$ inches.
   d. $x = 78$ inches.
2. If IQ scores have a normal probability distribution with a 100 average and 10 standard deviation,
   a. Specify the rate formula.
   b. What is the rate of occurrence at $x = 100$?

Continuous probability distributions, such as the normal distribution, are called probability density functions. Such functions give the rate of observation for the variable under study and can be obtained by differentiating cumulative distributions. The latter represents the total fraction of occurrences up to a certain point. We can demonstrate this with a problem in product failure.

Many products have a cumulative pattern of failure, which can be modeled as

$$F = 1 - e^{-(x/\mu)}$$

where F is the total fraction of a batch of product (say, light bulbs) that have failed within x time in use and μ is the average lifetime of the product. The derivative of this cumulative distribution, which represents the failure rate or probability density, is

$$\frac{dF}{dx} = D[1 - e^{-(x/\mu)}] = D[1] - D[e^{-(x/\mu)}] = 0 - e^{-(x/\mu)}(\ln e)\left(-\frac{1}{\mu}\right)$$

$$= \left(\frac{1}{\mu}\right)e^{-(x/\mu)}$$

For example, if a batch of light bulbs has an average life of four months, then the failure rate is

$$\frac{dF}{dx} = \left(\frac{1}{4}\right)e^{-(x/4)}$$

Initially, the light bulbs fail at the rate of $(\frac{1}{4})e^{-0} = \frac{1}{4}$, or 25 percent per month. However, 25 percent will not fail the first month since the failure rate declines continuously, as shown in Figure 14.8. For example, the failure rate at the end of the first month is

$$\left(\frac{1}{4}\right)e^{-(1/4)} = 19.5 \text{ percent}$$

Figure 14.8

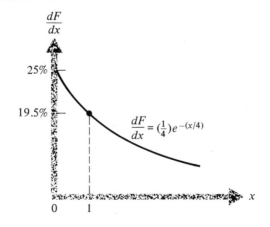

APPLICATIONS OF THE DERIVATIVE

Exercise 23

What is the light bulb failure rate at the following times:

    a. $x = 2$ months.
    b. $x = 4$ months.
    c. $x = 8$ months.

In the literature, this is called the negative exponential failure model. It turns out that for this model the failure rate is proportional to the fraction of "living" product, as we show now. Recall that the total fraction of failed product is $F = 1 - e^{-(x/\mu)}$. So the fraction of living product (L) is

$$L = 1 - F = 1 - (1 - e^{-(x/\mu)}) = e^{-(x/\mu)}$$

The failure rate, derived earlier, in terms of L is

$$\frac{dF}{dx} = \left(\frac{1}{\mu}\right) e^{-(x/\mu)} = \left(\frac{1}{\mu}\right) L$$

Since $(1/\mu)$ is a constant for a given product, failure rate for the negative exponential model is proportional to the fraction of living product. This explains the declining failure rate, since the fraction of living items from a given production batch continually declines over time.

Exercises 24

Find and graph the probability distribution for the following cumulative distributions:

1. $F = .1x$ $\{0 \le x \le 10\}$
2. $F = .2 + .4x$ $\{0 \le x \le 2\}$
3. $F = .1x + .02x^2$ $\{0 \le x \le 5\}$

14.8 SCIENCE AND ENGINEERING[3]

Science and engineering abound with differential calculus applications. Here we sketch examples relating to chemical reaction, heat transfer, and electric current flow.

Chemical Reaction

Chemical reactions are classified in terms of their rate of reaction. For example, if chemical A decomposes into chemicals B and C as

---

[3] This section is included in the spirit of revealing the widely varied applications of calculus, rather than to develop expertise in technical areas.

$$A \to B + C$$

in such a way that the reaction rate is proportional to the amount of A present (n_A), or

$$\frac{dn_A}{dx} = -kn_A$$

where k is a constant and x is the time, then the reaction is called *first order*. Such a rate occurs only if the following total function is true,

$$n_A = ae^{-kx}$$

where a is the initial amount of A. We can verify this by taking the derivative

$$\frac{dn_A}{dx} = aD[e^{-kx}] = a(e^{-kx} \ln e)(-k) = -k(ae^{-kx}) = -kn_A$$

Heat Transfer

Whenever there exists a difference in temperature between two regions separated by a wall of some sort, heat will flow through the wall from the warm to the cool area, as shown in Figure 14.9. This process is called *conduction*. So, heat flows from a room into the refrigerater; from the outside into an air-conditioned house; from a house to the outside in winter; and so on.

Refrigerating and heating devices are rated according to the rate at which they can take out or add heat to a system. Choosing the right size of such a unit depends, in part, on the rate of heat flow through the walls. This conduction rate through one wall is given by the fundamental law of heat transfer.

$$\frac{dH}{dx} = kA\Delta T$$

where

k is a constant depending on wall insulation
A is the wall area

Figure 14.9

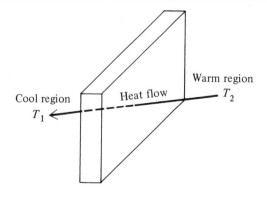

APPLICATIONS OF THE DERIVATIVE

$\Delta T = T_2 - T_1$ is the difference in temperature
H = BTU's of heat
x = time

Of course, in a complicated air conditioning or heating system, many "walls" must be considered to arrive at the proper specification.

As an example of the use of this law, consider the total conduction rate into a refrigerator if $\Delta T = 50°F$, $A = 60$ square feet, and $k = .1$.

$$\frac{dH}{dx} = (.1)(60)(50) = 300 \text{ BTU/hour}$$

This equation points out the importance of good insulation as measured by k. Had k been double, or .2, the conduction heat gain would have doubled.

Electric Current Flow

Electrical systems are described by the rate at which current flows through them. One such system, involving a battery and an induction coil, as illustrated in Figure 14.10, has a total current (C) given by the equation

$$C = \frac{V}{R} - \left(\frac{V}{R}\right)e^{-(Rx/L)}$$

where

V is the constant battery voltage
L is the constant induction coefficient
R is the constant resistance of the system
x is the time in seconds

From this, the rate of current flow can be found by differentiating

$$\frac{dC}{dx} = D\left[\frac{V}{R} - \left(\frac{V}{R}\right)e^{-(Rx/L)}\right] = D\left[\frac{V}{R}\right] - D\left[\left(\frac{V}{R}\right)e^{-(Rx/L)}\right]$$

Figure 14.10

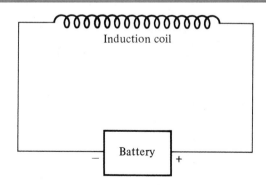

$$\frac{dC}{dx} = 0 - \left(\frac{V}{R}\right)e^{-(Rx/L)}(\ln e)\left(-\frac{R}{L}\right)$$
$$= \left(\frac{V}{L}\right)e^{-(Rx/L)}$$

The current flow rate is a negative exponential function starting out at V/L when $x = 0$ and diminishing with time—in fact, approaching zero as x approaches ∞. See Figure 14.11.

Figure 14.11

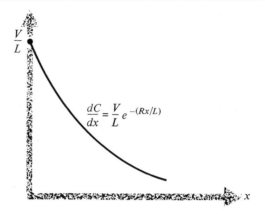

14.9 GRAPHING AID

Knowledge of where the derivative is positive and negative is an important aid when graphing unfamiliar functions. For example, consider the family of power functions $y = x^n$, where n is a positive integer. We haven't graphed such functions (except for the parabola $y = x^2$), so you should not be familiar with their shapes.

The derivative of power functions

$$D[x^n] = nx^{n-1}$$

reveals quite different graphs for even and odd integer values of n, as described in Table 1.

Since the absolute value of the derivative increases as n gets larger, the curves get steeper as n increases (the basic slope directions remained unchanged). See Figure 14.12.

Exercises 25

1. The logarithmic derivative rule is

$$D[\log_b x] = \frac{1}{x \ln b}$$

APPLICATIONS OF THE DERIVATIVE

Explain how this result helps graph logarithmic functions with different bases. Specifically, consider the following.
  a. Can such functions ever "turn down" as x increases?
  b. What is the effect of increasing the base b on the relative positions of different logarithmic functions?
2. Consider the function: $y = f(x) = xe^{-x}$
Find the derivative and use it to determine the "shape" of the function.

Table 1

Case	n is an even integer (call it ε)	n is an odd integer (call it θ)
$\dfrac{dy}{dx}$	$\varepsilon x^{\varepsilon-1}$ since $\varepsilon - 1$ is odd, we have εx^θ	$\theta x^{\theta-1}$ since $\theta - 1$ is even, we have θx^ε
Direction of slope $x < 0$	negative, since a negative number to an odd integer power is negative	positive ⎫
$x = 0$	zero	zero ⎬ since any number to an even integer power is at least zero
$x > 0$	positive	positive ⎭
Graphs	Negative slope / Positive slope / 0 slope	Positive slope / 0 slope / Positive slope

Figure 14.12

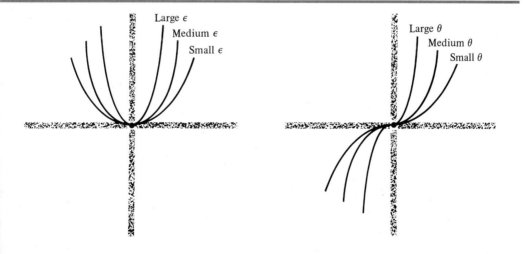

Large ϵ / Medium ϵ / Small ϵ

Large θ / Medium θ / Small θ

PROBLEMS

TECHNICAL TRIUMPHS

1. Earthquake intensity, Q, is the following function of the Richter scale reading, R.

   $$Q = f(R) = 10^R$$

   a. Determine $\dfrac{dQ}{dR}$.
   b. What is $\dfrac{dQ}{dR}$ for an earthquake with a Richter scale reading of 7?
   c. Graph Q and $\dfrac{dQ}{dR}$.
   d. Find $R = g(Q)$.
   e. Find $\dfrac{dR}{dQ}$.

2. The unemployment rate is defined as the number of unemployed people out of 1000 seeking employment. Check the business section to find the current unemployment rate.

3. Obese Observers has a 5-week crash diet that purports to take off fat (F), as given by the following function of time (t) in weeks:

   $$F = 5t + 2t^2 - \frac{t^3}{3}$$

   At what rate does it remove fat at
   a. 1 week.
   b. 3 weeks.
   c. 5 weeks.

4. The normal probability distribution for IQ scores with a mean of 100 and standard deviation of 15 is

   $$\frac{dy}{dx} = 266e^{-\left(\frac{x-100}{15}\right)^2} \quad \{55 \le x \le 145\}$$

   a. Graph the probability distribution.
   b. From your graph, what value of x results in the highest value of $\dfrac{dy}{dx}$?

5. Have you noticed the sign in front of McDonald's giving the total number of hamburgers sold since its inception? Assuming that the past sales growth patterns hold into the future, the number on the sign (H) in billions will be the following function of time (t) since January 1, 1974.

   $$H = f(t) = 13(1.15)^t$$

   a. If the above function holds in the future, at what rate (hamburgers per year) will they be dishing them out as of January 1, 2005.
   b. When will the hamburger sales rate equal 500 billion per year?

APPLICATIONS OF THE DERIVATIVE

6. Find the elasticity of demand function for a product with the following demand curve

   $Q = f(p) = 200 - 5P \quad \{10 \leq P \leq 40\}$

7. On the planet Jupiter, the gravity would be such that a body would fall y feet according to the following function of time, x, in seconds

   $y = f(x) = 40.1x^2$

   a. Find the Jupiter velocity function
   b. Find the Jupiter acceleration.

8. If the U.S. total debt, D, is forecasted to be the following function of time, t, in years since Jan 1, 2003.

   $D = f(t) = 4(1.1)^t \quad \text{(in trillions)}$

   a. What is $\dfrac{dD}{dt}$?
   b. When will the rate of debt growth equal 1 trillion?

9. The cumulative distribution function for the life of a certain calculator battery is

   $F = f(x) = 1 - e^{-\left(\frac{x}{100}\right)}$

   Use the software www.calc101.com to find the failure rate.

10. The pressure, P, in pounds per square inch, of a tire that ran over a nail, was determined to be the following function of time, x, in seconds, since the puncture

    $P = f(t) = \dfrac{32}{x}$

    a. What was the tire pressure 2 seconds after the puncture?
    b. What was the rate of pressure loss at that time?

CONFIDENCE BUILDERS

1. Joe NoMath is traveling to visit his girlfriend, Minny Sota. If the distance traveled (y) is the following function of time (t) in hours

   $y = t^2 + 50t$

   a. When will he get a speeding ticket (55 mph speed limit)?
   b. What is his acceleration?

2. A firm's management science team estimates that the cumulative probability distribution (F) of sales (x) for a new product is

   $F = -.002x^3 + .03x^2 \quad \{0 \leq x \leq 10\}$

   a. Determine the probability density function.
   b. What type of function is the probability density function? The cumulative probability distribution?
   c. Graph both distributions.

3. Weather forecasters predicted a big lake-effect snowstorm for Syracuse, with total snow (S) forecasted to be the following function of t, time in hours since the storm began.

$$S = f(t) = \frac{t^2}{2} - .1\frac{t^3}{3} \quad \{0 \leq t \leq 10\}$$

   a. What was the predicted snowfall rate 2 hours into the storm?
   b. What was the snowfall rate at the end of the storm? Does that make sense?
   c. When will the snowfall rate be 2.4 inches per hour?

4. The macroeconomic model

   $$MV = PT$$

   relates money in the economy (M), velocity at which money circulates (V), average price level (P), and the number of transactions or units sold (T) in the economy. Monetarists place heavy emphasis on this model to explain economic activity.
   a. Assuming V and T are constant, at what rate does price change with respect to money supply?
   b. Assuming P and T are constant, at what rate does velocity change with respect to money supply?

5. On page 305, we derived the following least-squares regression line for a mathematics quiz grade (Y) as a function of study hours (X) and math aptitude (Z).

   $$Y = f(X,Z) = 4.77 + 2.97X + 1.64Z$$

   a. Determine the two partial derivatives.
   b. What is the effect of an additional hour of study?
   c. What is the effect of an additional unit of mathematics aptitude?

6. Random numbers (x) over the interval, 0 to K, where K is a constant, have a cumulative probability distribution (F) of

   $$F = \frac{x}{K}$$

   a. Determine the probability density function. What type of function is it?
   b. Show that the area under the probability density function equals 1.

7. The sample size, n, required to estimate a population mean with 95% confidence is given by the following formula

   $$n = \frac{3.96\sigma^2}{e^2}$$

   where σ^2 is the population variance and e is the tolerable sampling error. Suppose $\sigma^2 = 10$.
   a. Determine $\frac{dn}{de}$.
   b. Is $\frac{dn}{de}$ positive or negative for all values of e?
   c. What does your answer to part b mean if one wants to reduce e ?

8. M. Presario will present "The Plumbers" at the upcoming concert in Madison Rectangle Garden. From past concerts he has been able to relate attendance (Q) to admission fee (P) as

APPLICATIONS OF THE DERIVATIVE

$Q = f(P) = 20{,}000e^{-.2P}$

a. Determine the elasticity and marginal revenue functions.
b. Show that elasticity equals 1 at the price that marginal revenue equals zero. What price is that?

9. We derived Blurry TV's demand curve on page 235 as

$Q = -4{,}000P + 2{,}200{,}000 \quad \{P \leq 300\}$

$Q = -10{,}000P + 4{,}000{,}000 \quad \{P > 300\}$

a. Find elasticity for each section of Blurry's kinked demand curve.
b. At what point(s) does elasticity equal 1?
c. Show that marginal revenue equals zero at the point(s) found in part b.
d. What is the range of elasticities found in the right ($P > 300$) section of the demand curve? What are the pricing implications of this?

10. A company has determined that the total cost (C) of producing x units of a certain product is

$C = \dfrac{x^3}{3} - 5x^2 + 100x + 200$

a. Determine the marginal cost function.
b. Approximately how much would the third unit cost? (Use the marginal cost function to approximate this.) Compare this result to the result obtained by using the total cost function.
c. For what value(s) of x would marginal cost equal 76?

MIND STRETCHERS

1. Consider an infantry platoon moving to a new position over an 8-hour period. Suppose that the distance (y) they move in miles is the following parabolic function of time since starting (t) in hours.

$y = at^2 + bt$ where a and b are constants

a. Determine the values of a and b if the platoon's initial velocity is 3 miles per hour and its acceleration (actually deceleration in this case) is $-\frac{1}{4}$.
b. How far will the unit move in the 8-hour period?
c. How long will it take for the unit to move 10 miles?

2. Star outfielder, Rich Money, hits well at the beginning of the year, improves throughout the early part of the year, but then runs into a slump that lasts until the end of the season. Rich's hitting pattern is captured by the function

$A = f(x) = -.2x^3 + .05x^2 + .2x + .3$

where A is his batting average (decimal) and x is the fraction of the season completed.
a. Find the rate of change of Rich's batting average.
b. At what point does his slump begin?

3. In section 7.2, we derived demand curves for S. Lumlord assuming hyperbolic and parabolic models as

Hyperbolic $\quad Q = f(P) = \dfrac{168{,}000}{80+P} \quad \{200 < P \leq 600\}$

Parabolic $\quad Q = g(P) = .004167P^2 + -6.0835P + 1{,}650 \quad \{200 < P \leq 600\}$

a. Determine the elasticity for each model.
b. For each case find the price at which elasticity equals one.
c. For each model find the marginal revenue function.
d. Show that marginal revenue equals zero where elasticity equals 1 for the parabolic case.
e. Does marginal revenue ever equal zero (finite price) for the hyperbolic case? Is marginal revenue positive or negative for values of P in the domain? What are the pricing implications of this result?

4. Margie Nall's family has the following total consumption (C) function of income (Y):

$$C = f(Y) = 4 + 3Y^{1/2}$$

a. Find the marginal consumption function.
b. Determine the marginal saving function.
c. At what point of increasing income does marginal saving become positive? Explain why total consumption is still greater than income at this point.

5. Consider the balance, B, of a savings account that gets continuous compounding over time, t, at an annual rate of i.
a. Find the rate of change of the balance with respect to time.
b. Find the relative rate of change by dividing the derivative by B.

6. Bull Rally, football player-turned-stockbroker, in a speech before the local garden club, observed that it was a good time to sell stocks since after a period of rapid increases in stock prices, recent increases have become smaller and smaller. Rose Bush, the club president, perplexed by the advice to sell while stock prices were still climbing, asked if it was really the second derivative of prices that was the basis for his caution. Bull paused, then, as he scurried out of the room, was heard to mumble, "I wish I had taken more math in college." See if you can answer Rose's question.

7. Two companies (A and B) have total sales since their inception given by the following functions of time, t.

$$f(t) = S_A(1 + g_A)^t$$
$$g(t) = S_B(1 + g_B)^t$$

Where S_A and S_B are the respective total sales since company inception as of Jan 1, 2004, and g_A and g_B are the respective annual growth rates of sales. If $S_A > S_B$ and $g_A < g_B$ and $t = 0$ on Jan 1, 2004.
a. Derive the formula that gives the time when the two total sales rates are equal.
b. At the point found in part a, what is the ratio (Company A divided by Company B) of total sales since inception.

CHAPTER 15

Optimization

15.1 INTRODUCTION

We have already flirted with the concept of optimization; for example, wanting to know the rent that results in the most revenue for S. Lumlord, and wishing to find the velocity that minimizes gasoline consumption. In general, optimization involves the search for extreme points (highest and/or lowest) on a function. In B.C. (Before Calculus) days, we only had the "exhaustion" method for this search. It isn't too swift. Now in A.D. (After Differentiation) time, this search is cream puff.

So that you can better appreciate the calculus methods presented later, we'll now review the "exhaustion" method. It involves substituting many arbitrarily chosen values of x to find the respective values of $f(x)$, and then hoping and praying that the extremes can be spotted. Let's illustrate with the polynomial function that we met on page 193.

$$y = f(x) = \frac{1}{3}x^3 - \frac{11}{4}x^2 + \frac{9}{2}x + 6$$

whose functional values and graph we repeat here (see Figure 15.1).

x	f(x)
0	6
1	$8\frac{1}{12}$
2	$6\frac{2}{3}$
3	$3\frac{3}{4}$
4	$1\frac{1}{3}$
5	$1\frac{5}{12}$
6	6

This method can only approximate the extreme points. For example, the highest $f(x)$ was observed at $x = 1$; however, had we tried $x = .98$ or $x = 1.05$, we might have found a higher point. The lowest point found corresponds to $x = 4$, with the point at $x = 5$ a close second. However, as the graph indicates, the lowest point on the function lies somewhere between $x = 4$ and $x = 5$. Only more exhaustion could pinpoint it better.

Figure 15.1

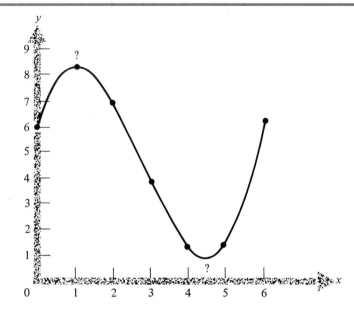

The exhaustion method can be quite frustrating and time consuming if the region of extreme points isn't known. For example, $f(x) = -x^2 + 3050x + 1500$ has a maximum at $x = 1525$. Can you imagine starting at $x = 1$, then $x = 2$, ... puff puff ... $x = 100$ ... and on and on. Why, you'd probably give up by $x = 500$ without getting close to the maximum. Even if you were smart enough to increment x by 100 each time, it would take a lot of effort to approximate the maximum.

Now let's advance to the A.D. era and bring the power of calculus to this problem. Both maximum and minimum points are characterized by zero slope (see Figure 15.1). So, by merely finding the derivative of a function, setting this derivative equal to zero and solving for x, we can quickly locate the extreme points. Carrying out these operations for our example function gives

$$D\left[\frac{1}{3}x^3 - \frac{11}{4}x^2 + \frac{9}{2}x + 6\right] = x^2 - 5.5x + 4.5$$

$$x^2 - 5.5x + 4.5 = 0$$

$$x = \frac{-(-5.5) \pm \sqrt{30.25 - 18}}{2}$$

$$= \frac{5.5 \pm 3.5}{2}$$

$$= 1 \text{ and } 4.5$$

So the maximum occurs at $x = 1$, and the minimum occurs at $x = 4.5$. Quick as that!

OPTIMIZATION

Learning Plan

The objective of this chapter is to optimize various real world functions. First, though, we must develop some additional theory. We begin by carefully observing the nature of high and low points on a function of one variable. Second derivatives are then introduced to distinguish between the two types of extremes (only because we knew the graph could we distinguish the extremes for our example problem). We wrap up the theory section by considering some special optimization cases and functions of two variables. Then comes the important task of applying our knowledge. This we do first for S. Lumlord, who has waited many chapters for the secret of how to fill his pockets to the fullest. Then we present applications relating to the optimization of profits, net returns, costs, revenues, happiness, statistical information, study time, and physical capacity. These all involve functions of one variable. We conclude the chapter with least-squares regression, an application of optimization for functions of more than one variable.

15.2 OPTIMIZATION THEORY

The first derivative is zero at both a maximum and a minimum point. For our example problem the graph made it obvious which was which. But how could we distinguish the two types of extremes without resorting to graphing or evaluating the function?

Careful observation of the slope in the vicinity of the two types of extremes will answer our question. In the vicinity of the maximum, as we ascend to go over the hill, we encounter positive, zero, then negative slope. In the vicinity of the minimum, as we trudge through the valley, we encounter negative, zero, then positive slope. These general slope characteristics are presented in Figure 15.2.

As can be clearly seen, the slope of the slope function (derivative of the derivative) differs for the two types of extremes. In the vicinity of a maximum, the slope function has a negative slope; in the vicinity of a minimum, the slope function has a positive slope, as indicated in the bottom level of graphs.

Before we finalize our procedure for locating extremes, let's investigate this new creature, the derivative of a derivative, which is called the *second derivative*.

Second Derivative

The derivative of x^3 is $3x^2$. Wouldn't it be discriminatory if $3x^2$ were denied the right of having a derivative? But, don't worry, it has one—$6x$. Starting with x^3 we had to take the derivative of a derivative (henceforth called taking the second derivative) to reach $6x$. Now, why stop a good thing? The derivative of $6x$ is 6, which would then be the third derivative of x^3. The derivative of 6 is zero, which is then the fourth derivative of x^3. Can you see that the fifth, sixth, and all succeeding derivatives of x^3 are zero?

There is no limit to the number of derivatives that one can take. Likewise, there is no limit to the respect one can obtain from glibly announcing the results of such. Can't you just imagine the rapid improvement of your image if you declare at the next cocktail party that the four hundred seventy-eighth derivative of x^3 is zero? Practically speaking, only the first and second de-

Figure 15.2

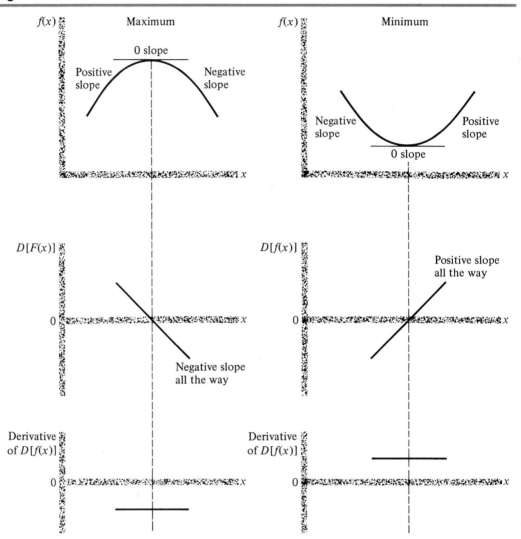

rivatives have widespread application. You should already be an expert on the first derivative. Now let's take a close look at the second derivative.

The symbolism for the second derivative is analogous to that used for the first derivative.

Symbol for first derivative	Corresponding symbol for second derivative
$\dfrac{dy}{dx}$	$\dfrac{d^2y}{dx^2}$
$f'(x)$	$f''(x)$
$D[f(x)]$	$D^2[f(x)] = D[D[f(x)]]$

OPTIMIZATION

All three symbols for the second derivative will be used in this text, since each will find situations where it conveys the most meaning or is the most useful. For example, to convey that the second derivative is of revenue with respect to price, the first symbol, or d^2R/dP^2, is the most useful. However, the $f''(x)$ symbol is most meaningful when evaluating a second derivative at a particular value of x. For example, if $f''(x) = 3x$, then $f''(2) = 3 \cdot 2 = 6$. But in using the derivative rules to find a second derivative, it is easier to use the last symbol, as we illustrate next.

Finding second derivatives is as simple as finding first derivatives. You can use all the rules that we developed in Chapter 13 as long as you take the derivative of the first derivative function. We can show this process using the last symbolism.

Symbol for second derivative		symbols for first derivative		derivative of	first derivative function		second derivative function
		↓ ↓		↘	↓		
$D^2[x^3]$	=	$D[D[x^3]]$	=		$D[3x^2]$	=	$6x$

Examples

1. $D^2[5] = D[D[5]] = D[0] = 0$
2. $D^2[3x] = D[D[3x]] = D[3] = 0$
3. $D^2[4x^2] = D[D[4x^2]] = D[8x] = 8$
4. $D^2[x^{-3}] = D[D[x^{-3}]] = D[-3x^{-4}] = 12x^{-5}$
5. $D^2[e^x] = D[D[e^x]] = D[e^x] = e^x$
6. $D^2[\ln x] = D[D[\ln x]] = D\left[\dfrac{1}{x}\right] = -x^{-2}$
7. $D^2[x^4 + 2^x] = D[D[x^4 + 2^x]] = D[4x^3 + 2^x \ln 2]$
   $= 12x^2 + 2^x(\ln 2)^2$
8. $D^2[x(\ln x)] = D[D[x(\ln x)]] = D\left[x \cdot \dfrac{1}{x} + (\ln x) \cdot 1\right]$
   $= D[1 + \ln x] = 0 + \dfrac{1}{x} = \dfrac{1}{x}$
9. $D^2\left[\dfrac{6}{e^x}\right] = D\left[D\left[\dfrac{6}{e^x}\right]\right] = D\left[\dfrac{e^x \cdot 0 - 6e^x}{e^{2x}}\right]$
   $= D\left[\dfrac{-6}{e^x}\right] = \dfrac{6}{e^x}$
10. $D^2[(1 + 3x)^{1/2}] = D[D[(1 + 3x)^{1/2}]] = D[\tfrac{3}{2}(1 + 3x)^{-(1/2)}]$
    $= -\tfrac{9}{4}(1 + 3x)^{-(3/2)}$

Exercises 1

1. $D^2[x^{3.5}] =$
2. $D^2[\log_{10} x] =$
3. $D^2[2^x] =$
4. $D^2[xe^x] =$
5. $D^2[x^{-2}] =$

6. $D^2[-4] =$
7. $D^2[10x + 5] =$
8. $D^2\left[\dfrac{2}{e^x}\right] =$
9. $D^2[x^3 + x2^x] =$
10. $D^2[(4x + 3)^{-(1/2)}] =$

Optimization Rule

Now that we have an understanding of the second derivative, we can incorporate it into a rule for finding and distinguishing maximum and minimum points. Recall that the slope of the first derivative is negative at a maximum (thus, the second derivative is negative); the slope of the first derivative is positive at a minimum (thus, the second derivative is positive). Putting it all together, we have the rule shown in Figure 15.3.

Let's try out this rule on our example function. We already know that the first derivative is zero when $x = 1$ and $x = 4.5$. Now let's find the second derivative and evaluate it for these values of x.

$$D^2[\tfrac{1}{3}x^3 - \tfrac{11}{4}x^2 + \tfrac{9}{2}x + 6] = D[x^2 - 5.5x + 4.5]$$
$$= 2x - 5.5$$

Figure 15.3

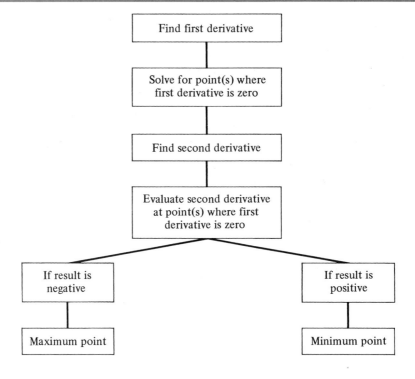

At $x = 1$, the second derivative equals $2 \cdot 1 - 5.5 = -3.5$. Since it is negative, a maximum occurs there. At $x = 4.5$, the second derivative is $2 \cdot 4.5 - 5.5 = +3.5$. Since it is positive, a minimum occurs there.

Exercises 2

Locate maximum and/or minimum points on the following functions.

1. $-x^2 + 10x + 4$
2. $\frac{1}{3}x^3 + \frac{3}{2}x^2 - 10x$
3. $x^3 - 12x$

Special Optimization Cases

Our optimization rule can detect extreme points for most functions. However, our rule has one gap (What if the second derivative is neither positive nor negative, but zero?), and it can't detect an extreme at the end of the domain. So, to be complete, we now consider these two special cases.

Inflection Points

Inflection points can be part maximum and part minimum, as illustrated in Figure 15.4. In part (a), the curve rises to a momentary peak, and then rises again. In part (b), the curve drops to a momentary low, and then drops further. Notice that both first and second derivatives are zero at these momentary extremes. *Note:* Points where $f''(x) = 0$ but $f'(x) \neq 0$ are also called inflection points; however we won't consider them here.

In order to distinguish the upward from the downward trending inflection points, the third derivative is necessary. A positive third derivative means an upward inflection point, while a negative third derivative means a downward inflection point (see figure). An example of a function with an inflection point is x^3. Its first derivative, $3x^2$, equals zero when $x = 0$, at which point the second derivative, $6x$, is also zero. Since the third derivative, 6, is positive, the function possesses an upward inflection point at $x = 0$.

Exercises 3

1. Graph the function $y = x^3$ to verify that it has an upward inflection point.
2. Find and identify the inflection point for the function $y = -x^3 + 3x^2 - 3x + 1$.

End-of-Domain Extremes

An extreme point can exist on a function where the slope is not zero. However, this can occur only at the ends of the domain. To illustrate, consider our example function, $y = f(x) = \frac{1}{3}x^3 - \frac{11}{4}x^2 + \frac{9}{2}x + 6$. Its first derivative, $x^2 - 5.5x + 4.5 = (x - 4.5)(x - 1)$, is positive for any $x > 4.5$, since both factors $(x - 4.5)$ and $(x - 1)$ would be positive. This means that $f(x)$ increases continuously

Figure 15.4

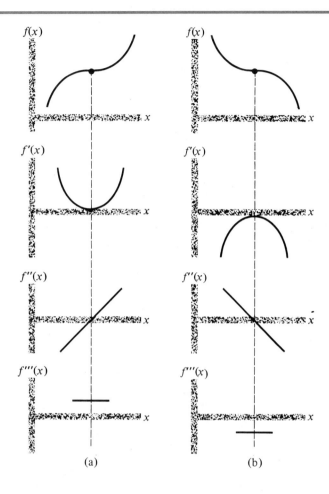

for $x > 4.5$. So, after some point, $f(x)$ must become larger than $f(1) = 8\frac{1}{12}$, which was the maximum in the $\{0 \le x \le 6\}$ domain.[1] We can locate this point by solving the following equation:

$$\tfrac{1}{3}x^3 - \tfrac{11}{4}x^2 + \tfrac{9}{2}x + 6 = 8\tfrac{1}{12}$$

A trial-and-error procedure reveals the solution to be $x = 6.25$. For domains stretching beyond this value of x, the right end point would be the highest point on the function. For example, if the domain were $\{0 \le x \le 8\}$, $f(8) = 36\frac{2}{3}$ would be the maximum.

Exercises 4

1. By graphing, find the approximate left domain edge such that $f(x) = f(4.5)$ for our example function.

---

[1]This wouldn't be true if the function had an upper limit. For example, $f(x) = 8 - 1/x$, for $x > 0$, always has positive slope and thus increases continuously; however, it never gets over 8.

OPTIMIZATION

2. Find the highest and lowest point on the function, $f(x) = x^3 - 12x$, over the domain, $\{-3 \le x \le 5\}$.

Optimization Rule for Functions of Two Variables

A function of two variables is a surface in three dimensions, such as the terrain on earth. Picture a mountain. At the peak (or at the bottom of the valley), the slope is temporarily zero in all directions. The analogy for functions is that both partial derivatives are zero at a maximum or at a minimum point.

$y = f(x, z)$

$\dfrac{\partial y}{\partial x} = 0$ and $\dfrac{\partial y}{\partial z} = 0$ at extreme points

There are tests to distinguish maximums from minimums for functions of two variables. However, they involve second partial derivatives and mixed partial derivatives. Rather than get involved in such headaches, you should just test the value of the function at and around each extreme point.

15.3 OVERVIEW OF OPTIMIZATION APPLICATIONS

Optimization theory has potential applications in almost every area of human existence because, as thinking beings, people are continually trying to do the best under the circumstances. To realize this potential, the situation must lend itself to mathematical formulation. To illustrate this, we will return to the saga of S. Lumlord, who has been waiting many chapters to find the rent that would fill his pockets to the fullest.

Riches at Last for S. Lumlord

Assuming S. Lumlord's demand function is $Q = 1000 - 2P$, where Q is the number of rented apartments, P is the rent per apartment, and his costs are $30,000 (fixed), plus $30 for each rented and $10 for each vacant apartment, let's find the rents that maximize revenue and profit. These optimum rents won't be the same, as you'll see.

The revenue (R) function is

$R = PQ = 1000P - 2P^2$ (see Figure 15.5)

Setting the first derivative equal to zero

$\dfrac{dR}{dP} = 1000 - 4P = 0$

and solving gives $P = \$250$, which yields a maximum revenue since the second derivative is negative (-4).

Figure 15.5

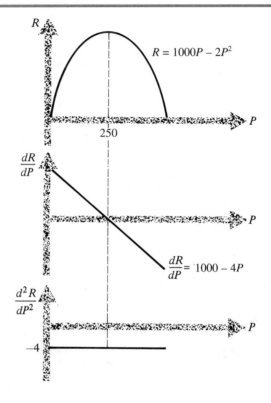

Profits (π) are revenues minus costs (see Figure 15.6).

We already know the revenue function. From the cost information given, we can find the total cost function as

$$C = 30{,}000 + 30Q + 10(600 - Q)$$
$$= 36{,}000 + 20Q$$

(*Note:* With 600 total apartments, $600 - Q$ is the number of vacancies.) Since $Q = 1000 - 2P$, we can express C as a function of P.

$$C = 36{,}000 + 20(1000 - 2P)$$
$$= 56{,}000 - 40P$$

Thus, the profit function is

$$\pi = (1000P - 2P^2) - (56{,}000 - 40P)$$
$$= -2P^2 + 1040P - 56{,}000$$

Setting the first derivative equal to zero

OPTIMIZATION

Figure 15.6

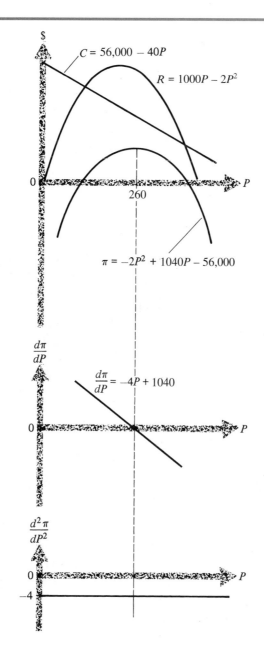

$$\frac{d\pi}{dP} = -4P + 1040 = 0$$

and solving gives $P = \$260$, which yields maximum profit since the second derivative is negative (-4).

To find the actual revenues and profits at these optimum rent levels, we must substitute into the respective total revenue and total profit functions as follows:

$R = f(P) = 1000P - 2P^2$

$f(250) = 1000(250) - 2(250)^2 = \$125{,}000$

$\pi = g(P) = -2P^2 + 1040P - 56{,}000$

$g(260) = -2(260)^2 + 1040(260) - 56{,}000 = \$79{,}200$

Exercises 5

1. Find total revenue at $260 rent and total profit at $250 rent. Compare these to the above optimum results.
2. Show that the magnitude of fixed costs does not affect the maximum profit rent level.
3. If the variable costs for rented apartments increase to $40, find the new maximum profit rent level.
4. If the demand curve was $Q = 1200 - 3P$, what rents would maximize revenue and profit (assume $30 variable cost per rented apartment)?

Applications of Optimization

For the remainder of this chapter, we illustrate the diverse applications of optimization theory classified into the following areas.

1. Profit maximization
2. Net return maximization
3. Cost minimization
4. Revenue maximization
5. Happiness maximization
6. Optimization in statistics
7. Student study optimization
8. Physical capacity maximization
9. Least-squares regression

To illustrate the applications in the first eight areas, we employ functions of one variable. Least-squares regression illustrates the optimization of multivariate functions.

15.4 PROFIT MAXIMIZATION

S. Lumlord provided us with an example of a maximum profit model. Now let's look further at such applications.

Profit (π) is defined as total revenue (R) minus total cost (C).

$\pi = R - C$

If R and C are functions of price (P) or output (Q), then π is a function of P or Q, and we can find the P or Q value that maximizes profit. First we'll focus on functions of P, so

OPTIMIZATION

$R = g(P) \qquad C = h(P)$

then

$\pi = f(P) = g(P) - h(P)$

Taking the first derivative of π with respect to P and setting it equal to zero gives

$D[f(P)] = D[g(P) - h(P)] = D[g(P)] - D[h(P)] = 0$

So

$D[g(P)] = D[h(P)]$

or

$\dfrac{dR}{dP} = \dfrac{dC}{dP} \qquad$ marginal revenue = marginal cost

So you can finally understand why your economics professor told you that profits are maximized where marginal revenue equals marginal cost. For example, S. Lumlord's marginal functions are

$\dfrac{dR}{dP} = 1000 - 4P \qquad \dfrac{dC}{dP} = -40$

which are equal when $1000 - 4P = -40$ or $P = \$260$.

In similar fashion, we can consider the variables as functions of Q, and derive the result that profits are optimized where

$\dfrac{dR}{dQ} = \dfrac{dC}{dQ}$

Exercises 6

Determine R and C as functions of Q for S. Lumlord and show

1. $Q = 480$ when $P = \$260$
2. $\dfrac{dR}{dQ} = \dfrac{dC}{dQ}$ when $Q = 480$

There is a possible fallacy to finding maximum profit points by setting marginal revenue equal to marginal cost, since these two derivatives would be equal at minimum profit points too! So before accepting any point as yielding maximum profits, you should confirm by showing that the second derivative of the profit function is negative. To do this, you need to start with the profit function itself (also the profit function is needed to determine the magnitude of the maximum profits), so it is wise to define and work with it throughout the optimization procedure.

CALCULUS

We illustrate profit maximization applications by helping Maxi set taxi fare, showing Creaky Airlines the most profitable No-Extras price, and proving that profits are maximized halfway between the breakeven points for the case of linear demand and cost curves. The basic situation and functions for each of these three cases were developed in chapter 7.

Maxi Taxi Company

On pages 244–245, we derived Maxi's profit to be the following function of Q, the number of cab riders.

$$\pi = -2.5Q^2 + 200Q - 3500$$

The first derivative of profit with respect to Q is

$$\frac{d\pi}{dQ} = -5Q + 200$$

Setting this equal to zero and solving for Q gives

$$-5Q + 200 = 0 \quad \text{so} \quad Q = 40$$

The negative second derivative

$$\frac{d^2\pi}{dQ^2} = -5$$

confirms that maximum profit occurs when 40 riders decide to hail Maxi's cab.

To be of maximum help to Maxi, we must go one step further and tell him what numbers to paint on the side of the taxi. We do this by solving Maxi's demand function, derived on page 244.

$$Q = 90 - .4P$$

for the value of P that results in $Q = 40$.

$$P = f(Q) = \frac{90 - Q}{.4}$$

$$f(40) = \frac{90 - 40}{.4} = 125¢ = \$1.25$$

So, Maxi, make it $1.25 in big big numerals.

Exercises 7

1. On page 245, we found the two breakeven points for Maxi to be 26 and 54 riders. Do you see any pattern between these results and the maximum profit point of 40 riders? Discuss.

2. Develop Maxi's profit as a function of P and find the optimum P. Compare the result to that arrived at by optimizing profit as a function of Q above. (*Note:* $C = 25Q + 3500$.)
3. Find the fare price that maximizes Maxi's revenue.

Creaky Airlines No-Extras Fare

On pages 232–233, we derived Creaky Airline's cost (C), regular seat revenue (R_R), and No-Extras seat revenue (R_{NE}) functions of No-Extras seat price (P) as

$$C = 6100 + 10P$$

$$R_R = 1P^2 + 10P$$

$$R_{NE} = -2P^2 + 180P$$

From these we can derive the profit function as

$$\pi = (R_R + R_{NE}) - C$$
$$= -P^2 + 180P - 6100$$

Maximizing this function gives

$$\frac{d\pi}{dP} = -2P + 180 = 0 \quad \text{so} \quad P = 90$$

The negative second derivative ($d^2\pi/dP^2 = -2$) confirms that a No-Extras price of $90 per seat maximizes total profit.

It is interesting to note that when this price is substituted in the No-Extras demand function, it results in no No-Extras passengers. Could that be a clue as to why the real airlines dropped this type of service in 1976? Updating to today's conditions, it could be argued that all airline seats are No-Extras!!

Exercises 8

The cost curve implicit in the above application was

$$C = F + v_R(P + 10) + v_{NE}(-2P + 180)$$

with parameter settings of

F = fixed costs = $5000

v_R = regular seat variable cost = $20

v_{NE} = No-Extras seat variable cost = $5

1. Show that the No-Extras pricing decision does not change if fixed costs jump to $5500 as a result of an increase in jet fuel price.

2. If v_R and v_{NE} change to be \$25 and \$2.50, respectively, what No-Extras seat price would maximize profits?
3. Find the total profit as a result of the price found above.

Profit Maximization-Breakeven Relationship

We stated the general case of linear demand and total cost functions in chapter 7 as:

$$Q = sP + i \quad \{s < 0\} \quad \text{so} \quad P = \frac{Q}{s} - \frac{i}{s}$$

$$C = F + vQ$$

See if you can derive the profit (π) function of Q and the break-even points (Q_{BE}).

$$\pi = \frac{1}{s}Q^2 - \left(\frac{i}{s} + v\right)Q - F$$

$$Q_{BE} = \frac{\left(\frac{i}{s} + v\right) \pm \sqrt{\left(\frac{i}{s} + v\right)^2 + \frac{4F}{s}}}{2/s}$$

Now let's find the output level that results in maximum profit, and compare it to the breakeven output levels.

$$\frac{d\pi}{dQ} = \frac{2}{s}Q - \left(\frac{i}{s} + v\right) = 0 \quad \text{when} \quad Q = \frac{\left(\frac{i}{s} + v\right)}{2/s}$$

The second derivative ($2/s$) is negative, since s, the slope of the demand curve, is negative. Thus, we have found the maximum profit point.

Since breakeven points are merely the maximum profit output plus or minus a certain amount, it follows that the maximum profit output lies midway between the two breakeven points for this case.

Exercise 9

Show that the maximum profit output level lies midway between the breakeven output levels for the case of linear demand and parabolic total cost functions.

15.5 NET RETURN MAXIMIZATION

Profit is a dirty word in some nonbusiness settings. However, its equivalent, net return, isn't. So let's see how net return can be maximized for a nonprofit university that wishes to get the most

OPTIMIZATION

out of its student recruiting effort and for individuals who want to structure their education so as to achieve the highest net lifetime earnings.

Student Recruiting

The Monkey Business School wishes to know the optimum number of days it should spend on the road recruiting unsuspecting students, assuming the following circumstances:

1. The number of recruits snared is a parabolic function of the number of days on the road. Ten and 100 days of recruiting would yield 40 and 310 additional applications (these would not have been made without the recruiting), respectively.
2. Recruiting cost is $100 per day.
3. Three-quarters of the applicants are accepted, and one-half of these enroll in the school.
4. The net additional tuition income (after subtracting all variable costs) is $200 per student.

Using the curve-fitting methods of Chapter 7, you should be able to fit the recruiting function as

$$R = 4.1d - .01d^2$$

where R is applications as result of recruiting, and d is days on the road. (*Note:* when $d = 0$, $R = 0$.) With acceptances (A) and enrollees (E) being the following functions of R

$$A = \tfrac{3}{4}R \qquad E = \tfrac{1}{2}A = \tfrac{3}{8}R$$

then total net additional tuition (T) from these recruits is

$$T = 200E = 200(\tfrac{3}{8}R) = 75R = 307.5d - .75d^2$$

With total recruiting cost (C) as

$$C = 100d$$

the net return (N) from the recruiting effort (net additional tuition minus recruiting cost) is

$$N = T - C = (307.5d - .75d^2) - (100d) = 207.5d - .75d^2$$

Optimizing this function gives

$$\frac{dN}{dd} = 207.5 - 1.5d = 0$$

$$d = \frac{207.5}{1.5} = 138.3 \quad \text{(say 138 days)}$$

Since the second derivative is -1.5, we have found a maximum net return. The 138 days of recruiting should result in the following additional applications, enrollees, and net tuition.

$$R = 4.1(138) - .01(138)^2 = 375$$

$$E = \tfrac{3}{8}R = 141$$

$$N = 207.5(138) - .75(138^2) = \$14{,}352$$

Exercises 10

1. If the recruiting costs were $150 per day, what level of recruitment would be best?
2. If 90 percent of the recruitees are accepted and 80 percent of these enroll, what level of recruitment is best?
3. This exercise deserves a part 3, so change one of the problem conditions and solve for the optimum recruitment level.

College and Big $

Professor Ray Search did a study and discovered that lifetime earnings (L) approximated the following function of years of college education (x)

$$L = 300{,}000 + 70{,}000 \ln(x + 1)$$

If the cost of attending college per year is $10,000, Professor Search would like to know the best advice for students planning their stay at his university.

Additional years of college increase lifetime earnings, but also increase the cost of obtaining such earnings. The net increase (N) is

$$N = L - 10{,}000x$$
$$= 300{,}000 + 70{,}000 \ln(x + 1) - 10{,}000x$$

Maximizing N reveals

$$\frac{dN}{dx} = 70{,}000 \left(\frac{1}{x+1}\right) - 10{,}000 = 0$$

$$\frac{1}{x+1} = \frac{10{,}000}{70{,}000} = \frac{1}{7}$$

So $x = 6$.

Exercise 11

Show that the second derivative is negative when $x = 6$.

Thus, six years of college (master's degree) yields the highest net lifetime earnings, which on average would be

$N = 300{,}000 + 70{,}000(\ln 7) - 10{,}000(6)$

$= \$376{,}220$

Exercises 12

1. How much did the author suffer (net lifetime earnings only) by spending eight years in college? (Now you know why I had to write this book!)
2. What would the optimum educational level be if a year of college cost $20,000?
3. If the following lifetime earnings function is true for students going to college in the years 2004–2007.

    $L = 400{,}000 + 45{,}000 \ln(x + 1)$

    please stay in college long enough to determine the optimum educational level (assume $20,000 college cost per year).

15.6 COST MINIMIZATION

In this section we consider two models, economic order quantity model and branch office model, where costs are minimized under the assumption of constant revenue.[2] Such models have found wide application in business.

Economic Order Quantity Model

The economic order quantity (EOQ) model addresses the question, How many orders of constant size should be placed per year for a given product? Its simplicity, while still capturing the delicate cost balance involved in inventory decisions, accounts for its wide use in business and even wider use in the classroom (if you graduate from a business school without having the EOQ model at least three times, apply for a tuition refund).

The model assumes (1) constant annual sales (S), spread evenly throughout the year, (2) n orders of constant size equally spaced throughout the year, and (3) no volume discounts. Notice that there are many different ordering alternatives: S units once a year, $S/2$ units every six months, $S/4$ units every quarter, and so on. The inventory levels for these alternatives are shown in Figure 15.7.

With few orders inventory holding costs for storage and financing are high. However, ordering costs for supervising the deliveries, placing the orders, bookkeeping, and so on, are low. With many orders, the reverse of low inventory and high ordering costs is true.

The model assumes that inventory holding costs are proportional to average inventory size, and ordering costs (which repeat for each order) are proportional to the number of orders. Product cost is constant since the same total number of units is ordered regardless of the ordering strategy. So the formulas for the various costs are:

---

[2] These models are really only profit maximization models in disguise. This can be seen from the definition of profit, $\pi = R - C$. If R is constant, making C as small as possible is equivalent to making π as large as possible.

Figure 15.7

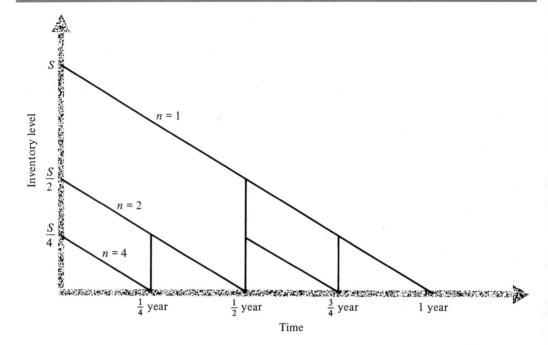

Holding cost $= HC = k\left(\dfrac{S}{2n}\right)$

Ordering cost $= OC = Kn$

Product cost $= PC = \rho$

where k, K, and ρ are constants and $S/2n$ is average inventory.

Total cost (C) is the sum of these three costs, as shown in Figure 15.8.

$C = PC + HC + OC$

$\quad = \rho + k\left(\dfrac{S}{2n}\right) + Kn$

We minimize C by first taking its derivative.

$\dfrac{dC}{dn} = D[\rho] + D\left[k\left(\dfrac{S}{2n}\right)\right] + D[Kn]$

$\quad = 0 - \dfrac{kS}{2n^2} + K$

At an extreme point, the derivative is zero.

$0 = \dfrac{-kS}{2n^2} + K$

OPTIMIZATION

Figure 15.8

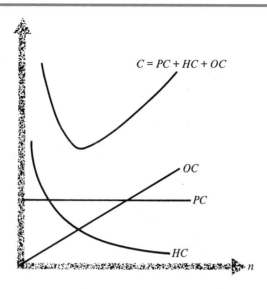

Solving for n,

$$\frac{kS}{2n^2} = K$$

$$n = \sqrt{\frac{kS}{2K}}$$

The second derivative

$$\frac{d^2C}{dn^2} = \frac{kS}{n^3}$$

is positive for all n, since k, S, and n^3 are all positive. Thus we have verified a minimum cost level when $n = \sqrt{kS/2K}$, with order size

$$Q = \frac{S}{n} = \frac{S}{\sqrt{kS/2K}} = \sqrt{\frac{2KS}{k}}$$

Example

A product with annual sales of 1,000,000 units, with $k = .1$ (holding costs are 10 percent of average inventory level) and $K = \$20$ (ordering cost per order) would have an EOQ solution

$$n = \sqrt{\frac{kS}{2K}} = \sqrt{\frac{.1(1,000,000)}{2(20)}} = \sqrt{\frac{100,000}{40}} = \sqrt{2500} = 50$$

$$Q = \frac{1,000,000}{50} = 20,000$$

So, the economic order quantity (the one that minimizes total cost and thus maximizes profit) is 20,000 units that should be ordered 50 times per year (approximately once a week).

Exercises 13

1. How do the following changes affect EOQ?
   a. A doubling of product cost.
   b. A doubling of the holding cost constant (k).
   c. A doubling of the ordering cost constant (K).
2. If both ordering and holding cost constants increased by same percentage, what effect would this have on EOQ?

Branch Office Model

The branch office model balances travel-related and office-maintenance-related costs to arrive at the best number of sales offices to serve a large geographic market. It assumes that a target sales level (call it $\$S$) could be generated with 1, 2, 3, 4, ... branch offices. With a small number of offices the salespeople will have extensive travel-related expenses to generate $\$S$ sales, but, of course, office-related costs will be low. With a large number of offices the reverse is true: low-travel and high-office costs.

Travel-related costs (T) and office-maintenance costs (θ) required to generate $\$S$ sales are functions of the number of branch offices (n), the form and specifics of which depend on the market particulars. To illustrate the model and solution method, we can use the following reasonable functions:

$$T = \frac{k}{n^{1/2}} = kn^{-1/2} \qquad \theta = Kn \qquad \text{where } k \text{ and } K \text{ are constants}$$

Total cost (C) of generating $\$S$ sales is the sum of the above functions, as shown in Figure 15.9.

$$C = T + \theta$$
$$C = f(n) = kn^{-1/2} + Kn$$

Since the target sales can be generated with many different branch office levels, the one that minimizes cost will maximize profit.

$$\frac{dC}{dn} = -\tfrac{1}{2}kn^{-3/2} + K$$

$$0 = -\tfrac{1}{2}kn^{-3/2} + K$$

$$n = \left(\frac{k}{2K}\right)^{2/3}$$

(See if you can supply the missing algebraic steps leading to the solution for n.) The second derivative

OPTIMIZATION

Figure 15.9

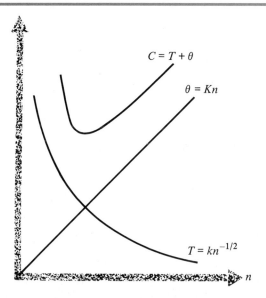

$$\frac{d^2C}{dn^2} = \tfrac{3}{4}kn^{-5/2} = \frac{3k}{4n^{5/2}}$$

is always positive, so $n = (k/2K)^{2/3}$ results in minimum cost.

Example

To generate $10 million sales in the United States market, the Tic-Tock Watch Company has estimated the following travel and office expense functions:

$$T = \frac{1{,}000{,}000}{n^{1/2}}$$

$$\theta = 45{,}000\, n$$

With $k = 1{,}000{,}000$ and $K = 45{,}000$, the optimum number of sales offices is

$$n = \left(\frac{k}{2K}\right)^{2/3}$$

$$= \left[\frac{1{,}000{,}000}{2(45{,}000)}\right]^{2/3} = 4.98 \quad (\text{say } 5)$$

and total selling cost is

$$C = 1{,}000{,}000(5)^{-1/2} + 45{,}000(5)$$

$$= 447{,}214 + 225{,}000$$

$$= 672{,}214$$

Exercises 14

1. Find the optimum number of sales offices for Tic-Tock if $k = \$3,000,000$ and $K = \$75,000$.
2. Determine the optimum number of sales offices if the travel and office expense functions were $T = 1,000,000n^{-1/3}$ and $\theta = 50,000n$.

15.7 REVENUE MAXIMIZATION

In this section we consider revenue maximization under the assumption of constant costs. These, too, are really profit maximization cases in disguise. (See if you can understand why by considering the profit equation for constant C.)

Constant cost is a reasonable assumption for many sporting events, for which the major costs are player-coaching salaries and stadium rental, which do not vary, regardless of whether 5,000 or 50,000 fans show up to cheer their heroes. Of course, some costs, such as for ushers, vary with attendance, but their insignificance compared to other costs allows these models to give good approximations to the optimum solutions.

Put yourself in the place of Cindy Kate, promotor of the Al Lee-Ken Dive heavyweight fight. Cindy's market research group, headed by Les Sense, an ex-fighter who retired after 100 straight knockouts (the last time it took three days to revive him) estimated that attendance (A) would be the following function of ticket price (P).

$$A = \frac{50,000}{P^2 + 1}$$

Les, in his usual cerebral manner, suggested that they charge $P = 0$ because that would maximize attendance, since

$$\frac{dA}{dP} = \frac{-50,000(2P)}{(P^2 + 1)^2} \quad \text{which equals 0 when } P = 0$$

At that point Les suffered his 101st straight knockout!

Cindy, a recent graduate of the Monkey Business School, knew better. She first had to define total ticket revenue (R) and maximize it (not attendance).

$$R = AP$$
$$= \left(\frac{50,000}{P^2 + 1}\right)P = \frac{50,000P}{P^2 + 1}$$
$$\frac{dR}{dP} = \frac{(P^2 + 1)D[50,000P] - (50,000P)D[P^2 + 1]}{(P^2 + 1)^2}$$
$$= \frac{(P^2 + 1)(50,000) - (50,000P)(2P)}{(P^2 + 1)^2}$$
$$= \frac{50,000P^2 + 50,000 - 100,000P^2}{(P^2 + 1)^2}$$

OPTIMIZATION

$$= \frac{50{,}000 - 50{,}000P^2}{(P^2 + 1)^2}$$

When the slope is zero,

$$0 = \frac{50{,}000(1 - P^2)}{(P^2 + 1)^2}$$

The numerator $50{,}000(1 - P^2)$ must equal 0, so $P = \$1$ maximizes revenue, which we confirm by the second derivative

$$\frac{d^2R}{dP^2} = \frac{(P^2 + 1)^2(-100{,}000P) - [(50{,}000)(1 - P^2)(2)(P^2 + 1)(2P)]}{(P^2 + 1)^4}$$

At $P = 1$, $(1 - P^2) = 0$ so the massive term on the right of the numerator is zero. The term on the left of the numerator is negative, and that divided by a positive number (any number raised to an even power is positive) yields a negative second derivative. Whew!

Exercises 15

1. If the attendance function were

    $$A = \frac{50{,}000}{P^{3/2} + 1}$$

    what would the optimum price be?

2. Consider the general attendance function

    $$A = \frac{50{,}000}{P^k + 1} \quad \text{where } k \text{ is a constant}$$

    Derive the optimum price in terms of k. Use the result to check your answer in exercise 1.

3. After making a fortune on the fight, Cindy decided to produce rock-and-roll concerts. Her first one featured Pelvis Essley. Assuming that attendance is the following function of price,

    $$A = 20{,}000e^{-.2P}$$

    what admission fee would maximize revenue? What would that maximum revenue be?

4. For the linear demand curve,

    $$Q = -sP + i \quad \{s > 0\}$$

    show that revenue is maximized at $P = i/2s$. (*Note:* This is the price where elasticity equals 1.)

15.8 HAPPINESS OPTIMIZATION

Until now we have been working with maximum profit models (explicitly or in disguise)! But profits are often not the sole basis for decision making. For example, college professors are said to get enough "psychic income" from the freedom of the academic environment to be willing to pass up jobs paying thousands more. (If you hear of such a job, please let me know!) In such cases people act to maximize overall "happiness," which is the result of a combination of monetary and nonmonetary factors. If, further, these factors can be developed into a happiness index, we can then optimize it. We'll consider two cases—a car velocity decision (with ecological implications) and an investment decision—to illustrate these ideas.

Car Velocity Decision

There you are embarked on a long, thruway car trip. On the one hand, getting to your destination quickly gives you happiness. On the other hand, driving fast is wasteful of gasoline, which on monetary and ecological grounds adds to your unhappiness.

In this case it is better to define a measure of unhappiness (U) and minimize it. Let's assume that only two factors—gasoline cost of the trip (C) and total travel time (T)—contribute to unhappiness, but not necessarily to the same degree. Thus, a dollar added cost doesn't give the same unhappiness as an added hour of travel. Reflecting this, we define U as

$$U = C + \lambda T$$

where λ is a constant that incorporates the different cost and time effects. For example, if your displeasure index remains unchanged ($\Delta U = 0$) by a $\frac{1}{2}$ hour longer trip if you could save \$1 in the process, then

$$\Delta U = 0 = \Delta C + \lambda(\Delta T)$$
$$0 = -1 + \lambda(+\tfrac{1}{2})$$
$$\lambda = 2$$

Now let's define C and T in terms of velocity (v). Assuming that C only reflects gasoline cost, then

$$C = c \cdot g$$
$$= c \cdot \frac{L}{m}$$

where

- c is cost of gas per gallon
- g is gallons needed
- L is total length of trip in miles
- m is miles per gallon of gas

OPTIMIZATION

You fitted the following gasoline mileage function (problem 7, page 246), which we assume holds true for your car.

$$m = .9v - .01v^2$$

So C as a function of v is

$$C = \frac{cL}{.9v - .01v^2}$$

Assuming a constant velocity, travel time (T) is merely L/v.
Finally, U as a function of v is

$$U = \frac{cL}{.9v - .01v^2} + \lambda \frac{L}{v}$$

We will minimize U in general, but first let's get a feel for the problem by solving for the specific case of $c = \$1$, $L = 100$ miles, and $\lambda = 1$.

$$U = \frac{100}{.9v - .01v^2} + \frac{100}{v}$$

$$\frac{dU}{dv} = \frac{-100(.9 - .02v)}{(.9v - .01v^2)^2} - \frac{100}{v^2}$$

$$= \frac{v^2(2v - 90) - 100(.9v - .01v^2)^2}{(.9v - .01v^2)^2(v^2)}$$

Setting the derivative equal to zero and multiplying both sides by the denominator gives

$$v^2(2v - 90) - 100(.81v^2 - .018v^3 + .0001v^4) = 0$$

Dividing by v^2 and simplifying gives

$$2v - 90 - 81 + 1.8v - .01v^2 = 0$$

$$-.01v^2 + 3.8v - 171 = 0$$

Finally, we solve for v using the quadratic roots formula.

$$v = \frac{-3.8 \pm \sqrt{(3.8)^2 - 4(-.01)(-171)}}{2(-.01)}$$

$$= 52 \text{ and } 328 \text{ miles per hour}$$

(*Note:* Since the m function doesn't hold for large v ($m < 0$ if $v > 90$), the 328 root is not meaningful.)

Exercises 16

1. By using the second derivative test, show that a velocity of 52 miles per hour minimizes unhappiness.
2. What are the travel time, gasoline cost, and unhappiness index for this optimum case? Recompute these for a 60 mile per hour trip, and compare to the optimum results.
3. If gas costs 160¢ per gallon and $\lambda = 2$, what velocity minimizes unhappiness for a 500 mile trip? What would travel time and gasoline cost be for a trip at this velocity?

We can now generalize this problem by not specifying values for the constants c, L, and λ. This will lead us to insights that were hidden in solving the specific case.

$$U = \frac{cL}{.9v - .01v^2} + \frac{\lambda L}{v}$$

Differentiating, forming a common denominator, and setting the result equal to zero yields

$$\frac{dU}{dv} = \frac{-cL(.9 - .02v)}{(.9v - .01v^2)^2} - \frac{\lambda L}{v^2}$$

$$\frac{-cL(.9 - .02v)v^2 - \lambda L(.9v - .01v^2)^2}{(.9v - .01v^2)^2 v^2} = 0$$

Multiplying by $(.9v - .01v^2)^2 v^2$, dividing by v^2, and simplifying gives

$$-.9cL + .02cLv - .81\lambda L + .018\lambda Lv - .0001\lambda Lv^2 = 0$$

$$-(.0001\lambda L)v^2 + (.02cL + .018\lambda L)v - (.9cL + .81\lambda L) = 0$$

$$-\lambda Lv^2 + (200cL + 180\lambda L)v - (9000cL + 8100\lambda L) = 0$$

The symbol L is factored and divided out.

$$-\lambda v^2 + (200c + 180\lambda)v - (9000c + 8100\lambda) = 0$$

Finally, the solution for v is brought to you by the quadratic roots formula.

$$v = \frac{-(200c + 180\lambda) \pm \sqrt{(200c + 180\lambda)^2 - 4(-\lambda)(-9000c - 8100\lambda)}}{2(-\lambda)}$$

You would need the second derivative and the values of c and λ to determine which roots produce minimum and maximum unhappiness levels.

The total trip distance, L, is conspicuous by its absence in the optimum velocity formula. This points out the value of generalization, since we might never have known that it had no relevance to optimum velocity if we confined our effort to working specific problems.

Exercises 17

1. Substitute the specific conditions that we employed earlier ($\lambda = 1$ and $c = 1$) into the general formula and show that it gives the specific solution ($v = 52$) found earlier.

2. If the gas mileage function is $m = 1v - .01v^2$, derive the general formula for the optimum velocity.

Investment Portfolio Selection

Investors selecting portfolios of common stock like high expected or average return (E), but dislike high risk or variability (V) on their investments. These tendencies are captured with the following happiness (H) function

$$H = \lambda E - V$$

where λ represents the increase in V that nullifies a one-unit increase in E. Thus, if $\lambda = 3$, increases of 3 in V and 1 in E leave H unchanged.

Portfolio theory enables the determination of the set of "efficient" portfolios from the set of all possible ones. Each efficient portfolio is characterized by E and V values, and together these points form the efficient frontier. As you would expect, one can only get higher average return if he or she is willing to accept higher risk. The efficient frontier incorporating this truism can be approximated by the parabolic function

$$V = (E - k)^2$$

where k represents the riskless return. For example, if one can earn 5 percent in a riskless savings account, then $k = 5$ since $V = 0$ when $E = 5$. Using $k = 5$, let's maximize H.

$$H = \lambda E - (E - 5)^2$$

$$\frac{dH}{dE} = \lambda - 2(E - 5) = (10 + \lambda) - 2E = 0$$

$$E = \frac{10 + \lambda}{2} = 5 + \frac{\lambda}{2}$$

Since the second derivative

$$\frac{d^2H}{dE^2} = -2$$

is negative, $E = 5 + (\lambda/2)$ results in maximum happiness. So, for example, if $\lambda = 2$, the efficient portfolio with $E = 6$ percent and $V = 1$ maximizes happiness.

On a V-E coordinate system, if happiness is held constant, then V is a linear function of E

$$V = -H + \lambda E$$

with slope of λ. Happiness increases moving in a southeast direction, as shown in Figure 15.10. Our maximization procedure in effect locates the greatest happiness line (for a given λ) tangent to the efficient frontier.

Figure 15.10

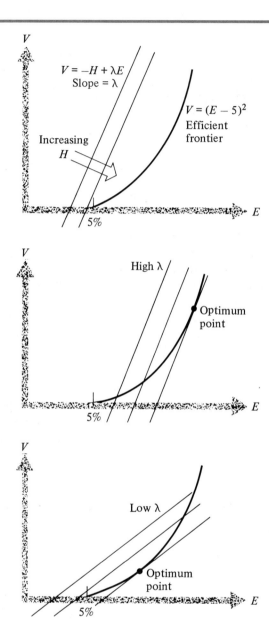

Exercises 18

1. Consider this problem for the general case of unspecified riskless return k. Show that the portfolio with $E = k + (\lambda/2)$ results in maximum happiness.
2. Suppose that your happiness would remain unchanged if V increased by 2 as long as E increased by 1. If the riskless return is 6 percent, determine the expected return and variability of the optimum portfolio.
3. If $V = (E - k)^3$, derive the formula for optimum investment return. Answer exercise 2 for this case.

15.9 OPTIMIZATION IN STATISTICS

Statistical theory includes many applications of optimization. Here, we consider two of them, relating to sample size determination and survey efficiency.

Sample Size Determination

In taking a sample survey of a dichotomous population, where each elementary unit is in one of two states (Republican or Democratic, drink Booze Beer or not), the sample size (n) needed to meet a specified precision goal is given by

$$n = f(p) = kp - kp^2$$

where k is a positive constant determined by the precision required, and p is the proportion of population having the characteristic of interest (e.g., proportion that drink Booze Beer).

The value of p is not known prior to the study; in fact the purpose of the study is to estimate p. So, how can the sample size be determined so as to get on with the study? One approach is to select the sample size that would be large enough to meet the precision goal under all conceivable situations for p. But this means that the maximum value of n, determined from the above sample size formula, be selected.

Taking the derivative of the sample size formula and setting the result equal to zero

$$\frac{dn}{dp} = k - 2kp = 0$$

$$p = \tfrac{1}{2}$$

reveals that the maximum sample size corresponds to $p = \tfrac{1}{2}$. (It is a maximum since the second derivative is $-2k$, a negative number.)

The magnitude of the maximum sample size is

$$f(\tfrac{1}{2}) = k\tfrac{1}{2} - k(\tfrac{1}{2})^2 = \tfrac{1}{4}k$$

Of course, if p is actually not $\tfrac{1}{2}$, then the sample will be larger than really needed, and the precision of the study will be greater than needed.

As shown in Figure 15.11, n is a parabolic function of p (you should verify this). So, if we know that p is restricted to a range of possible values, not including $\tfrac{1}{2}$, we can reduce the necessary sample size. For example, if we know that no more than 20 percent of the people prefer Booze Beer, we can determine the sample size for a Booze Beer preference poll based on the end-of-domain extreme point, or $f(.2)$, as shown in Figure 15.11.

Exercises 19

1. Horace Gallup needs your advice to choose a sample size for the upcoming presidential preference poll between Senator Graft and Governor Watergate. If $k = 8000$,
   a. What sample size should he employ?

Figure 15.11

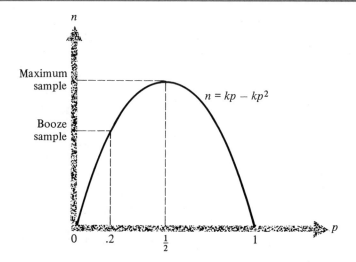

b. If Horace knows that Graft has a commanding lead (at least 70 percent of the electorate), what sample size should he take?

2. To estimate the proportion of the population with a certain rare disease (you are sure that it is less than .01), what sample size is needed, assuming $k = 1,000,000$?

Survey Efficiency

Increasing the sample size (n) of a survey increases both the information content (I) of the results and the cost (C) of the study. A measure of study efficiency (E), which relates the informational content to the cost of obtaining it, is defined by

$$E = \frac{I}{C}$$

Statistical theory has shown that

$$I = n^{.5}$$

Assuming a linear cost function

$$C = F + vn$$

where F is the fixed cost of designing the study, and v is variable cost per unit sampled, let's find the sample size that maximizes survey efficiency.

$$E = \frac{n^{.5}}{F + vn}$$

OPTIMIZATION

Taking the derivative

$$\frac{dE}{dn} = \frac{(F+vn)D[n^{.5}] - (n^{.5})D[F+vn]}{(F+vn)^2}$$

$$= \frac{(F+vn)(.5n^{-.5}) - (n^{.5})(v)}{(F+vn)^2}$$

$$= \frac{.5Fn^{-.5} + .5vn^{.5} - vn^{.5}}{(F+vn)^2}$$

$$= \frac{.5Fn^{-.5} - .5vn^{.5}}{(F+vn)^2}$$

and setting it equal to 0 (after multiplying numerator and denominator by 2) gives

$$\frac{Fn^{-.5} - vn^{.5}}{2(F+vn)^2} = 0$$

Simplifying, by multiplying both sides by the denominator, and then solving for n gives

$$Fn^{-.5} - vn^{.5} = 0$$

$$F - vn = 0$$

$$n = \frac{F}{v}$$

The second derivative is a real lulu, but with a little effort, we can show it to be negative when $n = F/v$.

$$\frac{d^2E}{dn^2} = D\left[\frac{Fn^{-.5} - vn^{.5}}{2(F+vn)^2}\right]$$

$$= \frac{2(F+vn)^2 D[Fn^{-.5} - vn^{.5}] - (Fn^{-.5} - vn^{.5})D[2(F+vn)^2]}{4(F+vn)^4}$$

$$= \frac{\overbrace{[2(F+vn)^2(-.5Fn^{-1.5} - .5vn^{-.5})]}^{A\text{ term}} - \overbrace{[(Fn^{-.5} - vn^{.5})4(F+vn)v]}^{B\text{ term}}}{4(F+vn)^4}$$

The denominator of this mess is positive. Let's now investigate the nature of the A and B terms in the numerator. The A term can be written as

$$A = -(F+vn)^2(Fn^{-1.5} + vn^{-.5})$$

This is negative for any positive n (do you know why?). The first factor in the B term is zero when $n = F/v$, as we show

$$Fn^{-.5} - vn^{.5} = F\left(\frac{F}{v}\right)^{-.5} - v\left(\frac{F}{v}\right)^{.5} = F^{.5}v^{.5} - F^{.5}v^{.5} = 0$$

so the entire B term is zero. Thus at $n = F/v$

$$\frac{d^2E}{dn^2} = \frac{\text{negative number} - 0}{\text{positive number}} = \text{negative number}$$

Thus, $n = F/v$ results in the maximum survey efficiency, which is

$$E = \frac{(F/v)^{.5}}{F + v(F/v)} = \frac{1}{2(Fv)^{.5}}$$

So, decreasing fixed or variable cost increases the maximum efficiency of the survey.

The total cost of the optimally designed survey is

$$C = F + v\left(\frac{F}{v}\right) = 2F$$

or twice the fixed cost.

Exercises 20

1. If a sample survey has a $500 fixed cost and a $5 variable cost per sampled unit,
   a. What sample size maximizes efficiency?
   b. How much would such a study cost?
2. If the cost function were $C = 1000 + 20n^{1/2}$, show that survey efficiency always increases with increases in sample size. If one had $2000 available for a study under these cost conditions, what sample size should be chosen in order to maximize efficiency?

15.10 STUDENT STUDY OPTIMIZATION

If you have trouble figuring out how much to study and how to allocate your study effort among the various courses, this section is just for you.

Right before a final exam period you probably spend all your time either studying or sleeping (sure, you must eat and do the unmentionables, but we can consider this time to be negligible). Suppose that total learning (L) is approximated by

$$L = H \cdot E$$
$$= (24 - S)E$$

where

H is study hours per day
S is sleep hours per day
E is study effectiveness

OPTIMIZATION

Some of you set sleep time equal to zero on those all-nighters. However, you realize that with less sleep, your study becomes less effective. The following effectiveness function captures this pattern:

$$E = S^{.5}$$

So, we can express L as a function of S:

$$L = f(S) = (24 - S)S^{.5}$$

Employing the optimization procedure

$$\frac{dL}{dS} = (24 - S)D[S^{.5}] + (S^{.5})D[(24 - S)]$$

$$= (24 - S)(\tfrac{1}{2}S^{-.5}) + (S^{.5})(-1)$$

$$= 12S^{-.5} - \tfrac{3}{2}S^{.5}$$

$$12S^{-.5} - \tfrac{3}{2}S^{.5} = 0$$

$$12 - \tfrac{3}{2}S = 0 \quad \text{or} \quad S = 8$$

reveals that eight hours of sleep is the optimum.

Exercises 21

1. Verify by taking the second derivative that eight hours of sleep results in maximum (not minimum) learning.
2. Consider the revised effectiveness function

   $$E = \ln S \quad \{S \geq 1\}$$

   Show that the optimum sleep time is found by solving

   $$\frac{24}{S} - \ln S = 1$$

   (*Note:* Since the solution of this equation is beyond the scope of this book, don't proceed to solve explicitly for S.)

Now consider the study allocation decision. There you are, trying to decide whether to study for the math final or for your other finals in Accounting Methods to Hide Overseas Bribes, Sadistics, and Financial Mismanagement. You know that every hour spent in studying math will help your grade in that course. However, since those hours will take study time away from the other courses, the other grades will suffer. After some thinking on the problem, you estimate the grade effects of studying x hours on mathematics are ($4 = A$)

$$M = 2 + .3x^{.5}$$
$$N = 3 - .01x$$

where M is math grade, and N is average grade in the other three non-math courses.

Since each course carries the same credit, your overall grade point average (G) is computed as

$$G = \frac{M + 3N}{4}$$

which is the following function of x:

$$G = \frac{(2 + .3x^{.5}) + 3(3 - .01x)}{4} = \frac{11}{4} + \frac{.3}{4}x^{.5} - \frac{.03}{4}x$$

As we show now G is maximized at 25 hours of mathematics study.

$$\frac{dG}{dx} = \frac{.3}{8}x^{-.5} - \frac{.03}{4} = 0$$

$$x^{-.5} = \frac{.06}{.3} \quad \text{so} \quad x = 25$$

The second derivative

$$\frac{d^2G}{dx^2} = -\frac{1}{2}\left(\frac{.3}{8}x^{-1.5}\right) = -\frac{.3}{16x^{1.5}}$$

is negative when $x = 25$, so a maximum is verified.

Exercises 22

1. What will your grade-point average be if you study 25 hours?
2. If you have four other courses, so that

$$G = \frac{M + 4N}{5}$$

   how much mathematics study will maximize grade-point average?
3. In general, the formula for grade-point average is

$$G = \frac{M + kN}{1 + k}$$

   where k is the number of non-math credits per math credit. Derive the general formula for hours of math study to maximize grade point average, in terms of k. Show that your formula works by applying it to exercise 2.
4. Construct M and N functions that reflect your own study situation, and maximize G.

15.11 PHYSICAL CAPACITY MAXIMIZATION

The physical capacity (area or volume) of any entity depends on its dimensions. However, by the choice of optimum dimensions, the capacity can be maximized, as illustrated by the following two examples.

A farmer has 200 yards of fencing that will be used to enclose three sides of a rectangular grazing area; the fourth side need not be fenced since it is a steep hill. If the two equal sides are x yards long, then the remainder $(200 - 2x)$ will be the length of the third fenced side, as shown in Figure 15.12.

The enclosed area (A), which is the product of length times width,

$$A = (x)(200 - 2x) = 200x - 2x^2$$

is maximized when

$$\frac{dA}{dx} = 200 - 4x = 0$$

$$x = 50$$

The negative second derivative (-4) confirms that we have found the dimensions that maximize grazing area.

Exercises 23

1. Find the dimensions and the area of the grazing region.
2. If the fourth side must be fenced, too, find the revised dimensions and area of the grazing region.

An open box is made from a 10-inch by 8-inch metal sheet by cutting out identical squares from the corners, as shown in Figure 15.13. The maximum volume box is desired.

The volume of the box is

$$V = (\text{length})(\text{width})(\text{height})$$
$$= (10 - 2x)(8 - 2x)(x)$$
$$= 4x^3 - 36x^2 + 80x$$

Figure 15.12

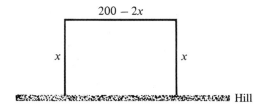

Figure 15.13

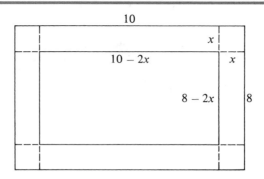

Maximizing V by setting the derivative equal to zero gives

$$\frac{dV}{dx} = 12x^2 - 72x + 80 = 0$$

$$3x^2 - 18x + 20 = 0$$

$$x = \frac{18 \pm \sqrt{324 - 240}}{6} \cong 1.5 \text{ and } 4.5$$

The second derivative is $24x - 72$, which is negative when $x = 1.5$ and positive when $x = 4.5$. So $x = 1.5$ results in the maximum volume box (see Figure 15.14).

Exercises 24

1. Show why $x = 4.5$ is a physical impossibility.
2. Find the dimensions and volume of the maximum volume open box that can be constructed from a 20 inch-by-10 inch sheet.

Figure 15.14

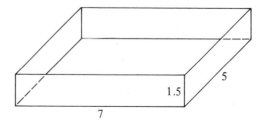

$V = 7 \cdot 5 \cdot 1.5 = 52.5$ cubic inches

OPTIMIZATION

15.12 LEAST-SQUARES REGRESSION

Least-squares regression finds the equation of the best fitting linear function through a set of data. We introduced the subject and solved for the unique best-fitting line using so-called normal equations on pages 216–218, 264–267, and 350–352. In those B.C. (Before Calculus) times, you were told "take our word for it" and "wait for calculus to see why." Now, in A.D. (After Differentiation) times, you FINALLY can see the light.

Here, we show how calculus, specifically optimization of functions of more than one variable, leads to the normal equations for determining the best-fitting line. We begin by stating that the best-fitting line is the one that has the minimum sum of squares of the deviations from the points to the line.

The deviation for a typical point, as shown in Figure 15.15, is

$$d = y - y_{LS}$$
$$= y - (sx + i) = y - sx - i$$

Squaring this deviation gives

$$d^2 = y^2 + s^2x^2 + i^2 - 2ysx - 2yi + 2sxi$$

Now, if we consider all n deviations and sum their squares, we get

$$D = \Sigma d^2$$
$$= \Sigma(y^2 + s^2x^2 + i^2 - 2ysx - 2yi + 2sxi)$$
$$= \Sigma y^2 + s^2(\Sigma x^2) + ni^2 - 2s(\Sigma xy) - 2i(\Sigma y) + 2si(\Sigma x)$$

The various sums are constants for a given set of data (n observations), as illustrated by the example on page 265. So, D is a function of two variables, s and i.

$$D = f(s, i)$$

Figure 15.15

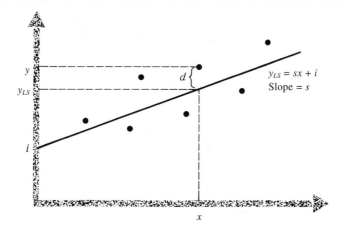

We minimize D by setting the partial derivatives equal to zero.

$$\frac{\partial D}{\partial s} = 2s(\Sigma x^2) - 2(\Sigma xy) + 2i(\Sigma x) = 0$$

$$\frac{\partial D}{\partial i} = 2ni - 2(\Sigma y) + 2s(\Sigma x) = 0$$

After some simplification we arrive at the so-called normal equations

$(\Sigma x^2)s + (\Sigma x)i = \Sigma xy$

$(\Sigma x)s + ni = \Sigma y$

which are solved using Cramer's rule to give the slope (s) and intercept (i) of the least-squares regression line as

$$s = \frac{n(\Sigma xy) - (\Sigma x)(\Sigma y)}{n(\Sigma x^2) - (\Sigma x)^2}$$

$$i = \frac{(\Sigma x^2)(\Sigma y) - (\Sigma x)(\Sigma xy)}{n(\Sigma x^2) - (\Sigma x)^2}$$

Exercise 25

Apply these formulas to the data on page 265 and find the least-squares regression line for math grade as a function of hours studied.

Better prediction of a dependent variable can be achieved by employing more independent variables in the regression model. In each case, though, optimization theory is needed to find the regression equation. For example, if we postulate a two-independent variable (x and z) model of the form

$$y = a + bx + cz$$

where a, b, and c are parameters to be found. The typical deviation from a data point to the least-squares plane is

$$d = y - y_{LS} = y - a - bx - cz$$

Squaring the n deviations and summing gives

$$\begin{aligned}D &= \Sigma d^2 \\ &= \Sigma y^2 + na^2 + b^2(\Sigma x^2) + c^2(\Sigma z^2) - 2a(\Sigma y) - 2b(\Sigma xy) - 2c(\Sigma yz) \\ & + 2ab(\Sigma x) + 2ac(\Sigma z) + 2bc(\Sigma xz)\end{aligned}$$

OPTIMIZATION

Exercise 26

Are you feeling brave? Take over and find the three partial derivatives $\partial D/\partial a$, $\partial D/\partial b$, and $\partial D/\partial c$; set them equal to zero; and simplify. You should get the three normal equations given on page 266.

PROBLEMS

TECHNICAL TRIUMPHS

1. Consider the function

   $$y = f(x) = x^3 - 2x^2 - 7x + 10$$

   a. Graph it over the domain $\{-2 \leq x \leq 3\}$.
   b. Determine the maximum and minimum points from your graph.
   c. Use calculus to determine the extreme points.
   d. Use the second derivative check to identify the maximum and minimum points.

2. Consider the linear demand function

   $$Q = f(P) = 100 - 5P \quad \{0 \leq P \leq 20\}$$

   a. What price maximizes total revenue?
   b. What is the maximum revenue level?
   c. Verify your answer in part b. with the second derivative check.

3. The Slo and Dirty (SAD) Railroad Company, operating a line between Slippery Rock and Turtle Creek, Pennsylvania, has determined that the cost per car (C) of running an x-car train is

   $$C = \frac{-36}{x} - (20 \ln x) + x + 100 \quad \{x \geq 1\}$$

   a. Find the extreme points.
   b. Use the second derivative check to determine the length of train that minimizes the cost per car.

4. Suppose total revenue (R) as a function of No-Extras ticket price (P) for Creaky Airlines is

   $$R = -2P^2 + 275P + 950$$

   a. Determine the marginal revenue function.
   b. What No-Extras ticket price maximizes revenue? What is this maximum revenue level?
   c. Verify your result by using the second derivative check.

5. A company estimates that the total cost (C) of obtaining sample information for accounting purposes is the following function of the sample size (n).

   $$C = 5n + \frac{2000}{n}$$

   a. Determine the marginal cost function.
   b. What sample size provides for the minimum cost?

6. In problem 4 on page 249, you derived the following total cost (C), total revenue (R), and total profit (π) functions of the number of thousands of tomato plants (x) planted on a one-acre plot.

$C = 100 + 500x$

$R = 10{,}000 \ln x$

$\pi = 10{,}000 \ln x - 500x - 100$

   a. How many plants maximize revenue?
   b. How many plants maximize profit? What would this profit be?

7. In problem 3 on page 249 you derived the Ivory Tower University revenue (R), cost (C), and profit (π) functions of tuition level (t) as

$R = 16{,}000t - 2t^2$

$C = 34{,}000{,}000 - 1{,}000t$

$\pi = -2t^2 + 17{,}000t - 34{,}000{,}000$

   a. What tuition maximizes revenue? What is this revenue?
   b. What tuition maximizes profit? What is this profit?
   c. Compare the maximum profit tuition to the breakeven tuition levels of $3219 and $5281.

8. Consider the function

$y = f(x) = x^4 - x^3 - x^2 \quad \{-1 \leq x \leq 3\}$

   a. Graph the function.
   b. Determine the points where maximums or minimums occur.
   c. From the second derivative check, distinguish the maximums from the minimums.

CONFIDENCE BUILDERS

1. Maximum Industries estimated that its total revenue (R) and total cost (C) functions of output (Q) are

$R = f(Q) = 10Q - .01Q^2$

$C = g(Q) = 30 + 5Q + .01Q^2$

   a. Determine Maximum's profit function.
   b. At what output level does marginal revenue equal marginal cost? Show that profits are maximized at this output level.
   c. Find the output level that maximizes revenue. Is this point different than the one that maximizes profit?
   d. Plot the revenue, cost, and profit functions on a single graph, showing the various maximization points.

2. Show that the peak of the normal probability distribution, given by the function,

$$f(x) = \left(\frac{1}{2.5\sigma}\right) e^{-.5\left(\frac{x-\mu}{\sigma}\right)^2}$$

   where μ (the mean) and σ (the standard deviation) are constants, occurs at the mean.

3. If Gangland Industries has the following revenue (R) and cost (C) functions of time (t) in years from now

OPTIMIZATION

$$R = 10(1.2)^t$$
$$C = 8(1.2)^t$$

show that profits are never maximized, and in fact continue to grow forever and ever.

4. On pages 212–214, we fit demand curves for S. Lumlord assuming hyperbolic, $f(P)$, and parabolic, $g(P)$, models

$$Q = f(P) = \frac{168{,}000}{80 + P} \qquad Q = g(P) = .004167P^2 - 6.0835P + 1{,}650$$

   a. For each model, determine the rent price that maximizes revenue.
   b. Assuming the total cost function derived in the beginning of the chapter

   $$C = 49{,}000 - 40P$$

   for each model, determine the rent price that maximizes profits.

5. The GO Gas Station charges 1.60 per gallon for its gas. At that price it sells 1000 gallons per day. By market research, it has determined that for every 1¢ reduction in price per gallon, it can sell an additional 100 gallons.
   a. Determine gallons sold per day (Q) as a function of price per gallon (P) in cents.
   b. What price maximizes revenue?
   c. If total cost (C) is the following function of Q

   $$C = 30{,}000 + 120Q \quad (\textit{Note: } C \text{ is in cents})$$

   determine the price that maximizes profit.

6. M. Presario is presenting "The Plumbers" in the upcoming concert in Madison Rectangle Garden. From past concerts he was able to relate attendance (Q) to ticket price (P) as

$$Q = 20{,}000e^{-.2P}$$

   a. If it costs $5000 to rent the "Garden," $5000 to pay the "Plumbers," and if the various other costs (ticket takers, cleaning up, etc.) are figured at $1 per person attending, what admission fee maximizes profit?
   b. What is this maximum profit?

7. Carmen Monoxide derived the revenue, cost, and resulting profit functions of fare price for the Fumes Bus Company. After taking all the proper derivatives, she found the fare price that would maximize profit. However, when she substituted that fare into the profit function, lo and behold, it showed that a negative profit (loss) would result. Carmen was puzzled and sought help from her co-workers, Smokey Hays and Morris Mog. Smokey ridiculed Carmen's mathematics and said that such a result was impossible. Morris disagreed, saying that Carmen's result was indeed possible. Who is right? Discuss and back up your position with a graphical presentation.

8. In problem 7 on page 246 you derived gasoline mileage (m) for a certain car to be the following function of velocity (v).

$$m = f(v) = -.01v^2 + .9v$$

   a. What velocity maximizes gasoline mileage?
   b. What is the maximum gasoline mileage level?

9. In problem 1 on page 248 you derived two attendance (A) functions for Syracuse Tiddler's Tiddly Wink games using Tillie Winkle's, $f(P)$, and Tedlee Winkler's $g(P)$, assumptions of functional relationship as

$$A = f(P) = \frac{5000}{1 + P^2} \qquad A = g(P) = \frac{3333}{\frac{1}{3} + P}$$

a. For each functional assumption, what ticket price maximizes revenue? What are the respective maximum revenue levels?
b. Do your results indicate that one of the functions is unrealistic?

10. In problem 2 on page 249, you derived attendance (Q) at the Al Lee—Willy Dive heavyweight championship fight to be the following function of ticket price (P):

$$Q = -500P + 40{,}000$$

a. What ticket price maximizes revenue?
b. What will the maximum revenue be?
c. If total costs are $500,000, what ticket price results in maximum profit? How much is this profit?
d. If Al Lee gets 10 percent of the revenue, what price maximizes profit?

11. A company has the choice of producing a product entirely on its own or subcontracting all or part of the production. Depending on how many units of its own labor (L) and machines (M) it uses, the total cost (C) including the subcontracting cost is given by

$$C = f(L, M) = L^2 - 30L + M^2 - 36M + LM + 2000$$

a. Determine the cost if the company subcontracts the entire operation.
b. What internal labor and machine inputs minimize the total cost?
c. What is the minimum cost?

12. Universe Book Encyclopedia has determined that its profits (π) are the following function of expenditures on personal selling (x), such as for salespeople, and expenditures on impersonal selling (y), such as for advertising—all variables are in millions of dollars.

$$\pi = -2x^2 - 4y^2 - xy + 18x + 20y + 10$$

a. Find the amounts of personal and impersonal selling that maximize profits.
b. What is that maximum profit level?

MIND STRETCHERS

1. A real estate investor has a property worth $1,000,000 now and expects the value to increase $80,000 per year. If the cost of capital is 5 percent,
   a. Determine the present value of the selling price n years away.
   b. When should the property be sold to maximize the present value?

2. By using optimization concepts show that the parabola $y = ax^2 + bx + c$, where a, b, and c are constants, opens upward if a is positive and downward if a is negative.

3. The New York Debts football team currently charges $30 per ticket and averages 40,000 fans per game. They are thinking about ways to increase revenue so that they can meet the fat payroll of such stars as Joe NoMath, Rich Mann, and Manny Bucks. When they charged $25 per ticket, attendance averaged 45,000. Costs are fairly stable regardless of the crowd size. Assuming a linear demand curve,
   a. What ticket price maximizes revenue?

b. If the team realizes about $5 per fan in refreshment stand and scorecard sales, what ticket price maximizes overall revenue?
c. Reanswer parts a and b assuming a parabolic demand curve of the form,

$$Q = aP^2 + bP.$$

4. Show that profit is maximized at the right end of the domain for the case of linear revenue ($R = PQ$) and linear cost ($C = F + vQ$) curves, where $P > v$.

5. A farmer has 1000 feet of fence to enclose three equal animal stalls, as shown in the figure.

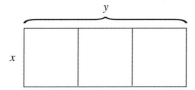

a. What should the dimensions of the stalls be in order to enclose the maximum area? What will the total area equal?
b. If the land is up against a steep cliff (thus, one of the y dimension sides need not be fenced), what should the dimensions of the stalls be in order to enclose the maximum area? What will the total area equal?

6. Suppose that for an inventory system, total cost (TC) is the sum of inventory carrying cost (IC) and ordering costs (OC).
   If $IC = a/n$ and $OC = bn$, where a and b are constants, and n is the number of orders per year,
   a. Show that the optimum solution (the value of n that minimizes TC) occurs at the point where $IC = OC$.
   b. Illustrate this system and the solution graphically.

7. If $f(x) = \ln x$
   a. Find the second derivative.
   b. Derive the formula that gives the nth derivative.

8. Tom A. Doe operates a fruit and vegetable stand. He can sell a box (100 pounds) of tomatoes at 50¢ per pound. For every 1¢ increase in price up to a price of 125¢ per pound, he would sell 3 pounds less. For every 1¢ over 125¢ per pound, he would sell one pound less. Assume the unsold tomatoes get overripe and are worthless.
   a. Find the demand curve.
   b. What price should Tom charge if he wishes to maximize revenue?
   c. How many pounds of tomatoes would Tom sell if he charged the price that maximized revenue?

9. Consider the case of linear demand of the form

$$Q = sP + i \quad \text{where } s \text{ and } i, \text{ respectively, are the slope and intercept}$$

and parabolic total cost of the form

$$C = F + aQ + bQ^2 \quad \text{where } F \text{ (fixed cost)}, a, \text{ and } b \text{ are constants}$$

a. Show that profits as a function of price (P) are

$$\pi = (s - bs^2)P^2 + (i - as - 2bsi)P - (F + ai + bi^2)$$

b. Show that profits are maximized when

$$P = \frac{i - s(a + 2bi)}{-2s(1 - bs)}$$

10. An importer charges $10 per ounce of perfume and sells 45,000 ounces at that price. However, the government is considering the imposition of a tax (t) on the perfume. The importer claims that for every dollar of tax, (selling price would be $10 + t$), he would lose 4500 ounces of sales. Assuming the importer's claim is true,
    a. What tax level would maximize the tax revenue to the government?
    b. What tax level would result in maximum product revenue (excluding taxes) for the importer?
    c. If the government imposes the tax that maximizes its revenue, what total tax revenue will it achieve? How many ounces of perfume will the importer sell under these conditions?

11. Al Pell operates a 1000 tree orchard. As of September 1, his trees yield an average of 300 pounds of apples, for which Al can get 20¢ per pound on the market. For each week that he waits, he can get another 40 pounds per tree; however, the price drops 1¢ per pound. Al's total cost would be $30,000 for a September 1 picking; however, for each week of delay, he incurs an additional $1000 cost.
    a. When should Al pick the apples in order to maximize revenue?
    b. When should he pick the fruit in order to maximize profit?

12. In the cold of winter you enjoy higher home temperature (T) but don't enjoy paying the higher cost (C) of achieving it. If $C = .001(T - 25)^3$ and you would get equal pleasure from an additional degree of temperature and an additional $5 in your pocket, what home temperature maximizes the happiness (H) function

$$H = T - \lambda C$$

13. Glamourpuss Toothpaste Company estimates that its total revenue (R) in millions of dollars and its total advertising budget (A) in millions of dollars are related as follows:

   for $\{1 \leq A \leq 8\}$ $R = 10 + 5(\ln A)$

   for $\{A > 8\}$ $\frac{dR}{dA} < 0$

   In addition to the advertising cost, Glamourpuss has one million dollars in fixed cost, plus the product cost which amounts to 20% of the revenue.
    a. How much of that "Look Ma, no cavities (no teeth either)" advertising should they do in order to maximize profits? Explain why.
    b. What is the maximum revenue that they can achieve? You must carefully explain your reasoning here.

CHAPTER 16

Introduction to Integral Calculus

16.1 INTRODUCTION

Integral calculus is the study of "going backwards" in the differentiation process. The reverse process to differentiation is called *antidifferentiation,* or *integration.* Could this be a godsend for all those who were doing differentiation all backwards? Could you now be a whiz at integration? Sorry, too bad, but no. It will take a good understanding of differential calculus, along with a grasp of how to find an area under a curve and an understanding of the fundamental theorem of calculus to be that whiz.

But what is this reverse process, and why undo what was so difficult to learn in the first place? Consider $F(x) = x^3$. Differentiating, we get $3x^2$. Now, suppose we began with $f(x) = 3x^2$ and wanted to know (heavens knows why) what function had this as its derivative. Seeing the above, one could say, "Ah hah, it is x^3 that you seek."

On first exposure the integration process might appear to have only academic interest. But wait. Differential calculus determines the rate of change of a total function. For example, differentiating the total revenue function $R = 5Q$ gives 5, which is the rate at which total revenue changes with respect to changes in quantity. But, what if you had rate of change information, yet wished to determine the total function? For example, if you knew that revenue changed by $5 for each additional item sold, by integration you could find the "total" function, or $5Q$, which has the given "rate" function.

The real world does contain applications where rate of change data is available, yet information about the total function is needed. Birth and death rates can be tabulated and estimated for the future, but estimates of the total future population might be desired. Sales rates are commonly forecast, but total sales over some extended period might be needed. Rates of snowfall or rainfall can be determined, but how much will the total snowfall or rainfall be? The rate of pollution entering a lake can be measured, but what is the important level of total pollution in the lake? Velocity is measured on one's speedometer, but what is the total distance traveled? This need to translate rate information to totals information exists in such diverse areas as atomic physics, statistical theory, business administration, public administration, ecology, and outer space exploration. On a more down-to-earth level this process could determine the amount of beer left in your mug given your sipping function. Now, if these application areas, especially the last one, don't impress you, and you still haven't figured out what differential calculus is all about, as your last act, it can help determine how long it would take to leap from the nearest tall building (certainly the most down-to-earth application)!

Learning Plan

The ability to apply integral calculus to everyday problems cannot occur until the student has successfully passed through the following three-stage learning process:

1. Learn differential calculus.
2. Understand theory for finding the area under a curve.
3. Understand the fundamental theorem of calculus.

Stage 1 should be achieved already. If not, go back to Chapter 12, but do not collect $200. The means to successfully complete stages 2 and 3 are included in this chapter. These stages are theoretical, but hopefully the reader, knowing that an understanding of this theory is a prerequisite to application ability, will bear with it.

16.2 AREA UNDER A CURVE

Integration and areas under curves are theoretically tied together. This point will be brought out with the fundamental theorem of calculus. Now, to prepare you for this occasion, geometric methods to approximate the area under a curve are presented.

Consider the function $y = f(x) = x^2$, and suppose we wish to determine the area under this curve from $x = 0$ to $x = 2$, as shown in Figure 16.1. The area of a rectangle, triangle, or even trapezoid can be easily determined from formulas in any geometry book. Unfortunately, the area here does not conform to any of these basic geometric shapes. However, we will be able to approximate the area under the curve by methods employing rectangular areas.

Figure 16.1

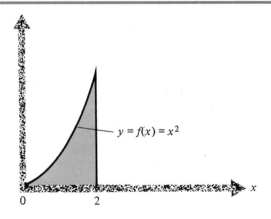

INTRODUCTION TO INTEGRAL CALCULUS

Figure 16.2

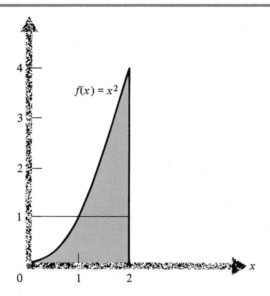

Midpoint Rectangle Method

Let us approximate the area under the curve by a single rectangle with base of two units and height equal to the value of the function at the midpoint of our interval, as shown in Figure 16.2. Recalling that the area of a rectangle is equal to the base times the height, we can calculate the area to be $2 \cdot f(1) = 2 \cdot 1 = 2$. Note that this single rectangle overestimates the area in the left side of the interval and underestimates the area in the right side. Overall, a net underestimate results. Possibly two rectangles, one approximating the left side and the other the right side, could do a better overall job of approximating the area under the curve. Suppose the two rectangles, as shown in Figure 16.3, are formed by dividing the domain into two equal subintervals, which form the bases of the rectangles, then taking the value of the function at the midpoint of each subinterval ($\frac{1}{2}$ and $\frac{3}{2}$) to serve as the heights of the rectangles. The left rectangle has a height of $f(\frac{1}{2}) = (\frac{1}{2})^2 = \frac{1}{4}$; the right rectangle has a height of $f(\frac{3}{2}) = (\frac{3}{2})^2 = \frac{9}{4}$. Thus, the left rectangle has an area of $1 \cdot \frac{1}{4} = \frac{1}{4}$; the right rectangle has an area of $1 \cdot \frac{9}{4} = \frac{9}{4}$, and the total area in both rectangles is 2.5.

Seeing that two rectangles approximate the area under the curve better than the single rectangle, one is tempted to try four rectangles to do even better. Again, using "midpoint" rectangles (see Figure 16.4), we divide the domain into four equal subintervals (base of $\frac{1}{2}$), with midpoints at $\frac{1}{4}$, $\frac{3}{4}$, $\frac{5}{4}$, and $\frac{7}{4}$, respectively. The total rectangular area (A), which is the sum of the individual rectangle areas, is

$$A = \tfrac{1}{2} \cdot f(\tfrac{1}{4}) + \tfrac{1}{2} \cdot f(\tfrac{3}{4}) + \tfrac{1}{2} \cdot f(\tfrac{5}{4}) + \tfrac{1}{2} \cdot f(\tfrac{7}{4})$$
$$= \tfrac{1}{2}[f(\tfrac{1}{4}) + f(\tfrac{3}{4}) + f(\tfrac{5}{4}) + f(\tfrac{7}{4})]$$
$$= \tfrac{1}{2}(\tfrac{1}{16} + \tfrac{9}{16} + \tfrac{25}{16} + \tfrac{49}{16}) = 2\tfrac{5}{8} = 2.625$$

One senses that the total rectangular area is getting closer and closer to the actual area under the curve. But for further insight into this, let us consider eight rectangles. With eight equal

Figure 16.3

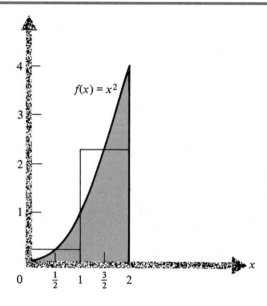

Figure 16.4

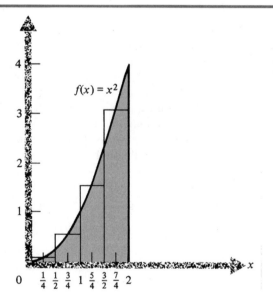

subintervals between 0 and 2, each rectangle has a base of $\frac{1}{4}$ unit and an altitude determined at the appropriate midpoint. Thus,

$$A = \tfrac{1}{4}[f(\tfrac{1}{8}) + f(\tfrac{3}{8}) + f(\tfrac{5}{8}) + \cdots + f(\tfrac{15}{8})]$$
$$= \tfrac{1}{4}(\tfrac{1}{64} + \tfrac{9}{64} + \tfrac{25}{64} + \cdots + \tfrac{225}{64})$$
$$= 2\tfrac{21}{32} = 2.656$$

Exercise 1

Verify that 16 rectangles, constructed according to the midpoint rectangle method, would have a total area of 2.664.

The results of our midpoint rectangular approximation method are summarized as follows:

Number of rectangles	Total area of rectangles
1	2.0
2	2.5
4	2.625
8	2.656
16	2.664

One can see that the total area seems to be approaching some limit, which appears to be equal to the actual area under the curve. However, it is impossible to tell from the information above the precise limiting point. Is it 2.67, 2.7, 2.7012, 2.8, or what?

Now, consider the following "mysterious operation:"

$$\int_0^2 x^2 \, dx = \frac{x^3}{3} \bigg|_0^2 = \frac{8}{3} - 0 = 2\frac{2}{3} = 2.666\ldots$$

It tells us that the *exact* area under the function $f(x) = x^2$ between $x = 0$ and $x = 2$ is equal to $2\frac{2}{3}$. Do not worry about how this mysterious operation was performed—it is beyond your current mathematical prowess. But take note that integral calculus was employed, and in a matter of seconds the exact area under a curve was determined.[1]

One interesting thing to note about the mystery operation is that the "answer" involved the function $x^3/3$, which we know has a derivative of x^2, the function bounding the region whose area was sought.

Until now rectangles have been formed with the altitude being the value of the function at the midpoint of each interval. But would rectangles formed in some other way, in the limit as the number of rectangles approaches infinity, still provide the area under the curve? The answer is yes, and the following demonstration consists of showing that the two "worst possible" methods provide the area; therefore, other "better" methods would do likewise.

"Worst" Methods—Smallest and Largest Rectangles

Consider any function represented by $y = f(x)$ over the interval $x = a$ to $x = b$, as shown in Figure 16.5. As we have done previously, let us divide the interval into n equal subintervals, each of width $\Delta x = (b - a)/n$. If y_k is the smallest value of $f(x)$ in the kth interval, and Y_k is the largest value of $f(x)$ in the kth interval, one can form both a smallest and a largest rectangle for each subinterval by taking y_k and Y_k as the respective altitudes. Thus, in the kth interval, the smallest

---

[1] It is interesting to note that the rectangular approximation method was close to the actual area with as few as 16 rectangles. However, only in the limit, as the number of rectangles goes to infinity, would that method provide the exact area under the curve.

Figure 16.5

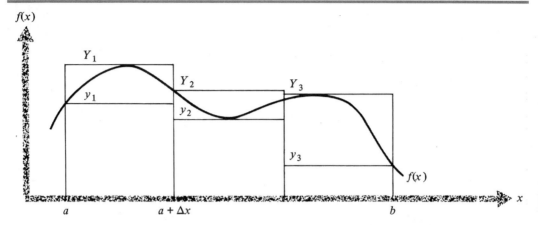

rectangle will have area of $\Delta x \cdot y_k$, and the largest rectangle will have area of $\Delta x \cdot Y_k$. If S is the total area of all the smallest rectangles, and L is the total area of all the largest rectangles, then

$$S = \Delta x \cdot y_1 + \Delta x \cdot y_2 + \Delta x \cdot y_3 + \cdots + \Delta x \cdot y_n = \Delta x \left(\sum_{k=1}^{n} y_k \right)$$

$$L = \Delta x \cdot Y_1 + \Delta x \cdot Y_2 + \Delta x \cdot Y_3 + \cdots + \Delta x \cdot Y_n = \Delta x \left(\sum_{k=1}^{n} Y_k \right)$$

Figure 16.5 points out that in each subinterval the smallest rectangle encompasses less area than does $f(x)$, while the largest rectangle encompasses more area than $f(x)$. Consequently, the following relationship holds

$$S \leq T \leq L$$

where T is the true area under $f(x)$.

To get a sense for these two "worst" methods for finding areas under curves, let's return to our example function, $y = f(x) = x^2$ for $\{0 \leq x \leq 2\}$.

One can see on Figure 16.3 that the smallest rectangle for each subinterval would be formed by using the height or value of the function at the left end of the subinterval, while the largest rectangle for each subinterval would be formed by using the height or value of the function at the right end of the subinterval. Using two rectangles ($\Delta x = 1$), then

$$S = \Delta x \left(\sum_{k=1}^{2} y_k \right) = 1(0 + 1) = 1$$

$$L = \Delta x \left(\sum_{k=1}^{2} Y_k \right) = 1(1 + 4) = 5$$

Now let us observe the effect, on S and L, of increasing the number of subintervals. Suppose the number of subintervals were doubled so two exist where one did before. Consider Figure 16.6, which illustrates the effect on the first subinterval in Figure 16.5. A small area (dotted region) was gained by S, while L lost a small area (cross-hatched region). If each of the two new

INTRODUCTION TO INTEGRAL CALCULUS

Figure 16.6

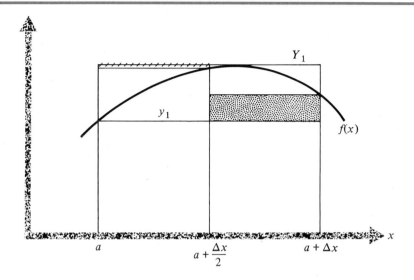

subdivisions were further subdivided, then S would gain a little more and L would lose a little more area.

We illustrate this effect on our example function, $y = f(x) = x^2$ for $\{0 \leq x \leq 2\}$ by doubling the number of subintervals to 4 ($\Delta x = \frac{1}{2}$). Then, S and L are (see Figure 16.4)

$$S = \Delta x \left(\sum_{k=1}^{4} y_k \right) = \tfrac{1}{2}(0 + \tfrac{1}{4} + 1 + \tfrac{9}{4}) = \tfrac{14}{8}$$

$$L = \Delta x \left(\sum_{k=1}^{4} Y_k \right) = \tfrac{1}{2}(\tfrac{1}{4} + 1 + \tfrac{9}{4} + 4) = \tfrac{30}{8}$$

Notice how S got bigger and L got smaller than they were with 2 subintervals.

Exercise 2

Find S and L for 8 subintervals.

And on and on the process goes. As the number of subdivisions increases, S gets larger while L gets smaller. Yet, by construction S can't be larger than T and L can't be smaller than T. Therefore, both S and L must approach the true area under the curve.

$$\lim_{n \to \infty} [S] = T$$

$$\lim_{n \to \infty} [L] = T$$

Since these two worst methods both give the area under the curve in the limit as $n \to \infty$, any other method of forming rectangles will do likewise. Consider one such method that uses intermediate rectangles.

Intermediate Rectangle Method

With this method the interval is again subdivided into n equal-width sub-intervals. But in determining the altitude of each rectangle, suppose one takes any x value in the subinterval—call this value of x for the kth subinterval c_k. The location of c_k within the kth subinterval does not matter. It could be the midpoint, a point randomly determined, and so on; it is sometimes on the left side, other times on the right side of the subinterval, and so on.

The "intermediate rectangle" method is illustrated in Figure 16.7.

Notice that each intermediate rectangle has an area at least as great as the corresponding smallest rectangle, but no greater than the corresponding largest rectangle. Consequently, I, the total area of all the intermediate rectangles, defined as

$$I = \sum_{k=1}^{n} f(c_k) \cdot \Delta x$$

is related to S and L as

$$S \leq I \leq L$$

We have already seen that both S and L approach the true area, T, as n approaches infinity. Since I is sandwiched in between S and L, it follows that it too must approach the true area.

$$\operatorname*{limit}_{n \to \infty} [I] = \operatorname*{limit}_{n \to \infty} \left[\sum_{k=1}^{n} (f(c_k) \cdot \Delta x) \right] = T$$

This result is crucial to the development of the fundamental theorem of calculus, which now follows.

Figure 16.7

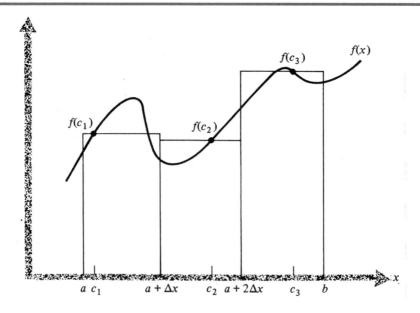

16.3 FUNDAMENTAL THEOREM OF CALCULUS

The fundamental theorem of calculus is indeed fundamental to an understanding of calculus. It serves as the bridge between the two main branches of calculus—differential and integral. An understanding of this theorem allows one to view integral calculus as just a logical extension of differential calculus, and the process of integration as just the reverse process to differentiation.

The fundamental theorem states that if $F(x)$ is some function, the derivative of which is $f(x)$, and if both functions are defined over the interval from $x = a$ to $x = b$, then

$$\int_a^b f(x)\, dx = F(b) - F(a)$$

For those who wish it repeated in English, a little patience is urged in as much as the following exposition has a good chance (not 100 percent, to be sure) of providing the reader with the understanding needed.

Consider a continuous function of x, called $F(x)$, which possesses a continuous derivative function, called $f(x)$. Thus,

$$D[F(x)] = f(x)$$

For example, if $F(x) = 3x^2$, then

$$D[F(x)] = D[3x^2] = 6x = f(x)$$

Now consider Figure 16.8, which shows both $F(x)$ and $f(x)$ over some interval $x = a$ to $x = b$. Again, we divide the interval into n equal subintervals, each of width $\Delta x = (b - a)/n$. The kth subinterval is bounded on the left by x_{k-1} and on the right by x_k, where $x_k = a + k(\Delta x)$. Vertical lines from each subinterval boundary divide the area under both $F(x)$ and $f(x)$ into n smaller areas. Now, it is necessary to invoke a theorem, which we are in no position to prove, but which should make sense. This theorem, called the mean value theorem of calculus, states that for any subinterval, where k stands for the number of the subinterval,

$$\frac{F(x_k) - F(x_{k-1})}{x_k - x_{k-1}} = f(c_k) \quad \text{where} \quad \{x_{k-1} \leq c_k \leq x_k\}$$

To understand this theorem, let's focus on the kth subinterval portrayed in Figure 16.9.

Notice that $F(x_k) - F(x_{k-1})$ is the ordinate change or Δy when one connects the endpoints of $F(x)$ in the kth subinterval. Of course, $x_k - x_{k-1}$ is the change in x or Δx over the subinterval. Thus, the ratio of the left side of the theorem is the change in y divided by the change in x, or our old friend, the slope of the secant line connecting the endpoints of the subinterval.

The theorem states that this slope is equal to $f(c_k)$, where c_k is some value of x in the kth subinterval. But since $f(x)$ is the derivative of $F(x)$, it follows that the straight line (secant line) connecting the endpoints of $F(x)$ has a slope that is equaled by some tangent line along the function $F(x)$ somewhere in the subinterval. In this case it looks as if c_k is about two-thirds of the way into the subinterval. In general c_k could occur anywhere in the subinterval, or there even could be more than one value of c_k, as illustrated in Figure 16.10.

Let's illustrate this theorem by returning to our example function, $y = f(x) = x^2$ (see Figure 16.3). Using two subintervals over the domain $\{0 \leq x \leq 2\}$, the table on page 590 gives the points (c_k) where the secant and tangent slopes are equal. (Note $\dfrac{dy}{dx} = 2x$).

Figure 16.8

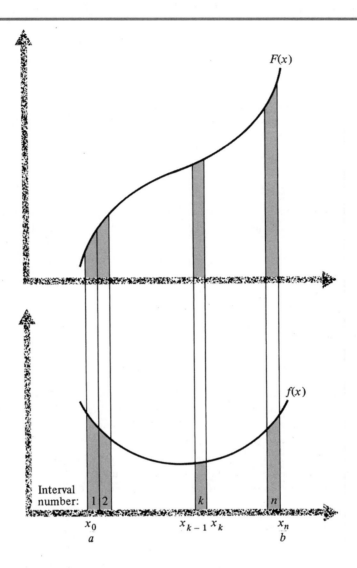

Subinterval	Secant Slope $\left(\dfrac{\Delta y}{\Delta x}\right)$	Finding c_k
0–1	$\dfrac{1-0}{1-0} = 1$	$2x = 1$ when $x = \dfrac{1}{2}$
1–2	$\dfrac{4-1}{2-1} = 3$	$2x = 3$ when $x = \dfrac{3}{2}$

In this case, we found secant and tangent slopes equal at the midpoints of the subintervals. Of course, such midpoint locations need not be so for other functions.

INTRODUCTION TO INTEGRAL CALCULUS

Figure 16.9

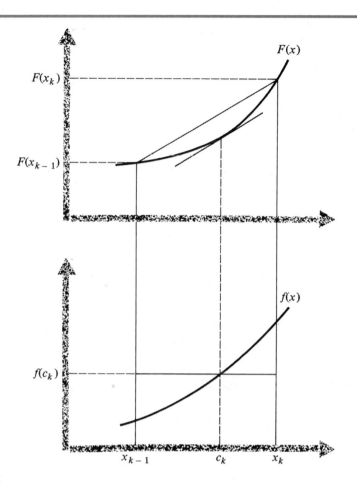

Figure 16.10

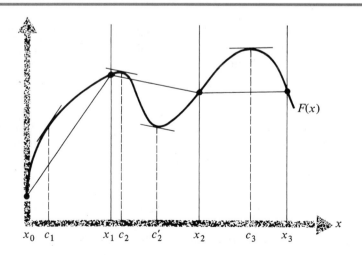

Exercises 3

For those who find this theorem hard to believe, consider the following functions

1. $y = F(x) = x^{1/2}$
2. $y = F(x) = e^x$
3. $y = F(x) = 2^{-x}$

over the subinterval from $x = 0$ to $x = 1$. For each case find the slope of the straight line connecting the endpoints. Then find the derivative function. Then locate the point where the derivative equals the slope of the line connecting the end points.

If you think the above examples and exercises were rigged, draw any continuous function (make it weird) with a derivative throughout the interval and use an "eyeball" approach to determine where the derivative equals the slope of the line connecting the end points.

Another form of the mean value theorem is

$$F(x_k) - F(x_{k-1}) = f(c_k) \cdot \Delta x$$

This was obtained from the original form by multiplying both sides by $(x_k - x_{k-1}) = \Delta x$. Now, since this theorem applies to any subinterval, let's write a statement of it for every one of the n subintervals.

Interval	Statement of mean value theorem
1	$F(x_1) - F(x_0) = f(c_1) \cdot \Delta x$
2	$F(x_2) - F(x_1) = f(c_2) \cdot \Delta x$
3	$F(x_3) - F(x_2) = f(c_3) \cdot \Delta x$
$\vdots$	$\vdots$
n	$F(x_n) - F(x_{n-1}) = f(c_n) \cdot \Delta x$

Summing up the left sides of the n equations gives

$$[F(x_1) - F(x_0)] + [F(x_2) - F(x_1)] + [F(x_3) - F(x_2)] + \cdots + [F(x_n) - F(x_{n-1})]$$

Observe that there is a lot of canceling going on; in fact, only two terms remain so that the sum of the left sides is equal to $F(x_n) - F(x_0)$, or $F(b) - F(a)$.

Now focus on the sum of the right sides

$$f(c_1) \cdot \Delta x + f(c_2) \cdot \Delta x + f(c_3) \cdot \Delta x + \cdots + f(c_n) \cdot \Delta x$$

Observe that c_1 is some value of x in the first subinterval and thus $f(c_1)$ is some value of $f(x)$ in the first subinterval. Thus, the product $f(c_1) \cdot \Delta x$ is the area of an "intermediate type" rectangle, and the sum of these areas is what we earlier called I. Thus,

$$F(b) - F(a) = \sum_{k=1}^{n} f(c_k) \cdot \Delta x = I$$

Letting n get larger allows a better approximation to areas under functions. With our pen-

INTRODUCTION TO INTEGRAL CALCULUS

chant for doing things BIG, why not let n go to infinity? Formally, this means taking the limit of all sides of the equation as n goes to infinity

$$\lim_{n \to \infty} [F(b) - F(a)] = \lim_{n \to \infty} \left[\sum_{k=1}^{n} f(c_k) \cdot \Delta x \right] = \lim_{n \to \infty} [I]$$

The left side is unaffected by the value of n since $F(b)$ and $F(a)$ only depend on the values of a and b. From the discussion on page 588, we know that the far right side limit is equal to T, the true area under $f(x)$ between $x = a$ and $x = b$. Thus,

$$F(b) - F(a) = \sum_{k=1}^{n \to \infty} f(c_k) \cdot \Delta x = T$$

Believe it or not we have just completed the development of the fundamental theorem. However, we could use a symbolism that better reflects our arrival in the world of infinite sums. By letting n go to infinity, we have caused the number of rectangles to approach infinity, and consequently the width of each to approach (but not equal) zero. We have also "boxed in" the value of c_k to approach the value of x at that point. Using $\int$ as the symbol for an infinite summation and dx as the symbol for the infinitesimal (mighty small) width, we have

Old symbol $\quad \sum_{k=1}^{n \to \infty} f(c_k) \cdot \Delta x$

New preferred symbol $\quad \int_{a}^{b} f(x)\, dx$

We are finished. The fundamental theorem is demonstrated in all its glory,

$$F(b) - F(a) = \int_{a}^{b} f(x)\, dx$$

$$= \text{area under } f(x) \text{ between } a \text{ and } b$$

where $D[F(x)] = f(x)$.

So in suspense-thriller fashion, we have found an area, not under $F(x)$ as you probably figured, but rather under $f(x)$, the unlikely suspect. Summing up, the fundamental theorem states that the area under $f(x)$, the rate function which is also called the *integrand function,* between two points a and b can be determined once one finds $F(x)$, called the *total function,* and evaluates it at a and at b. The overall process is called *integration*. The process of finding $F(x)$ is called *indefinite integration*. This, of course, is the heart of integration, since finding $F(x)$ allows for finding the area under $f(x)$ between any two points.

16.4 ANTIDIFFERENTIATION

The indefinite integration form of the fundamental theorem

$$\int f(x)\, dx = F(x)$$

points out the antidifferentiation nature of integration. Since

$$D[F(x)] = f(x)$$

then

$$\int D[F(x)]\,dx = F(x)$$

Beginning with $F(x)$, taking the derivative $D[F(x)]$, then integrating $\int D[F(x)]\,dx$, one arrives back at the starting point. In other words integration has undone the operation of differentiation. To bring out this point further let $y = F(x)$. Then $dy/dx = D[F(x)]$ and

$$\int \frac{dy}{dx}\,dx = y$$

Now, y as the variable symbol wasn't sacred. We could have $S = F(x)$ where S stands for total sales, or $P = F(x)$ where P stands for total pollution, and so on. In these cases

$$\int \frac{dS}{dx}\,dx = S$$

where dS/dx is sales rate,

$$\int \frac{dP}{dx}\,dx = P$$

where dP/dx is pollution rate, and in general

$$\int \frac{d\,(\text{any symbol})}{dx}\,dx = \text{any symbol}$$

Thus, the fundamental theorem allows us to convert rate of change information into total information.

The Mystery Operation Revisited

If you haven't succumbed to the suspense, you are now ready to understand the mystery operation performed on page 585. There, in trying to find the area under $f(x) = x^2$, the following was stated with no perceptible explanation:

$$\int_0^2 x^2\,dx = \frac{x^3}{3}\Big|_0^2 = \frac{8}{3} - 0 = \frac{8}{3}$$

Ah hah, but now we know full well that this is merely an application of the fundamental theorem. The clues give it away. Observe that the derivative of $x^3/3$ is x^2, so $x^3/3$ is the total function. The fraction 8/3 is the total function value at the upper limit of 2; 0 is the total function value at the lower limit of 0. Figure 16.11 shows this graphically.

INTRODUCTION TO INTEGRAL CALCULUS

Figure 16.11

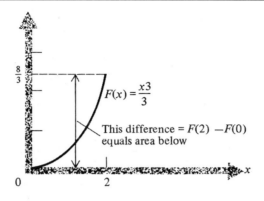

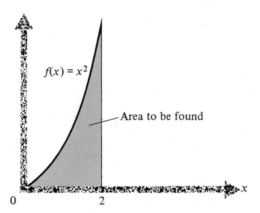

This new notation, or $F(x)|_a^b$, which we will use often, means carry out the operation $F(b) - F(a)$. Thus $\frac{x^3}{3}\Big|_0^2$, where $F(x) = \frac{x^3}{3}$, means to find $F(2) - F(0) = \frac{2^3}{3} - \frac{0^3}{3} = \frac{8}{3} - 0 = \frac{8}{3}$.

Area Determination by Fundamental Theorem

Now for those who still find it hard to believe the advertized claims of the fundamental theorem, let's test it in finding the area under three functions, two of which can be alternatively found by geometric methods (and thus a check on the results).

1. Area under $f(x) = 2$ from $x = 1$ to $x = 4$
2. Area under $f(x) = x$ from $x = 0$ to $x = 3$
3. Area under $f(x) = e^x$ from $x = 0$ to $x = 1$

The first area is simply that of the rectangle with altitude of 2 and base of $(4 - 1) = 3$, or 6. This is shown graphically in Figure 16.12. Since the derivative of $2x$ is 2, the fundamental theorem states:

$$\int_1^4 2\, dx = 2x \Big|_1^4 = 8 - 2 = 6$$

And lo and behold, we get the correct answer.

Exercises 4

Find the area under the function, $y = f(x) = 2$ from $x = 1$ to $x = 6$

  a. geometrically
  b. using the fundamental theorem

The area under $f(x) = x$ can be found geometrically since it forms a triangular region (see Figure 16.13). A triangle has area of $\frac{1}{2}$ base times the altitude, or in this case, $\frac{1}{2}(3)(3) = 4.5$. Recognizing that the derivative of $x^2/2$ is x, the fundamental theorem provides the same result:

$$\int_0^3 x\,dx = \left.\frac{x^2}{2}\right|_0^3 = \frac{9}{2} - 0 = 4.5$$

Exercises 5

Find the area under the function, $y = f(x) = x$ from $x = 0$ to $x = 5$

  a. geometrically
  b. using the fundamental theorem

Figure 16.12

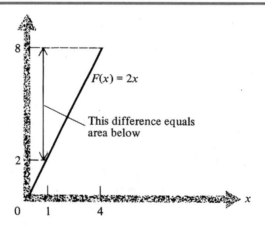

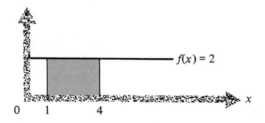

Figure 16.13

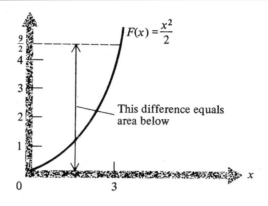

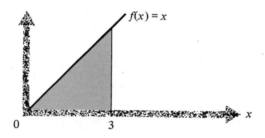

In the above examples we were checking out the fundamental theorem to see if it can really find the area under a function. And indeed it did. However, for those cases, the sophistication of calculus was not necessary. The next example is different. Without the fundamental theorem, the task of finding the area would be very difficult.

Consider $f(x) = e^x$. It does not form a familiar geometric shape (see Figure 16.14). Only integral calculus can provide us with the exact area.

Recall that the derivative of e^x is e^x itself. Thus,

$$\int_0^1 e^x \, dx = e^x \Big|_0^1 = e^1 - e^0$$

$$= e - 1$$

which is the desired area.

The Constant of Integration

The observant reader should have noticed that no unique total function exists. Referring to example 1 above, $F(x) = 2x + 1$ meets the requirement of having a derivative of 2, so does $F(x) = 2x - 3$, or in general $F(x) = 2x +$ any constant, since the derivative of any constant is zero. Thus, the total function for $f(x) = 2$ is $F(x) = 2x + C$, where C is any constant. But then does the existence of an infinite number of total functions wreak havoc with the ability to find areas? Intuitively, it would seem so, but fortunately, it doesn't. Let's use this example to investigate why. Integrating with the constant included in the total function, we get (see Figure 16.15).

Figure 16.14

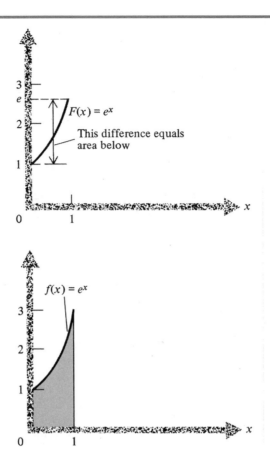

$$\int_1^4 2\,dx = 2x + C \Big|_1^4 = (8 + C) - (2 + C)$$
$$= 6 + C - C$$
$$= 6$$

The constant of integration conveniently (lucky for us) cancels out. Graphically, we can see that C merely raises (or lowers if it is negative) the entire total function by a constant amount. Thus, the relative positions or differences between two ordinates are the same.

Exercise 6

Redo Exercise 5 for the revised function, $y = f(x) = x + 10$. Use a graphical approach to show why you got the same answer.

The reader should verify that the constant of integration would not affect the areas found in the other examples.

INTRODUCTION TO INTEGRAL CALCULUS

Figure 16.15

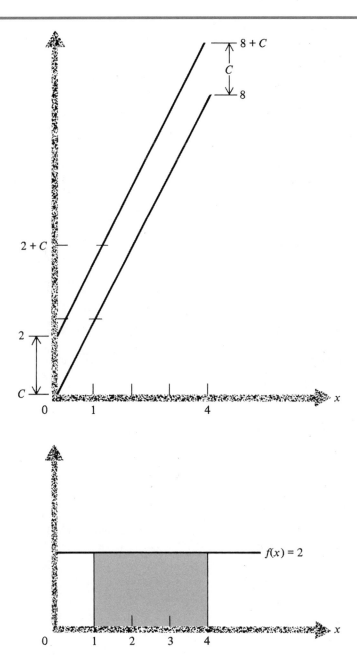

This chapter has given us a solid introduction to integral calculus. Now we are ready to move on to chapter 17, where we derive integration rules and introduce various integration applications.

PROBLEMS

TECHNICAL TRIUMPHS

1. Find the derivative of the following functions:

   $y = f(x) = x^4 + 10$

   $y = g(x) = x^4 + 100$

   Do the constants in these functions have any effect on the derivative?

2. Approximate area under the function, $y = f(x) = x^2$ for $\{2 \leq x \leq 4\}$, using two
   a. mid-point rectangles
   b. smallest rectangles
   c. largest rectangles
   d. repeat the above questions, using 4 rectangles.

3. Consider the function

   $y = f(x) = x^3$ for $\{0 \leq x \leq 1\}$

   a. Approximate the area under this curve using 4 mid-point rectangles.
   b. Use 4 smallest and 4 largest rectangles.
   c. Integral calculus would find the exact area to be $\frac{1}{4}$. Compare your various approximations to that.

4. a. Use geometric methods to find the area under

   $y = f(x) = 2x$ for $\{0 \leq x \leq 3\}$

   b. Notice that the derivative of $F(x) = x^2$ is $2x$. So find $F(3) - F(0)$. Explain why this is equal to your answer in part a.

CONFIDENCE BUILDERS

1. Sonny Wadders owns a lakeside property as shown by the accompanying sketch. Approximate his lot area by 2, then 4, then 8 midpoint rectangles.

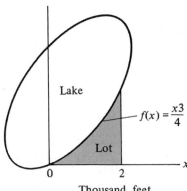

INTRODUCTION TO INTEGRAL CALCULUS

2. Consider the function

   $y = f(x) = x^3 + 3x$ for $\{0 \leq x \leq 4\}$

   Using 4 subintervals
   a. Find the area under the curve using mid-point rectangles.
   b. Find all the c_k, or the points in each subinterval where the tangent slope equals the secant line slope.
   c. Use the $f(c_k)$ to form rectangles and approximate the area under the curve.

3. The velocity, v, of a body falling frictionless in Earth's gravity, is the following function of time, x, in seconds.

   $v = f(x) = 32x$

   a. Find the area under the function for $\{0 \leq x \leq 5\}$.
   b. Find the antiderivative.
   c. What does the antiderivative represent in our physical world?
   d. Use the antiderivative to find the total distance a body falls in 5 seconds.

MIND STRETCHERS

1. Matt Rugg needs to carpet his study with the curved windowed wall as depicted in the accompanying sketch. If the cost is $2 per square foot, how much will the bill be? (*Hint:* The integral of a sum of functions is the sum of the integrals.)

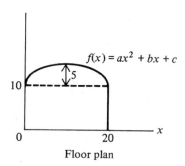

Floor plan

2. The Hoboken Hookeys play hockey in an area with the sketched shape. What is the total ice area? (See hint in problem 1 above.)

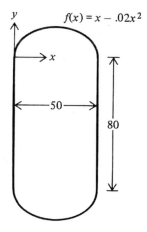

3. The area under a probability distribution must be equal to 1. Show which of the following qualify as a probability distribution.
   a. $f(x) = e^x$ from $x = 1$ to $x = 2$.
   b. $f(x) = .08x$ from $x = 0$ to $x = 5$.
   c. $f(x) = 1$ from $x = 2$ to $x = 4$.

CHAPTER 17

Integration Methods and Their Application

17.1 INTRODUCTION

In this chapter we will develop and apply the rules of integration. These rules enable the efficient transformation from rate of change information to total information, which comes in handy in the real world where it is often easier to determine or estimate rates of change than totals.

The fundamental theorem of calculus developed in the last chapter serves as our basis for developing rules of integration. Recall that it states that the integral of a rate function $f(x)$, which represents the area under that function between the points $x = a$ and $x = b$, is equal to the difference of the total function $F(x)$ evaluated at those two points, or

$$\int_a^b f(x) = F(x)\Big|_a^b = F(b) - F(a)$$

where $D[F(x)] = f(x)$

The fundamental theorem provides no direct way to determine $F(x)$, it only points out the need to find the missing total function. Fortunately, it does provide one big fat clue as to the nature of $F(x)$; namely, that the unknown total function has $f(x)$, the rate function, as its derivative.

Developing the rules of integration is more challenging than developing the rules of differentiation. This results from the lack of a direct way to integrate as compared to the direct method of differentiation. Recall that beginning with the definition of a derivative as the limit of a difference quotient, one could take any function, do the proper algebraic simplification, take the proper limits, and eventually arrive at the derivative. Granted that someone might get stuck along the way for lack of algebraic or limit-taking prowess; nevertheless the process is straightforward. Do this, do that, and the answer is yours. Unfortunately, no universal starting point and direct process exists for integration. All we know is that we must find a function whose derivative is the function we have. Therefore, success in the development of integration rules requires a good understanding of differentiation.

It is possible (even likely) that the reader is perplexed at this point. What with the fundamental theorem, rate functions, total functions, and areas under curves all twirling about, and integration rules looming on the horizon, it is not easy to see the interrelationships or the everyday applications that can come from this. So let's pause for some "put it all together," or literally "integrating" vignettes in the life of Ben Dover.

Ben is traveling on the thruway at a constant velocity of 60 miles per hour to visit his sweetheart Polly Ethyl Lynn. Even Ben knows that if he travels x hours at 60 miles per hour, he will cover a distance of 60 times x, or $60x$ miles. Note that the derivative of $60x$ is equal to 60, so $60x$ is the total function, or $F(x)$, for the rate function $f(x) = 60$. Figure 17.1 shows that the area under the rate function from time zero to time x is $60x$, the total distance traveled. This is also the difference of the total function, $F(x) - F(0)$, as in the fundamental theorem.

Symbolically, with y representing distance, v representing velocity, and x representing time,

$$v = \frac{dy}{dx} = 60$$

$$y = \int \frac{dy}{dx} dx = \int 60\, dx = 60x$$

In this case the integration "rule" to transform the rate $f(x) = 60$ to a total was simply to multiply by x. We shall see later that this rule generalizes and is called the *power rule*.

Now suppose that Ben's longing for Polly causes him to go faster and faster according to the velocity function $v = 60 + 2x$. How far will he travel in two hours?

Figure 17.1

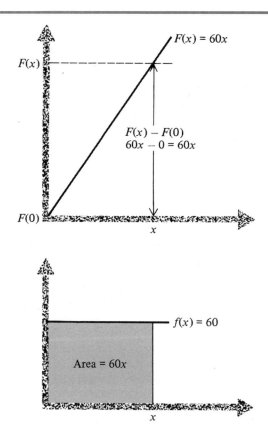

INTEGRATION METHODS AND THEIR APPLICATION

Determining distance traveled as velocity × time, as in the previous case, is no longer possible since the velocity is continually changing. If, however, we focus on short time intervals during which velocity doesn't change much, we can approximate the answer. Consider breaking the two hours into four $\frac{1}{2}$-hour periods. During the first half-hour, maintenance of the initial velocity of 60 would result in a distance of 30 miles. Of course, Ben went further than 30 miles, since he was going faster than 60 after the first split second. Notice that 30 is the area of the first rectangle in Figure 17.2. Ben begins the second half-hour at $60 + 2 \cdot \frac{1}{2} = 61$ miles per hour. Maintaining that speed would result in traveling a distance of 30.5 miles in a half-hour. Again, of course, he went more than that. During the third and fourth half-hour periods he would travel 31 and 31.5 miles, respectively, had he maintained the velocity at the beginning of each period. All the half-hour distances represent areas of "smallest" rectangles as defined in the previous chapter. The total area of these rectangles, which is our approximation to total distance, is 123 miles, which, of course, is an underestimate of the truth. By choosing periods of one-quarter hour, this process would provide a total area closer to the truth.

If you have the impression that we are going through the same process as we did in the previous chapter, you are absolutely right. This method will give us the exact total distance only in the limit as the number of rectangles goes to infinity. And this distance will be the area under the velocity function.

Consider the area under the velocity function up to time x. The reader can verify geometrically that this trapezoidal area is equal to $60x + x^2$. For $x = 2$, this formula gives an area of 124 miles, which is the actual distance traveled by Ben. Inasmuch as $60x + x^2$ has $60 + 2x$ or the rate function as its derivative, it is the total function. Unfortunately, we cannot always determine total functions geometrically. The next case will point this out and thus emphasize the need for integration rules to do this job.

While Ben is moving at 60 miles per hour (88 feet per second), he sees a deer run on to the road and slams on the brakes. Under what conditions will he avoid the crash if his velocity function is

$v = 88 - 3x^2$

Figure 17.2

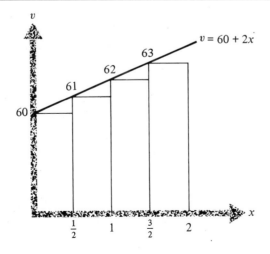

Figure 17.3

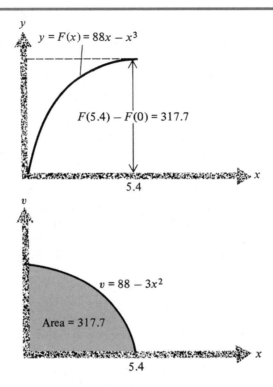

where v is in feet per second and x is in seconds? This rate function, as displayed in Figure 17.3, reflects the fact that velocity drops slowly at first, since momentum must be overcome, then drops faster and faster.

By now it should be apparent that total distance traveled is the area under the rate function, which is, by the fundamental theorem, the difference of two points on the total function. But it isn't easy to find a function whose derivative is equal to $88 - 3x^2$. Fortunately, the rules of integration to be developed in this chapter make it relatively simple. Knowledge of them would permit one to find the total function as $88x - x^3$, which, the reader might check, does have the rate function as its derivative.

Solving for the total distance before stop, we first find the stopping time by setting $v = 0$.

$3x^2 = 88$

$x = 5.4$ seconds

Integrating to find the area under the rate function up to $x = 5.4$, we get

$$y = \int_0^{5.4} \frac{dy}{dx} dx$$

$$= \int_0^{5.4} (88 - 3x^2)\, dx = 88x - x^3 \Big|_0^{5.4} = [88(5.4) - 5.4^3] - 0$$

$$= 317.7 \text{ feet}$$

INTEGRATION METHODS AND THEIR APPLICATION 607

So, Ben, if you were more than 317.7 feet away when you hit those brakes, you'll live to learn the integration rules that now follow.

Learning Plan

In this chapter we first develop some basic integration rules—mainly by recalling the derivative rules and working backward. Once we have amassed a working set of these rules, we apply them to problems of sales forecasting, weather forecasting, market research, highway design, probability and statistics, and economic theory. When we get stuck in a capital budgeting application with an integral that we can't handle, we stop. Finding integration rules for complicated functions will have to wait until the next chapter.

We finish this chapter by showing how that neat computer software, www.calc101.com, finds integrals. Since your professor probably won't let you do exams with such software, you should use it as a check on your own "brainware" methods. At some point when we get to really advanced integrals in the next chapter, we will need the software in order to avoid punting!

Each solved application is followed by related exercises. Use these exercises to obtain immediate feedback on whether you understand the application well enough to be able to extend your knowledge to a similar situation.

Because this chapter rests so heavily on differential calculus, don't be afraid to go back to Chapters 12–15 and solidify your knowledge of derivatives. It goes without saying that you must have a good knowledge of the introductory integral calculus material in Chapter 16.

17.2 HEURISTIC DEVELOPMENT OF INTEGRATION RULES

One way to meet the integration challenge is to differentiate many functions, because in doing so the reverse process of integration is also accomplished. For example, this tack would go

If	Then	Thus
$F(x) = x^3 + C$	$D[F(x)] = 3x^2$	$\int 3x^2 \, dx = x^3 + C$
$F(x) = x \ln x + C$	$D[F(x)] = 1 + \ln x$	$\int (1 + \ln x) \, dx = x \ln x + C$
$F(x) = \dfrac{x+2}{e^x} + C$	$D[F(x)] = -(x+1)e^{-x}$	$\int -(x+1)e^{-x} \, dx = \dfrac{x+2}{e^x} + C$

Certainly this is not an efficient method to develop the rules of integration. The next logical step would be to differentiate important general classes of functions and thereby learn to integrate the resulting general classes of functions. For this refer to the differentiation rules on page 481. For example, these rules tell us that the derivative of the power function x^n is nx^{n-1}. Thus, $\int nx^{n-1} \, dx = x^n + C$. But this raises the question of how to integrate just x to the n power, $\int x^n \, dx = ?$ An educated guess for the total function, knowing that the power rule of differentiation gives a derivative with x raised to one lower power, is $F(x) = x^{n+1} + C$. But $D[F(x)] =$

$(n + 1)x^n$. Indeed, we must revise $F(x)$ so that the constant $(n + 1)$ will cancel out upon differentiation. Recalling that the derivative of a constant times a function is the constant times the derivative of the function, it is apparent that $F(x) = [1/(1 + n)]x^{n+1} + C$ will do the trick. So, our first integration rule, called the power rule, is

$$\int x^n \, dx = \frac{x^{n+1}}{n + 1} + C$$

The power rule works for all real values of n except -1. For this value of n, the power rule gives $\int x^{-1} \, dx = (x^0/0) + C = \infty$. Knowing that integration gives an area under a curve and knowing that the hyperbola $f(x) = x^{-1}$ does indeed have a finite area between any two points with positive x coordinates, one senses something is awry. Certainly no power function has x^{-1} as its derivative, but possibly some other type of function does. The mathematicians of the eighteenth century who discovered this dilemma searched long and hard for this missing function, and when they realized that no such function existed, they did something about it. They gave birth to or, as we say, mathematically defined a brand new function, the derivative of which is x^{-1}, and which represents the area under x^{-1}. This new function turned out to have all the properties of a logarithmic function. Unfortunately, they couldn't choose the base of this logarithmic function; it turned out to be the unwieldly constant 2.718. . . . But don't despair—this is e, a number we have already met and learned to love. Recall that base e logarithms are called natural logarithms and are given the special symbol $\ln x$. Thus, our second integration rule is

$$\int x^{-1} \, dx = \log_e x + C = \ln x + C$$

Values for $\ln x$ can be found in Table 1 in the Appendix. These values, which represent the area under $f(x) = x^{-1}$ from $x = 1$ to $x = x$ (see Figure 17.4), were obtained by area approximation methods similar to those employed in the previous chapter. With the table one can evaluate the definite integral. For example,

Figure 17.4

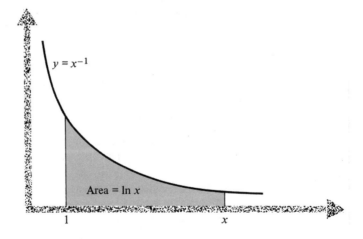

$$\int_2^5 x^{-1}\,dx = \ln x + C \Big|_2^5 = \ln 5 - \ln 2$$
$$= 1.609 - .693$$
$$= .916$$

which represents the area under x^{-1} from $x = 2$ to $x = 5$.

Consider integrating the general exponential function, $\int b^x\,dx = ?$ Differentiating b^x gives us $b^x \ln b$. As we did for the power function, we can cancel the constant obtained upon differentiating by starting with a function having the inverse of that constant. Thus

$$D\left[\left(\frac{1}{\ln b}\right)b^x\right] = \left(\frac{1}{\ln b}\right)b^x \ln b = b^x$$

So our third integration rule is

$$\int b^x\,dx = \frac{b^x}{\ln b} + C$$

As a special case of this for the base e

$$\int e^x\,dx = e^x + C \quad \text{since} \quad \ln e = 1$$

At this point we can integrate x^n, for any real n including -1, and b^x for any real positive b. However, we cannot integrate a constant times these functions (i.e., $5x^2$ or $-3\cdot 2^x$) or a sum of such functions (i.e., $x^{-4} + e^x$). We can now fill these gaps by developing the general rules for a constant times any function and for a sum of functions.

Consider $f(x) = k \cdot g(x)$, where $g(x)$ is any function of x and k is any constant. For expository purposes suppose $g(x)$ takes on only positive values. Then (see Figure 17.5) each ordinate of $f(x)$ stands k times as high as the corresponding ordinate of $g(x)$. For example, if $g(x) = x^2$ and $f(x) = 5x^2$, at $x = 2$, $g(2) = 4$ while $f(2) = 5\cdot 4 = 20$. Recognizing that $\int g(x)\,dx$ represents the area under $g(x)$ and that each point on $f(x)$ is k times as high as the point on $g(x)$, it is apparent that

Figure 17.5

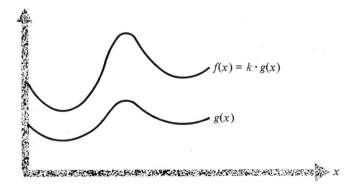

Figure 17.6

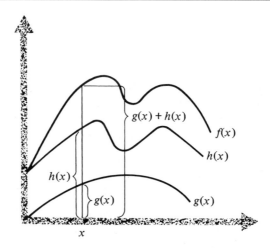

the area under $f(x)$ is k times the area under $g(x)$. Writing this in terms of integrals, we have our fourth integration rule, which holds for all possible k and $g(x)$.

$$\int k \cdot g(x)\, dx = k \cdot \int g(x)\, dx$$

Now consider a sum of two functions: $f(x) = g(x) + h(x)$. As shown in Figure 17.6 each ordinate of $f(x)$ stands $g(x) + h(x)$ units above the axis. So the area under $f(x)$ is equal to the area under $g(x)$ plus the area under $h(x)$. Since these areas correspond to integrals, we have

$$\int [g(x) + h(x)]\, dx = \int g(x)\, dx + \int h(x)\, dx$$

This rule can be generalized to the case of a sum of many functions. The integral of a sum of n functions is the sum of the integrals of the n functions. Symbolically, our fifth integration rule is

$$\int [f_1(x) + f_2(x) + \ldots + f_n(x)]\, dx = \int f_1(x)\, dx + \int f_2(x)\, dx + \ldots + \int f_n(x)\, dx$$

We summarize the integration rules developed thus far in Table 1.

The following examples make use of these rules. The reader is encouraged to identify the integration rule that allows each step and to complete the exercises in the same manner.

Examples

1. $\int x^4\, dx = \dfrac{x^5}{5} + C$

2. $\int 3x^{5/2}\, dx = 3\int x^{5/2}\, dx = 3\left(\dfrac{x^{7/2}}{7/2}\right) + C$

INTEGRATION METHODS AND THEIR APPLICATION

Table 1 Integration Rules

Rule 1	$\int x^n \, dx = \dfrac{x^{n+1}}{n+1} + C$	For any n except -1
Rule 2	$\int x^{-1} \, dx = \int \dfrac{dx}{x} = \ln x + C$	
Rule 3	$\int b^x \, dx = \dfrac{b^x}{\ln b} + C$	For positive values of b When $b = e$, $\int e^x \, dx = e^x + C$
Rule 4	$\int kg(x) \, dx = k \int g(x) \, dx$	For any constant k
Rule 5	$\int [f_1(x) \, dx + f_2(x) \, dx + \cdots + f_n(x)] \, dx = \int f_1(x) \, dx + \int f_2(x) \, dx + \cdots + \int f_n(x) \, dx$	

3. $\int \dfrac{5}{x} \, dx = 5 \int \dfrac{dx}{x} = 5 \ln x + C$

4. $\int 3^x \, dx = \dfrac{3^x}{\ln 3} + C$

5. $\int (3 + x) \, dx = \int 3 \, dx + \int x \, dx$

$$= 3 \int x^0 \, dx + \int x \, dx$$

$$= (3x + C_1) + \left(\dfrac{x^2}{2} + C_2\right)$$

$$= 3x + \dfrac{x^2}{2} + C$$

6. $\int \left(\dfrac{6}{x^2} - 3.7x^2\right) dx = \int 6x^{-2} \, dx - \int 3.7x^2 \, dx$

$$= 6 \int x^{-2} \, dx - 3.7 \int x^2 \, dx$$

$$= (-6x^{-1} + C_1) + \left(-3.7 \dfrac{x^3}{3} + C_2\right)$$

$$= -6x^{-1} - \dfrac{3.7}{3} x^3 + C$$

7. $\int (10^x + x^{10}) \, dx = \int 10^x \, dx + \int x^{10} \, dx$

$$= \left(\dfrac{10^x}{\ln 10} + C\right) + \left(\dfrac{x^{11}}{11} + C_2\right)$$

$$= \dfrac{10^x}{\ln 10} + \dfrac{x^{11}}{11} + C$$

8. $\int \left(x^{-3} + \frac{1}{x} + 2^x \right) dx = \int x^{-3} dx + \int \frac{1}{x} dx + \int 2^x dx$

$= \left(\frac{x^{-2}}{-2} + C_1 \right) + (\ln x + C_2) + \left(\frac{2^x}{\ln 2} + C_3 \right)$

$= -\frac{1}{2} x^{-2} + \ln x + \frac{2^x}{\ln 2} + C$

Exercises 1

1. $\int -4x^{1/2} dx =$

2. $\int (e^x + 2^x) dx =$

3. $\int \left(-10 - \frac{3}{x} \right) dx =$

4. $\int (2x^{-3} + 4(3)^{-5} - 7x^{-1}) dx =$

The rules of integration, mechanical in themselves, gain relevance inasmuch as they allow for the solution of practical problems, a scattering of which are now presented.

17.3 SOME APPLICATIONS OF INTEGRATION

Sales Forecasting

Shock Inc. has just introduced their new electric ear muffs. The sales manager wishes to forecast total sales of the product; however, he is better able to forecast sales rates. He knows that he can count on sales of 130,000 units per year from a market already cultivated—via the recent introduction of electric knee warmers. Additional sales will be difficult to make at first, but the new market sales rate should be 30,000 per year initially with growth continuing so that at any time the rate is double that of the previous year. His forecast is captured by the following sales rate function:

$$\frac{dS}{dx} = 100,000 + 30,000 \cdot 2^x$$

where dS/dx is muffs sales per year, and x is time from now in years. What will total sales be in the first two years? When will the 1,000,000th sale occur, as a big advertising campaign is planned to herald this event?

INTEGRATION METHODS AND THEIR APPLICATION

Figure 17.7

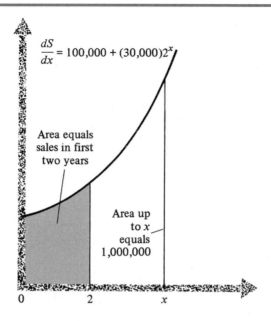

Solution

Total sales, S, which appears as an area under the sales rate function, is found by integrating the sales rate function. As shown in Figure 17.7, what is specifically required here is to find the area from $x = 0$ to $x = 2$ and the value of x such that the area to that point is 1,000,000. Our solution process is to determine the total function and then to perform the proper definite integrations.

$$S = \int \frac{dS}{dx} dx$$

$$= \int (100{,}000 + 30{,}000 \cdot 2^x) \, dx$$

$$= 100{,}000 \int x^0 \, dx + 30{,}000 \int 2^x \, dx$$

$$= 100{,}000x + 30{,}000 \left(\frac{2^x}{\ln 2} \right) + C$$

$$= 100{,}000x + 43{,}290 \cdot 2^x + C$$

Note: $\dfrac{30{,}000}{\ln 2} = \dfrac{30{,}000}{.693} = 43{,}290$

The constant of integration can be determined by noting that $S = 0$ when $x = 0$. Thus,

$$0 = 100{,}000 \cdot 0 + 43{,}290 \cdot 2^0 + C$$

So

$$C = -43{,}290$$

and

$$S = F(x) = 100{,}000x + 43{,}290 \cdot 2^x - 43{,}290$$

Now, by substituting values of x in $F(x)$ we can find total sales from time zero up to that point. Thus, total sales over the first two years would be

$$S = F(2) = 100{,}000 \cdot 2 + 43{,}290 \cdot 2^2 - 43{,}290$$
$$= 200{,}000 + 43{,}290(2^2 - 1)$$

Alternatively, we could have gotten this result without finding the constant of integration (in fact, by completely ignoring it). This approach rests on the fact that C cancels out upon use of the fundamental theorem of calculus, as shown below:

$$F(x) = 100{,}000x + 43{,}290 \cdot 2^x + C$$
$$F(2) - F(0) = (100{,}000 \cdot 2 + 43{,}290 \cdot 2^2 + C)$$
$$- (100{,}000 \cdot 0 + 43{,}290 \cdot 2^0 + C)$$
$$= (100{,}000 \cdot 2 + 43{,}290 \cdot 2^2) - (100{,}000 \cdot 0 + 43{,}290 \cdot 2^0)$$
$$= 200{,}000 + 43{,}290(2^2 - 1)$$

So whenever we use the fundamental theorem of calculus, we can ignore C in the total function as shown below:

$$F(2) - F(0) = 100{,}000x + 43{,}290 \cdot 2^x \Big|_0^2$$
$$= (100{,}000 \cdot 2 + 43{,}290 \cdot 2^2) - (100{,}000 \cdot 0 + 43{,}290 \cdot 2^0)$$
$$= 200{,}000 + 43{,}290(2^2 - 1)$$
$$= 329{,}870$$

We take this approach in solving for the time when total sales will equal 1,000,000.

$$F(x) - F(0) = 1{,}000{,}000 = 100{,}000x + 43{,}290 \cdot 2^x \Big|_0^x$$
$$= (100{,}000x + 43{,}290 \cdot 2^x) - (43{,}290)$$
$$= 100{,}000x + 43{,}290(2^x - 1)$$

Solving for x requires a trial-and-error procedure. When $x = 3$, total sales equal 603,030; when $x = 4$, total sales are 1,049,350. Thus, an x of slightly under 4 (or, 3.9 to the nearest decimal place) is the answer. This means that the big advertising campaign should begin in late November of the fourth year.

Exercises 2

1. Recalculate this problem assuming that the "already established" market was only 50,000 units per year but that the "new" market sales rate was forecast to triple each year.
2. Recalculate this problem if the sales rate function is

$$\frac{dS}{dx} = 130{,}000 + 60{,}000x^2$$

Weather Forecasting

The weather bureau forecasts a snowstorm for the Syracuse area, which is not unusual. The forecast for total accumulation is difficult, but can be determined from snow intensity (inches per hour) information, which is easier to estimate. They forecast the storm to begin at 8:00 A.M. ($x = 0$), to intensify quickly, reach its peak intensity at 10:30 A.M., remain heavy for a while, and then diminish, ending at 1:00 P.M. At the peak of the storm, a snowfall rate of 1.25 inches per hour is forecasted. If a parabolic function of the form

$$\frac{dS}{dx} = ax^2 + bx + c$$

where

$\frac{dS}{dx}$ is rate of snowfall (inches per hour)

x is time (in hours) elapsed since storm beginning

a, b, and c are constants

is assumed to describe the snowfall rate pattern, how much snow should be forecasted?

Solution

Using the curve-fitting methods of Chapter 7, you should arrive at the following snowfall rate function:

$$\frac{dS}{dx} = x - .2x^2$$

Total snowfall, the area shown in Figure 17.8, is found by integrating the rate function from $x = 0$ to $x = 5$.

$$S = \int \frac{dS}{dx} dx = \int (x - .2x^2)\, dx$$

$$= \int x\, dx - .2 \int x^2\, dx = \frac{x^2}{2} - .2\frac{x^3}{3} + C$$

Figure 17.8

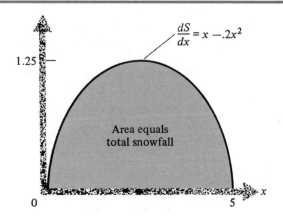

Total snowfall $= \int_0^5 \frac{dS}{dx} dx = \frac{x^2}{2} - .2\frac{x^3}{3} \Big|_0^5$

$= \frac{25}{2} - .2\left(\frac{125}{3}\right)$

$= 4.17$ inches

Exercises 3

1. Redo this problem for an eight-hour storm with a peak snowfall rate of 1 inch per hour occurring at the four-hour point.
2. If the snowfall rate during the first four hours of the intensifying period of a storm is

   $$\frac{dS}{dx} = .1 + .4x$$

   and a snow emergency is called when 3 inches have fallen, when will the snow emergency be called?

Market Research

Hifalootin Inc., which operates exclusive boutiques, is trying to decide whether to open a shop in East Podunk. The town has been growing rapidly in population; but, in the words of Hifalootin's president, I. M. Rich, "Most of it is merely riffraff." In the future the population is expected to continue growing, but at a slower rate. Reflecting this pattern, the company forecasts the population growth rate of the "people who count"—loaded (L) customers—as

$$\frac{dL}{dx} = \frac{5000}{x} \quad \text{with } x = 1 \text{ currently}$$

The company estimates that there are now 40,000 loaded customers currently in town, but 50,000 are needed to make the business profitable. The company is willing to open provided the necessary population will be there within three years. Should it plan to open?

INTEGRATION METHODS AND THEIR APPLICATION   617

Figure 17.9

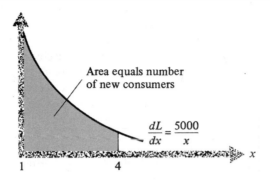

Solution

The designated area under the rate function in Figure 17.9 is the number of new customers in town. So, we must integrate over the first three years to see if more or less than 10,000 newcomers arrive. An alternative solution method would be to determine the proper constant of integration for the total function so as to include all loaded customers, and then to see if the total exceeds 50,000. Both approaches follow.

$$L = \int \frac{dL}{dx} dx = \int \frac{5000}{x} dx = 5000 \ln x + C$$

$$\text{New customers} = \int_1^4 \frac{dL}{dx} dx = 5000 \ln x \Big|_1^4$$

$$= 5000 \ln 4 - 0$$

$$= 5000(1.386) = 6930$$

Since total loaded customers equals 40,000 when $x = 1$,

$$40,000 = 5,000 \ln 1 + C$$

Thus, the constant of integration is 40,000. As such the total function is

$$L = 5,000 \ln x + 40,000$$

So when $x = 4$, we get $L = 46,930$.

Inasmuch as the relevant population falls short of 50,000, Hifalootin should not open the store.

Exercises 4

1. Determine when the relevant population will reach 50,000.
2. If the growth rate was higher, as given by

$$\frac{dL}{dx} = \frac{6000}{x}$$

should Hifalootin plan to open?

Highway Design

What should the length of the acceleration lane to a superhighway be? Certainly, it should be long enough to allow acceleration from zero to 60 miles per hour (88 feet/second) for even the most sluggish car (mine) under bad traffic and weather conditions. If this worst case is a constant acceleration of 2 feet/second^2 (my car can actually do better, but this low number will include the possibility of deliberately low acceleration so as to wait for an opening in the traffic pattern), how long should the acceleration lane be?

Solution

Recall from Chapter 14 that acceleration (a) is the derivative of the velocity function, and velocity (v) is the derivative of the distance (y) function.

$$v = \int \frac{dv}{dx} dx = \int 2 \, dx = 2x + C$$

Since $v = 0$ when $x = 0$, in the equation above $C = 0$.

$$y = \int \frac{dy}{dx} dx = \int v \, dx = \int 2x \, dx = x^2 + C$$

Since $y = 0$ when $x = 0$, then $C = 0$.

Therefore, the velocity and distance functions are $v = 2x$ and $y = x^2$. The velocity will reach 88 feet per second after 44 seconds (solving the velocity function for x). When $x = 44$, the distance function shows that the car will have traveled $44^2 = 1936$ feet, or about $\frac{1}{3}$ mile, which represents the minimum length necessary for the acceleration lane (see Figure 17.10).

Exercises 5

1. Determine the necessary length of the acceleration lane if the worst acceleration were 3 feet per second per second.
2. An existing acceleration lane is only 880 feet long. What is the worst constant acceleration (from 0 to 88 feet per second) that it can accommodate?

Probability and Statistics

Earlier we developed differential calculus applications of probability (section 14.7) and statistics (section 15.9). Here, we will build on those applications by developing some integral calculus applications in probability and statistics.

If $P(x)$ is a continuous probability density function for a random variable, then, by definition, the area under this function must equal 1.

$$\int_{-\infty}^{+\infty} P(x) \, dx = 1$$

Averages-Central Tendency: Median and Mean

Two so-called "averages," the median and the mean, have been defined in terms of integrals to measure the central tendency of a probability distribution.

INTEGRATION METHODS AND THEIR APPLICATION

Figure 17.10

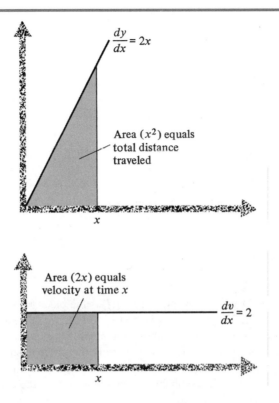

The median symbolized as *M*, is the point where half the values are below it and half are above it. In area terms, half the probability distribution's area is below the median and half the area is above the median. To find the median, one integrates the probability distribution function as

$$\int_{\infty}^{M} P(x)\, dx = \frac{1}{2}$$

and solves for *M*.

One can see that the middle value is the median. As such, extreme values, either high or low, have no effect on the median.

The mean, symbolized by μ, is defined by the integral

$$\text{Mean} = \mu = \int_{-\infty}^{\infty} x \cdot P(x)\, dx$$

Since each value of *x* is "weighted" by its probability, the mean is affected by extreme values.

Variability: Variance and Standard Deviation

The variance, symbolized by σ^2, is a measure of the variability of the values of *x*. It is defined by the integral

Figure 17.11

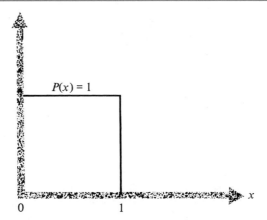

$$\text{Variance} = \sigma^2 = \int_{-\infty}^{\infty} (x - \mu)^2 \cdot P(x)\, dx$$

Notice that the "deviation" from the mean, or $(x - \mu)$, is the basis for this measure. As values of x get further from the mean, the effect would be to increase σ^2. By squaring the deviation, its magnitude, and not whether it is below or above the mean, is relevant.

The standard deviation, symbolized by σ, is merely the positive square root of the variance.

Consider the uniform probability density function, as shown in Figure 17.11, with $P(x) = 1$ for $\{0 \le x \le 1\}$. Let's show it to be a probability density function (thus has an area of 1) and find its median, mean, variance, and standard deviation.

Solution

Geometrically the area under the uniform distribution is a square with area of $1 \cdot 1 = 1$. This is verified by integration[1]

$$\int_0^1 1\, dx = x \Big|_0^1 = 1 - 0 = 1$$

The median is found as

$$\int_0^M 1\, dx = 1x \Big|_0^M = \frac{1}{2}$$

$$M - 0 = \frac{1}{2} \quad \text{so} \quad M = \frac{1}{2}$$

The mean is

$$\mu = \int_0^1 x \cdot 1\, dx = \frac{x^2}{2} \Big|_0^1 = \frac{1}{2} - 0 = \frac{1}{2} = .5$$

---

[1] Note that the limits of integration are zero and 1 rather than $-\infty$ and $+\infty$. This is because $P(x) = 0$ for $\{0 > x > 1\}$, so that $\int_{-\infty}^{0} P(x)\, dx = \int_{1}^{+\infty} P(x)\, dx = 0$

INTEGRATION METHODS AND THEIR APPLICATION

The variance is

$$\sigma^2 = \int_0^1 (x - .5)^2 \cdot 1 \, dx$$

$$= \int_0^1 (x^2 - x + .25) \, dx$$

$$= \int_0^1 x^2 \, dx - \int_0^1 x \, dx + \int_0^1 .25 \, dx$$

$$= \frac{x^3}{3} - \frac{x^2}{2} + .25x \Big|_0^1$$

$$= \tfrac{1}{3} - \tfrac{1}{2} + \tfrac{1}{4} - 0 = \tfrac{1}{12}$$

Thus the standard deviation is

$$\sigma = \sqrt{\sigma^2} = \sqrt{\frac{1}{12}} = .289$$

Now let's determine the median, mean, and variance for the slightly more complicated probability density function

$$P(x) = .02x \quad \text{for} \quad \{0 \leq x \leq 10\}$$

The median is found by solving

$$\int_0^M .02x \, dx = \frac{1}{2}$$

$$.02 \frac{x^2}{2} \Big|_0^M = \frac{1}{2}$$

$$.01 M^2 = \frac{1}{2} = .5$$

So $M = \sqrt{50} = 7.07$

The mean is found by the following integration:

$$\mu = \int_0^{10} x \cdot P(x) \, dx = \int_0^{10} x \, (.02 \, x) \, dx$$

$$= .02 \int_0^{10} x^2 \, dx$$

$$= .02 \frac{x^3}{3} \Big|_0^{10}$$

$$= \frac{20}{3} - 0 = \frac{20}{3}$$

The variance is found by the following integration:

$$\sigma^2 = \int_0^{10} (x - \mu)^2 P(x)\, dx = \int_0^{10} \left(x - \frac{20}{3}\right)^2 (.02x)\, dx$$

which after a great deal of algebra becomes

$$\sigma^2 = \int_0^{10} \left(.02x^3 - \frac{.8}{3}x^2 + \frac{8}{9}x\right) dx$$

Exercises 6

1. Complete the integration to find the variance. Then find the standard deviation.
2. Consider $P(x) = .05 + .45x$ for $\{0 \le x \le 2\}$. Verify that it is a probability density function and determine its mean, median, and variance.
3. Show that $P(x) = 2e^x$ for $\{0 \le x \le 1\}$ is not a probability density function. For what value of the constant k is $P(x) = ke^x$ for $\{0 \le x \le 1\}$ a probability density function? Find the median of this function.

Economic Theory

It is established economic theory that marginal consumption decreases as one's income rises. A marginal consumption function of the form

$$\frac{dK}{dx} = 1 - ax^n$$

where

  K is consumption in thousands
  x is income in thousands
 a, n are constants

reflects this pattern.

Consider the case of Juana Moore who (1) would consume $2 thousand worth of goods even if she had no income, (2) would consume at a 90 percent marginal rate if her income were $1 thousand, and (3) would consume at an 85.9 percent marginal rate if her income were $2 thousand. Determine Juana's consumption function. What salary would Juana have to earn in order to break even?

Solution

We must first find Juana's marginal consumption function. Then we must integrate this function to get total consumption.

$.9 = 1 - a \cdot 1^n$

$.859 = 1 - a \cdot 2^n$

INTEGRATION METHODS AND THEIR APPLICATION

Solving for a and n gives $a = .1$ and $n = .5$. Thus, the marginal consumption function is

$$\frac{dK}{dx} = 1 - .1x^{.5}$$

Integrating this to get the total consumption function

$$K = \int (1 - .1x^{.5}) \, dx = \int 1 \, dx - .1 \int x^{.5} \, dx$$

$$= x - .1\left(\frac{x^{1.5}}{1.5}\right) + C$$

K is 2 when $x = 0$; therefore, $C = 2$ and

$$K = x - \frac{1}{15}x^{1.5} + 2$$

Setting consumption equal to income, we find that Juana must earn $9655 to break even.

$$x = x - \frac{1}{15}x^{1.5} + 2$$

$$\frac{1}{15}x^{1.5} = 2$$

$$x = 30^{2/3} = \$9.655 \text{ thousand, or } \$9655$$

Exercise 7

Consider this problem if Juana would (1) consume $1 thousand even with no income, (2) consume at an 80 percent marginal rate if her income were $1 thousand, and (3) consume at a 60 percent marginal rate if her income were $4 thousand.

Capital Budgeting

Matt Tress, the owner of a bedding company, is considering a capital investment. Recently he purchased a machine that forms the springs for his "Kinck" size beds. Now he learns that a more efficient machine is available. To obtain the new machine, he'll have to put up $10,000 plus the proceeds from selling the older machine.

Matt notes that the different machines make springs of identical quality; however, since less maintenance is needed, the new machine does produce savings. With the passage of time, though, the maintenance savings should diminish. By a careful study of both machines, Matt estimates the savings (S) rate as the following function of time (x) in years from now

$$\frac{dS}{dx} = 6000\left(\frac{x}{2} + 1\right)^{-2}$$

where x is time in years and S is savings in dollars.

Matt bases his investment decisions on the "payback" method. Specifically, if the savings incurred in the first four years exceed the initial investment ($10,000), he invests. Considering all this, what should Matt do?

Solution

We must integrate the savings rate function over the first four years to determine if total savings is more or less than $10,000. The total function is found as

$$S = \int \frac{dS}{dx} dx = \int 6000 \left(\frac{x}{2} + 1\right)^{-2} dx$$

$$= 6000 \int \left(\frac{x}{2} + 1\right)^{-2} dx$$

Notice that this integrand does not fit exactly into any of our five integration rules. It does bear a resemblance to the power function, so let's try that one.

$$= 6000 \frac{\left(\frac{x}{2} + 1\right)^{-2+1}}{-2 + 1} + C = -6000 \left(\frac{x}{2} + 1\right)^{-1} + C$$

Since we are a little unsure of our method here, we should check the result by differentiating. Using the chain rule with $u = (x/2) + 1$, we get

$$\frac{dS}{dx} = \left(\frac{dS}{du}\right)\left(\frac{du}{dx}\right) = (-6000 \cdot -1u^{-2})\left(\frac{1}{2}\right)$$

$$= 3000 \left(\frac{x}{2} + 1\right)^{-2}$$

Whoops, we did not get back to the rate function. In other words, we did a no-no above when we integrated with the power rule. Now what should we do? The answer is obvious. We better go back to the salt mines and dig up some advanced integration rules so that Matt can make the decision. We will do that in the next chapter.

17.4 INTEGRATION USING www.calc101.com

We have made extensive use of the neat software, www.calc101.com, already. Recall that we used it for matrix operations (See pages 352–355) and differential calculus (See pages 485–489). Looking at it's home page on page 353, you can see that the software allows one to click on "integrals" (used in this chapter) and various "stored integrals" (used in the next chapter).

Let's illustrate this software's use with the simple function, $y = f(x) = x^2$. Assuming that you are "cheap" like me and don't pay for a password, you would still get the first and last lines of the solution, as shown in Figure 17.12.

I must admit that the software developer, George Beck, was kind enough to give me a

Figure 17.12

Step-by-Step Integration

Click "Do it" to get your answer; to see all the steps, get a *password* fast!

Plot your function!

home | FAQs | info@calc101.com | polynomial multiplication | polynomial division

graphs *new!* | matrix algebra *new!* | derivatives | **passwords**

Use the right input format to get good results.	Wrong	Right
Use square brackets [] for functions.	cos(x), sqrt x, arcsinx, sin[ln x]	cos[x], sqrt[x], arcsin[x], sin[ln[x]]
Use parentheses () for grouping.	[x+1]^3, e^[x+1], x^2/[1+x^3]	(x+1)^3, e^(x+1), x^2/(1+x^3)
Use spaces or * for multiplication.	xe^-x, xsin[pix], bx^2	x e^-x, x sin[pi x], b x^2

enter your password

Do the integral of
x^2

with respect to
x

[DO IT ▶]

Sorry, the password is not valid.

For your convenience, here are the first and last lines of the solution:

$\int x^2 \, dx$

$= \frac{x^3}{3} + C$

Exercises 8

Use www.calc101.com to find the integrals for the following rate functions:

1. $f(x) = x^{-2}$
2. $f(x) = \ln x$
3. $f(x) = x^4 + x^3$
4. $f(x) = 2^x$

PROBLEMS

TECHNICAL TRIUMPHS

1. Integrate the following functions. In each case, specify the integration rule(s) used.
   a. $f(x) = x^{-4}$
   b. $f(x) = e^x + x^e$
   c. $f(x) = \dfrac{-10}{x} + 10^x$

2. Find the definite integral for $x = 1$ to $x = 2$ for the three cases in problem #1 above.

3. If Ben Dover's velocity function was $v = 65$
   a. Find the total distance function.
   b. How far will Ben travel in 2 hours?
   c. If his girl friend lives 200 miles away, how long will it take Ben to arrive?
   d. Graphically illustrate your answer to the above questions.
   e. If Ben's velocity function was

   $v = 65 + 2x$

   reanswer the above questions.

4. If Ben sees a deer on the road ahead of him and his braking velocity function is

   $v = 72 - 2x^2$

   a. How long will it take for him to stop?
   b. How many feet will he travel during the breaking period?

5. Sales of Worldcon have been declining according to the following rate function

   $\dfrac{dS}{dx} = 10 - 2x$

   where S is sales in millions and x is time in years. ($x = 0$ in 1998).

INTEGRATION METHODS AND THEIR APPLICATION

a. Graph the rate function.
b. What does the area under that graph from $x = 1$ to $x = 3$ represent?
c. When the sales rate drops to 0, the firm declared bankruptcy. When did that occur?
d. What was the total sales until bankruptcy?

6. Consider the Calculus III rocket whose velocity function on lift-off was

$$v = 2x^2$$

where v is velocity in miles per minute and x is time in minutes.
a. Determine the total distance function.
b. How high was the rocket after 3 minutes?
c. How many miles did the rocket go up from 1 minute to 4 minutes into the flight?
d. At what time will the rocket be 50 miles above ground?

7. Use www.calc101.com to find the integrals of the following functions.
a. $f(x) = x^4 + x^3$
b. $f(x) = 7e^x$
c. $f(x) = \dfrac{5}{x} + 3x^{-2}$

8. A probability distribution has an area of 1 under the rate function. Does $f(x) = x^2$ for $\{0 \le x \le \sqrt{3}\}$ qualify as a probability distribution?

9. The marginal revenue and marginal cost functions for a firm are

$$\frac{dR}{dP} = 100 - 4P$$

$$\frac{dC}{dP} = 90 - 2P$$

a. Find the total revenue and total cost functions if $R = 1000$ and $C = 2000$ when $P = 1$.
b. Find the break-even price level.

10. A 10-hour snowstorm is approaching Syracuse, NY (whats new!). The rate of snowfall is expected to be

$$\frac{ds}{dx} = 2x - .2x^2$$

How much snow will fall?

CONFIDENCE BUILDERS

1. "Lefty" McGoo, president of Left-Handed Liberation, is sick and tired of working for the right-handed establishment, so he decided to go into business for himself to manufacture left-handed screwdrivers. Lefty believes that the rate of sales should be about 80,000 per year initially, then grow continuously to 87,000 a year from now and 100,000 two years from now. If a parabolic function is assumed to approximate the sales rate as a function of time.
a. What will total sales in the first three years be?
b. When will Lefty sell the one millionth screwdriver (approximate)?

2. The death rate from lung cancer in the United States can be approximated by the following function of time.

$$\frac{dL}{dx} = 300{,}000 + 20{,}000x$$

where L is deaths because of lung cancer and x is time in years since 1990.
  a. Plot and describe this rate function.
  b. Approximately how many lung cancer deaths occured in the 1990s?
  c. If this pattern continues, how many such deaths are forecast for the years 2000–2009?

3. Fad Products has just introduced its new product, the Lulu Loop. Market research forecasted that it would sweep the country in a four-month period, then die out in the next three months. Their thinking is reflected in the following sales rate function:

$$\frac{dS}{dx} = 1 + \frac{x}{2} \quad \text{for} \quad \{0 \le x \le 4\}$$

$$\frac{dS}{dx} = 7 - x \quad \text{for} \quad \{4 < x \le 7\}$$

$$\frac{dS}{dx} = 0 \quad \text{for} \quad \{x > 7\}$$

where x is time in months and S is sales in thousands of units.
  a. What are total sales going to be if this forecast holds up?
  b. If their breakeven point is 10,500 units, when should this occur?

4. If marginal cost is a parabolic function of output, determine the total cost of producing 10,000 units if

Output level (thousands)	Marginal cost ($)
0	3
2	2
10	4

5. A recent Buffalo blizzard began at 8:00 A.M., intensified, reached a peak at 11:00 A.M., and then diminished until 2:00 P.M., at which time it ended. The following rate function approximated the pattern of intensity of the storm

$$\frac{dS}{dx} = \frac{x^2}{3} \quad \text{for} \quad \{0 \le x \le 3\}$$

$$\frac{dS}{dx} = 6 - x \quad \text{for} \quad \{3 < x \le 6\}$$

where x is time in hours ($x = 0$ at 8:00 A.M.) and S is snow in inches.
  a. Graph the rate function.
  b. How much snow fell by 10:00 A.M.?
  c. When did the city call a snow emergency (this occurs when six inches are on the ground)?
  d. How much snow fell during the entire storm?

6. The president of a national women's liberation group is anxious to forecast future membership totals. With 1 million current members, she expects the rate of new memberships to be

$$\frac{dM}{dx} = .1 \ln x$$

where x is time in years ($x = 1$ now) and M is members in millions.
  a. Graph the rate function.
  b. If her forecast is correct, how many members will be in the group ten years from now?

INTEGRATION METHODS AND THEIR APPLICATION

7. Ella Fant, a 300-pound woman, is interested in Slim Pickins. She knows that she can catch him only if she can slim down to a svelte 250 pounds in 6 weeks. If Ella goes on a diet and loses fat (F) at a rate given by the following function of time (t) in weeks

$$\frac{dF}{dt} = t^2 - 10t + 26$$

will there be wedding bells for Ella and Slim?

8. In Chapter 6 we considered the following Gangland and C.A.R.T.E.L. sales rate functions of time (t) from now (we didn't know these as rates then):

Gangland $\quad \dfrac{dS}{dt} = 10(1.2)^t \quad t = 0$ now for both

C.A.R.T.E.L. $\quad \dfrac{dS}{dt} = 50(1.1)^t$

Back then we determined that the Gangland sales rate would catch up to C.A.R.T.E.L.'s in about 19 years. Now, see if you can determine when the total sales of Gangland from now on will catch up to the corresponding total sales of C.A.R.T.E.L.

9. If Juana Moore's marginal consumption function is

$$\frac{dK}{dx} = 1 - .15x^5$$

where K is total consumption in thousands and x is income in thousands.
a. Find her total consumption function if she consumes 15 when $x = 0$.
b. What is her break-even income level?

10. Use www.calc101.com to integrate the following functions
a. $f(x) = x^3 + x^2 + x + x^{-1}$
b. $f(x) = 4 \cdot 3^x + 2^x$
c. Find the total area under these rate functions between $x = 1$ and $x = 3$.

MIND STRETCHERS

1. A new company, beginning with no customers at time zero, gains new customers according to the monthly rate $dC/dx = 20 - x$, where x in months is less than 20. For the first four months of operation, they lose no established customers. Thereafter, they lose five previously established customers per month.
a. How many customers will they have after four months? After six months?
b. When will they have 100 customers?
c. When will they have the maximum number of customers?

2. The acceleration (a) of moon-bound rockets in the first few minutes after lift-off can be approximated by the following function of time (x) in minutes:

$$a = kx^2$$

where k is a constant. If the rocket is 16 miles high after 2 minutes, how high will it be after 3 minutes?

3. A harried taxpayer estimates that her taxes will grow 10 percent per year. If she pays taxes for 40 years and initially pays $1000, use calculus to estimate the total she will pay. Compare this answer to the one obtained by summing the appropriate geometric progression.

4. Hot-Rod Rodney claims that from a stopped position his car, with constant acceleration, can reach a velocity of 60 miles per hour, or 88 feet per second, in exactly 10 seconds. If his claim is true,
   a. How far will he travel during this feat?
   b. What will his velocity be after he has gone 110 feet?

5. Hot-Rod Rodney is at the beginning of the acceleration lane to Route 3.1416 and Felicia Freeway is on the main highway. Their initial positions and velocities are as shown in the diagram.

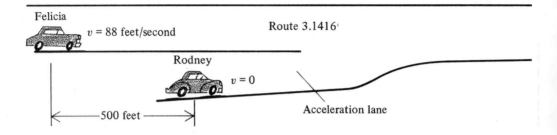

Without checking for traffic Rodney accelerates at a constant 8 feet per second^2 until he reaches a velocity of 60 miles per hour, or 88 feet per second, at which time he swings from the acceleration lane to the main highway. Felicia doesn't see Rodney's car and continues at a constant velocity of 88 feet per second. Describe the split second that Rodney enters the main highway. Is there a crash? If not, which car is first and how many feet separate the vehicles? (Neglect car length and Rodney's lateral movement in the calculations.)

6. Consider the above problem for the following conditions:
   (1) initial separation distance of 400 feet,
   (2) Rodney darts onto the main highway after traveling 196 feet,
   (3) Felicia sees Rodney and hits the brakes three seconds after his starting time, and,
   (4) Felicia's velocity function during the braking period is $v = 88 - 3x^2$. What happened?

7. Suppose the birth (B) and death (D) rates in the United States are forecasted to be

$$\frac{dB}{dx} = 3,500,000 - 50,000x$$

$$\frac{dD}{dx} = 2,000,000 + 25,000x$$

where x is time in years ($x = 0$ now). If the current population is 300 million, when will the point of ZPG (zero population growth) occur? What will the population be then?

8. Gypyou Co. has just introduced its newest product. Market researchers forecast the sales rate to climb initially, but then to decline to zero (as consumers realize they were taken) according to the function

$$\frac{dS}{dt} = 20 + 3t - \frac{1}{5}t^2$$

where $\frac{dS}{dt}$ is sales rate in units per month, and t is time in months ($t = 0$ now).
   a. When will the sales rate be highest?
   b. What will total sales for the product be?

9. Consider the probability density function

$$P(x) = x \quad \{0 \leq x \leq X\}$$

INTEGRATION METHODS AND THEIR APPLICATION

   a. Find X for $P(x)$ to be a probability density function.
   b. What is the median?
   c. Find the mean.
   d. Find the variance.

10. A Syracuse snowstorm is forecast to have a snowfall rate as given by the following function (s is total snowfall in inches and x is the total number of hours since the beginning of the storm):

$$\frac{ds}{dx} = x - .1x^2$$

   a. How much snow will fall in the entire storm?
   b. At what time will the storm be most intense (snow falling at the greatest rate)?

CHAPTER 18

Advanced Integral Calculus

18.1 INTRODUCTION

We now embark on the last chapter of this book. If you got to this point, even slightly bruised and bloodied, you deserve a round of applause. But, like a marathon runner with 1 mile to go, the biggest challenge could be ahead. So don't celebrate yet. Luckily, the previous two chapters provided you with a solid foundation that should propel you through the more advanced work here.

Learning Plan

At the end of Chapter 17, we were stymied by an integral that could not be found by the simple rules known to us then. We make up this deficiency here by developing more advanced integration rules.

Two major integration rules, called change of variable and integration by parts, will be developed in this chapter. Once accomplished, we can return to help Matt Tress solve his capital investment problem. But that is just the lead-off application hitter. The rest of the applications lineup includes the big hitters of learning curves, job training, physical fitness, city planning (garbage dump forecasting, and ecology (water and air pollution).

At some point, we will still get stymied by integrals for which we don't have handy-dandy rules. But don't fear, we then bring out the neat software, www.calc101.com, from the bull pen to save the game.

We end the chapter with a section on multiple integration. That may sound ominous. But it isn't. Just as we can take a second derivative or take a partial derivative, we can handle them. You'll see.

18.2 HEURISTIC DEVELOPMENT OF SOME MORE INTEGRATION RULES

The two methods of integration presented now, namely, (1) change-of-variable integration and (2) integration by parts, will enable the reader to handle Matt's investment decision as well as many other applications.

Change-of-Variable Integration

The change-of-variable method involves changing the variable of integration so that the integrand shapes up into a function of the new variable, which can then be integrated by already established rules. For example, if $f(x)$ is some complicated function for which the integral is not obvious, defining a new variable (u) as a function of x so as to transform the integrand to the u variable as shown here.

$$\int f(x)\, dx = \int g(u)\, du$$

This transformation will help us only if we know how to integrate $g(u)$. But in many cases, that will indeed by possible.

But how can this wonderful shaping up of the integrand occur? Isn't it too much to expect? To understand the answers to these pressing questions, we must begin back in differential calculus, which along with the fundamental theorem is our source of inspiration.

Consider our struggle to differentiate $y = f(x) = (x^2 - 3)^{1.5}$. Recall that we had to define a new variable, u, and use the chain rule. If $u = x^2 - 3$, then $y = u^{1.5}$, so the chain rule gives

$$\frac{dy}{dx} = \left(\frac{dy}{du}\right)\left(\frac{du}{dx}\right) = (1.5u^{.5})(2x)$$

$$= (1.5(x^2 - 3)^{.5}(2x)$$

Consequently,

$$\int 1.5(x^2 - 3)^{.5} 2x\, dx = (x^2 - 3)^{1.5} + C$$

For our understanding of integration it is important to realize that differentiating by the chain rule insures that some function in the result is the derivative of another function in the overall expression (here $2x$ is the derivative of $x^2 - 3$).

If confronted with the function $1.5(x^2 - 3)^{.5} 2x$ to integrate without the benefit of having found it by differentiation, one could be in a bind; that is, unless one noticed that $2x$ is the derivative of $x^2 - 3$. This clue should ring the chain rule bell, which should start the wheels turning to the change-of-variable method of integration. The procedure would be: let

$$u = x^2 - 3$$

then

$$\frac{du}{dx} = 2x \quad \text{and} \quad du = 2x\, dx$$

(The derivative symbol, that is, du/dx, has, until now, been treated as a whole. However, it can be treated as two "differentials," du and dx, bearing the relationship $du \div dx = (du/dx)$, or the value of the derivative.)

So we accomplish the change of variable as

# ADVANCED INTEGRAL CALCULUS

$$1.5 \int (x^2 - 3)^5 (2x\, dx) = 1.5 \int u^5\, du$$

This expression is in power rule form, which can easily be integrated to give

$$= 1.5 \left(\frac{u^{1.5}}{1.5} \right) + C$$

$$= u^{1.5} + C$$

And, finally, replacing u by its equal, or $x^2 - 3$,

$$= (x^2 - 3)^{1.5} + C$$

Notice that the constant 1.5 didn't interfere with our ability to integrate, inasmuch as it could be brought out from the integrand. But what if the original integrand had the constants 1.5 and 2 multiplied out as $\int 3(x^2 - 3)^5 x\, dx$? Since x is no longer the derivative of $x^2 - 3$, are we out of luck? No, because the necessary constant of 2 can always be put in front of x to get $2x$ as long as we compensate by inserting a $\frac{1}{2}$ elsewhere in the integrand. So,

$$\int 3(x^2 - 3)^5 x\, dx = \int 3(x^2 - 3)^5 \left(\frac{1}{2} \right)(2x\, dx)$$

$$= \int 1.5(x^2 - 3)^5 (2x\, dx)$$

To generalize, as long as du/dx lacks only the constant multiplier needed to effect a change of variable, the integrand can be multiplied and divided by that constant and the change of variable method can be used.

Now consider

$$\int \frac{2x + 1}{x^2 + x}\, dx$$

At first glance it looks impossible. But note the interesting link between the numerator and the denominator. The derivative of the denominator equals the numerator. Ah, hah! We are back in business with the change of variable method. Let $u = x^2 + x$. Then $du/dx = 2x + 1$ and consequently $du = (2x + 1)\, dx$. So, we have

$$\int \frac{(2x + 1)\, dx}{x^2 + x} = \int \frac{du}{u}$$

$$= \ln u + C = \ln(x^2 + x) + C$$

Note that as long as the numerator is a constant times $2x + 1$ or $k2x + k$, we can use the change of variable method here. Thus,

$$\int \frac{k2x + k}{x^2 + x}\, dx = \int \frac{k(2x + 1)}{x^2 + x}\, dx = k \int \frac{(2x + 1)\, dx}{x^2 + x}$$

$$= k \ln(x^2 + x) + C$$

However,

$$\int \frac{2x+3}{x^2+x} \, dx$$

could not be integrated by this method. Do you understand why?

Now consider

$$\int e^{-x} \, dx$$

One is tempted to say, "Ah, hah, the integral is $e^{-x} + C$ since $\int e^x \, dx = e^x + C$. But upon checking the derivative of e^{-x} using the chain rule with $u = -x$,

$$\frac{dy}{dx} = \left(\frac{dy}{du}\right)\left(\frac{du}{dx}\right)$$
$$= (e^u)(-1)$$
$$= -e^{-x}$$

we don't get back to $+e^{-x}$.

Of course, had e^{-x} been the integral of e^{-x}, its derivative should have been e^{-x}. It wasn't, but the fact that the chain rule was used in checking should indicate the use of the change of variable method. Letting $u = -x$, so $du = -1 \, dx$, we have

$$\int e^{-x} \, dx = \int e^u(-1)(-1 \, dx)$$
$$= -1 \int e^u \, du$$
$$= -e^u + C$$
$$= -e^{-x} + C$$

Check to see that $D[-e^{-x} + C]$ is e^{-x}.

Now see if you can follow these three examples and carry the ball alone for the three exercises.

Examples

1. $\displaystyle\int 10^{x^2} x \, dx = \int 10^u \left(\frac{1}{2}\right) \overbrace{(2x \, dx)}^{du} \quad$ where $\quad u = x^2$

$$= \frac{1}{2} \int 10^u \, du$$

$$= \frac{1}{2}\left(\frac{10^u}{\ln u}\right) + C$$

$$= \frac{1}{2}\left(\frac{10^{x^2}}{\ln x^2}\right) + C$$

ADVANCED INTEGRAL CALCULUS

2. $\displaystyle\int \frac{5 \ln x}{x} dx = 5 \int u \overbrace{\left(\frac{dx}{x}\right)}^{du}$ where $u = \ln x$

$\displaystyle = 5 \int u \, du$

$\displaystyle = 5 \left(\frac{u^2}{2}\right) + C$

$\displaystyle = \frac{5(\ln x)^2}{2} + C$

3. $\displaystyle\int \left(\frac{x^3}{3} - e^x\right)^3 (x^2 - e^x) \, dx = \int u^3 \overbrace{(x^2 - e^x) \, dx}^{du}$ where $u = \dfrac{x^3}{3} - e^x$

$\displaystyle = \int u^3 \, du$

$\displaystyle = \frac{u^4}{4} + C$

$\displaystyle = \frac{\left(\dfrac{x^3}{3} - e^x\right)^4}{4} + C$

Exercises 1

1. $\displaystyle\int \frac{x^{-2} + 4}{x^3} \, dx =$

2. $\displaystyle\int \frac{1}{x \ln x} \, dx =$

3. $\displaystyle\int x e^x (e^x)(x + 1) \, dx =$

Integration by Parts

As expected, the integration-by-parts method has its inception in differential calculus. Recall the product rule for differentiation. If

$$y = u \cdot v \quad \text{where} \quad u = g(x) \quad \text{and} \quad v = h(x)$$

then

$$\frac{dy}{dx} = u \frac{dv}{dx} + v \frac{du}{dx}$$

If we integrate both sides,

$$\int \frac{dy}{dx} dx = \int u \frac{dv}{dx} dx + \int v \frac{du}{dx} dx$$

$$\int dy = \int u\, dv + \int v\, du$$

$$y = u \cdot v = \int u\, dv + \int v\, du$$

Thus,

$$\int u\, dv = u \cdot v - \int v\, du$$

Since u and v are functions of x, the above expression can be put completely in terms of x.

$$\int f(x)\, dx = i(x) - \int j(x)\, dx$$

where

$$f(x) = g(x) \cdot D[h(x)]$$
$$i(x) = g(x) \cdot h(x)$$
$$j(x) = h(x) \cdot D[g(x)]$$

A good question now is, What's all this for? The answer is that this theoretical result allows one to trade the determination of a monster integral for the chance to work on less mind-boggling ones. So, if we can't integrate $f(x)$ directly, we can do it indirectly by determining $i(x)$ and integrating $j(x)$. The indirect method assumes we can define certain functions and carry out certain integrations, which are best explained by example. Consider

$$\int xe^x\, dx$$

First, note that the methods previously learned do not help. Trying integration by parts, we must first set

$$\int xe^x\, dx = \int u\, dv$$

There are two ways to do this:

$u = e^x$ and $dv = x\, dx$
$u = x$ and $dv = e^x\, dx$

The first way gives

ADVANCED INTEGRAL CALCULUS

$$\int \overset{u}{\overset{\downarrow}{(e^x)}}\overset{dv}{\overset{\downarrow}{(x\,dx)}} = uv - \int v\,du$$

This points out the need to find v, du, and $\int v\,du$.

$$v = \int dv = \int x\,dx = \frac{x^2}{2}$$

$$du = D[e^x] \cdot dx = e^x\,dx$$

$$\int v\,du = \int \frac{x^2 e^x}{2}\,dx = \text{Huh!!}$$

This integral is worse than the original beast. Let's try the second way.

$$\int \overset{u}{\overset{\downarrow}{(x)}}\overset{dv}{\overset{\downarrow}{(e^x\,dx)}} = uv - \int v\,du$$

$$v = \int dv = \int e^x\,dx = e^x$$

$$du = D[x] \cdot dx = 1\,dx = dx$$

$$\int v\,du = \int e^x\,dx = e^x$$

Thus,[2]

$$\int xe^x\,dx = xe^x - e^x + C = e^x(x-1) + C$$

The reasons for success are that both dv and $v\,du$ were easily integrated. This trading of a complex integral for two simpler ones is the essence of the method.

Now consider $\int x \ln x\,dx$. The two ways to assign u and dv here are

$u = x$ and $dv = \ln x\,dx$

$u = \ln x$ and $dv = x\,dx$

Looking ahead, if we choose the first way, how in the heck would we get v (actually this integral itself requires an integration by parts)? The second way doesn't seem to present any pitfalls, so let's try it.

---

[2] Both intermediate integrations for the integration-by-parts method have a constant of integration, which we don't show. The constant of integration that results from $\int dv$ cancels out in a subsequent step (see if you can verify this for our example problem). The constant of integration from $\int v\,du$ doesn't cancel; however, we wait until the final result to show it.

$$\int (\ln x)x \, dx = uv - \int v \, du$$

$$v = \int dv = \int x \, dx = \frac{x^2}{2}$$

$$du = D[\ln x] \, dx = \frac{1}{x} dx$$

Thus

$$\int (\ln x)x \, dx = (\ln x)\frac{x^2}{2} - \int \frac{x^2}{2} \cdot \frac{1}{x} dx$$

$$= \frac{x^2 \ln x}{2} - \int \frac{x}{2} dx$$

$$= \frac{x^2 \ln x}{2} - \frac{x^2}{4} + C$$

Having seen two examples of integration by parts, one can appreciate that success depends mainly on two factors:

1. Ability to integrate $\int dv$ to obtain v, and
2. Obtaining $\int v \, du$, which is simpler to integrate than $\int u \, dv$.

Now see if you can understand solved exercise 1. The next two exercises are all yours.

Exercises 2

1. $\int \overbrace{\ln x}^{u} \overbrace{dx}^{dv} = (\ln x)x - \int x \cdot \frac{1}{x} dx$

    $= x(\ln x - 1) + C$

2. $\int x^2 e^x \, dx =$ (*Hint:* Use integration by parts twice.)

3. $\int \log_b x \, dx =$

This ends our development of integration rules. Those presented are by no means a complete set; however, they should suffice for most managerial applications. A good question now is, What would one do if one met a hairy integral that could not be handled by the methods described herein? Well, the unfortunate person could do one of five things:

1. Give up,
2. Approximate the area under the curve, since this is what the integral represents physically.
3. Consult an advanced mathematics text for advanced integration methods,

ADVANCED INTEGRAL CALCULUS   641

   4. Consult a table of integrals which contain thousands of solved integrals,
   5. Use the www.calc101.com software.

In the next section, we will take the fifth option and illustrate that software.

18.3 INTEGRATION USING www.calc101.com SOFTWARE

The home page for www.calc101.com is shown on page 353. Before proceeding, you should look under "stored integrals" and click on "substitution" and later on "integration by parts". You will then see many example integrals and be shown how to input their formulas.

With the knowledge of the above examples, you are now ready to help Matt Tress solve his investment decision problem (See page 623). Matt's rate function is

$$\frac{ds}{dx} = 6000 \left(\frac{x}{2} + 1\right)^{-2}$$

You would input this equation as

6000 ((x/2) + 1) ∧ −2

as shown in Figure 18.1.

The software gives the indefinite integral as

$$\frac{-24{,}000}{x+2} + C$$

At first glance, that result looks different than our book's answer on page 645. But a little algebra will reveal that both answers are equivalent

$$-12{,}000 \left(\frac{x}{2} + 1\right)^{-1} = \frac{-12{,}000}{\left(\frac{x+2}{2}\right)} = \frac{-24{,}000}{x+2}$$

Now, let's do the integration by parts problem stated on page 638 which we solved on page 639 as

$$\int x e^x \, dx = e^x(x - 1) + C$$

If you input the formula in the software as

x * e ∧ x

and furthermore input a valid password (you'll need to pay for that.), you'll not only get the ex-

Figure 18.1

Step-by-Step Integration

Click "Do it" to get your answer; to see all the steps, get a *password* fast!

Plot your function!

home | FAQs | info@calc101.com | polynomial multiplication | polynomial division

graphs *new!* | matrix algebra *new!* | derivatives | **passwords**

Use the right input format to get good results.	Wrong	Right
Use square brackets [] for functions.	cos(x), sqrt x, arcsinx, sin[ln x]	cos[x], sqrt[x], arcsin[x], sin[ln[x]]
Use parentheses () for grouping.	[x+1]^3, e^[x+1], x^2/[1+x^3]	(x+1)^3, e^(x+1), x^2/(1+x^3)
Use spaces or * for multiplication.	xe^-x, xsin[pix], bx^2	x e^-x, x sin[pi x], b x^2

enter your password

Do the integral of
`6000*((x/2)+1)^-2`

with respect to
`x`

▶ DO IT ▶

Sorry, the password is not valid.

For your convenience, here are the first and last lines of the solution:

$$\int \frac{6000}{(\frac{x}{2}+1)^2} \, dx$$
$$= -\frac{24000}{x+2} + C$$

To see all the steps, please buy a password!

If you forgot or lost your password, email info@calc101.com and mention your name.

ADVANCED INTEGRAL CALCULUS

act same solution we found by hand, but you'll also get the various instructive steps leading up to the integration by parts solution (See Figure 18.2).

Now let's try the software on an integrand that stumped us earlier. Refer to page 636 where we realized that we didn't have the rules needed to integrate

$$\int \frac{2x + 3}{x^2 + x} \, dx$$

By inputting this integrand and a valid password, you would get a long series of steps using a partial fractions method and a final result of

$$\int \frac{2x + 3}{x^2 + x} \, dx = 3 \log x - \log(x + 1) - 2 + C$$

where log represents the natural logarithm or ln.

You can verify its correctness by showing that the derivative of the answer does equal the integrand.

The partial fractions method is beyond the scope of our book; so it is nice to have software that "knows it."

Now, even www.calc101.com has its limitations. If you asked it to do the following integral

$$\int x e^x \ln x \, dx$$

it would politely print out, "Sorry, calc101 cannot do this integral." For such cases, you would need to consult a table of integrals.

Exercises 3

Use www.calc101.com to do the following indefinite integrations.

1. The 3 substitution examples on page 636.
2. The 3 integration by parts exercises on page 640.

18.4 APPLICATIONS OF INTEGRATION

Now we'll return to Matt Tress's decision problem. Our understanding of the change-of-variable and the integration-by-parts methods, as well as the vast software integration prowess, will enable us to solve Matt's as well as other everyday problems.

Matt Tress's Capital Budgeting Problem Revisited

Recall that Matt's investment problem reduced to integrating the following:

$$S = 6000 \int \left(\frac{x}{2} + 1\right)^{-2} dx$$

Figure 18.2

Step-by-Step Integration

Click "Do it" to get your answer; to see all the steps, get a *password* fast!

Plot your function!

home | FAQs | info@calc101.com | polynomial multiplication | polynomial division

graphs *new!* | matrix algebra *new!* | derivatives | **passwords**

Use the right input format to get good results.	Wrong	Right
Use square brackets [] for functions.	cos(x), sqrt x, arcsinx, sin[ln x]	cos[x], sqrt[x], arcsin[x], sin[ln[x]]
Use parentheses () for grouping.	[x+1]^3, e^[x+1], x^2/[1+x^3]	(x+1)^3, e^(x+1), x^2/(1+x^3)
Use spaces or * for multiplication.	xe^-x, xsin[pix], bx^2	x e^-x, x sin[pi x], b x^2

enter your password

Do the integral of
x*e^x

with respect to
x

[DO IT ▶]

---

solving integrals step-by-step

$\int x\, e^x\, dx$

Integrate by parts using
$u = x$, $dv = e^x\, dx$,
$du = 1\, dx$, $v = e^x$.

$= e^x x - \int e^x\, dx$

The integral of e^x is e^x.

$= e^x x - e^x + C$

Factor for another way to see the result.

$= e^x (x - 1) + C$

ADVANCED INTEGRAL CALCULUS

Using the change-of-variable method, with $u = \left(\dfrac{x}{2} + 1\right)$ so $du = \tfrac{1}{2} dx$, gives

$$S = 6{,}000 \int u^{-2} \cdot 2 \cdot \dfrac{1}{2} dx = 12{,}000 \int u^{-2} du$$

$$= 12{,}000 \left(\dfrac{u^{-1}}{-1}\right) + C = -12{,}000 \left(\dfrac{x}{2} + 1\right)^{-1} + C$$

Total forecasted savings in the first four years is found by taking the definite integral over the interval from zero to 4.

$$F(4) - F(0) = (-12{,}000 \cdot 3^{-1}) - (-12{,}000 \cdot 1^{-1})$$

$$= -4{,}000 + 12{,}000$$

$$= 8{,}000$$

Matt should not invest in the new machine, since it does not meet his payback standard requiring a $10,000 savings in four years.

Exercises 4

1. What is the upper limit of total savings for the project?
2. Determine when the project will pay back the initial investment of $10,000.
3. Will the project meet the four-year payback standard if its savings rate is

$$\dfrac{dS}{dx} = 7000 \left(\dfrac{x}{4} + 1\right)^{-3}$$

Learning Curve: Job Training

A certain industrial job has a 100-day training period. On the first day the trainee is given various tests and thus spends no time training. Thereafter, he is trained on the job, and his production rate gets higher and higher as he learns the job. Learning theory suggests a logarithmic function like the following to represent the training situation.

$$\dfrac{dP}{dx} = K + k \log_{10} x$$

where

    dP/dx is daily production rate in units per day
        x is time in days $\{x \geq 1\}$
        K, k are constants

If the average trainee begins ($x = 1$) at a zero rate and ends ($x = 100$) at a 100 rate, how much will he produce in the entire training period?

Solution

First, the specific rate function must be determined by finding K and k from the specified conditions. Then, the rate function must be integrated over the interval 1 to 100 to obtain total production (see Figure 18.3).

Since $dP/dx = 0$ when $x = 1$ and $dP/dx = 100$ when $x = 100$, we have

$$0 = K + k \log_{10} 1$$
$$100 = K + k \log_{10} 100$$

Solving, we get $K = 0$ and $k = 50$. Thus

$$\frac{dP}{dx} = 50 \log_{10} x$$

Integration by parts is required for this rate function.

$$\int 50 \log_{10} x \, dx = 50 \int \log_{10} x \, dx$$

$$= 50 \left[(\log_{10} x)x - \int x \frac{1}{x \ln 10} \, dx \right]$$

$$= 50 \left[x \log_{10} x - \frac{x}{\ln 10} \right] + C$$

Total production is then

$$F(100) - F(1) = 50 \left[\left(100 \log_{10} 100 - \frac{100}{\ln 10} \right) - \left(1 \log_{10} 1 - \frac{1}{\ln 10} \right) \right]$$

$$= 50 \left(200 - \frac{99}{\ln 10} \right) = 7{,}850$$

Figure 18.3

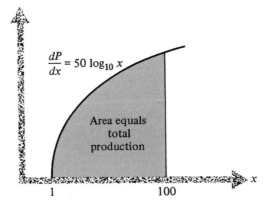

ADVANCED INTEGRAL CALCULUS

Exercise 5

Consider the following production rate function:

$$\frac{dP}{dx} = 5 + \ln x \quad \text{for} \quad \{x \geq 1\}$$

How much production will occur over the first 100 days?

Physical Fitness

Phil Z. Calfitness is a push-up nut. Initially, he can do them at a rate of 60 per minute. As he tires, the rate drops, and he is down to 42 per minute by the end of 1 minute. The function

$$\frac{dP}{dx} = \frac{K}{(x+1)^n}$$

where dP/dx is the push-up rate (number per minute), K, n are constants, and x is time in minutes reflects this pattern of push-ups. How many push-ups can Phil do in 1 minute? How long would it take him to do 100?

Solution

The specific rate function must be determined. Then it must be integrated over the interval $x = 0$ to $x = 1$. Finally, the value of x up to which the area is 100 needs to be located. (See Figure 18.4).

The two points given on the rate function allow determination of K and n as follows:

$$60 = \frac{K}{(0+1)^n} \quad \text{and} \quad 42 = \frac{K}{(1+1)^n}$$

Solving simultaneously gives $K = 60$ and $n = \frac{1}{2}$. So

$$\frac{dP}{dx} = \frac{60}{(x+1)^{1/2}}$$

Figure 18.4

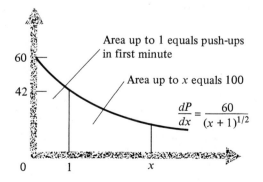

The change of variable method is necessary to integrate this rate function. Letting $u = x + 1$, then $du = dx$, thus,

$$P = \int \frac{dP}{dx} dx = 60 \int (x + 1)^{-1/2} dx = 60 \int u^{-1/2} du$$

$$= 120 u^{1/2} + C$$

$$= 120(x + 1)^{1/2} + C$$

Total push-ups during the first minute are

$$F(1) - F(0) = 120 \cdot 2^{1/2} - 120 \cdot 1^{1/2}$$

$$= 120(2^{1/2} - 1)$$

$$= 49.7 \text{ or } 49 \text{ complete ones}$$

Total push-ups in x seconds is $F(x) - F(0)$. Setting this equal to 100 and solving for x gives

$$100 = F(x) - F(0)$$

$$= 120(x + 1)^{1/2} - 120 \cdot 1^{1/2}$$

$$= 120(x + 1)^{1/2} - 120$$

A little algebra and a little luck should reveal that $x = (121/36) - 1$, or 2.36 minutes.

Exercise 6

Consider this problem if Phil's initial rate is 100 per minute, but he is down to 25 per minute as of the end of the first minute.

City Planning: Garbage Dump Forecasting

The city of Ohdore must plan for its future garbage disposal needs. The existing city dumps can hold an additional 10,000,000 cubic feet of garbage. Other dumps are difficult to obtain (not many people like them next door), so careful forecasting is necessary to find out when they will be required. A study over the past several years revealed that garbage disposal rate is an exponential function of time.

$$\frac{dG}{dx} = ke^{.1x}$$

where

> dG/dx is the garbage disposal rate in cubic feet per year
> x is time in years ($x = 0$ now)
> k is a constant

ADVANCED INTEGRAL CALCULUS

Currently ($x = 0$), the garbage rate is 1,000,000 cubic feet per year. How long do the city fathers have to obtain additional dump sites?

Solution

The value of k must be determined from the initial conditions. Then the garbage rate function must be integrated to find that time, x, such that the area under the curve to that point is 10,000,000 (see Figure 18.5).

Since $dG/dx = 1,000,000$ when $x = 0$, then $1,000,000 = k \cdot e^0 = k$. Consequently, the problem reduces to finding x such that

$$10,000,000 = \int_0^x 1,000,000 e^{.1x}\, dx$$

$$= 1,000,000 \int_0^x e^{.1x}\, dx$$

Integration here requires the change of variable method. Letting $u = .1x$, then $du = .1\, dx$, so

$$10^6 \int e^{.1x}\, dx = 10^6 \int e^{.1x}\left(\frac{1}{.1}\right).1\, dx$$

$$= 10^7 \int e^u\, du$$

$$= 10^7 e^u + C$$

$$= 10^7 e^{.1x} + C$$

Then

$$10,000,000 = 10^7 = F(x) - F(0)$$

$$= 10^7 e^{.1x} - 10^7 e^0$$

$$= 10^7(e^{.1x} - 1)$$

Adroit use of algebra will reveal that $x = (\ln 2)/.1 = 6.93$ years.

Figure 18.5

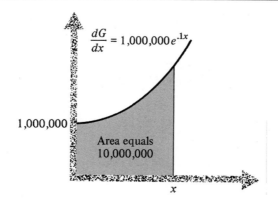

Exercises 7

1. Determine what the current garbage rate would have to be in order for the city to have 10 years use of the current site.
2. If the garbage rate were

$$\frac{dG}{dx} = 1{,}000{,}000 e^{.15x}$$

how long could the current site be used?

Ecology: Water Pollution

Lake Murky has the distinction of having several large chemical companies on its shores. The companies have had a tendency to dump mercury in the lake, but recently (after government suits) their public-spirited managements have undertaken steps to reduce the flow of mercury into the lake. Consequently, the rate of mercury flow (dM/dx) in pounds per year is forecasted by pollution experts to decline over time (x) in years from now, as the various filtering projects come to fruition, according to the function

$$\frac{dM}{dx} = 75(x+1)e^{-x}$$

If it is estimated that 100 pounds are already in the lake and 300 pounds would be disastrous to marine life, are the steps taken by the firms sufficient?

Solution

If $F(x)$ is the total amount of mercury in the lake at time x, then

$$F(x) = 100 + \int_0^x 75(x+1)e^{-x}\, dx$$

The question is whether $F(x)$ can exceed 300 pounds or alternatively can

$$\int_0^x 75(x+1)e^{-x}\, dx$$

be greater than 200 pounds.

This integral is a natural for the by-parts method. Letting $u = x + 1$ and $dv = e^{-x}\, dx$, then $du = dx$ and $v = -e^{-x}$.

$$75 \int (x+1)e^{-x}\, dx = 75[(x+1)(-e^{-x}) - \int -e^{-x}\, dx]$$

$$= 75[x+1)(-e^{-x}) - e^{-x}] + C$$

$$= 75(x+2)(-e^{-x}) + C$$

ADVANCED INTEGRAL CALCULUS

The definite integral can never have a value greater than 150, as shown now,

$$75(x+2)(-e^{-x})\Big|_0^x = 75(x+2)(-e^{-x}) - 75(2)(-e^0)$$

$$= 150 - \frac{75(x+2)}{e^x}$$

since the second term is positive for all positive x. Thus, the lake will be saved.

Exercises 8

1. How much mercury will be dumped in the lake during the first year? During the second year?
2. How much mercury will be dumped in the lake in the first three years?

Ecology: Air Pollution

Car engine warming is a source of air pollution. Cold engines operate inefficiently, with the result being incomplete combustion and introduction of carbon monoxide into the air. As the engine warms, its efficiency increases and carbon monoxide discharge decreases. Suppose that for a certain engine on a 20° day the following functions represent the carbon monoxide discharge rate (dc/dx) in liters per minute during the time (x) in minutes of the warm-up period.

$$\frac{dc}{dx} = 10 - 5x \quad \text{for} \quad \{x \leq 1\} \text{ minutes}$$

$$\frac{dc}{dx} = \frac{3}{x} + 2 \quad \text{for} \quad \{x > 1\} \text{ minutes}$$

How many liters of carbon monoxide are discharged into the air during a three-minute warm up?

Solution

This is a different kind of an integration problem inasmuch as we have two different rate functions. However, it should pose no problem. The total carbon monoxide emission is the area under the two rate functions. To determine this area we will have to integrate each rate function over its domain.

$$\int_0^1 (10 - 5x)\,dx = 10x - \frac{5x^2}{2}\Big|_0^1$$

$$= 10 - \frac{5}{2} = 7.5$$

$$\int_1^3 \left(\frac{3}{x} + 2\right) dx = 3\ln x + 2x\Big|_1^3$$

$$= (3\ln 3 + 6) - (0 + 2) = 3\ln 3 + 4$$

Figure 18.6

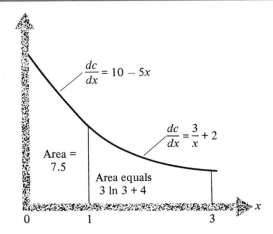

The total emission, which is the area under the rate functions as shown in Figure 18.6, is equal to $11.5 + 3 \ln 3$, or 14.8 liters.

Exercise 9

Use the initial and one-minute emission rates from the above rate function to determine a negative exponential function of the form

$$\frac{dc}{dx} = ae^{-dx}$$

where a and d are constants, and use it to approximate total emission in the first three minutes.

Probability—Exponential Distribution

The exponential probability density function is

$$\frac{dP}{dx} = \frac{1}{\mu} e^{-\frac{x}{\mu}}$$

where μ is the mean.

A total probability (which is also the area under the curve) of 1 is necessary for any function to be a probability density function. Let's see if the exponential function meets this test.

$$\int_0^\infty \frac{1}{\mu} e^{-\frac{x}{\mu}} dx = \frac{1}{\mu} \int_0^\infty e^{-\frac{x}{\mu}} dx$$

If we let $u = -\frac{x}{\mu}$, then $du = -\frac{1}{\mu} dx$; so $dx = -\mu\, du$. Thus, we transform our integral to

ADVANCED INTEGRAL CALCULUS

$$\frac{1}{\mu}\int_0^\infty e^{-\frac{x}{\mu}}\,dx = \frac{1}{\mu}\int_0^\infty e^u(-\mu\,du) = -1\,e^u\Big|_0^\infty$$

Substitution $u = -\dfrac{x}{\mu}$ back in, we get

$$= -1\,e^{-\frac{x}{\mu}}\Big]_0^\infty$$

$$= -1(0 - 1)$$

$$= 1$$

Let's consider a specific case. Suppose Neverready batteries have a life described by an exponential distribution with a mean life of 100 hours, or

$$\frac{dP}{dx} = -\frac{1}{100}e^{-\frac{x}{100}}$$

Using www.calc101.com, we input this function as

$(-1/100) * e \wedge -(x/100)$

The software returns the indefinite integral

$$-e^{-\frac{x}{100}} + C$$

We can now answer various questions. For example, what is the probability that the batteries will last longer than the mean time. For this, we find the definite integral from 100 to ∞, or

$$-e^{-\frac{x}{100}}\Big]_{100}^\infty = -e^{-\infty} - (-e^{-1})$$

$$= 0 + \frac{1}{e} = \frac{1}{e} = .368$$

Exercises 10

Find the probability that the batteries will

a. last more than 200 hours.
b. fail before 10 hours.

18.5 MULTIPLE INTEGRATION

We will consider multiple integration for two cases

1. single independent variable, or $y = f(x)$
2. two independent variables, or $z = f(x, y)$

Single Independent Variable

This case is analogous to taking a second derivative. Recall that if we began with the total distance (y) function that a body falls per second (x) in frictionless Earth gravity

$$y = f(x) = 16x^2$$

Taking the first derivative gives velocity, v

$$v = \frac{dy}{dx} = 32x$$

and then taking the second derivative gives acceleration, a

$$a = \frac{dv}{dx} = 32$$

If we started with a known acceleration function, we could go in reverse and integrate twice to get the total distance function, or

$$\iint 32 \, dx$$

We would do this multiple integration one at a time by first doing the "inner" integral

$$v = \int \left[\int 32 \, dx \right] dx = \int (32x + C) \, dx$$

The constant of integration would have to be determined. Since $v = 0$ when $x = 0$, then $C = 0$. So, we could then find the second integral

$$y = \int 32x \, dx = 16x^2 + C$$

Again $C = 0$, since $v = y = 0$ when $x = 0$. So finally we have

$$y = 16x^2$$

Exercise 11

If the acceleration in Earth's gravity, assuming friction is

$a = 20$

Find the total distance function.

Now let's see how a second derivative used as a check for a maximum could be the basis for finding the total revenue function by using two integrations.

ADVANCED INTEGRAL CALCULUS

Suppose Orange Computers has the following second derivative of revenue (R) as a function of price (P) for its laptop computers.

$$\frac{d^2R}{dP^2} = -100$$

Integrating this function gives marginal revenue

$$\frac{dR}{dP} = \int -100 \, dP = -100\,P + C$$

If total revenue was maximized when $P = 1000$, then $\frac{dR}{dP} = 0$ when $P = 1000$, so $C = 100{,}000$. Thus

$$\frac{dR}{dP} = -100\,P + 100{,}000$$

Then, the total revenue function can be found by a second integration

$$R = \int (-100\,P + 100{,}000) \, dP = -50P^2 + 100{,}000\,P + C$$

The constant of integration could be found knowing one point on the revenue function.

Exercise 12

If the second derivative check for verifying a maximum profit (π) is the following function of price

$$\frac{d^2\pi}{dP^2} = -10\,P$$

and total profit was maximized when $P = 20$, and $\pi = 1000$ when $P = 5$, find the total profit function.

Two Independent Variables

This case involves integrating functions of two independent variables, which symbolically is

$$\iint f(x, y) \, dx \, dy$$

Recall that integrating a function of one independent variable is equivalent to find the area under the one-dimension curve. So, you wouldn't be surprised to learn that integrating a function of two independent variables is equivalent to finding the volume under the two dimensional surface. We can illustrate that with the relatively simple function

Figure 18.7

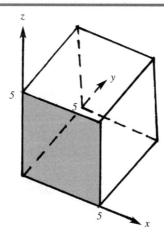

$$z = f(x, y) = \int 5x^0 y^0 \, dx \, dy = \int 5 \, dx \, dy$$

This function represents a horizontal surface 5 units high above the $x - y$ plane, as shown in Figure 18.7.

Carrying out the "inner" integration first

$$\int_0^5 5 \, dx = 5x \Big|_0^5 = 25 - 0 = 25$$

This represents the shaded area in Figure 18.7. Then carrying out the outer integration

$$\int_0^5 25 \, dy = 25y \Big|_0^5 = 125 - 0 = 125$$

gives the volume of the entire cube. This latter integration can be thought of as summing an infinite number of areas (like the shaded one) in the y direction.

Now let's do the following more complicated multiple integration.

$$\int_0^1 \int_0^1 xy \, dx \, dy$$

When integrating with respect to one variable, we hold the other one constant (like we did with partial derivatives).

Carrying out the inner integration on x first

$$\int_0^1 \left[\int_0^1 x \, dx \right] y \, dy = \int_0^1 \frac{x^2}{2} \Big|_0^1 y \, dy$$

$$= \int_0^1 \left(\frac{1}{2} - 0 \right) y \, dy$$

$$= \int_0^1 \frac{1}{2} y \, dy$$

ADVANCED INTEGRAL CALCULUS

Now carrying out the integration on y

$$\frac{1}{2}\int_0^1 y\,dy = \frac{1}{2}\frac{y^2}{2}\Big]_0^1 = \frac{1}{4} - 0 = \frac{1}{4}$$

The final result of $\frac{1}{4}$ represents the volume under the function xy when each variable ranges from 0 to 1.

Exercise 13

Find the volume

$$\int_0^1 \int_0^1 e^x \cdot y\,dx\,dy$$

Probability-Exponential Distribution Application

Let's see an interesting application of multiple integration. One multivariate exponential probability density function is

$$P(x) = e^{-x}e^{-y}$$

A legitimate probability distribution would have the volume of 1.

$$\int_0^\infty \int_0^\infty e^{-x}e^{-y}\,dx\,dy = 1$$

Let's see.
  Integrating first on x, then y

$$\int_0^\infty \left[\int_0^\infty e^{-x}\,dx\right]e^{-y}\,dy = \int_0^\infty \left\{-e^{-x}\Big|_0^\infty\right\}e^{-y}\,dy$$

$$= \int_0^\infty (-0 - (-1))\,e^{-y}\,dy$$

$$= \int_0^\infty 1e^{-y}\,dy$$

$$= -e^{-y}\Big|_0^\infty$$

$$-(0 - (-1)) = 1$$

Yes indeed it is a probability distribution.
  More generally, multivariate exponential probability distributions are described by

$$P(x, y) = \mu_x \mu_y\, e^{-\frac{x}{\mu_x}} e^{-\frac{y}{\mu_y}}$$

where μ_x is the mean of the x's and μ_y is the mean of the y's.

Let's apply this case to a real business situation. Suppose an executive during a day spends time out of her office for business calls (x minutes) and in her office in staff meetings (y minutes). If $\mu_x = 50$ and $\mu_y = 20$, we can find the probability she can have $x \leq 100$ and $y \geq 10$ by

$$\int_{10}^{\infty} \int_0^{100} e^{-\frac{x}{100}} e^{-\frac{y}{10}} \, dx \, dy$$

Finding the inner integral on x

$$\int_0^{100} e^{-\frac{x}{100}} \, dx = -\frac{1}{100} e^{-\frac{x}{100}} \Big|_0^{100} = -\frac{1}{100}(e^{-1} - 1)$$

$$= \frac{1}{100}(1 - e^{-1})$$

Next integrating with respect to y

$$\int_{10}^{\infty} \frac{1}{100}(1 - e) e^{-\frac{y}{10}} \, dy = \frac{1}{100}(1 - e) \frac{-1}{10} e^{-\frac{y}{10}} \Big|_{10}^{\infty}$$

$$= -\frac{1}{1000}(1 - e^{-1})(e^{-\infty} - e^{-1})$$

$$= -\frac{1}{1000}(1 - e^{-1})(0 - e^{-1})$$

$$= \frac{1}{1000}\left(\frac{1 - e^{-1}}{e}\right)$$

Exercise 14

Find the probability that the executive spends less than 50 minutes for business calls and more than 20 minutes for staff meetings.

18.6 THE END!

We made it! We finished!! Hopefully, the book's conversational style, humor (yes, bad jokes), silly names, and especially the many real world applications made the learning easier.

I won't invoke the dreaded word "Again," which my three-year old twin grandchildren, Seth and Liana, always say (When will one-year-old Jessica join the chorus?) after I finish reading them a book (even if it is the fifth straight reading). Rather, I suggest you return to sections as and when needed in your classroom work and career.

Good Luck!!

ADVANCED INTEGRAL CALCULUS

PROBLEMS

TECHNICAL TRIUMPHS

1. Integrate the following
   a. $\ln(x^2 + x)$
   b. $x \cdot 10^x$
   c. $\left(\dfrac{x^2}{2} + e^x\right)^4 (x + e^x)$

2. Use www.calc101.com to integrate the functions in problem #1.

3. Find the area under the curves in problem #1 from $\{1 \le x \le 2\}$

4. Carry out the following multiple integrations
   a. $\iint x^3 y^{-4} \, dx \, dy$
   b. $\iint e^x \ln y \, dx \, dy$

5. Use www.calc101.com to do problem #4. (Note: In each case, you'll need to do two integrals. For each, just do a single variable integration by omitting the other variable. Later, put that other variable back)

6. Show that the area under the following exponential probability distribution is 1, or

$$\int_0^\infty e^{-x} \, dx = 1$$

7. If a second derivative check for a cost (C) minimization reveals

$$\frac{d^2C}{dx^2} = 10$$

and costs are minimized when $x = 50$, find the cost function (including a constant of integration).

8. Find the volume under the multivariate function

$$z = f(x, y) = x^2 y^2 \quad \text{for} \quad \begin{array}{l}\{0 \le x \le 1\} \\ \{0 \le y \le 1\}\end{array}$$

CONFIDENCE BUILDERS

1. If the rate of grade (G) improvement for a mathematics exam as a function of hours of time studied (x) is

$$\frac{dG}{dx} = 20 + 5 \ln x$$

Find $G = f(x)$ if $G = 70$ when $x = 10$.

2. Consider the exponential probability distribution

$$P(x) = \frac{1}{10} e^{-\frac{x}{10}} \quad \text{for} \quad \{0 \le x \le \infty\}$$

a. Verify that it is a probability distribution by showing the area under the curve is 1.
b. What is the probability that $x \leq 5$?
c. Find the median.
d. Find the mean

3. If the height (z) of the dome over a sports complex is

$$z = 100 - .01x^2 - .01y^2 \quad \begin{array}{l} \{0 \leq x \leq 100\} \\ \{0 \leq y \leq 100\} \end{array}$$

find the volume of the complex.

4. Consider the probability density function

$$P(x) = (x + 1)^{1/2} \quad \text{for} \quad \{0 \leq x \leq 2.953\}$$

a. Verify that it is a probability distribution.
b. Find $P(x \leq 2)$.
c. Find the median.
d. Find the mean.

5. Dr. Smock's research indicates that the function

$$\frac{dW}{dx} = 50 \cdot 2^{-x/52}$$

reasonably reflects the wetting rate of babies in their first year, where x is time in weeks and dW/dx is wetting rate in "wets" per week. How many diaper changes would be necessary in the first year if a parent desired to keep the baby as dry as possible? (Neglect other baby "happenings.")

6. Consider the exponential probability distribution

$$P(x) = \frac{1}{\mu} e^{-x/\mu} \quad \text{for} \quad \{0 \leq x \leq \infty\}$$

where μ is a constant.
a. Show that it is a probability density function by verifying that the area under the curve is 1.
b. Show that a random variable with this probability density function has a mean of μ. (*Note:* as $x \to \infty$, $xe^{-x/\mu} \to 0$)
c. Consider the specific case of $\mu = 2$. Find the median and variance of the random variable.

7. In 1972 the fertility rate in the United States dropped below the replacement level for the first time. (A fertility rate at the replacement level would eventually result in zero population growth.) Were the nation to sustain the recent low fertility rate indefinitely, the population rate of change (dP/dt) in millions per year would be the following function of time (t) in years after 1975 (1975 population was 215 million).

$$\frac{dP}{dt} = 2e^{-.02t}$$

a. What was the initial (1975) population rate of change?
b. Determine total population as a function of t.
c. What will the population be in the year 2010?
d. At what point will the population be highest? What will this maximum population level be?

ADVANCED INTEGRAL CALCULUS

MIND STRETCHERS

1. Gidget Gadget Co. has found that the production rate of the average trainee during the training period can be approximated by the function

$$\frac{dP}{dx} = a + c(x + 1)^{1/2}$$

where x is time in days, P is units of production, and a and c are constants. If the initial production rate averages 5 units per day and the rate at the end of 8 days averages 11 units per day, how many units will the average trainee produce in the 35-day training period?

2. An agency forecasted a farm population decline in the United States in the next 25 years, with an increase thereafter according to the functions

$$\frac{dP}{dx} = -.5x^{-.5} \quad \text{for} \quad \{x \le 25\}$$

$$\frac{dP}{dx} = .01x \quad \text{for} \quad \{x > 25\}$$

where x is time in years ($x = 0$ now) and P is population in millions. If the current farm population is 15 million,
   a. What should the population be nine years from now?
   b. When will it be down to 12 million?
   c. What is the lowest farm population expected?
   d. When will the population be back to the current level?

3. General Southmoreland, after assessing the military situation in the 99-year war between East and West Wetnam, concluded that casualties on our side, now running at an annual rate of 10,000 will increase "slightly" in linear fashion to a rate of 20,000 per year two years from now as we "completely smash" the enemy's offensive. After that, the annual casualty (C) rate will drop "drastically" over the next two years according to the function of time (x)

$$\frac{dC}{dx} = 10,000(4 - x)e^{-(x-2)} \quad \text{for} \quad \{2 < x \le 4\}$$

as the war is concluded and won. If the general's forecast is correct, how many casualties can we expect during the four-year period of "victory"?

4. On page 230, we derived Miles Runner's velocity (v) in miles per minute as the following function (we have rounded the constants for ease of computation) of time (t) in minutes.

$$v = f(t) = .02t^2 - .1t + .3$$

   a. Will he finish the mile run in under four minutes?
   b. What will his time to finish be (approximate)?

Ray Cez likes to start the mile race at a fast pace, but then he tires, as given by his velocity function

$$v = g(t) = .3e^{-.1t}$$

   c. Will Ray break the four-minute mile?
   d. What will his finish time be?

World record holder Roger Rannister typically starts the mile race slowly, but continuously speeds up throughout the race, as described by his velocity function

$$v = h(t) = ke^{.1t}$$

where k is a constant.

e. What must his initial velocity (k) be in order to finish in four minutes?

5. Suppose the annual rate of warranty costs (W) associated with a specific production run are estimated by the following function of time (x) in years since production.

$$\frac{dW}{dx} = 1{,}000{,}000xe^{-.5x}$$

a. When will warranty cost rates reach a peak?
b. What are total warranty costs forecasted to be over the first four years?

6. The rate at which a certain flu virus spreads through a large community is

$$\frac{dV}{dx} = 50 \cdot 10^x \quad \text{for } \{x \le 3\}$$

$$\frac{dV}{dx} = 50{,}000e^{-(x-3)} \quad \text{for } \{x > 3\}$$

where x is weeks, and V is people with the virus.

If it takes one week to be cured of this virus,
a. How many people would have had the virus by the end of three weeks?
b. How many have it at the end of three weeks?
c. How many people would have had it by the end of the sixth week?
d. At what time was there a maximum number of people with the virus?

7. Consider the probability density function

$$P(x) = x^{\frac{1}{2}}$$

over the domain $\{0 \le x \le X\}$.
a. What must X be for $P(x)$ to be a probability distribution?

Given the value of X found in part a,
b. Find the median
c. Find the mean
d. Find the variance

Appendix

Table 1 Natural (base e) and Common (base 10) Logarithms

x	$\ln x$	$\log_{10} x$	x	$\ln x$	$\log_{10} x$	x	$\ln x$	$\log_{10} x$
0.1	−2.303	−1.000	3.5	1.253	0.544	6.8	1.917	0.833
0.2	−1.609	−0.699	3.6	1.281	0.556	6.9	1.932	0.839
0.3	−1.204	−0.523	3.7	1.308	0.568	7.0	1.946	0.845
0.4	−0.916	−0.398	3.8	1.335	0.580	7.1	1.960	0.851
0.5	−0.693	−0.301	3.9	1.361	0.591	7.2	1.974	0.857
0.6	−0.511	−0.222	4.0	1.386	0.602	7.3	1.988	0.863
0.7	−0.357	−0.155	4.1	1.411	0.613	7.4	2.001	0.869
0.8	−0.223	−0.097	4.2	1.435	0.623	7.5	2.015	0.875
0.9	−0.105	−0.046	4.3	1.459	0.633	7.6	2.028	0.881
1.0	0.000	0.000	4.4	1.482	0.643	7.7	2.041	0.886
1.1	0.095	0.041	4.5	1.504	0.653	7.8	2.054	0.892
1.2	0.182	0.079	4.6	1.526	0.663	7.9	2.067	0.898
1.3	0.262	0.115	4.7	1.548	0.672	8.0	2.079	0.903
1.4	0.336	0.146	4.8	1.569	0.681	8.1	2.092	0.908
1.5	0.405	0.176	4.9	1.589	0.690	8.2	2.104	0.914
1.6	0.470	0.204	5.0	1.609	0.699	8.3	2.116	0.919
1.7	0.531	0.230	5.1	1.629	0.708	8.4	2.128	0.924
1.8	0.588	0.255	5.2	1.649	0.716	8.5	2.140	0.929
1.9	0.642	0.279	5.3	1.668	0.724	8.6	2.152	0.934
2.0	0.693	0.301	5.4	1.686	0.732	8.7	2.163	0.940
2.1	0.742	0.322	5.5	1.705	0.740	8.8	2.175	0.944
2.2	0.788	0.342	5.6	1.723	0.748	8.9	2.186	0.949
2.3	0.833	0.362	5.7	1.740	0.756	9.0	2.197	0.954
2.4	0.875	0.380	5.8	1.758	0.763	9.1	2.208	0.959
2.5	0.916	0.398	5.9	1.775	0.771	9.2	2.219	0.964
2.6	0.956	0.415	6.0	1.792	0.778	9.3	2.230	0.968
2.7	0.993	0.431	6.1	1.808	0.785	9.4	2.241	0.973
2.8	1.030	0.447	6.2	1.825	0.792	9.5	2.251	0.978
2.9	1.065	0.462	6.3	1.841	0.799	9.6	2.262	0.982
3.0	1.099	0.477	6.4	1.856	0.806	9.7	2.272	0.987
3.1	1.131	0.491	6.5	1.872	0.813	9.8	2.282	0.991
3.2	1.163	0.505	6.6	1.887	0.820	9.9	2.293	0.996
3.3	1.194	0.519	6.7	1.902	0.826	10.0	2.303	1.000
3.4	1.224	0.531						

This table, in conjunction with the laws of logarithms, can be used to determine the logarithms of larger numbers. For example, $\ln 20 = \ln 2 \cdot 10 = \ln 2 + \ln 10 = .693 + 2.303 = 2.996$.

Table 2 Powers of e and 10

x	e^x	10^x	x	e^x	10^x
0.1	1.105	1.259	2.6	13.464	398.107
0.2	1.221	1.585	2.7	14.880	501.187
0.3	1.350	1.995	2.8	16.445	630.957
0.4	1.492	2.512	2.9	18.174	794.328
0.5	1.649	3.162	3.0	20.086	1000.000
0.6	1.822	3.981	3.1	22.198	1258.925
0.7	2.014	5.012	3.2	24.533	1584.893
0.8	2.226	6.310	3.3	27.113	1995.262
0.9	2.460	7.943	3.4	29.964	2511.886
1.0	2.718	10.000	3.5	33.115	3162.278
1.1	3.004	12.589	3.6	36.598	3981.072
1.2	3.320	15.849	3.7	40.447	5011.872
1.3	3.669	19.953	3.8	44.701	6309.573
1.4	4.055	25.119	3.9	49.402	7943.282
1.5	4.482	31.623	4.0	54.598	10000.000
1.6	4.953	39.811	4.1	60.340	12589.254
1.7	5.474	50.119	4.2	66.686	15848.932
1.8	6.050	63.096	4.3	73.700	19952.623
1.9	6.686	79.433	4.4	81.451	25118.864
2.0	7.389	100.000	4.5	90.017	31622.777
2.1	8.166	125.893	4.6	99.484	39810.717
2.2	9.025	158.489	4.7	109.947	50118.723
2.3	9.974	199.526	4.8	121.510	63095.734
2.4	11.023	251.189	4.9	134.290	79432.823
2.5	12.182	316.228	5.0	148.413	100000.000

This table, in conjunction with the exponent laws, can be used to determine higher (or negative) powers. For example,

$e^7 = e^5 \cdot e^2 = (148.413)(7.389) = 1096.624$

$10^{-.3} = 1/10^{.3} = 1/1.995 = .501$

Index

ABC "Wide World of Sports," 65
Abscissa, 136
Acapulco divers, 65
Acceleration
   of automobile, 230–231, 560–563
   in earth's gravity, 518–520
   of hot rod, 521
   in moon's gravity, 520
Addition
   of equations, 278–282
   of fractions, 34–35
   of matrices, 316–319
   of terms in arithmetic progression, 59–62, 173–174
   of terms in geometric progression, 69
Algebra "magic," 51
Algebra quiz, 24, 52
Annuities, 97–104
   retirement, 101–102
   sinking fund, 99
   sports salaries, 103
Antidifferentiation, 593–594. *See also* Integration
Area under a curve
   approximation by rectangles, 582–588
   as given by fundamental theorem of calculus, 595–599
   for probability distribution, 618
Arithmetic progression, 58–59
   applications of, 62–67
   common difference, 59
   general expression, 58
   sum of terms, 59–62, 173–174
Associative axioms, 16
Asymptote, 167
Average
   mean, 618–619
   median, 618–619
Average rate of change. *See* Rate of change
Axioms, 15–16
Axis, 136

Banking. *See also* Compound interest
   balance of savings account, 86–89
   money supply, 111
Base
   of logarithmic function, 185
   number representation, 47–49

Basic solution of linear programming problem, 388
Basis, 391
   revision of, 394–396
Binomial expansion, 38–39
Bivariate normal distribution, 174–175
Branch office model, 556–558
Breakeven analysis
   with linear revenue and cost, 243–244
   with parabolic revenue and linear cost, 244–245
   relationship to maximum profit point, 550
Breakeven point. *See* Breakeven analysis

Calculus basics
   programmed learning of, 445–451
Calculus I rocket, 520–521
California budget plan, 514–516
Capital budgeting applications, 104–107
Capital budgeting applications, 104–107, 623–624, 643–645
Car financing decision, 114–115
Car rental model, 8–9, 255, 299–300
Cartesian coordinate system, 136
Chain letters, 76–77
Chain rule, 418, 477–481
Circle
   applications of, 176
   graph of, 168
   identification, 166
Classification using a matrix, 310–311
Closure axioms, 16, 34
Column vector, 313
Common denominator, 34
Common stock. *See also* Stock market
   betas, 159
   valuation, 115–117
Commutative axioms, 16, 34
Commutative, matrix multiplication is not, 325–327
Complex numbers, 14, 15
Compound growth, 84–85
Compounding, 84
   continuous, 442–444
Compound interest, 85–89
   balance, 86–89
   effective interest rate, 87
   formula, 89
   process of, 86–87

Compound period, 86
  effect of changing, 88
Computer solution
  differential calculus, 485–489
  integral calculus, 641–643
  input-output problem, 353–354
  linear programming problem, 413–417
  matrix algebra, 352–359
Constant of integration, 597–599
Constraints, 369, 375–378
Consumption functions
  individual, 504–507
  aggregate for economy, 508
  marginal consumption, 622–623
Continuity, 444–445
Corner point theorem, 385–388
Cost
  average, 64
  fixed, 155, 240
  marginal, 547
  total, 155, 240, 554
  unit, 173
  variable, 155, 209, 240–243
Cost of capital, 104
Cost functions
  fitting to real cases, 240–242
  linear, 243, 244–245
  minimization of, 553–556
  parabolic, 240
  polynomial, 209
  total, 155
Cramer, 293
Cramer's rule, 293–305
  generalization, 298–299
  for 2×2 systems, 292–295
  for 3×3 systems, 295–298
  use in solving applications, 299–305
Crude oil production, 220–221
Curve fitting
  basic models, 211
  exponential, 222–225
  hyperbolic, 212, 215–216, 227–229
  least-squares, 217, 264–267, 304–305, 350–352, 573–575
  linear, 211–212
  logarithmic, 225–227
  parabolic, 213–214, 229–231
  parameters, 210–214

Decimal, 10
Demand curves, 232–235
  constant, 236
  hyperbolic, 238–239
  linear, 153–154, 206, 221–222, 234–235
  parabolic, 196, 213–214
Demand, elasticity of. *See* Elasticity of demand
Demand and supply equilibrium, 262–263, 302–303
Dependent equations, 283–284

Depreciation of car, 224
Depreciation methods, 107–110
  double-declining balance, 109–110
  straight line, 110, 155–157
  sum-of-years' digits, 107–109
Derivative
  applications of, 497–529
  chain rule, 477–481
  computer determination, 485
  of constant, 463–464
  of constant times a function, 471–472
  definition, 452
  of exponential functions, 469–470
  as graphing aid, 528–529
  of logarithmic functions, 467–469
  notation for, 452
  partial, 483–485
  of power functions, 464–466
  of product of functions, 475–476
  programmed learning of, 445–451
  of quotient of functions, 476–477
  rules, 481
  second, 537–540
  of sum or difference of functions, 473–475
Descarte, Rene, 137, 143
Determinant
  of 2×2 matrix, 293–295
  of 3×3 matrix, 264–266, 295–298
  use in Cramer's rule, 293–298
  zero, 266
Diagonal matrix, 312
Dice, 196
Difference quotient, 440
Differentiation, 453. *See also* Derivative
Direct proof, 17–18
Discounts, volume, 62
Distribution. *See* Probability distribution
Distributive axiom, 16
Division of fractions, 35
Domain of a function, 127
Double-declining balance depreciation, 109–110

e
  base of natural logarithms, 441
  base of some exponential functions, 183–185
  defined as limit, 441–442
  table of powers of, 000
Earnings (lifetime)
  daily pay doubling case, 70, 154
  resulting from college education, 552
Earthquakes, 47
Economic order quantity, 195
Effective interest rate, 92
Elasticity of demand
  arc, 498
  constant, 503–504
  definition, 497
  for linear demand, 499–500

INDEX

point, 498
  relationship to marginal revenue, 500–502
  unitary, 175–176, 430–431, 503
Electric
  current flow, 527–528
Element
  of matrix, 310
  of set, 11
  of product matrix, 320–325
Elementary row operations (ERO's)
  nature of, 338
  use in finding inverses, 338–341
  use in linear programming, 396–399
Elimination method for solving simultaneous equations, 277–282
Ellipse
  bivariate normal application, 174–175
  graph of, 165, 169–172
  identification, 166
End-of-domain extremes, 541–542
Equality, 25
  axioms, 16
  as balance, 26
  of matrices, 312
  of numbers, 25–26
Equations, 25
  adding and subtracting, 278–282
  dependent, 283–284
  independent, 283–284
Equilibrium
  demand and supply, 262–263, 302–303
  market share, 348–349
Excel (Microsoft) software, 352, 356–359, 413–417
Exchange equations, 392–393
Exponential functions, 177–180
  applications of, 180–185, 194–195
  constant percentage change, 222–223
  derivative of, 469–470
  general formula, 177
  graph of, 178–180
  integral of, 609
  as model, 193–194, 222–225
  relationship to logarithmic functions, 190–191
Exponential probability distribution, 183–184, 552–553
Exponents, 28–30
  laws of, 31

Factorial, 35
Factoring, 30, 32
Farming applications, 571–572
Feasible solution space, 370
Fermat, Pierrede
  graphical methods, 143
  last theorem, 4–5
Fixed cost, 155, 240–242
Football mathematics, 125–128
Fractions, 32–38
  addition and subtraction of, 34

common denominator, 34
compound, 36
multiplication and division of, 35
sign of, 33
simplification of, 33
Functions, 125–128
  classification of, 139
  definition of, 105
  domain of, 109, 127
  equation of, 130
  exponential, 150–158
  graph of, 135–138
  linear, 145–153
  logarithmic, 185–188
  mixed, 194
  multivariate, 132–133
  numerical, 129–130
  notation for, 111–112, 133–135
  objective (in linear programming), 369
  polynomial, 193–194
  power, 190–192
  range of, 129–130
  univariate, 131–132
Fundamental theorem of calculus, 589–593
Future value, 94, 101, 154–155

Gambling scheme, 72–73
Gangland–C.A.R.T.E.L. competition, 181, 194
Gasoline
  mileage problem, 230–231, 255, 300–301
Geometric progression, 67–70, 255
  applications of, 70–78
  common multiple, 68
  general expression, 68
  sum of terms, 69
Gold, Metlzer problem, 91
Government spending, 197, 514–517
Grade point average, 569–570
Graph of a function, 137–138
  exponential, 177–180
  linear, 150–151
  logarithmic, 185–188
  mixed, 194
  polynomial, 193–194
  power, 191–192
Graph of a quadratic equation, 166
Graph of a relation, 138
Gravitational constant, 518–520
Growth companies, 194

Half-plane, 373–375
Harrod–Domar model, 509–510
Higher derivatives, 537
Highway design, 618
Hyperbola
  application of, 175–176
  demand, 238–239
  elasticity of, 603

Hyperbola (*Continued*)
  graph of, 211, 216
  identification, 166
  as model, 211–212, 227–229
  slope of, 227–229
Hyperplane, 259

Identification of quadratic, 166
Identity axioms, 16
Identity matrix
  definition, 313
  multiplication by, 328–330
Implicit representation
  linear equations, 257–259, 276
  quadratic equations, 162–164
Income tax
  federal tax table, 162, 221
  as segmented linear function, 161, 221–222
Independent equations, 283–284
Indirect proof, 17–18
Inequality
  as constraint, 369
  formulation in real world, 375–378
  graph of, 373–375
  manipulation of, 372
  sense, 371
  strong, 371
  weak, 371
Inflection point, 541
Input-output model, 267–269, 345–346
  computer solution, 353–354
Instantaneous rate of change. *See* Rate of change
Insurance applications, 220
Integers, 12
Integral
  definite, by fundamental theorem, 219
  indefinite, 593
  notation for, 593
Integrand function, 593
Integration
  applications of, 612–624, 643–653
  by change of variable, 634–637
  by computer, 624–626
  constant of, 597–599
  of constant times function, 609–610
  definition, 593
  of exponential functions, 609
  multiple, 653–657
  by parts, 637–640
  of power functions, 608
  rules for, 611
  of sum (difference) of functions, 610
Intercept, i, 147–148
Interest. *See* Compound interest
Internal Revenue Service, 161–162
Inventory. *See also* Economic order quantity
  classification, 311, 318
Inverse axioms, 16, 27

Inverse of a matrix
  determination by elementary row operations, 338–341
  theory of, 332–338
  use in solving linear systems, 342–345
Investment in the economy, 508–509
Investments problem, 260–261, 269, 301, 306, 308–309, 322–325, 361–363, 367–369
Irrational numbers, 13

Laws of exponents, 31
Laws of logarithms, 44
Learning curves
  academic, 188–189
  job, 225–226, 645–647
Least-squares regression, *See* Regression, least squares
Leibnitz, 14
Leontief input-output model. *See* Input-output model
Limits
  applications, 441–442
  laws, 440
  notation, 439
Limiting ratios, 393–394, 397
Limit process
  for finding derivatives, 439–441
  for finding e, 441
Line. *See* Linear function
Linear function
  applications of, 153–162, 219–222
  explicit formulation of, 145
  graph of, 147, 150–151
  implicit formulation of, 245, 257–259
  intercept, 147, 229
  as model, 187–188
  multivariate, 152–153
  point-slope formula, 149–150
  as process, 130
  segmented, 161–162
  slope of, 148–150, 219–220
Linear inequality. *See* Inequality
Linear programming
  algebraic solution method, 388–399
  basic solution, 391
  basis, 391
  constraints, 369, 375–378
  corner point theorem, 385–388
  exchange equations, 392–393
  feasible solution space, 383–385
  format, 369
  graphical solution, 342
  objective function, 350, 369, 396
  problem formulation, 255, 322–325
  simplex method of solution, 399–404
  slack variables, 389–391
Linear programming applications, 378–383
  diet planning (minimization case), 364–367, 380–381, 410–413
  fuel blending, 367–369, 381–383
  investments, 322–325, 361–363, 407–409

INDEX

production scheduling, 363–364, 379–380, 409–410
vacation planning, 358–361, 378–379, 404–407
Linear systems, 285
  applications of, 260–270
  computer solution
    Microsoft Excel, 356–359
    www.calc101.com, 352–356
  formulation of, 255
  general 2×2 solution, 286–287
  general 3×3 solution, 288–291
  general solution by Cramer's rule, 298–299
  matrix representation of, 307–309
  order, 259
  solution, 259–260, 284–286
  solution of applications by Cramer's rule, 299–305
  solution of applications by matrix algebra, 342–352
  solution patterns, 292
Liquidity trap, 198–199
Loan
  payments, 112–114
Logarithmic functions, 185–188
  applications of, 188–191
  derivative of, 467–469
  general formula, 185
  graph of, 186–187
  as model, 225–227
  relationship to exponential functions, 190–191
  slope of, 160, 195
Logarithms, 42–47
  definition, 42
  base b, 42
  base e (natural logarithms), 46, 608–609
  conversion of base, 46
  laws of, 44
  tables of
    base e, 000
    base 10, 000

Manhattan, sale of in 1624, for $24, 90
March madness, 78
Marginal cost, 547, 554
Marginal propensity to consume, 505–507
Marginal propensity to save, 505–507
Marginal revenue, 473, 495–496, 500–502
  relationship with elasticity, 500–502
Market research, 616–617
Market share
  equilibrium, 348–349
  forecast using Markov chains, 346–349
Markov chain analysis of market share, 346–349
Mathematical model, 7–9
Matrix. *See also* Matrix algebra
  determinant of. *See* Determinant
  diagonal, 313
  elementary, 313, 332
  elements of, 310
  equality, 312
  identity, 313
  inverse, 309, 331–341

notation, 307
null, 314
order of, 310
square, 313
transpose, 311–312
unity, 314
Matrix algebra. *See also* Matrix, Chapter 10
  addition, 316–319
  applications of, 345–352
  combining addition (subtraction) and constant multiplication, 319
  constant multiplication, 315–316
  flexibility, 344–345
  multiplication, 319–327
  subtraction, 316–319
  use in solving linear systems, 342–344
Maximization
  of grades, 568–570
  of happiness, 560–564
  of net return, 550–553
  of objective function. *See* Linear programming
  of physical capacity, 571–573
  of profit, 546–550
  of revenue, 558–559
Maxi Taxi case, 244–245, 548–549
Mean, 618–619
Mean value theorem of calculus, 589
Median, 618–619
Melody Tent case, 62–64, 147
Metric system (temperature conversion), 157–158
Microeconomic applications, 497–507
Microsoft (Excel) software, 352, 356–359, 413–417
Minimization
  of cost, 553–558
  of objective function, 364–367, 369, 380–381, 410–413
Mixed functions, 194
Model, mathematical, 7–9
Moore's law of computing power, 73
Money supply, 111
Mortgage payments, 117–119
Multiplication
  of equation by constant, 277–278
  of fractions, 35
  of matrices, 319–327
  of matrix by constant, 315–316
Multiplication of matrices, 319–327
  condition necessary for multiplication, 285
  by identity matrix, 328–329
  by inverse matrix, 293–294, 329–330
  not commutative, 325–327
  product matrix elements, 320–325
Multiplier, 508–509
Multivariate function, 132–133

NASA, 58
Natural logarithms, 46, 608–609
  table of, 000
Natural numbers, 12

INDEX

NBA draft, 66
NCAA basketball tournament, 78
Negative exponents, 29–30
Net present value, 105
Normal equations, 265–267, 350–352, 573–575
Normal probability distribution
    bivariate, 174–175
    as model of IQ scores, 184–185
    as model of physical characteristics, 523
Notation for functions, 133–135
Null matrix, 314
Numbers, 12–15
    complex, 14
    integer, 12
    irrational, 13
    natural, 12
    rational, 12–13
    real, 13–14
    representation of, 47–49

Objective function, 369
    value of, 387, 417
Operations, 14
Optimization, linear constrained. *See* Linear programming
Optimization of function of one variable
    applications of, 546–572
    theory for, 537–543
Optimization of functions of two variables, 497–499, 543, 573–574
Order of a matrix, 259, 310
Ordinate, 136
Origin, 136

Parabola
    applications of, 229–231
    graph of, 211
    identification, 166
    as model, 209, 229–231
    slope reversal, 229–
Parallel lines, 285
Parameters, 210
Partial derivative, 483–485
Parts, integration by, 637–641
Pay problem, 70
π (Pi), 13, 58
Pivoting, 401–402
Plane, 258
Point-slope formula, 149–150
Pollution related applications, 650–652
Polynomial functions, 193–194
Population growth, 77, 224
Portfolio selection, 563–564
Postage cost function, 160–161
Power function, 190–192
    derivative of, 464–466, 481
    general formula, 191
    graph of, 192
    integral of, 507–508
Power rule of differentiation, 464–466, 481

Power rule of integration, 507–508
Powers of e and 10, table of, 000
Present value
    application to capital budgeting, 104–105
    definition, 93
    net, 105
    relation to future value, 181
    summing stream of, 95
Probability
    continuous, 523–525
    discrete, 522–523
    as long-run rates, 449
Probability distribution, 620
    area under, 618
    discrete, 522
    exponential, 183–184, 652–653
    multivariate exponential, 657–658
    normal, 184–185
Product failure, 000
Product rule of differentiation, 475–476, 481
Production applications, 263, 303–304
Profit functions
    and breakeven analysis, 242–246
    development of, 231–232
    fitting to real cases, 243–246
Profit maximization
    breakeven relationship to, 550
    Creaky Airlines, 549–550
    Maxi Taxi Co., 548–549
Progressions. *See* Arithmetic progressions *or* Geometric progressions
Promotion system, 67
Proofs, 16
    direct, 17
    indirect, 17–18
Public administration applications, 514–518
Pythagoras, 13

Quadratic equations, 162–172
    applications of, 172–176
    general equation, 163
    graph of, 166
    identification, 166
    quadratic roots formula, 164–165
Quadratic roots formula, 164–165
    use in finding roots of quadratic, 165, 245

Racing (car) application, 521–522
Racing (foot) applications, 229–230
Random variable. *See also* Probability distribution
    mean of, 618–621, 652–653
    median of, 618–621
    standard deviation of, 619–622
    variance of, 000
Random walk, 176
Range of a function, 127
Rate of change
    applications as sports statistics, 510–514
    average, 430–432

INDEX

concept of, 426
instantaneous, 432–433
practical examples of, 427
as slope of line, 428–429
Rational numbers, 12–13
Real numbers, 13
Regression, least squares, 217, 264–267, 304–305, 350–352, 573–575
Relation, 128
Revenue functions, 235–240
   for constant demand, 205
   for constant price, 154–155
   fitting to real cases, 235–241
   for hyperbolic demand, 238–240
   for linear demand, 172–173, 206, 237, 243
   maximization of, 558–559
   for parabolic demand, 196
Richter scale for earthquake measurement, 47
Rocket propulsion application, 433–439
Row operations, elementary. *See* Elementary row operations
Row vector, 313
Rule of 72, 5–6, 74–76

Sales forecasting, 73–74, 181–183, 612–615
Sample size determination, 565–566
Sample survey efficiency, 566–568
Saving account balance. *See* Compound interest
Savings functions of individuals, 504–507
Scatter diagram, 264
Science applications
   chemical reaction, 525–526
   electric current flow, 527–528
   heat transfer, 526–527
Scientific notation, 49–51
Secant line, 433
   slope of, 434
   use in limit process to find derivative, 387–398
Second derivative, 463–464, 537–540
   notation for, 538
   test for optimum, 540–541
Segmented functions, 161
Sets, 11
Simplex method, 399–404
   applications of, 404–413
   basis solution, 400
   computer solution, 413–417
   departing variable, 401–
   entering variable, 400–401
   initial tableau, 399–400
   limiting ratios, 393–394
   pivoting: revise tableau, 401–402
   recycling, 402–403
   slack variables, 389–390
Simultaneous equations, 40–42
Simultaneous equations solution methods. *See also* Linear systems
   elimination, 277–282

graphical, 276
substitution, 40–42, 211–214
Sinema Theater case, 131–135
Sinking fund, 99
Sky diving, 65–66
Slack variables, 389–390
Slope
   as aid in model choice, 208–209
   concept of, 179–180
   of curves, 433
   of linear function, 148
   as rate of change, 376, 378–382
   of secant line, 397–398, 434–439
   symbol (s) for, 148
   of tangent line, 435, 438
S. Lumlord saga, 207–208, 489–490, 495–496, 543–546
Solution of linear systems
   Cramer's rule, 299–305
   elimination method, 277–282
   graphical method, 276
   matrix algebra, 342–352
   substitution method, 276–277
Solver (on Excel) software (for linear programming), 413–417
Sports applications
   annuities for salaries, 103
   batting averages, 512–514
   earned run averages, 511–512
   football mathematics, 125–128
   foot racing, 229–230
   sports statistics, 510–514
Square matrix, 313
Standard deviation, 619–622
Statistics applications
   mean, 618–619
   median, 618–619
   sample size determination, 565–566
   standard deviation, 184–185, 619–622
   survey efficiency, 566–568
   variance, 619–622
Step function, 160–161
Stock market
   valuation of stock, 115–117
Straight line. *See* Linear function
Straight line depreciation, 110, 155–157
Strong inequality, 371
Substitution method for solving simultaneous equations, 40–42, 276–277
Subtraction
   of equations, 278–282
   of fractions, 34–35
   of matrices, 316–319
Sum-of-years'-digits depreciation, 107–109
Supply equilibrium with demand, 262–263, 302–303

Tableau. *See* Simplex method
Tangent line, 435
   slope of, 435, 438

Tax
  equalization rate, 516–518
  federal income, 161–162, 221–222
Temperature
  cricket thermometer, 158–159
  Fahrenheit–Centigrade conversion, 157–158
Theorems, 16, 18
Total cost functions. *See* Cost functions
Total function, 426, 593
Transpose of a matrix, 311–312

Uniform probability distribution, 620–623
Unitary elasticity, 503–504
Unit cost, 173
Unity matrix, 314
Univariate function, 131–132
Unknown
  matrix notation, 307–308
  relationship to variable in linear system, 260

Variable, 260
  dependent, 265
  independent, 265–266
  random, 618–622
  relationship to unknown in linear system, 260
  slack, 389–391
Variable cost, 155, 209–213, 240–243
Variance, 619–622
Vector
  column, 13
  row, 313

Velocity
  of calculus I rocket, 433–439
  of car, 230–231, 560–563, 603–607, 618
  of falling body, 518–520
  of hot rod, 521–522
  on moon, 518–520
  during race, 229–230
  as rate of change, 428–431
  as rate to be summed to find total distance, 603–607
  walking pace, 189–190
Volume, 195

Walking pace, 189–190
Weak inequality, 371
Weather forecasting, 615–616
www.calc101.com, 352–356, 485–489, 641–644

x-axis, 136
x-coordinate, 136

y-axis, 136
y-coordinate, 136
Yield curves, 226–227
y-intercept, i, 147–148

Zero determinant, 297–298
Zero exponent, 29